Results and Problems in Cell Differentiation

42

Series Editors
D. Richter, H. Tiedge

Philipp Kaldis (Ed.)

Cell Cycle Regulation

With 26 Figures, 1 in Color, and 9 Tables

Philipp Kaldis, PhD
National Cancer Institute, NCI-Frederick
1050 Boyles Street
Bldg. 560
Frederick, MD 21702-1201
USA

ISSN 0080-1844
ISBN-10 3-540-34552-3 Springer Berlin Heidelberg New York
ISBN-13 978-3-540-34552-7 Springer Berlin Heidelberg New York

Library of Congress Control Number: 2006925965

This work is subject to copyright. All rights are reserved, whether the whole or part of the material is concerned, specifically the rights of translation, reprinting, reuse of illustrations, recitation, broadcasting, reproduction on microfilm or in any other way, and storage in data banks. Duplication of this publication or parts thereof is permitted only under the provisions of the German Copyright Law of September 9, 1965, in its current version, and permission for use must always be obtained from Springer. Violations are liable for prosecution under the German Copyright Law.

Springer is a part of Springer Science+Business Media

springer.com

© Springer-Verlag Berlin Heidelberg 2006
Printed in Germany

The use of registered names, trademarks, etc. in this publication does not imply, even in the absence of a specific statement, that such names are exempt from the relevant protective laws and regulations and therefore free for general use.

Cover design: *Design & Production* GmbH, Heidelberg
Typesetting and Production: LE-TEX Jelonek, Schmidt & Vöckler GbR, Leipzig

Printed on acid-free paper   31/3150/YL – 5 4 3 2 1 0

Preface

The cell cycle is tightly regulated on many different levels to ensure properly controlled proliferation. In the last 20 years, through the contributions of many laboratories, we have gained insight into many important aspects of the regulation of the cell cycle and its relation to cancer, which culminated in the 2001 Nobel Prize being awarded to Leland Hartwell, Tim Hunt, and Paul Nurse. In the investigations of cell cycle regulation, it has been essential to use different model systems from yeast to mouse, where the results from one system have led to advances in another system. Recently, studies have been done using more complex organisms like the mouse, which has taught us much about redundancy and flexibility in the regulation of the cell cycle. Some of the (even fundamental) results from yeast or mammalian cell lines had to be revised since they were not completely applicable to complex animal systems. It is a major challenge to keep an open mind when new results overthrow established dogmas, especially since some of the dogmas have never been backed by convincing experiments. This book will provide an updated view of some of the most exciting areas of cell cycle regulation.

The chapters of this book have been written by experts in the cell cycle field and cover topics ranging from yeast to mouse and from Rb to sterility. In the first chapter Moeller and Sheaff review recent results regarding G1 phase control, which might suggest that depending on the context or cell type, the G1 phase control could be different. The second chapter by Teer and Dutta deals with the regulation of DNA replication during the S phase. They discuss the origin of replication complex, MCMs, and how they are controlled by different factors. The next chapter, by Yang and Zou, reviews checkpoints and the response to DNA damage, followed by a chapter by Hoffmann, which deals with protein kinases that are involved in the regulation of the mitotic spindle checkpoint. The regulation of the centrosome cycle is discussed in the chapter by Mattison and Winey. In the sixth chapter Reed reviews the regulation of the cell cycle by ubiquitin-mediated degradation. The next chapter, by Dannenberg and Te Riele, deals with the Rb family and its control of the cell cycle using in vivo systems. Lili Yamasaki reviews the relations between cancer and the Rb/E2F pathway in the eighth chapter and Hiroaki Kiyokawa then discusses interactions of senescence and cell cycle control. Aleem and Kaldis follow with new concepts obtained by studying mouse models of cell cycle regulators. In the eleventh chapter Bernard and Eilers review the functions of Myc in the control of cell growth and proliferation. The book concludes with a chapter by Rajesh and Pittman, who discuss the relations of cell cycle regulators and mammalian germ cells.

The future challenges in cell cycle research will be to integrate our knowledge coming from different systems, extend it to tumorigenesis in humans, and use all this information to design clinically relevant studies. This cannot happen in one step or overnight and will necessitate a lot of effort. It will continue to require broad-based basic research, along with the development of relevant animal models. These animal models need to recapitulate human diseases as closely as possible. Currently, many questions remain regarding animals being good models for human diseases. Nevertheless, more effort needs to be expended in developing better animal models before conclusions can be drawn. It is obvious that without appropriate animal models we will have to continue to test newly developed drugs in clinical trials without knowing the potential outcome. This is a time-consuming and risky procedure, which has been going on for too long a time. The future of cell cycle research is bright and the results of such studies will hopefully influence the battle against cancer.

This book could not have been completed without the outstanding contributions from the authors and I would like to thank them all for their valuable effort. In addition, I thank the members of the Kaldis lab as well as Michele Pagano for encouragement and support. I also acknowledge the support of Ursula Gramm, Sabine Schreck (Springer, Heidelberg), and Michael Reinfarth (Le-TeX GbR, Leipzig) for editorial managing and production of this book.

March 2006   Philipp Kaldis

Contents

G1 Phase: Components, Conundrums, Context
Stephanie J. Moeller, Robert J. Sheaff 1
1   Introduction 1
2   Arrival of the Cycle 2
2.1   Discrete Events during Division 2
2.2   Maintaining Order 3
2.3   Cell Cycle Machinery 4
3   G1 Progression in Cultured Cells 5
3.1   Coordinating Cell Growth and Division 6
3.2   Information Integration 7
3.3   The Cyclin-Cdk Engine 8
3.4   Removing Impediments: Inactivating Rb 9
3.5   Removing Impediments: Inactivating $p27^{kip1}$ 10
3.6   Preparing for the Future 11
4   Ablating G1 Regulators in Mice 12
4.1   Cyclin D-Cdk4/6 12
4.2   Cyclin E/Cdk2 14
4.3   G1 Targets 16
5   Implications and Future Directions 19
5.1   Conundrums 19
5.2   G1 in Context 20
6   Conclusions 23
References 24

Regulation of S Phase
Jamie K. Teer, Anindya Dutta 31
1   Introduction 31
2   Origins of Replication 32
2.1   Genome Replicator Sequences 32
3   Pre-Replication Complex 35
3.1   ORC 35
3.2   Cdt1 37
3.3   Cdc6 38

3.4	MCM2-7	40
3.5	Geminin	41
3.6	Summary	42
4	Pre-Initiation Complex	43
4.1	Mcm10	43
4.2	Cdc45	44
4.3	Dbf4/Cdc7	45
4.4	GINS	46
4.5	DPB11	47
4.6	Summary	47
5	S-phase Regulation and Cancer	49
6	Conclusion	50
References		52

Checkpoint and Coordinated Cellular Responses to DNA Damage
Xiaohong H. Yang, Lee Zou 65

1	Introduction	65
2	Sensing DNA Damage and DNA Replication Stress	66
2.1	Recruitment of ATR to DNA	66
2.2	DNA Damage Recognition by the RFC- and PCNA-like Checkpoint Complexes	69
2.3	Processing of DNA Lesions	71
2.4	MRN Complex and Activation of ATM and ATR	73
3	Transduction of DNA Damage Signals	74
4	Regulation of Downstream Cellular Processes	76
4.1	Regulation of the Cell Cycle	77
4.2	Regulation of DNA Replication Forks	78
4.3	Regulation of DNA Repair	79
4.4	Regulation of Telomeres	80
5	Interplay between Checkpoint Signaling and Chromatin	81
6	Perspectives	82
References		83

Protein Kinases Involved in Mitotic Spindle Checkpoint Regulation
Ingrid Hoffmann 93

1	Introduction	93
2	The Spindle Assembly Checkpoint	94
3	Regulation of the Spindle Checkpoint by Protein Kinases	95
3.1	Bub1	95
3.2	BubR1	98
3.3	Aurora B	99
3.4	Mps1	101
3.5	Mitogen-activated protein kinase	102
4	The Spindle Checkpoint and Cancer	102

5	Conclusions	104
References		104

The Centrosome Cycle
CHRISTOPHER P. MATTISON, MARK WINEY 111

1	Introduction	111
1.1	History	111
1.2	Microtubule Organizing Centers	112
1.3	Centrosome Functions	112
1.4	Centrosome Dysfunction and Cancer/Disease	113
1.5	Centrosome Structure	113
2	The Centrosome Cycle	114
2.1	Introduction	114
2.2	Centrosome Duplication	116
2.3	Centrosome Maturation	126
2.4	Centrosome Separation	130
2.5	Licensing of Centrosome Duplication	133
2.6	Post-Mitosis Return to G1	133
3	Conclusion	134
References		135

The Ubiquitin-Proteasome Pathway in Cell Cycle Control
STEVEN I. REED 147

1	Introduction	147
2	The Ubiquitin-Proteasome Pathway	148
3	Protein-Ubiquitin Ligases in the Cell Cycle Core Machinery	149
3.1	APC/C Protein-Ubiquitin Ligases	151
3.2	APC/C Substrates and Biology	154
3.3	APC/C and Meiosis	156
3.4	SCF Protein-Ubiquitin Ligases	156
3.5	SCF Substrates and Biology	157
3.6	Regulation of SCF Activity	162
4	Checkpoint Control	163
5	Atypical Roles of Proteasomes and Ubiquitylation	166
6	Deubiquitylating Enzymes	167
7	Conclusions	167
References		169

The Retinoblastoma Gene Family
in Cell Cycle Regulation and Suppression of Tumorigenesis
JAN-HERMEN DANNENBERG, HEIN P. J. TE RIELE 183

1	Cancer and Genetic Alterations	183
2	The pRb Cell Cycle Control Pathway: Components and the Cancer Connection	184

3	Regulation of E2F Responsive Genes by pRb	185
4	The Retinoblastoma Gene Family	187
4.1	Rb Gene Family Members	187
4.2	pRb Family Protein Structure	187
4.3	Similar and Distinct Functions of the pRb Protein Family	188
4.4	pRb Family Mediated Regulation of E2F by Cellular Localization	190
4.5	Regulation of E2F Mediated Gene Expression	190
4.6	The pRb Family and the Cellular Response Towards Growth-Inhibitory Signals	192
5	The pRb and p53 Pathway in Senescence and Tumor Surveillance	193
5.1	Replicative Senescence	193
5.2	Tumor Surveillance	195
6	Interconnectivity between the pRb and p53 Pathway	196
7	The *Rb* Gene Family in Tumor Suppression in Mice	199
7.1	Mechanistic Insights in the Tumor Suppressive Role of the Rb Gene Family	205
8	Role of *p107* and *p130* in Human Cancer	207
9	The Retinoblastoma Gene Family in Differentiation and Tumorigenesis	208
9.1	A Link between Pax, bHLH and Pocket Proteins in Differentiation and Tumorigenesis	209
9.2	Pax and bHLH Proteins in Retina and Pulmonary Epithelium Development	209
10	Conclusion	210
References		211

Modeling Cell Cycle Control and Cancer with pRB Tumor Suppressor
Lili Yamasaki 227

1	Introduction and Background	227
1.1	Epidemiology	227
1.2	Modeling Human Cancer in the Mouse	228
2	The Universality of the Cell Cycle	230
3	The pRB Tumor Suppressor Pathway	231
3.1	The Discovery of pRB	231
3.2	Upstream Regulators of pRB	232
3.3	Phenotype of Mice Lacking pRB Family Members	233
3.4	pRB Regulates Growth and Differentiation	236
4	The E2F/DP Transcription Factor Family	237
4.1	E2F Target Genes and Repression	237
4.2	Mice Deficient in E2F Family Members	238
5	Cyclin-dependent Kinases and their Inhibitors	240
5.1	Deregulation of Cyclins, Cdks and CKIs in Human Tumors	240

5.2	Mice Deficient in Cyclins, Cdks and CKIs	241
6	Links Between the pRB and p53 Tumor Suppressor Pathway	243
7	Murine Models of Retinoblastoma	245
8	Revising Cell Cycle Models	246
References		248

Senescence and Cell Cycle Control
HIROAKI KIYOKAWA 257

1	Senescence	257
2	Role of the p53 Pathway in Senescence	258
3	Role of the Rb Pathway in Senescence	260
4	The Role of the INK4A/ARF Locus in Senescence	262
5	Mouse Cells vs. Human Cells: Roles of Reactive Oxygen Species and Telomere Attrition	263
6	Conclusions	266
References		266

Mouse Models of Cell Cycle Regulators: New Paradigms
EIMAN ALEEM, PHILIPP KALDIS 271

1	Introduction	271
2	History of the Cell Cycle Model	273
2.1	The Concept of Mammalian Cell Cycle Regulation	273
2.2	Lessons from Yeast	273
2.3	Human Cdc2, Cdk2 and Cyclin E	275
2.4	G1 Phase in Mammalian Cultured Cells	276
3	Mouse Models of Cell Cycle Regulators	279
3.1	Targeting of Individual Cell Cycle Regulators Results in Embryonic Lethality	279
3.2	Sterility	281
3.3	Mouse Models with Hematopoietic Defects	287
3.4	Mouse Models with Pancreatic Defects	289
3.5	Placental Defects and Endoreduplication	291
4	Tumorigenesis in Mouse Models of Cell Cycle Regulators	294
4.1	Pituitary Tumors	294
4.2	Skin Cancer and Melanoma	298
4.3	Breast Cancer	298
4.4	Ovarian Tumors	300
5	New Functions for Old Players	301
5.1	Cdc2 Regulates S Phase Entry	301
5.2	p27 Regulates the Rho Pathway	303
6	Genetic Interaction and Functional Complementation of Cell Cycle Regulators	304
6.1	Interactions of Cyclin D1 and p27	304

6.2	Functional Complementation of Cdc2 and Cdk2 in G1/S Phase Transition	305
6.3	Functional Cooperation Between Cdk2, Cdk4 and p27	307
6.4	Compensation Between the D-type Cyclins	308
6.5	Interactions Between Cdk4 and Cdk6	309
6.6	Cyclin E Can Functionally Compensate for Cyclin D1	310
7	Implications of Data from Cell Cycle Mouse Models to Human Cancer	310
7.1	Cdk2 in Human Tumors and in Tumor Cell Lines	311
8	Conclusions	312
References		314

Control of Cell Proliferation and Growth by Myc Proteins
SANDRA BERNARD, MARTIN EILERS 329

1	Introduction	329
2	Mechanisms of Myc Action	332
3	Targets	334
4	Checkpoints and Apoptosis	336
5	Conclusions	337
References		338

Cell Cycle Regulation in Mammalian Germ Cells
CHANGANAMKANDATH RAJESH, DOUGLAS L. PITTMAN 343

1	Introduction	343
2	Cell Cycle Regulatory Genes Required for Initiation and Maintenance of Meiosis	353
3	Transcriptional and Translational Factors	355
4	Cell Signaling	356
5	Cytoplasmic and Apoptotic Factors	357
6	Cell Cycle Regulation during Prophase I	358
7	Future Perspectives	360
References		361

Subject Index 369

G1 Phase: Components, Conundrums, Context

Stephanie J. Moeller[1] · Robert J. Sheaff[2] (✉)

[1]Corporate Research Materials Laboratory, 3M Center, Building 201-03-E-03, St. Paul, MN 55144-1000, USA

[2]University of Minnesota Cancer Center, MMC 806, 420 Delaware Street SE, Minneapolis, MN 55455, USA
sheaf004@tc.umn.edu

Abstract A eukaryotic cell must coordinate DNA synthesis and chromosomal segregation to generate a faithful replica of itself. These events are confined to discrete periods designated synthesis (S) and mitosis (M), and are separated by two gap periods (G1 and G2). A complete proliferative cycle entails sequential and regulated progression through G1, S, G2, and M phases. During G1, cells receive information from the extracellular environment and determine whether to proliferate or to adopt an alternate fate. Work in yeast and cultured mammalian cells has implicated cyclin dependent kinases (Cdks) and their cyclin regulatory partners as key components controlling G1. Unique cyclin/Cdk complexes are temporally expressed in response to extracellular signaling, whereupon they phosphorylate specific targets to promote ordered G1 progression and S phase entry. Cyclins and Cdks are thought to be required and rate-limiting for cell proliferation because manipulating their activity in yeast and cultured mammalian cells alters G1 progression. However, recent evidence suggests that these same components are not necessarily required in developing mouse embryos or cells derived from them. The implications of these intriguing observations for understanding G1 progression and its regulation are discussed.

1
Introduction

"All theory is grey, life's golden tree alone is green."

Johann Wolfgang von Goethe

Ever since the cell was designated the fundamental unit of living organisms, efforts have been increasingly devoted to solving the mystery of its propagation. Physical observation in diverse systems, from simple unicellular bacteria to complex multicellular animals, revealed that this process involves duplicating cellular contents followed by division into two identical cells (Nurse 2000a).

Cell cycle theory is a generalized conceptual framework for describing how a eukaryotic cell copies itself by coordinating an increase in mass, chromosome replication/segregation, and division (Mitchison 1971). Over the past

3 decades, the machinery controlling these processes has been identified and organized into a description of cell cycle progression. Now that the field has its Nobel Prize, one might assume that the picture is largely complete and only details remain. A broader perspective, however, reminds us that those who ignore the history of scientific advancement are often doomed *not* to repeat it. That the cell cycle field will be no exception is evidenced by surprising new observations hinting that it might be time to start a new canvas.

This chapter will first undertake an examination of how cell cycle theory developed, which reveals the rationale for G1 phase and its role in cell division. We next lay out in broad strokes the current understanding of molecular events controlling G1 progression in mammalian cells. Principles and generalizations underlying this model will be explicitly identified and discussed, with particular emphasis on how they are now being called into question by recent experimental data analyzing cell cycle regulators in mice. Ultimately, we hope to illustrate how accumulating evidence provides hints of a richer and more complex picture of G1 phase waiting to be discovered.

2
Arrival of the Cycle

Discovery of cell division marked the birth of cell cycle research (Nurse 2000b). Subsequent investigations identified two major events during this process, mitosis and DNA replication, and demonstrated they occur at different times and in a particular order. The existence of gap phases and why they separate these key events has long been appreciated, but molecular mechanisms defining transitions between them could not be investigated until cell cycle machinery was identified.

2.1
Discrete Events during Division

Physical observation of animal cell duplication identified discrete events during this process, the most dramatic being condensation of thread-like structures shortly before cell division (Flemming 1965). We now know this period as mitosis, when the chromosomes segregate and are equally distributed to the mother and daughter cell. Subsequent work revealed chromosomes contain the hereditary material, are composed of DNA, and are duplicated at a defined period occurring before cell division (Nurse 2000a). These initial observations suggested that cell duplication is divided into discrete periods or phases, an organizing principle distinguishing bacteria from eukaryotic cells. Molecular mechanisms are therefore required to coordinate these processes in time and space.

G1 Phase: Components, Conundrums, Context

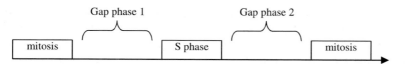

Fig. 1 Temporal separation of S and M phases in a typical cell cycle. DNA replication (S-phase) and cell division (mitosis, M phase) are separated by distinct gap phases

Physical and temporal separation of DNA synthesis (S-phase) and mitosis (M phase) implies existence of gap phases separating these events (Fig. 1). Gap phase 1 (G1) is defined as the period from end of mitosis to initiation of DNA synthesis. Gap phase 2 (G2) separates end of DNA synthesis from initiation of mitosis (Mitchison 1971). Time spent in G1 varies between cell types and in different situations, but in mammalian cells it usually accounts for a significant amount of total cycling time. A typical mammalian cell might require 24 h to make a copy of itself and spend half this time in G1. However, in some specialized situations such as early development, G1 is absent and cells go directly from M phase to synthesizing DNA (Murray and Hunt 1993). These extremes provide important clues about why separating the end of mitosis from initiation of DNA synthesis is sometimes necessary and desirable. In such cases it becomes important to understand how this period is traversed, but before discussing this issue, the relationship between distinct cell cycle phases must be further defined.

2.2
Maintaining Order

Continuity through multiple cell divisions requires that each new daughter receive a complete and accurate copy of the genome. Chromosomes must be duplicated once and only once before mitosis; conversely, mitosis must be completed before DNA replication is re-initiated (Fig. 2) (DePamphilis

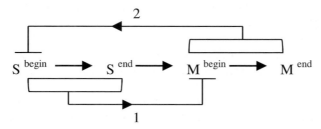

Fig. 2 Checkpoint control of S and M phase initiation. In pathway 1, ongoing DNA replication transmits a signal that blocks beginning of M phase (M^{begin}). In pathway 2, ongoing mitosis transmits a signal that blocks start of S-phase (S^{begin})

2003). Cells also continually monitor for and repair the inevitable DNA damage occurring throughout the division cycle (Kastan and Bartek 2004). In all these situations order is maintained by checkpoints, wherein initiation of later events is dependent on successful completion of earlier ones (Hartwell 1974; Hartwell and Weinert 1989). Temporally and spatially separate events are linked via signaling components, which transmit information to elicit desired responses (Nurse 2000b). By monitoring and linking events required for cell division and repair, checkpoints help maintain genomic integrity essential for survival and continuation of the cell lineage.

Checkpoints represent an elegant solution to the problem of ordering DNA synthesis and cell division, while at the same time raising additional questions. What drives progression through the cell cycle, and how is this process regulated? These controls are distinct from machinery replicating DNA and dividing the cell, which must receive instructions to initiate and complete these tasks properly. Addressing such thorny issues required a paradigm shift from observation of cell duplication to analysis of molecular events. Breakthroughs came from disparate but ultimately complementary approaches: biochemical analysis of S to M phase cycling reproduced in a cell free system derived from frog oocytes, generation and analysis of yeast mutants defective in cell division control, and analysis of protein expression patterns in sea urchin extracts (Nurse 1990; Nasmyth 2001). These seminal investigations (along with other important contributions) led inexorably to identification of critical cell cycle machinery.

2.3
Cell Cycle Machinery

Recognition that specific protein catalysts are responsible for diverse cellular processes such as fermentation (late 1800s) suggested that cell growth and proliferation would be similarly controlled (Nurse 2000a). Division of the cell cycle into temporally ordered, discrete steps implied different proteins regulate specific cell cycle transitions (Hartwell 1974). If so, then factors advancing cell cycle progression might be rate limiting (Nurse 1975). These concepts gave birth to the idea of a cell cycle engine that both drives and controls progression through the division cycle (Murray and Hunt 1993).

Biochemical and genetic approaches in different systems converged to identify what we now know as the cell cycle machinery. A key discovery was that nuclear division in frog oocytes is controlled by a "maturation promoting factor", or MPF (Masui and Markert 1971). Around the same time, genetic screens identified yeast mutants defective in cell division or prematurely entering mitosis (Hartwell et al. 1973; Nurse et al. 1976). Rate limiting components of S to M phase cycling were eventually isolated from frog egg extracts, and the *Deus ex machina* turned out to be a kinase in association with a regulatory subunit called cyclin (Evans et al. 1983; Lohka et al. 1988).

These cyclin-dependent kinases (Cdks) transfer gamma-phosphate from ATP to a specific protein substrate (Morgan 1995). However, the kinase subunit alone is inactive because the bound ATP is not properly oriented, and access of the protein substrate is blocked by a section of Cdk called the T-loop (DeBondt et al. 1993). These impediments are removed by association with cyclin and T-loop phosphorylation by the multicomponent Cdk-activating kinase (CAK) (Russo et al. 1996).

The key to ordering and controlling cell cycle progression is thought to lie in periodic expression of different cyclins, which associate with Cdks at defined intervals and determine their specificity (Murray and Hunt 1993). These unique cyclin–Cdk complexes must phosphorylate specific substrates at the proper time to drive controlled progression through the cell cycle. After completing their task, complexes are disassembled and cyclin degraded as a prerequisite for subsequent steps (Murray et al. 1989). Temporal order is achieved and maintained by linking cyclin expression to completion of previous events, then regulating activity of the resulting cyclin–Cdk complex. Controlling complexes can be accomplished by removing activating modifications, inhibitory phosphorylation of the Cdk subunit, or tight binding of Cdk inhibitory proteins (CKIs) (Morgan 1995). Cdk activity can also be modulated by altering its location and/or accessibility to substrates, although these regulatory mechanisms are less well characterized (Murray 2004). Together, this molecular circuitry provides a mechanistic explanation of cell cycle progression during G1.

Basic underlying principles derived from these investigations are: 1) cell cycle machinery is evolutionarily conserved, 2) transitions between cell cycle phases are catalyzed by Cdks, 3) cell cycle machinery is highly regulated, and 4) cell cycle components are an obvious target in proliferative diseases like cancer (Murray and Hunt 1993).

3
G1 Progression in Cultured Cells

If John Donne were a developmental biologist, he might have penned: "In multicellular organisms no cell is an island, entire of itself; each must be responsive to the external environment". Cells receive specific signals to survive, nutrients to grow, and additional signals to proliferate. After each division a G1 phase cell must re-evaluate its overall situation and determine whether continued proliferation is desirable and feasible (Pardee 1974). Although precisely how cell cycle machinery regulates G1 progression remains poorly understood, a generally accepted working model has been constructed from investigations in many different experimental systems. It posits that unique G1 cyclin/Cdks are temporally expressed in response to extracellular signaling (Sherr and Roberts 1995). These complexes phosphorylate specific substrates to promote required events and remove negative impediments to G1 progression.

3.1
Coordinating Cell Growth and Division

A non-proliferating cell maintains a relatively constant size by establishing homeostasis between cellular processes such as protein synthesis and degradation (Neufeld and Edgar 1998). In contrast, conservation of mass requires that a proliferating cell at some point duplicate its cellular contents (i.e. grow) to maintain cell size; otherwise, it will become progressively smaller and smaller until survival is untenable. This problem could be avoided by exactly doubling cell components before each division, or by a stochastic process averaging the required mass increase over several division cycles. Although at some level proliferation must be coordinated with an increase in mass, manipulating this relationship is crucial for development of multicellular organisms (Su and O'Farrell 1998a,b).

DNA replication and segregation can occur much faster than mass increases, so a newly formed daughter cell must grow to become competent for S-phase (Saucedo and Edgar 2002). Although growth is not rigorously confined to a specific period like DNA synthesis and mitosis, much of the necessary mass increase in mammalian cells occurs during its lengthy G1. Consistent with these ideas, depriving cultured cells of growth factors or amino acids causes a reduction in the rate of protein synthesis and cell cycle arrest in G1. This result implies existence of a G1 checkpoint linking cell growth with cell cycle progression, as in yeast (Campisi et al. 1982; Rupes 2002). A sizing mechanism, such as overall increase in mass (reflected in protein synthesis) or production of a specific molecule(s), could determine when a critical size threshold is reached.

It is encouraging to see several recent reports re-invigorating the controversy about whether mammalian cells contain an active sizing mechanism. Rate of growth and division appears to be two separable and independently controlled processes in rat Schwann cells, because reductions in cell volume require several division cycles to re-establish homeostasis (Conlon and Raff 2003). In this case, size was determined by the net effect of how much growth and division occurred. In contrast, a number of other cell types (e.g. human, mouse, and chicken erythoblasts and fibroblasts) respond to size alterations by compensatory shortening of the subsequent G1 phase (Dolznig et al. 2004). These results provide evidence of a G1 size threshold that adjusts length of the next cell cycle to maintain balance between growth and division.

Additional work is clearly required to explain the differing conclusions reached in these two studies. One possibility is that generating cultured cell lines compromises or alters the link between growth and proliferation; alternatively, the extent or mechanics of coordination may vary depending on cell type or situation. Regardless, identifying cells in which a sizing mechanism is operational means that experiments can now be designed to identify its molecular components.

3.2
Information Integration

G1 phase of the cell cycle is organized around the concept of a restriction point (R point; called START in yeast) (Hartwell et al. 1973; Pardee 1974; Blagosklonny and Pardee 2002). Before this G1 checkpoint, the cell receives and interprets information from a variety of internal and external sources. A decision is then made whether or not to continue with the cycle and initiate another round of cell division. If conditions are not appropriate for proliferation, or the cell receives orders to adopt an alternative fate, it withdraws from the cycle into a G0 resting state. It can remain in this position until proliferative conditions are re-established, or initiate an alternative program resulting in differentiation, senescence, or apoptosis (Fig. 3).

The idea of a restriction point arose from analyzing how newly generated mammalian fibroblasts respond to nutrient and growth factor starvation (Zetterberg et al. 1982). If serum is removed up to an experimentally determined point, cells halt cell cycle progression in G1 phase. Upon serum re-addition, completion of the cell division cycle is significantly extended compared with continually fed cells. Thus, starvation not only blocks cell cycle progression, but causes cells to exit the cycle and enter G0. However, if serum is removed after this point, cells continue through the cycle unhindered (Zetterberg and Larsson 1985). Subsequent analysis identified other criteria that differ before and after this period in G1. Up until the R point, cells stop cycling in response to low concentrations of cyclohexamide (a pro-

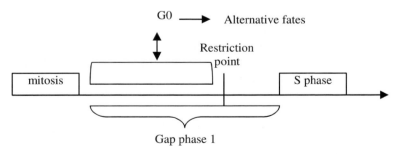

Fig. 3 Restriction point in G1 phase. The restriction point describes a position at which the cell irreversibly commits to completing the division cycle. Up until the R point the cell can withdraw to a quiescent state called G0. It can re-enter the cycle if conditions for proliferation are favorable, or pursue an alternative fate

tein synthesis inhibitor), while after the R point they are resistant (Pardee 1989). These observations suggested a molecular switch (such as an unstable protein) might define R point control (Zetterberg and Larsson 1991).

3.3
The Cyclin-Cdk Engine

G1 progression is promoted and controlled by cyclin/Cdk complexes, so they are often described as engines driving this process. Yeasts have only one Cdk (originally Cdc28; now called Cdk1), while 11 distinct versions have been identified in mammalian cells (van den Heuvel and Harlow 1994). Cdks accomplish their overall mission by promoting positive events, overcoming negative impediments, and policing themselves. In mammalian cells passage through G1 is controlled by ordered expression of the D and E type cyclins, which associate with Cdk4/6 and Cdk2/3, respectively (Fig. 4) (Sherr 1994). There are three members of the cyclin D family and two of cyclin E, each of which is expressed in a tissue-specific manner (Murray 2004). Current understanding of their regulation and function has emerged largely from the study of how cultured mammalian cells respond to serum starvation/refeeding.

When an asynchronous population of proliferating mammalian cells is deprived of serum, those located in G1 phase before the R point initiate a concerted shutdown of Cdk activity (Zetterberg and Larsson 1991; Sherr and Roberts 1995). Cyclin expression is inhibited and its destruction promoted. Any remaining cyclin/Cdk complexes are inhibited by phosphorylating Cdk and/or association of tight binding inhibitors (Sherr and Roberts 1995). Cells located after the R point when serum is removed complete the cycle and then exit G1 by similar mechanisms. In order to re-enter the cell cycle Cdk inhibition must be reversed. Refeeding G0 cells provides nutrients, growth factors, and mitogens, resulting in rapid activation of cell surface receptors and downstream signaling pathways like Ras/Map (also called Erk) kinase.

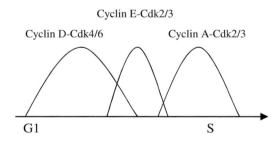

Fig. 4 Model of cyclin/Cdk activity during G1 phase. Ordered G1 progression in cultured cells involves temporal and transient expression of different cyclins, which bind their Cdk partners and determine specificity. The resulting complexes phosphorylate specific substrates required for regulated movement through the cycle

Activated Map kinase translocates to the nucleus, where it phosphorylates specific targets to promote transcription of genes required for growth, cell cycle progression, and upcoming S-phase (Alberts et al. 1994; Frost et al. 1994). Early mRNAs are induced within 30 min of refeeding cells and are insensitive to protein synthesis inhibitors, indicating components required for their production are already present. In contrast, late mRNAs are sensitive to these inhibitors because they depend on unstable products of early response genes. Identifying molecular events controlling this transcriptional program was essential for further defining R point control.

In fibroblasts and many other cell types a key consequence of activating Ras/Map kinase is rapid upregulation of cyclin D1 transcription; cyclin D1 protein then associates with Cdk4/6 and initiates G1 progression (Winston and Pledger 1993; Albanese et al. 1995). The Map kinase pathway also influences cyclin D1 localization, its association with Cdk4/6, and activation of the complex by the Cdk-activating kinase (CAK). These multiple levels of regulatory control help ensure cyclin D-Cdk4/6 is not inappropriately activated (Roussel et al. 1995). Mitogen dependence is maintained in part because cyclin D1 is a very unstable protein degraded by the ubiquitin/proteasome system (Matsushime et al. 1992; Diehl et al. 1997). Removing serum before the R point inhibits cyclin D transcription, resulting in rapid disappearance of cyclin D protein and subsequent exit from the cell cycle.

3.4
Removing Impediments: Inactivating Rb

A main target of activated cyclin D-Cdk4/6 is the retinoblastoma protein (Rb), so called because it was first identified as a tumor suppressor whose function is lost in a rare form of childhood retinal cancer (Friend et al. 1987). Rb siblings include p130 and p107, and this family occupies a central position in G1 control (Weinberg 1995). Rb acts in part as a repressor inhibiting members of the E2F transcription factor family (Bartek et al. 1996). E2Fs associate with Dp1 or Dp2 to form an active transcription factor complex upregulating a wide array of gene products required for growth, cell cycle progression, and upcoming S-phase (Stevaux and Dyson 2002). Rb can inhibit E2F-Dp complexes in a number of ways, including sequestration away from DNA and/or by forming active repressor complexes blocking DNA accessibility (Liu et al. 2004). This latter function is accomplished in part by recruiting histone deacetylases that alter chromatin structure (Harbour and Dean 2000).

In addition to its well-characterized role inhibiting E2F, Rb interacts with many different proteins and clearly regulates other processes in addition to transcription. It helps block global protein synthesis in response to nutritional deprivation by inhibiting expression of RNA polymerases I and III, which are responsible for synthesizing ribosomal RNAs needed for protein production (White 1994). Dual regulation of growth and cell cycle progres-

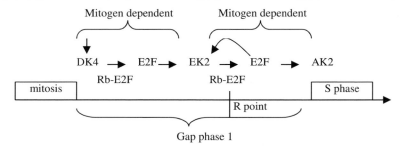

Fig. 5 Control of G1 progression by cyclin/Cdk complexes. Mitogens generate cyclin D-Cdk4 which phosphorylates Rb to release E2F. E2F transcriptionally upregulates cyclin E, which in association with Cdk2 inactivates additional Rb to generate more cyclin E. This positive feedback loop may represent the switch to mitogen independence (DK4: cyclin D/Cdk4; EK2: cyclin E/Cdk2; AK2: cyclin A/Cdk2)

sion by Rb may help coordinate these two processes during division or development.

As expected, Rb is highly regulated during the cell cycle. It is underphosphorylated (i.e. hypophosphorylated) in G0 cells and so binds E2F-Dp1 to prevent transcription (Weinberg 1995). Re-feeding generates active cyclin D1-Cdk4/6 that specifically phosphorylates Rb at a subset of available sites (Chen et al. 1989). Activated E2F-Dp1 then upregulates cyclin E, which associates with its partner Cdk2 and further phosphorylates Rb at distinct sites (Fig. 5) (Dynlacht et al. 1994). The resulting spike in E2F-Dp1 activity causes a burst of cyclin E synthesis and functional cyclin E/Cdk2 required for G1 progression (Ohtani et al. 1995). This positive feedback loop may represent the transition to mitogen independence during G1 (Hatakeyama et al. 1994). The burst is transient because cyclin E/Cdk2 marks its own cyclin subunit for ubiquitination by SCF^{Fbw7} and subsequent degradation by the proteasome (Clurman et al. 1996). Continued Rb inactivation during this period likely contributes to E2F-Dp1 dependent synthesis of cyclin A necessary for upcoming S-phase (Stevaux and Dyson 2002). As cells proceed through the cycle, Rb is dephosphorylated to reset the system (Buchkovich et al. 1989).

3.5
Removing Impediments: Inactivating p27^{kip1}

The Cdk inhibitor p27^{kip1} (p27) is an anti-mitogenic gene activated in response to serum starvation of proliferating cells (Sherr and Roberts 1995). It participates in cell cycle exit and helps maintain the G0 state by ensuring that cyclin/Cdk complexes remain inactive. High p27 levels in quiescent cells establish an inhibitory threshold that must be reduced for cell cycle re-entry (Roberts et al. 1994; Sherr and Roberts 1995). Cyclin D1–Cdk4/6 plays an important role in this process by sequestering p27 away from cyclin E-Cdk2 in early G1 (Reynisdottir et al. 1995).

In a satisfying twist, cyclin E/Cdk2 phosphorylates p27 later in G1 and targets it for recognition by SCF (Skp2, Csk1, Cul1), a ubiquitin ligase complex marking p27 for proteasomal degradation (Sheaff et al. 1997; Tsvetkov et al. 1999). Thus, p27 elimination may contribute to the rapid burst of cyclin E/Cdk activity and transversal of the R point. Recent evidence in support of this switch-like behavior comes from the discovery that Skp2 stability is controlled by APC^{Cdh1} (Bashir et al. 2004). This is a form of the APC/C (anaphase promoting complex/cyclosome) that operates as a major ubiquitin ligase in M phase. APC^{Cdh1} remains active into G1 and targets Skp2 for degradation, eventually running out of substrates and turning on itself. Skp2 protein can then accumulate and so p27 is degraded in a cyclin E/Cdk2 dependent manner. Thus, APC^{Cdh1} may contribute to maintenance and timing of G1 progression.

3.6
Preparing for the Future

The main function of cyclin D1-Cdk4/6 is inactivating Rb because these complexes are dispensable in an Rb-/- background (Lukas et al. 1995). In contrast, cyclin E/Cdk2 has other G1 targets since it is still required under these conditions. In addition to helping remove negative impediments such as Rb and p27, cyclin E/Cdk2 carries out various tasks required for upcoming S and M phases. It helps license replication origins to ensure DNA is only duplicated once per cell cycle. This process involves assembly of a pre-replicative complex (PRC) after mitosis, which then recruits MCM (minichromosome maintenance) helicases onto the DNA (Coverley et al. 2002; Diffley and Labib 2002). Cyclin E/Cdk2 may then participate in subsequent origin firing during S-phase (Woo and Poon 2003).

Cyclin E/Cdk2 has also been implicated in modulating chromatin structure. Histone H1 continues to serve as a substrate for purified cyclin E/Cdk2, but the physiological relevance of this reaction remained enigmatic. Recent evidence suggests cyclin E/Cdk2 phosphorylates histone H1 in cells, destabilizing its interaction with chromatin (Contreras et al. 2003). In addition, cyclin E/Cdk2 influences chromatin structure by phosphorylating Rb and altering its association with histone deacetylases (HDAC) at E2F promoters (Takaki et al. 2004). Finally, cyclin E/Cdk2 helps mediate centrosome duplication in preparation for upcoming M phase (Hinchcliffe et al. 1999). These microtubule organizing centers will relocate to opposite ends of the nucleus during mitosis, after which microtubules will create the spindles separating replicated chromosomes to daughter cells (Alberts et al. 1994).

4
Ablating G1 Regulators in Mice

If cyclin/Cdk complexes are engines driving and controlling G1 progression, then deleting their genes should result in very early lethality because cells cannot proceed through the cycle. In the absence of cyclin D-Cdk4/6 cells should not respond to mitogens or inactivate Rb, and thus fail to initiate the E2F transcriptional program required for G1 progression, S-phase, and division. Cyclin E/Cdk2 should also be absolutely required, since it controls the G1/S-phase transition by promoting key events required for cell cycle progression and DNA replication.

Similarly, deleting negative impediments such as Cdk inhibitors and Rb was predicted to result in very early lethality due to disruption of G1 timing. In the absence of Rb cells would be expected to inappropriately activate E2F/Dp1 and prematurely upregulate transcription of genes promoting G1 progression and S-phase entry. If p27 sets a threshold controlling S-phase entry, its absence should compromise the critical G1 to S-phase transition.

4.1
Cyclin D-Cdk4/6

Cdk4 and Cdk6 are closely related kinases associating with D-type cyclins to initiate G1 progression in response to proliferative signals (Sherr 1994). This idea arose from extensive work in cultured cells showing: 1) overexpressed cyclin D1 shortens G1 phase, suggesting activity of cyclin D1/Cdk4 is rate limiting for G1 progression, 2) cyclin D1 overexpression overcomes a G1 arrest caused by DNA damage or the unfolded protein response, 3) microinjected cyclin D1 antibodies, cyclin D1 antisense, inhibitory peptides derived from p16^{INK4a}, and small drug inhibitors all result in G1 arrest (Sherr and Roberts 1999).

Mice lacking Cdk4 are viable and display proliferative defects in a limited range of endocrine cell types, indicating Cdk4 is dispensable for proliferation in most situations (Tsutsui et al. 1999; Moons et al. 2002). Likewise, Cdk6-/- mice are also viable and for the most part develop normally (Malumbres et al. 2004). Cdk6 is preferentially expressed in hematopoietic cells, and its absence leads to delayed G1 progression in lymphocytes but not mouse embryo fibroblasts (MEFs) (Meyerson et al. 1992). Viability of single knockouts and their normal cell proliferation was initially thought to reflect compensation by the remaining family member.

Cdk4/6 double knockouts have now been generated and reveal that the above interpretation is only partially correct (Malumbres et al. 2004). Embryos lacking Cdk4 and Cdk6 die during late stages of embryogenesis due to severe anemia. However, they display normal organogenesis and most cell types continue to proliferate. In fact, embryonic fibroblasts (MEFs) derived from these animals can be immortalized (Malumbres et al. 2004). Quiescent MEFs lacking Cdk4/6 still respond to serum stimulation and enter S-phase with normal kinetics, albeit with lower efficiency. Maintenance of mitogen responsiveness in the absence of Cdk4/6 was quite surprising, especially in light of their reduced Rb phosphorylation and delayed expression of cyclin E and cyclin A. There is some evidence that Cdk2 partially compensates for the absence of Cdk4 and Cdk6, since its reduction by siRNA inhibited proliferation of knockout but not wild-type MEFs. Nevertheless, it appears full Rb modification is not necessary for G1 progression during mouse development, arguing against a stringent coupling of Cdk activation with initiation of DNA synthesis.

As activators of Cdk4/6, the D-type cyclins (1–3) are also viewed as essential links between environmental signals and control of cell proliferation (Sherr 1994). Mice lacking individual D-type cyclins are viable but display tissue specific phenotypes, suggesting they partially compensate for each other. Cyclin D1 knockout mice exhibit neurological abnormalities during development, and have hypoplastic retinas and mammary glands (i.e. underdeveloped tissues due to a decreased number of cells) (Fantl et al. 1995). Cyclin D2-/- females are sterile and males have hypoplastic testes (Sicinski et al. 1996). They also display cerebellar abnormalities, impaired proliferation of B lymphocytes, and hypoplasia of pancreatic Beta cells (Solvason et al. 2000). These phenotypes are similar to those occurring in Cdk4-/- mice, indicating a genetic link consistent with their biochemical partnership. Likewise, mice lacking cyclin D3 display defects in T lymphocyte development similar to Cdk6-/- mice (Sicinska et al. 2003). Analysis of double cyclin D knockouts indicated additive effects and failed to reveal any novel phenotypes (Ciemerych et al. 2002). These results suggested that one D type cyclin might be sufficient for development and viability, similar to budding yeast where two out of the three G1 Cln-type cyclins can be deleted (Richardson et al. 1989).

Triple cyclin D1-3 knockout mice were generated to directly test whether D-type cyclins are required for development and viability (Kozar et al. 2004). These animals survive until mid/late gestation and die due to heart abnormalities and anemia. Cause of death suggests that D-type cyclins are required for expansion of hematopoietic stem cells. Nevertheless, the majority of mouse tissues develop in their absence, indicating D-type cyclins are not required for proliferation of most mammalian cell types. Consistent with this prediction, MEFs lacking all D-type cyclins proliferate in culture (Kozar et al. 2004). Remarkably, they still exit the cell cycle when serum starved and re-enter upon refeeding, although increased mitogens are required. Levels of other cell cycle regulators such as cyclin E and A are unaffected and Rb is still phosphorylated. As was observed in Cdk4/6 knockouts, cells lacking all D-type cyclins appear to rely at least partially on Cdk2 activity to shoulder the burden of Rb phosphorylation (Malumbres 2004).

During mouse development D-type cyclins and Cdk4/6 are only necessary in a few select compartments. While surprising, these results are consistent with earlier observations that cyclin D or Cdk4 ablation affects post-embryonic growth but not embryonic development in *Caenorhabditis elegans* and *Drosophila* (Datar et al. 2000; Meyer et al. 2002). Cells derived from knockout mice lacking Cdk4/6 or all forms of cyclin D can clearly proliferate and respond to mitogens, in apparent contradiction to previous work suggesting that cyclin D-Cdk4/6 is ubiquitously required (Sherr and Roberts 1995). It therefore remains unclear how mitogen responsiveness is linked to cell cycle progression. The cell culture derived model of G1 progression posits that cyclin D generates cyclin E in order to connect extracellular signals and control of S-phase entry. In support of this interpretation, all cyclin D1-/- mice phenotypes were rescued by inserting cyclin E into the cyclin D1 gene locus (Geng et al. 1999). This explanation is now called into question because cells lacking D type cyclins or Cdk4/6 still express cyclin E and enter S-phase (Kozar et al. 2004). Based on these surprising results, it is necessary to reconsider whether a linear series of interdependent events initiated by cyclin D-Cdk4/6 lies at the heart of G1 progression.

4.2
Cyclin E/Cdk2

Cdk2 and Cdk3 are related kinases associating with E and A type cyclins to drive G1 progression and initiate S-phase (Sherr, 1994; Morgan 1995). Work in cultured cells showed that dominant negative Cdk2, microinjection of Cdk2 antibodies, Cdk2 antisense, and Cdk2 inhibitors (protein and small molecule) all block DNA synthesis and cell proliferation (Sherr and Roberts 1999). Because the Cdk3 gene is inactivated in commonly used mouse strains, targeting Cdk2 is sufficient to determine whether these complexes are truly necessary (Ye et al. 2001). The fact this experiment was only recently performed attests to the widespread assumption that it would be uninformative because the predicted outcome (early embryonic lethality) was so obvious.

Remarkably, Cdk2-/- mice are viable and survive up to 2 years (Berthet et al. 2003; Ortega et al. 2003). Embryonic fibroblasts lacking Cdk2 exhibit relatively normal proliferation, with a slight delay in S-phase entry (Berthet et al. 2003). Cdk2-/- cells enter crisis earlier than wild-type yet can still be immortalized, albeit somewhat less efficiently. Their response to DNA damage appears normal. While Cdk2 is not essential for mitotic cell division of most if not all cell types, it is necessary for completion of prophase 1 during meiotic cell division in male and female germ cells. This requirement explains sterility of both male and female knockouts. Experiments are underway to try and explain the apparent differential Cdk dependencies in cultured cells and those derived from Cdk2 knockout mice. Elimination of a conditional Cdk2 allele in immortal MEFs did not affect proliferation, arguing against developmental plasticity (Ortega et al. 2003). Instead, compensation (probably by cyclin A/Cdc2) might explain the benign phenotype of Cdk2-/- animals (Berthet et al. 2003).

Cyclins E1 and E2 are closely related and associate with Cdk2/3 to drive G1 progression and prepare for S-phase (Sherr 1994; Geng et al. 2001). Cyclin E/Cdk2 activity is thought to be required because blocking its function with microinjected cyclin E1 antibodies, cyclin E1 antisense, inhibitory peptides derived from p27 or p21, and small drug inhibitors all result in G1 arrest (Sherr and Roberts 1999). Cyclin E/Cdk2 activity also appears at least partially rate limiting for G1 progression because cyclin E overexpression shortens G1.

Mice lacking either cyclin E1 or E2 develop normally and are viable, although some cyclin E2-/- males exhibit defective spermatogenesis (Geng et al. 2003). Again, compensation was invoked to explain viability and the relatively benign phenotypes. The double cyclin E1/2 knockout results in mid-gestational embryonic lethality (Geng et al. 2003). This dramatic effect is curious given viability of mice lacking Cdk2, and suggests that cyclin E performs essential Cdk2 independent functions. Surprisingly, cause of death is not a failure of embryonic cell proliferation, but rather placental defects arising from severely compromised endoreplication of trophoblast giant cells and megakaryocytes (Geng et al. 2003; Parisi et al. 2003). This point was driven home by tetraploid rescue of trophoblast endoreplication in the cyclin E double knockouts, which resulted in normal development of late gestational embryos. Thus, like Cdk2/3, the E type cyclins are dispensable for mouse development.

MEFs lacking both cyclin E1 and E2 proliferate under conditions of continuous cell cycling (Sherr and Roberts 2004). Rb is still phosphorylated, possibly by cyclin A/Cdk2. However, serum starved cells lacking both cyclin E1 and E2 are unable to re-enter the cell cycle upon re-feeding despite normal induction of cyclin A, cyclin A/Cdk2 kinase activity, and Rb phosphorylation, indicating cyclin E performs a unique function(s) during this period (Geng et al. 2003). The molecular basis of S-phase entry varies depending on whether cells come from G0 or M. During continuous cycling the MCM helicase binds to origins immediately after exit from mitosis and in the absence of cyclin E/Cdk2 activity (Mendez and Stillman 2000). In contrast, MCM is displaced from chromatin in G0 cells and hence must be reloaded for S-phase to occur (Depamphilis 2003). Further analysis of serum starved/re-fed MEFs lacking cyclin E1 and E2 revealed a failure to incorporate MCM proteins onto DNA origins, consistent with the previously discussed role for cyclin E/Cdk2 in this process (Geng et al. 2003). This explanation is also supported by a requirement for cyclin E in MCM loading during *Drosophila* endoreplication cycles (Su and O'Farrell 1998).

Consequences of ablating cyclin E/Cdk2 in mice are clearly different from *C. elegans* and *Drosophila*, where cyclin E is required for development

(Knoblich et al. 1994; Fay and Han 2000). In fact, cyclin E inactivation in *Drosophila* blocks all mitotic cycles and endocycles (Follette et al. 1998). In contrast, analysis of mammalian cells derived from knockout mice indicates E-type cyclins are critically required in only a few select compartments. The situation is further confused by an extensive data set demonstrating that cyclin E is required and rate limiting in many types of cultured cells. This fundamental variation among species raises a host of intriguing questions about the role of cyclin E and its putative targets in different experimental systems, and suggests care should be taken when extrapolating mouse results to humans (Sherr and Roberts 2004). Together, these observations highlight limitations of current models and emphasize the need to re-evaluate generalizations about how G1 progression is controlled.

4.3
G1 Targets

The main roles of cyclin D-Cdk4/6 are Rb inactivation and sequestration of the cyclin E/Cdk2 inhibitor p27 (Sherr and Roberts 1999). More recently, cyclin D-Cdk4/6 has also been implicated in phosphorylation of SMAD3 to inhibit transcriptional complexes responsible for cell growth inhibition by the TGF beta family (Matsuura et al. 2004). In each of these cases cyclin D-Cdk4/6 removes an impediment to cell cycle progression. Similarly, cyclin E/Cdk2 inactivates Rb and its own inhibitor p27 to alleviate inhibitory thresholds in G1. Therefore, ablating either of these impediments in the mouse was predicted to cause inappropriate proliferation detrimental to development and/or survival.

Mice lacking Rb die at about day 14.5 of gestation (Clarke et al. 1992; Jacks et al. 1992; Lee et al. 1992). Death appeared to result from a compromised hematopoietic system and defective neurogenesis. Rb-/- cells derived from these environments exhibited apoptotic and proliferative defects that could be partially abrogated by deleting E2F1 in the Rb knockout (Tsai et al. 1998), consistent with the idea that Rb plays an important and essential role regulating E2F in the animal.

However, recent work has revealed a much more surprising explanation of Rb-/- lethality (Wu et al. 2003). During very early development cells make a decision whether to become part of the inner cell mass that eventually forms the embryo, or commit to become extraembryonic cells (such as trophoblasts) that help establish the placenta needed for proper development (Alberts et al. 1994). Rb loss causes hyperproliferation of these extraembryonic trophoblast cells, resulting in severe disruption of placenta architecture required for viability. Remarkably, Rb-/- embryos supplied with a wild-type placenta were carried to term and only died after birth (Wu et al. 2003). The animals exhibited no defects in the hematopoietic or nervous systems, suggesting that earlier phenotypes might be the result of non-cell autonomous effects. These results are quite similar in outcome to partial rescue of placenta defects in the cyclin E knockouts.

Rb is therefore largely dispensable during embryogenesis; cells still maintain control of G1 progression and respond appropriately to extracellular signaling. Given the presumed importance of controlling E2F activity during the G1 phase, it will be interesting to see if Rb family members p107 or p130 assume this burden. These results are puzzling because the original analysis of MEFs derived from Rb knockout mice indicated that restriction point control was compromised (Herrera et al. 1996). Rb-/- cells displayed increased levels of E2F dependent transcripts and premature synthesis of cyclin E, resulting in a shorter G1 phase and smaller cells that grew faster than wild-type cells. They were less responsive to mitogen removal and failed to G1 arrest in response to cyclohexamide, consistent with the idea that Rb makes an important contribution to R point control and hence G1 progression. However, it now seems possible that these earlier results are non-cell autonomous and arise from placenta defects.

Given the ability of embryogenesis to proceed in the absence of cyclin D-Cdk4/6 and Rb, it seems reasonable to ask whether E2F is a critical component in this process. E2F/Dp complexes activate or repress transcription of target genes depending on presence or absence of Rb (Frolov and Dyson 2004). Thus, loss of E2F transcription would be expected to severely compromise a cell's ability to proceed through G1. The E2F family has seven members, six of which have been individually ablated in the mouse (Trimarchi and Lees 2002). Phenotypes are tissue specific and indicate extensive functional overlap amongst family members. However, the E2F1-3 triple knockout prevents proliferation of primary mouse embryonic fibroblasts, consistent with a requirement for E2F activity during normal development (Wu et al. 2001). These observations indicate E2F transcriptional activity is required for development both in the mouse and in cultured cells. However, there appear to be fundamental differences in how its activity is positively and negatively regulated in different systems.

If E2F activity is essential, then ablating its Dp partner should have similar consequences. Of the two Dp family members, Dp1 has been knocked out in mice and causes early embryonic death (Kohn et al. 2004). This result is consistent with a basic requirement for E2F/Dp transcriptional activity. However, closer examination revealed that Dp1-/- lethality results from a failure of extraembryonic cell lineages to develop and replicate DNA properly. Putting Dp1-/- stem cells in wild-type blastocysts partially rescued this phenotype, as was the case with both cyclin E and Rb knockouts (Kohn et al. 2004). These surprising results suggest Dp1 is dispensable for development of most tissues. While it is possible that Dp1 has no role in the embryo, its RNA and protein levels are highly expressed during this period. Dp2 might compensate for Dp1 absence, or Dp1-/- embryos could be rescued by wild-type embryonic cells in a non-cell autonomous manner. Both these possibilities are currently under investigation (Kohn et al. 2004).

p27 is thought to establish a critical threshold of Cdk inhibition that must be overcome before cells can proceed through G1 and initiate DNA synthesis (Sherr and Roberts 1995). Ablating p27 was therefore expected to result in deregulated Cdk activity, accelerated G1 transit, and inappropriate S-phase entry. Surprisingly, p27-/- mice are viable and develop normally without any gross morphological or histological defects (Fero et al. 1996; Kiyokawa et al. 1996; Nakayama et al. 1996). Mice lacking p27 are approximately 33% larger than wild-type littermates due to an overall increase in cell number, while heterozygotes are intermediate in size. Although the reason for these size differences is unclear, it indicates the importance of precisely controlling p27 protein levels (Fero et al. 1998). Primary p27-/- MEFs do not exhibit increased Cdk activity or deregulated G1 progression, and they still respond to both mitogenic and antimitogenic signals. Some of the Cdk regulatory roles of p27 in the knockout animals are now supplied by the Rb family member p130 (Coats et al. 1999).

The p27 inhibitory threshold is overcome in part by cyclin E/Cdk2 phosphorylation of p27 at T187, which marks it for ubiquitination and degradation by the proteasome (Sherr and Roberts 1999). However, mice expressing a $p27^{T187A}$ mutant in place of wild-type are still viable and develop normally (Malek et al. 2001). Cells derived from these animals still proliferate, proceeding through G1 and initiating S-phase despite an inability to downregulate $p27^{T187A}$ at the G1/S-phase transition. If p27 establishes an inhibitory threshold, overcoming it does not appear to be a prerequisite for G1 progression and S-phase entry.

Current understanding of p27's role in the cell is based largely on its ability to inhibit cyclin/Cdk complexes. However, increasing evidence suggests p27 performs Cdk independent functions that may better explain its effects on cell fate determination and tumor suppression, as well as phenotypes of the p27 knockout mouse. We recently described a novel cytoplasmic role for p27 regulating the Ras/Map kinase pathway (Moeller et al. 2003). Mitogen stimulation of serum starved cells activates receptor tyrosine kinases, which recruit the adaptor protein GRB2 (growth factor receptor bound protein 2). GRB2 uses its SH3 domains to bind the guanine nucleotide exchange factor SOS, which in turn activates Ras and hence the Map kinase cascade. We found that in response to mitogen stimulation p27 is exported from the nucleus and binds the GRB2 SH3 domain, thereby preventing its interaction with SOS. In a similar type of analysis the Roberts group recently demonstrated cytoplasmic p27 also regulates the RhoA GTPase to influence cell migration (Besson et al. 2004). Although the significance of these observations remains to be seen, we have preliminary data that p27 targeting of GRB2 is disrupted in many different types of breast cancer cells (unpublished data). These observations are consistent with a growing chorus arguing that Cdk deregulation is not necessarily a prerequisite for tumorigenesis (Tetsu and McCormick 2003).

5
Implications and Future Directions

Current cell cycle theory provides a mature and well-supported explanation of G1 progression. Its basic premise is that extracellular signaling initiates temporal activation of unique cyclin/Cdk complexes, which phosphorylate specific substrates to drive G1 progression and prepare for upcoming S-phase (Sherr and Roberts 1999). This model explains many experimental results, and new investigations are continually being designed to confirm its predictive powers. Explanations are only as informative as the questions being addressed, however, and the present paradigm is derived in large part from analyzing proliferation of cultured cells. Observations inconsistent with current thinking inevitably arise in any field, often due to technological advances and/or investigation of different experimental systems. They can be marginalized or rationalized up to a point, but facts are persistent things; eventually their implications must be considered to advance understanding. Unanticipated consequences of ablating G1 regulators in mice suggest we are now at such a juncture in the cell cycle field.

5.1
Conundrums

Extensive analysis of cultured cell proliferation indicates that cyclin/Cdk complexes are required for G1 progression (Sherr and Roberts 1999). In contrast, ablating these same cyclin/Cdk complexes in mice does not prevent embryogenesis or cell proliferation, indicating they are not required for G1 progression (Sherr and Roberts 2004). As Robert Louis Stevenson wrote in *Catriona* (the sequel to *Kidnapped*), "I could see no way out of the pickle I was in no way so much as to return to the room I had just left." With adversity comes opportunity, however, and three general approaches to resolving this dilemma – rationalization, revision, or reinterpretation – each has important implications for re-conceptualizing G1 progression.

Rationalizing conflicting data with established theory is quite popular when an extensive body of work is called into question. During early stages of embryonic development cells cycle between S and M phases without intervening gaps, so G1 cyclin/Cdk complexes might be unnecessary at this stage (Alberts et al. 1994). However, this explanation cannot explain why significant proliferation and development still occurs normally even after G1 phase is introduced. Redundancy, compensation, and/or developmental plasticity have all been invoked to explain how extensive mouse embryogenesis takes place without G1 regulators (Murray 2004; Pagano and Jackson 2004; Sherr and Roberts 2004). Although primacy of cyclin/Cdk complexes is maintained by such maneuvers, they are nevertheless difficult to rationalize with basic tenets of G1 control. Cyclins are thought to determine Cdk substrate specificity that is essential for regulated G1 progression (Murray and Hunt 1993). However, this idea must be reconsidered if cyclins and/or Cdks are interchangeable and can readily substitute for one another. The temporal order and timing of cyclin/Cdk activity is also presumed critical during G1, and so it is difficult to envision how progression occurs normally without the complete complement of cyclins and Cdks. Finally, if compensation or redundancy permit normal proliferation during embryogenesis, why did these mechanisms fail to rescue proliferation when cyclins and/or Cdks were inhibited in wild-type culture cells?

A more radical solution is re-interpretation of a data set and its implications. The non-essential nature of G1 regulators during embryogenesis is reminiscent of a checkpoint that lies dormant until needed (Kastan and Bartek 2004). Perhaps cyclin/Cdk complexes represent a type of stress response, stopping and starting the cell cycle in response to low nutrients or other challenges (Pagano and Jackson 2004). Such controls may not be needed in an evolutionarily proscribed environment. This hypothesis is attractive because establishing cells in culture is a form of stress, and G1 regulators appear necessary under these conditions. However, it also assumes existence of a completely different mechanism, as yet unidentified, responsible for normal operation of G1. Such an unlikely situation could conceivably arise if cell cycle control in complex environments were subject to its own uncertainty principle, i.e. any attempt to study it alters the process by inducing a stress response and cyclin/Cdk intervention. The Achilles heel of this idea is that immortalized cell lines can be established from knockout mice lacking G1 regulators (Sherr and Roberts 2004). Such cells would not be expected to proliferate if they lack the ability to mount a viable stress response.

5.2
G1 in Context

Between the extremes of rationalization and re-interpretation, it may be possible to steer a middle course. Disrupting G1 regulators generally has little direct effect on embryo development, yet often results in lethality shortly thereafter. Post-gastrulation embryos are quite similar in size and morphology regardless of species, suggesting that their appearance is governed by pre-established rules (Follette and O'Farrell 1997). After this point, however, specific growth programs must be implemented to generate the substantial size variation observed among adult organisms of different species. One interesting possibility is that the physiological demarcation between embryo development and its subsequent growth arises in part from fundamental differences in molecular mechanisms controlling cell proliferation during these two stages.

Cell duplication obviously takes place during embryogenesis, but the overriding concern is generating a body plan for assembling the massive numbers of cells making up the organism (Follette and O'Farrell 1997). The transition between early developmental patterning and subsequent growth of the embryo might involve a switch between different molecular mechanisms controlling cell division. Data from knockout experiments indicate that E2F is required in both situations, and thus its activity is probably controlled by this putative molecular switch.

Because embryogenesis occurs normally despite the absence of G1 machinery thought to regulate E2F, other components controlling its function must exist. During early development cells are continuously exposed to nutrients and growth factors that drive an increase in mass, so these signals in conjunction with a minimum size threshold are possible candidates to regulate cell division (Neufeld and Edgar 1998). Once the embryo is formed, it might be advantageous to transfer control of E2F activity (and hence cell division) to cyclin/Cdk complexes, which can then drive the necessary increase in cell number (Fig. 6).

Several lines of evidence provide support for this model. E2F activation during embryogenesis does not appear to be rate limiting for cell proliferation because deleting Rb has minimal adverse effects (Wu et al. 2003). A more likely candidate for establishing overall division rate is a minimal size threshold. E2F activation may only become rate limiting after the switch to growth

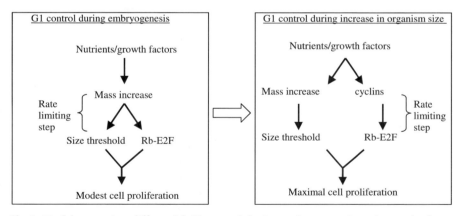

Fig. 6 Model proposing differential G1 control during embryogenesis and growth phase. On the left, embryonic cell proliferation is controlled by an increase in mass and passage of a minimal size threshold. On the right, extensive cell proliferation required for growth of the embryo is controlled by cyclins and Cdks. The need to increase mass and pass a size threshold is still present but no longer the rate limiting step for cell division

phase in order to unleash the cell's maximum proliferative capability. Consistent with this hypothesis, manipulating G1 machinery in cultured wild-type cells clearly affects cell cycle progression (Sherr and Roberts 1995). Furthermore, in cases where ablating G1 regulators is not lethal, final mouse size is often altered. p27-/- mice develop normally but are 33% larger than wild-type, while those lacking individual D-type cyclins, Cdk4, or Cdk2 are smaller than normal littermates (Sherr and Roberts 2004). These effects on organism size are the result of changes in total cell number, consistent with the idea traditional cell cycle machinery influences this parameter.

How might cyclin/Cdk independent control of E2F be generated during embryogenesis? New daughter cells must suppress S-phase entry until the necessary mass increase has been achieved. Rb contains distinct activities that can specifically inhibit cell cycle progression (by targeting E2F) and/or increases in cell mass (by targeting components of the protein synthesis machinery) (White 2004). After division Rb could preferentially associate with E2F and block cell cycle progression, while its effects on growth are muted due to lower levels of protein synthesis components (Fig. 7). As nutrients and growth factors increase overall mass and the cell approaches its minimum size threshold, levels of the mass target become sufficient to compete with E2F for binding Rb. The resulting switch would free E2F, initiate cell cycle progression, and suppress further increases in cell mass (Fig. 7). Transcription factor TFIIIB is a potential candidate for the mass machinery target because it specifically upregulates RNA polymerase III, which generates essential components required for protein synthesis (White 2004).

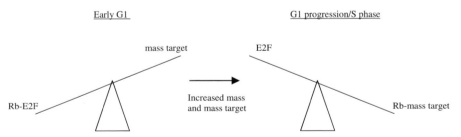

Fig. 7 Hypothetical Rb-mediated switch between increasing cell mass and G1 progression during embryogenesis. Rb negatively regulates both cell cycle progression (by inhibiting E2F) and increases in cell mass (by inhibiting components involved in overall protein synthesis). After embryonic cells exit mitosis Rb might preferentially inhibit E2F because levels of its mass machinery target are low. Cell cycle progression is therefore inhibited while cell mass increases. Once a cell traverses the minimum size threshold, levels of the mass machinery target become sufficient to compete with E2F for binding Rb. As a result E2F is released and drives G1 progression, while further increases in cell mass are suppressed

Existence of different G1 control mechanisms might explain why wild-type cells arrest when G1 cyclin/Cdk complexes are inhibited, yet cells derived from knockouts lacking the very same cyclin/Cdks proliferate normally. Immortalization selects for cell survival and expansion under culture conditions. This situation is most similar to the rapid expansion of cell number following embryogenesis, and thus should favor maintaining cyclin/Cdk control of E2F. In marked contrast, cells derived from cyclin/Cdk knockout mice are forced to utilize the less favorable cyclin/Cdk independent pathway during immortalization. Consistent with this explanation, cells lacking G1 regulators are often harder to immortalize than their wild-type counterparts (Sherr and Roberts 2004). There are several possible reasons the cyclin/Cdk independent pathway might fail to compensate when G1 regulators are targeted in wild-type cells: 1) the two pathways cannot exist simultaneously, 2) the two pathways cannot operate simultaneously, 3) the cyclin/Cdk independent pathway is actively selected against, or 4) components regulating the cyclin/Cdk independent pathway are compromised during immortalization.

The possible existence of distinct mechanisms controlling G1 progression during development has numerous important implications that should be pursued. Cell lines lacking G1 regulators will be quite useful for identifying components of the putative cyclin/Cdk independent pathway. We must also consider whether one or the other of these mechanisms predominates in specific situations or cell types such as stem and cancer cells (Burdon et al. 2002; Tetsu and McCormick 2003). Finally, additional methods of controlling G1 progression may have evolved in response to the increasing complexity of multicellular organisms. If such unanticipated diversity in G1 control exists, it might provide new therapeutic opportunities for discriminating between normal and diseased states.

6
Conclusions

The goals of this chapter were to describe the current understanding of G1 control, and illustrate how surprising results from knockout mice suggest a richer and more complex picture of this period in the cell cycle. Cyclin/Cdk complexes have long been described as the component evolutionarily selected to drive and control G1 progression. The internal logic and predictive capabilities of this model implied universality, but that assumption now appears unwarranted because G1 cyclin/Cdks are not required during mouse development. Nevertheless, the rationale for having G1 remains unchanged: integrating information, increasing cell mass, and deciding cell fate. Evolution may simply have found it necessary to develop additional mechanisms of G1 control optimized for situations unique to multicellular organisms. This hypothesis warns of over-reliance on a reductionist approach, and argues that controls have to be considered in context. Now is not the time to rest on our (or others) laurels, but instead to move the field forward by recognizing and critically examining the limitations of current ideas. As Daniel J. Boorstin recognized, "The greatest obstacle to discovery is not ignorance–it is the illusion of knowledge."

Acknowledgements The authors thank Lucas Nacusi for comments on the manuscript. S.J.M. was supported by a Susan G. Komen Dissertation Award and R.J.S. by an Ellison Foundation New Scholars Award.

References

Albanese C, Johnson J, Watanabe G, Eklund N, Vu D, Arnold A, Pestell RG (1995) Transforming p21^{ras} mutants and c-Ets-2 activate the cyclin D1 promoter through distinguishable regions. J Biol Chem 270:23589–23597

Alberts B, Bray D, Lewis J, Raff M, Roberts K, Watson J (1994) Molecular biology of the cell. Garland Publishing, New York

Bartek J, Bartkova J, Lukas J (1996) The retinoblastoma protein pathway and the restriction point. Curr Opin Cell Biol 8:805–814

Bashir T, Dorrello NV, Amador V, Guardavaccaro D, Pagano M (2004) Control of the SCF^{Skp2}-^{Cks1} ubiquitin ligase by the APC/C^{Cdh1} ubiquitin ligase. Nature 428:190–193

Besson A, Gurian-West M, Schmidt A, Hall A, Roberts JM (2004) p27^{Kip1} modulates cell migration through the regulation of RhoA activation. Genes Dev 18:862–876

Berthet C, Aleem E, Coppola V, Tessarollo L, Kaldis P (2003) Cdk2 knockout mice are viable. Curr Biol 13:1775–1785

Blagosklonny MV, Pardee AB (2002) The restriction point of the cell cycle. Cell Cycle 1:103–110

Buchkovich K, Duffy LA, Harlow E (1989) The retinoblastoma protein is phosphorylated during specific phases of the cell cycle. Cell 58:1097–1105

Burdon T, Smith A, Savatier P (2002) Signalling, cell cycle and pluripotency in embryonic stem cells. Trends Cell Biol 12:432–438

Campisi J, Medrano EE, Morreo G, Pardee AB (1982) Restriction point control of cell growth by a labile protein: evidence for increased stability in transformed cells. Proc Natl Acad Sci USA 79:436–440

Chen PL, Scully P, Shew JY, Wang JY, Lee WH (1989) Phosphorylation of the retinoblastoma gene product is modulated during the cell cycle and cellular differentiation. Cell 58:1193–1198

Ciemerych MA, Kenney AM, Sicinska E, Kalaszczynska I, Bronson RT, Rowitch DH, Gardner H, Sicinski P (2002) Development of mice expressing a single D-type cyclin. Genes Dev 16:3277–3289

Clarke AR, Maandag ER, van Roon M, van der Lugt NM, van der Valk M, Hooper ML, Berns A, te Riele H (1992) Requirement for a functional Rb-1 gene in murine development. Nature 359:328–330

Clurman BE, Sheaff RJ, Thress K, Groudine M, Roberts JM (1996) Turnover of cyclin E by the ubiquitin–proteasome pathway is regulated by Cdk2 binding and cyclin phosphorylation. Genes Dev 10:1979–1990

Coats S, Whyte P, Fero ML, Lacy S, Chung G, Randel E, Firpo E, Roberts JM (1999) A new pathway for mitogen-dependent Cdk2 regulation uncovered in p27^{Kip1}-deficient cells. Curr Biol 9:163–173

Conlon I, Raff M (2003) Differences in the way a mammalian cell and yeast cells coordinate cell growth and cell-cycle progression. J Biol 2:7

Contreras A, Hale TK, Stenoien DL, Rosen JM, Mancini MA, Herrera RE (2003) The dynamic mobility of histone H1 is regulated by cyclin/Cdk phosphorylation. Mol Cell Biol 23:8626–8636

Coverley D, Laman H, Laskey RA (2002) Distinct roles for cyclins E and A during DNA replication complex assembly and activation. Nat Cell Biol 4:523–528

Datar SA, Jacobs HW, de la Cruz AF, Lehner CF, Edgar BA (2000) The *Drosophila* cyclin D-Cdk4 complex promotes cellular growth. Embo J 19:4543–4554

De Bondt HL, Rosenblatt J, Jancarik J, Jones HD, Morgan DO, Kim SH (1993) Crystal structure of cyclin-dependent kinase 2. Nature 363:595–602

DePamphilis ML (2003) The "ORC cycle": a novel pathway for regulating eukaryotic DNA replication. Gene 310:1–15

Diehl JA, Zindy F, Sherr CJ (1997) Inhibition of cyclin D1 phosphorylation on threonine-286 prevents its rapid degradation via the ubiquitin-proteasome pathway. Genes Dev 11:957–972

Diffley JF, Labib K (2002) The chromosome replication cycle. J Cell Sci 115:869–872

Dolznig H, Grebien F, Sauer T, Beug H, Mullner EW (2004) Evidence for a size-sensing mechanism in animal cells. Nat Cell Biol 6:899–905

Dynlacht BD, Flores O, Lees JA, Harlow E (1994) Differential regulation of E2F transactivation by cyclin/Cdk2 complexes. Genes Dev 8:1772–1786

Evans T, Rosenthal ET, Youngblom J, Distel D, Hunt T (1983) Cyclin: a protein specified by maternal mRNA in sea urchin eggs that is destroyed at each cleavage division. Cell 33:389–396

Fantl V, Stamp G, Andrews A, Rosewell I, Dickson C (1995) Mice lacking cyclin D1 are small and show defects in eye and mammary gland development. Genes Dev 9:2364–2372

Fay DS, Han M (2000) Mutations in cye-1, a *Caenorhabditis elegans* cyclin E homolog, reveal coordination between cell-cycle control and vulval development. Development 127:4049–4060

Fero ML, Rivkin M, Tasch M, Porter P, Carow CE, Firpo E, Polyak K, Tsai LH, Broudy V, Perlmutter RM, Kaushansky K, Roberts JM (1996) A syndrome of multiorgan hyperplasia with features of gigantism, tumorigenesis, and female sterility in p27^{Kip1}-deficient mice. Cell 85:733–744

Fero ML, Randel E, Gurley KE, Roberts JM, Kemp CJ (1998) The murine gene p27^{Kip1} is haplo-insufficient for tumour suppression. Nature 396:177–180

Flemming W (1965) Contributions to the knowledge of the cell and its vital processes. J Cell Biol 25:1–69

Follette PJ, O'Farrell PH (1997) Connecting cell behavior to patterning: lessons from the cell cycle. Cell 88:309–314

Follette PJ, Duronio RJ, O'Farrell PH (1998) Fluctuations in cyclin E levels are required for multiple rounds of endocycle S phase in Drosophila. Curr Biol 8:235–238

Friend SH, Horowitz JM, Gerber MR, Wang XF, Bogenmann E, Li FP, Weinberg RA (1987) Deletions of a DNA sequence in retinoblastomas and mesenchymal tumors: organization of the sequence and its encoded protein. Proc Natl Acad Sci USA 84:9059–9063

Frolov MV, Dyson NJ (2004) Molecular mechanisms of E2F-dependent activation and pRB-mediated repression. J Cell Sci 117:2173–2181

Frost JA, Alberts AS, Sontag E, Guan K, Mumby MC, Feramisco JR (1994) Simian virus 40 small t antigen cooperates with mitogen-activated kinases to stimulate AP-1 activity. Mol Cell Biol 14:6244-6252

Geng Y, Whoriskey W, Park MY, Bronson RT, Medema RH, Li T, Weinberg RA, Sicinski P (1999) Rescue of cyclin D1 deficiency by knockin cyclin E. Cell 97:767-777

Geng Y, Yu Q, Whoriskey W, Dick F, Tsai KY, Ford HL, Biswas DK, Pardee AB, Amati B, Jacks T, Richardson A, Dyson N, Sicinski P (2001) Expression of cyclins E1 and E2 during mouse development and in neoplasia. Proc Natl Acad Sci USA 98:13138-13143

Geng Y, Yu Q, Sicinska E, Das M, Schneider JE, Bhattacharya S, Rideout WM, Bronson RT, Gardner H, Sicinski P (2003) Cyclin E ablation in the mouse. Cell 114:431-443

Harbour JW, Dean DC (2000) The Rb/E2F pathway: expanding roles and emerging paradigms. Genes Dev 14:2393-2409

Hartwell LH, Mortimer RK, Culotti J, Culotti M (1973) Genetic control of the cell division cycle in yeast: genetic analysis of cdc mutants. Genetics 74:267-286

Hartwell LH (1974) *Saccharomyces cerevisiae* cell cycle. Bacteriol Rev 38:164-198

Hartwell LH, Weinert TA (1989) Checkpoints: controls that ensure the order of cell cycle events. Science 246:629-634

Hatakeyama M, Herrera RA, Makela T, Dowdy SF, Jacks T, Weinberg RA (1994) The cancer cell and the cell cycle clock. Cold Spring Harb Symp Quant Biol 59:1-10

Herrera RE, Sah VP, Williams BO, Makela TP, Weinberg RA, Jacks T (1996) Altered cell cycle kinetics, gene expression, and G1 restriction point regulation in Rb-deficient fibroblasts. Mol Cell Biol 16:2402-2407

Hinchcliffe EH, Li C, Thompson EA, Maller JL, Sluder G (1999) Requirement of Cdk2-cyclin E activity for repeated centrosome reproduction in *Xenopus* egg extracts. Science 283:851-854

Jacks T, Fazeli A, Schmitt EM, Bronson RT, Goodell MA, Weinberg RA (1992) Effects of an Rb mutation in the mouse. Nature 359:295-300

Kastan MB, Bartek J (2004) Cell-cycle checkpoints and cancer. Nature 432:316-323

Kiyokawa H, Kineman RD, Manova-Todorova KO, Soares VC, Hoffman ES, Ono M, Khanam D, Hayday AC, Frohman LA, Koff A (1996) Enhanced growth of mice lacking the cyclin-dependent kinase inhibitor function of p27^{Kip1}. Cell 85:721-732

Knoblich JA, Sauer K, Jones L, Richardson H, Saint R, Lehner CF (1994) Cyclin E controls S phase progression and its down-regulation during *Drosophila* embryogenesis is required for the arrest of cell proliferation. Cell 77:107-120

Kohn MJ, Leung SW, Criniti V, Agromayor M, Yamasaki L (2004) Dp1 is largely dispensable for embryonic development. Mol Cell Biol 24:7197-7205

Kozar K, Ciemerych MA, Rebel VI, Shigematsu H, Zagozdzon A, Sicinska E, Geng Y, Yu Q, Bhattacharya S, Bronson RT, Akashi K, Sicinski P (2004) Mouse development and cell proliferation in the absence of D-cyclins. Cell 118:477-491

Lee EY, Chang CY, Hu N, Wang YC, Lai CC, Herrup K, Lee WH, Bradley A (1992) Mice deficient for Rb are nonviable and show defects in neurogenesis and haematopoiesis. Nature 359:288-294

Liu H, Dibling B, Spike B, Dirlam A, Macleod K (2004) New roles for the RB tumor suppressor protein. Curr Opin Genet Dev 14:55-64

Lohka MJ, Hayes MK, Maller JL (1988) Purification of maturation-promoting factor, an intracellular regulator of early mitotic events. Proc Natl Acad Sci USA 85:3009-3013

Lukas J, Bartkova J, Rohde M, Strauss M, Bartek J (1995) Cyclin D1 is dispensable for G1 control in retinoblastoma gene-deficient cells independently of Cdk4 activity. Mol Cell Biol 15:2600-2611

Malek NP, Sundberg H, McGrew S, Nakayama K, Kyriakides TR, Roberts JM (2001) A mouse knock-in model exposes sequential proteolytic pathways that regulate p27^{Kip1} in G1 and S phase. Nature 413:323–327

Malumbres M, Sotillo R, Santamaria D, Galan J, Cerezo A, Ortega S, Dubus P, Barbacid M (2004) Mammalian cells cycle without the D-type cyclin-dependent kinases Cdk4 and Cdk6. Cell 118:493–504

Masui Y, Markert CL (1971) Cytoplasmic control of nuclear behavior during meiotic maturation of frog oocytes. J Exp Zool 177:129–145

Matsushime H, Ewen ME, Strom DK, Kato JY, Hanks SK, Roussel MF, Sherr CJ (1992) Identification and properties of an atypical catalytic subunit (p34PSK-J3/cdk4) for mammalian D type G1 cyclins. Cell 71:323–334

Matsuura I, Denissova NG, Wang G, He D, Long J, Liu F (2004) Cyclin-dependent kinases regulate the antiproliferative function of Smads. Nature 430:226–231

Mendez J, Stillman B (2000) Chromatin association of human origin recognition complex, cdc6, and minichromosome maintenance proteins during the cell cycle: assembly of prereplication complexes in late mitosis. Mol Cell Biol 20:8602–8612

Meyer CA, Jacobs HW, Lehner CF (2002) Cyclin D-Cdk4 is not a master regulator of cell multiplication in *Drosophila* embryos. Curr Biol 12:661–666

Meyerson M, Enders GH, Wu CL, Su LK, Gorka C, Nelson C, Harlow E, Tsai LH (1992) A family of human cdc2-related protein kinases. EMBO J 11:2909–2917

Mitchison JM (1971) The biology of the cell cycle. Cambridge University Press, London

Moeller SJ, Head ED, Sheaff RJ (2003). p27^{Kip1} inhibition of GRB2-SOS formation can regulate Ras activation. Mol Cell Biol 23:3735–3752

Moons DS, Jirawatnotai S, Parlow AF, Gibori G, Kineman RD, Kiyokawa H (2002) Pituitary hypoplasia and lactotroph dysfunction in mice deficient for cyclin-dependent kinase-4. Endocrinology 143:3001–3008

Morgan DO (1995) Principles of Cdk regulation. Nature 374:131–134

Murrary A, Hunt T (1993) The cell cycle: an introduction. Freeman New York

Murray AW (2004) Recycling the cell cycle: cyclins revisited. Cell 116:221–234

Murray AW, Solomon MJ, Kirschner MW (1989) The role of cyclin synthesis and degradation in the control of maturation promoting factor activity. Nature 339:280–286

Nakayama K, Ishida N, Shirane M, Inomata A, Inoue T, Shishido N, Horii I, Loh DY (1996) Mice lacking p27^{Kip1} display increased body size, multiple organ hyperplasia, retinal dysplasia, and pituitary tumors. Cell 85:707–720

Nasmyth K (2001) A prize for proliferation. Cell 107:689–701

Neufeld TP, Edgar BA (1998) Connections between growth and the cell cycle. Curr Opin Cell Biol 10:784–790

Nurse P (1975) Genetic control of cell size at cell division in yeast. Nature 256:547–551

Nurse P (1990) Universal control mechanism regulating onset of M-phase. Nature 344:503–508

Nurse P (2000a) The incredible life and times of biological cells. Science 289:1711–1716

Nurse P (2000b) A long twentieth century of the cell cycle and beyond. Cell 100:71–78

Nurse P, Thuriaux P, Nasmyth K (1976) Genetic control of the cell division cycle in the fission yeast *Schizosaccharomyces pombe*. Mol Gen Genet 146:167–178

Ohtani K, DeGregori J, Nevins JR (1995) Regulation of the cyclin E gene by transcription factor E2F1. Proc Natl Acad Sci USA 92:12146–12150

Ortega S, Prieto I, Odajima J, Martin A, Dubus P, Sotillo R, Barbero JL, Malumbres M, Barbacid M (2003) Cyclin-dependent kinase 2 is essential for meiosis but not for mitotic cell division in mice. Nat Genet 35:25–31

Pagano M, Jackson PK (2004) Wagging the dogma; tissue-specific cell cycle control in the mouse embryo. Cell 118:535–538

Pardee AB (1974) A restriction point for control of normal animal cell proliferation. Proc Natl Acad Sci USA 71:1286–1290

Pardee AB (1989) G1 events and regulation of cell proliferation. Science 246:603–608

Parisi T, Beck AR, Rougier N, McNeil T, Lucian L, Werb Z, Amati B (2003) Cyclins E1 and E2 are required for endoreplication in placental trophoblast giant cells. Embo J 22:4794–4803

Reynisdottir I, Polyak K, Iavarone A, Massague J (1995) Kip/Cip and Ink4 Cdk inhibitors cooperate to induce cell cycle arrest in response to TGF-beta. Genes Dev 9:1831–1845

Richardson HE, Wittenberg C, Cross F, Reed SI (1989) An essential G1 function for cyclin-like proteins in yeast. Cell 59:1127–1133

Roberts JM, Koff A, Polyak K, Firpo E, Collins S, Ohtsubo M, Massague J (1994) Cyclins, Cdks, and cyclin kinase inhibitors. Cold Spring Harb Symp Quant Biol 59:31–38

Roussel MF, Theodoras AM, Pagano M, Sherr CJ (1995) Rescue of defective mitogenic signaling by D-type cyclins. Proc Natl Acad Sci USA 92:6837–6841

Russo AA, Jeffrey PD, Patten AK, Massague J, Pavletich NP (1996) Crystal structure of the p27Kip1 cyclin-dependent-kinase inhibitor bound to the cyclin A-Cdk2 complex. Nature 382:325-331

Rupes I (2002) Checking cell size in yeast. Trends Genet 18:479–485

Saucedo LJ, Edgar BA (2002) Why size matters: altering cell size. Curr Opin Genet Dev 12:565–571

Sheaff RJ, Groudine M, Gordon M, Roberts JM, Clurman BE (1997) Cyclin E/Cdk2 is a regulator of $p27^{Kip1}$. Genes Dev 11:1464–1478

Sherr CJ (1994) G1 phase progression: cycling on cue. Cell 79:551–555

Sherr CJ, Roberts JM (1995) Inhibitors of mammalian G1 cyclin-dependent kinases. Genes Dev 9:1149–1163

Sherr CJ, Roberts JM (1999) Cdk inhibitors: positive and negative regulators of G1-phase progression. Genes Dev 13:1501–1512

Sherr CJ, Roberts JM (2004) Living with or without cyclins and cyclin-dependent kinases. Genes Dev 18:2699–2711

Sicinska E, Aifantis I, Le Cam L, Swat W, Borowski C, Yu Q, Ferrando AA, Levin SD, Geng Y, von Boehmer H, Sicinski P (2003) Requirement for cyclin D3 in lymphocyte development and T cell leukemias. Cancer Cell 4:451–461

Sicinski P, Donaher JL, Geng Y, Parker SB, Gardner H, Park MY, Robker RL, Richards JS, McGinnis LK, Biggers JD, Eppig JJ, Bronson RT, Elledge SJ, Weinberg RA (1996) Cyclin D2 is an FSH-responsive gene involved in gonadal cell proliferation and oncogenesis. Nature 384:470–474

Solvason N, Wu WW, Parry D, Mahony D, Lam EW, Glassford J, Klaus GG, Sicinski P, Weinberg R, Liu YJ, Howard M, Lees E (2000) Cyclin D2 is essential for BCR-mediated proliferation and CD5 B cell development. Int Immunol 12:631–638

Stevaux O, Dyson NJ (2002) A revised picture of the E2F transcriptional network and RB function. Curr Opin Cell Biol 14:684–691

Su TT, O'Farrell PH (1998a) Chromosome association of minichromosome maintenance proteins in *Drosophila* endoreplication cycles. J Cell Biol 140:451–460

Su TT, O'Farrell PH (1998b) Size control: cell proliferation does not equal growth. Curr Biol 8:R687–R689

Takaki T, Fukasawa K, Suzuki-Takahashi I, Hirai H (2004) Cdk-mediated phosphorylation of pRB regulates HDAC binding in vitro. Biochem Biophys Res Commun 316:252–255

Tetsu O, McCormick F (2003) Proliferation of cancer cells despite Cdk2 inhibition. Cancer Cell 3:233–245

Trimarchi JM, Lees JA (2002) Sibling rivalry in the E2F family. Nat Rev Mol Cell Biol 3:11–20

Tsai KY, Hu Y, Macleod KF, Crowley D, Yamasaki L, Jacks T (1998) Mutation of E2f-1 suppresses apoptosis and inappropriate S phase entry and extends survival of Rb-deficient mouse embryos. Mol Cell 2:293–304

Tsutsui T, Hesabi B, Moons DS, Pandolfi PP, Hansel KS, Koff A, Kiyokawa H (1999) Targeted disruption of CDK4 delays cell cycle entry with enhanced p27^{Kip1} activity. Mol Cell Biol 19:7011–7019

Tsvetkov LM, Yeh KH, Lee SJ, Sun H, Zhang H (1999) p27^{Kip1} ubiquitination and degradation is regulated by the SCF^{Skp2} complex through phosphorylated Thr187 in p27. Curr Biol 9:661–664

van den Heuvel S, Harlow E (1993) Distinct roles for cyclin-dependent kinases in cell cycle control. Science 262:2050–2054

Weinberg RA (1995) The retinoblastoma protein and cell cycle control. Cell 81:323–330

White RJ (2004) RNA polymerase III transcription and cancer. Oncogene 23:3208–3216

Winston JT, Pledger WJ (1993) Growth factor regulation of cyclin D1 mRNA expression through protein synthesis-dependent and -independent mechanisms. Mol Biol Cell 4:1133–1144

Woo RA, Poon RY (2003) Cyclin-dependent kinases and S phase control in mammalian cells. Cell Cycle 2:316–324

Wu L, Timmers C, Maiti B, Saavedra HI, Sang L, Chong GT, Nuckolls F, Giangrande P, Wright FA, Field SJ, Greenberg ME, Orkin S, Nevins JR, Robinson ML, Leone G (2001) The E2F1-3 transcription factors are essential for cellular proliferation. Nature 414:457–462

Wu L, de Bruin A, Saavedra HI, Starovic M, Trimboli A, Yang Y, Opavska J, Wilson P, Thompson JC, Ostrowski MC, Rosol TJ, Woollett LA, Weinstein M, Cross JC, Robinson ML, Leone G (2003) Extra-embryonic function of Rb is essential for embryonic development and viability. Nature 421:942–947

Ye X, Zhu C, Harper JW (2001) A premature-termination mutation in the *Mus musculus* cyclin-dependent kinase 3 gene. Proc Natl Acad Sci USA 98:1682–1686

Zetterberg A, Larsson O (1985) Kinetic analysis of regulatory events in G1 leading to proliferation or quiescence of Swiss 3T3 cells. Proc Natl Acad Sci USA 82:5365–5369

Zetterberg A, Larsson O (1991) Coordination between cell growth and cell cycle transit in animal cells. Cold Spring Harb Symp Quant Biol 56:137–147

Zetterberg A, Engstrom W, Larsson O (1982) Growth activation of resting cells: induction of balanced and imbalanced growth. Ann N Y Acad Sci 397:130–147

Regulation of S Phase

Jamie K. Teer[1,2] · Anindya Dutta[1,2] (✉)

[1]Biological and Biomedical Sciences Program, Harvard Medical School,
Boston, MA 02115, USA
ad8q@virginia.edu

[2]Dept. Of Biochemistry, University of Virginia, Charlottesville, VA 22908, USA
ad8q@virginia.edu

Abstract Regulation of DNA replication is critical for accurate and timely dissemination of genomic material to daughter cells. The cell uses a variety of mechanisms to control this aspect of the cell cycle. There are various determinants of origin identification, as well as a large number of proteins required to load replication complexes at these defined genomic regions. A pre-Replication Complex (pre-RC) associates with origins in the G1 phase. This complex includes the Origin Recognition Complex (ORC), which serves to recognize origins, the putative helicase MCM2-7, and other factors important for complex assembly. Following pre-RC loading, a pre-Initiation Complex (pre-IC) builds upon the helicase with factors required for eventual loading of replicative polymerases. The chromatin association of these two complexes is temporally distinct, with pre-RC being inhibited, and pre-IC being activated by cyclin-dependent kinases (Cdks). This regulation is the basis for replication licensing, which allows replication to occur at a specific time once, and only once, per cell cycle. By preventing extra rounds of replication within a cell cycle, or by ensuring the cell cycle cannot progress until the environmental and intracellular conditions are most optimal, cells are able to carry out a successful replication cycle with minimal mutations.

1
Introduction

DNA replication is fundamentally critical, and yet rather problematic for all life. Cells must be prepared to replicate the entire genome, and must do so in a concerted, rapid, and efficient fashion. Failures in this process not only yield potentially damaging mutations, but may also hinder proper genome segregation between offspring. This problem becomes even more pronounced with increasing size of a given genome. Cells have a variety of mechanisms to ensure that the proper environment exists to carry out replication. These mechanisms can be divided into proper identification of appropriate origins of replication, and subsequent loading of the replication machinery itself. We will focus our discussion on the initiation of replication in eukaryotes, and how it can be used by the cells to control S-phase progression.

2
Origins of Replication

2.1
Genome Replicator Sequences

Theoretically, replication would be carried out most efficiently by starting from many different evenly spaced sites. Such a model might assume, however, that specific, conserved origins of replication exist. Studies on origins in various systems support the early theory of a *replicator* sequence that marks the origin of replication, and an *initiator* protein that binds this sequence and recruits downstream factors required for replication (Jacob et al., 1963). The earliest eukaryotic replicator was found in *Saccharomyces cerevisiae*, and was named ARS for autonomously replicating sequence (Stinchcomb et al., 1979; Struhl et al., 1979). Study of the ARS1 locus in budding yeast revealed conserved sequence blocks that were essential for replication, including an 11 bp consensus seqeunce (A element) and several other B elements (Marahrens and Stillman, 1992). These replicator sequences were used later to identify the putative initiator proteins: the Origin Recognition Complex [ORC] (see below). In *Schizosaccharomyces pombe,* two 30-55 base pair elements essential for replication were discovered in origin ars3002, and similar sequences were found in other ars regions (Dubey et al., 1996). Although the yeasts seem to have high sequence conservation from one replicator to the next, determining such consensus replicators in higher eukaryotes has been more difficult.

Studies in *Xenopus laevis* egg extracts have not identified a consensus replicator sequence. On the contrary, early results indicate the lack of sequence specificity in replicating regions (Hyrien and Mechali, 1992; Hyrien and Mechali, 1993; Mahbubani et al., 1992). Recent studies show that while the ORC proteins may prefer AT rich DNA stretches, they show no preference between defined origin sequences and control sequences in vitro, even with varying ORC concentration (Vashee et al., 2003). Such random origin selection may, however, be a function of the early embryogensis system. When origin selection in the rDNA locus was studied at different times in development, increasing origin specificity was seen as development progressed (Hyrien et al., 1995). In early stages, origin selection was random, but when rDNA gene expression began in late blastula and early gastrula stages, initiation frequency decreased in the transcribed regions. This effectively limited initiation to the intergenic regions. Similar results were observed in *Drosophila* embryos (Sasaki et al., 1999). Interestingly, when intact mammalian nuclei were added to *Xenopus* extracts, they initiated replication at specific sites. Disrupting the nuclei before incubation ablated this specificity (Gilbert et al., 1995). Additionally, intact mammalian nuclei isolated before a certain time in the G1 phase also failed to initiate specifically (Wu and Gilbert, 1996). Taken together, these results indicate that metazoans do seem to initiate replication at specific sites, but this specificity may be determined not by sequence, but by other influences from local chromatin and nuclear environments.

The picture is also complicated in mammalian systems. Early work in chinese hamster ovary cells revealed a replication origin in the dihydrofolate reductase (DHFR) locus (Heintz and Hamlin, 1982; Heintz et al., 1983). Although this origin firing was originally thought to be highly sequence specific, later two dimensional gel electrophoresis showed that origins fire in a broad zone (55 kb) throughout the intergenic region [but not in the DHFR gene itself] (Dijkwel and Hamlin, 1995; Vaughn et al., 1990). These observations argue against a defined sequence specificity. A different study, however, used nascent strand abundance assays on the same DHFR region to demonstrate only two to three major initiation sites, again raising the possibility of sequence specificity (Burhans et al., 1990; Kobayashi et al., 1998). Recently, a study using early labeled fragment hybridization (ELFH) showed that the earliest nascent strands could hybridize to many clones from different areas along the intergenic region, suggesting that replication can be initiated from many different sites (Dijkwel et al., 2002). Additionally, a deletion mapping experiment showed that replication could initiate from the intergenic region in the absence of the major sites, and even in the absence of 90% of the region (Mesner et al., 2003). These studies indicate the regions of potential origin firing in higher eukaryotes may not be determined simply by sequence, but by other factors.

Like Chinese hamster cells, few origins of replication are known in human cells. One of the earlier defined origins of replication is at the β-globin locus. A bidirectional origin was found to exist in the 2 kb region between the δ and β globin genes, and deleting this region abrogated the bidirectional activity (Kitsberg et al., 1993). This result suggested a sequence element may be present in this region to direct origin firing. Furthermore, an ectopically inserted β-globin locus promoted initiation at the ectopic site (Aladjem et al., 1998). Deletion mapping of the β-globin locus showed several sequences critical for replication initiation at the locus, in both ectopic and native locations (Aladjem et al., 1998; Wang et al., 2004). Similar results have been observed at the lamin B2 locus [1.2 kb] (Paixao et al., 2004), the hamster DHFR locus [5.8 kb] (Altman and Fanning, 2004), and the c-myc locus [2.4 kb] (Liu et al., 2003): ectopically inserted sequences can confer origin activity, and deletion of specific elements eliminates such activity. Unfortunately, the sequence elements do not seem to be identical, and no consensus sequences have emerged. There does seem to be an important role of AT rich sequences, as these are often found in critical deleted regions. Supporting this idea, an essential AT rich element in the lamin B2 locus can substitute for the AT-rich element in hamster ori-β locus (Altman and Fanning, 2004). One should note that experiments showing sequence specificity generally measure origin firing by PCR of nascent strands, while studies supporting sequence independent origin firing use ELFH and two dimensional gel electrophoresis. The possibility exists that different methodologies may have different effects on the results.

In addition to sequence effects, many studies have implicated transcription in selection of replication origins. In the DHFR locus, transcription of DHFR itself is required for origin firing activity, and yet origins do not fire in the gene (Kalejta et al., 1998; Saha et al., 2004). In yeast, evidence exists for transcriptional correlation with replication (Muller et al., 2000), but this may be limited to few specific sites, as a genomic microarray study failed to see a good correlation (Raghuraman et al., 2001). Many studies have shown a link between early origin firing and transcriptional activity by looking at replication of developmentally regulated genes, as well as genes from asymmetrically active alleles [reviewed in (Goren and Cedar, 2003)]. In the latter case, the active alleles are replicated much earlier than the silenced alleles. Additionally, replication studies using human (Jeon et al., 2005; Woodfine et al., 2004) and *Drosophila* (Macalpine et al., 2004; Schubeler et al., 2002) genome tiling microarrays show a positive correlation between early origin activity, gene density, and transcriptional activity. Recent results in *Drosophila* indicate that histone hyperacetylation is important for ORC recruitment, although induced hyperacetylation did not affect transcription (Aggarwal and Calvi, 2004). These studies indicate that, in higher eukaryotes, an environment generated by transcription allows for efficient origin firing. One might imagine that the open chromatin environment for transcription would also benefit replication, linking the two different activities.

Although controversy exists as to whether higher eukaryotic origins are sequence dependent or not, the complexity of these organisms may allow a reality that lies somewhere in the middle. In budding yeast, sequence specificity in the form of well established consensus replicators seems to be the primary determinant of origin locations. However, as organism complexity increases, so does the complexity of origin determination. In metazoans, a consensus origin sequence has not yet been identified. Many reports show that certain sequence elements are important for firing, but that these elements for the most part do not share common primary sequence, or even overall features, aside from AT rich sequence preference. Initiator proteins show no preference for sequence, but may show some preference instead for structure [to be discussed later] (Remus et al., 2004), indicating that these essential elements exist to provide a favorable environment for initiator loading. Other factors may also affect chromatin structure, especially during transcription. The positive correlation between origin firing and transcriptional activity in higher eukaryotes suggests that the more open chromatin structure not only allows efficient transcription, but efficient replication initiation as well. It is also plausible that different origins may have differential influences on their activity. Nearby transcriptional activity may be important for one origin, whereas critical sequence elements are important for another. In summary, origin selection in eukaryotes is defined by the proper environment for initiator binding, whether defined solely by DNA sequence elements, structural elements, chromatin structure itself, or a combination of effects. Thus, the early theory of a *replicator* still holds true today. The increasing complexity of higher eukaryotes simply means that the defining elements of a *replicator* are themselves more multifaceted.

3
Pre-Replication Complex

3.1
ORC

The identification of *replicator* sequences in S. cerevisiae opened the field of DNA replication in eukaryotes. One of the first critical discoveries stemming from this work was the identification of the proposed *initiator* proteins. The consensus A element of the ARS sequence was used to identify a six subunit complex termed ORC, or Origin Recognition Complex (Bell and Stillman, 1992). Mutations in the A element that prevent ORC binding also prevent replication from the mutated ARS, (Bell and Stillman, 1992; Rowley et al., 1995) supporting the idea that budding yeast ORC is the protein *initiator* responsible for recognizing specific *replicator* sequences. ORC is highly conserved, with homologues identified in *A. thaliana, S. pombe, D. melanogaster, X. laevis, M. musculus, H. sapiens*, and others (Carpenter et al., 1996; Dhar and Dutta, 2000; Gavin et al., 1995; Gossen et al., 1995; Leatherwood et al., 1996; Masuda et al., 2004; Muzi-Falconi and Kelly, 1995; Pinto et al., 1999; Quintana et al., 1997; Quintana et al., 1998; Tugal et al., 1998). The ORC subunits have been shown to form a functional complex in *D. melanogaster* (Chesnokov et al., 1999), *S. pombe* (Moon et al., 1999), *X. laevis* (Gillespie et al., 2001), and *H. sapiens* (Dhar et al., 2001a; Vashee et al., 2001).

As a replicative *initiator*, ORC should be able to recognize the *replicator* sequences. ORC has been shown by many to bind DNA, and this binding is dependent on ATP and the ATP binding functions of ORC (Bell and Stillman, 1992; Chesnokov et al., 2001; Gillespie et al., 2001; Seki and Diffley, 2000). Specific *replicator* sequence association has been observed in *S. cerevisiae* and *S. pombe*. Work in the latter organism has revealed that Orc4 dictates the specificity via a newly defined AT hook region (Kong and DePamphilis, 2001; Lee et al., 2001). ORC from higher eukaryotes, however, does not seem to have the same sequence specificity. Drosophila ORC showed little preference for chorion gene sequences compared to controls, but showed a much better preference for negative supercoiled DNA compared to relaxed or linear, suggesting secondary structure is more important than sequence for *initiator/replicator* interactions in metazoans (Remus et al., 2004). Similarly, human ORC shows no preference for known origins compared to random DNA sequences, but does show a slight preference for AT rich DNA (Vashee et al., 2003). While ORC may be responsible for DNA binding, the mechanism of such binding becomes unclear with increasing organism complexity, perhaps due to the increasing complexity of factors affecting origin selection.

Although the intricacies of the ORC-DNA interaction are not fully understood, the general role of ORC in replication is now accepted. Studies in *Xenopus* egg extracts have demonstrated that ORC is required to load Cdt1 and Cdc6, themselves factors required for replication initiation [discussed further below] (Coleman et al., 1996; Maiorano et al., 2000). This dependence of replication factor recruitment on ORC helps to explain the lethality of all ORC subunit deletions in yeast. As the foundation of replication initiation complexes, the role of ORC seems to be critical for downstream functions.

As ORC is a key player in defining and recruiting a replication complex to an origin, it is an important potential target for controlling replication. In yeasts, the ORC remains bound to chromatin throughout the cell cycle. However, in *S. cerevisiae*, Orc2 and Orc6 are phosphorylated by S-phase cyclin/Cdk1 during the G1/S transition. This ORC phosphorylation was found to be part of a mechanism to limit origin firing activity to only once per cell cycle; when Orc2 and Orc6 phosphorylation site mutants were introduced with constitutively active Cdc6 and MCM proteins (see Sects. 3.3 and 3.4, respectively), rereplication was observed (Nguyen et al., 2001; Wilmes et al., 2004). Similarly, *S. pombe* Orc2 is phosphorylated, which may be due to its similar interaction with Cdk1/cyclin B in the G2 phase. This interaction serves to prevent rereplication without an intervening mitosis, again ensuring only one replication event per cell cycle takes place (Wuarin et al., 2002). Phosphorylation is well studied as a molecular switch to regulate protein activity through a variety of mechanisms and ORC phosphorylation gives the cells a reversible way to prevent replication firing.

In higher eukaryotes, phosphorylation of ORC subunits is also observed. In *Xenopus* systems, phosphorylation of ORC by cyclin A dependent kinase activity disrupts ORC chromatin association (Findeisen et al., 1999). Similarly, mammalian Orc1 is phosphorylated. In Chinese hamster ovary cells, Orc1 interacts with cyclin A/Cdk1, which leads to the phosphorylation of Orc1. Inhibiting this phosphorylation with drugs allows Orc1 to rebind chromatin, indicating that phosphorylation is important for chromatin release in mitosis (Li et al., 2004). In human cells, Orc1 is also phosphorylated in vivo (our unpublished results) and in vitro by cyclin A/Cdk2 (Mendez et al., 2002). This phosphorylation seems to be required for Skp2 mediated ubiquitination of Orc1 (Mendez et al., 2002). Similarly, hamster Orc1 is also ubiquitinated. However, the nature and effect of these ubiquitination events is different. In hamster cells, Orc1 seems to be mono- and di-ubiquitinated, which causes its release from chromatin in S-phase until M-G1 (Li and DePamphilis, 2002).

In humans, several studies show that Orc1 is polyubiquitinated and then degraded by the proteasome during S-phase, (Fujita et al., 2002; Mendez et al., 2002; Tatsumi et al., 2003) although this observation may result from proteolysis after lysis (Ritzi et al., 2003). Although not degraded during the cell cycle, hamster Orc1 is increasingly sensitive to proteasomal degradation when artificially released to the cytoplasm (Li and DePamphilis, 2002). In *Drosophila* embryos, Orc1 is degraded in M and early G1 phases by the APC/fzr complex (Araki et al., 2003). Despite the apparent contradictions, which may simply result from differences between organisms or even cell types, Orc1 binding to chromatin can be regulated in higher eukaryotes. This regulation allows the cells to control replication initiation at the basic level of the ORC, ensuring inappropriate replication initiation has little, if any, chance of success.

3.2
Cdt1

Cdt1 was first identified as a Cdc10 regulated gene in *S. pombe*. This gene was found to be cell cycle regulated, and important for replication (Hofmann and Beach, 1994). It associates with Cdc6 and is required for loading of the MCM complex in several model systems (Maiorano et al., 2000; Nishitani et al., 2000; Tanaka and Diffley, 2002). Its own chromatin loading is dependent on ORC (Maiorano et al., 2000), supporting the idea that ORC serves as a foundation that recruits downstream factors for replication. Cdt1 protein itself is regulated during the cell cycle, not by transcription, but by proteasome dependent degradation in S-phase (Hofmann and Beach, 1994; Nishitani et al., 2001). A study in *C. elegans* implicated the ubiquitin ligase Cul-4 in Cdt1 degradation; when Cul-4 is absent, Cdt1 is stabilized in S-phase, and massive re-replication is observed (Zhong et al., 2003). In humans, the SCF^{Skp2} complex is implicated in the destruction of Cdt1, and is dependent upon phosphorylation by cyclin-dependent kinases [Cdks] (Li et al., 2003b; Sugimoto et al., 2004), although mutations in Cdt1 that disrupt association with Skp2 still permit degradation of Cdt1 in S-phase (Takeda et al., 2005). Recent work in *Xenopus* egg extracts indicates that the degradation of Cdt1 after replication initiation, together with its inhibition by the protein geminin (discussed later), limit replication to a single round per cell cycle (Arias and Walter, 2004; Li and Blow, 2004). As Cdt1 is required for pre-RC loading, its inhibition by either geminin interaction or degradation will help prevent further pre-RC formation, and thus, further replication initiation. Consistent with this, overexpression of Cdt1 alone leads to extensive re-replication in human cells (Vaziri et al., 2003).

In addition to its function as a replication licensing factor, Cdt1 has recently been implicated in preventing replication initiation after DNA damage in human cells. Cdt1 levels were found to be profoundly decreased after UV irradiation, and an E3 ligase, Cul4A-Roc1-Ddb1, was responsible for signaling this degradation via the proteasome (Higa et al., 2003; Hu et al., 2004). Interestingly, a separate study implicated the SCF^{Skp2} complex in the radiation induced degradation of Cdt1 (Kondo et al., 2004). It remains to be resolved which ubiquitin ligase is primarily responsible for both the cell cycle dependent modifications and the DNA damage induced modifications.

3.3
Cdc6

Cdc6 was originally identified in *S. cerevisiae* as a protein essential for cell cycle progression (Hartwell et al., 1974), and was thereafter shown to have an early DNA synthesis defect (Hartwell, 1976). Cdc6 interacts with ORC, forming a complex with an extended nuclease protected DNA footprint (Cocker et al., 1996; Liang et al., 1995). Furthermore, Cdc6 expression is required for MCM loading in budding yeast. Interestingly, phosphorylation of Cdc6 by B-type cyclin/Cdk complexes prevented the loading of Cdc6, illustrating a powerful way for the cells to regulate pre-RC formation (Donovan et al., 1997; Tanaka et al., 1997). By phosphorylating Cdc6 in S and G2/M, its activity was limited, preventing inappropriate origin firing in mitosis. ScCdc6 was also found to be marked for degradation at the G1/S transition by Clb/Cdc28 and the Cdc4/Cdc34/Cdc53 ubiquitination machinery, adding a further layer of regulation (Drury et al., 1997; Elsasser et al., 1999). A similar gene, Cdc18, was identified in *S. pombe*, and is also required for S-phase. Indeed, its overexpression in *S. pombe* resulted in severe rereplication. Additionally, its protein levels cycle, with maximum levels present during the G1/S transition (Muzi-Falconi et al., 1996; Nishitani and Nurse, 1995). Cdc18 is phosphorylated upon entry into S-phase, causing its rapid degradation (Jallepalli et al., 1997). Not only is Cdc18 required for MCM binding, but it seems to promote this binding in anaphase, supporting the idea that pre-RCs are formed in mitosis (Kearsey et al., 2000). Cdc18 and Cdc6 were later shown to be homologues.

Cdc6 is found in higher eukaryotes as well. Homologues have been identified in *Xenopus*, humans and others (Coleman et al., 1996; Saha et al., 1998; Williams et al., 1997). Using the *Xenopus* egg extract system, it was found that Cdc6 binding to chromatin is dependent on ORC, and is required for MCM2-7 loading, thus implicating a sequential assembly of pre-RC components (Coleman et al., 1996). Similar results were observed in a human cell free extract (Stoeber et al., 1998). Human Cdc6 is partially cell cycle regulated; it is under the control of the E2F transcription factor, which is responsible for promoting expression of numerous genes required for proliferation (Ohtani et al., 1998; Yan et al., 1998). However, unlike yeasts, human Cdc6 may not be degraded at the G1/S transition, but has been found to be exported from the nucleus in a phosphorylation dependent manner (Delmolino et al., 2001; Fujita et al., 1999; Jiang et al., 1999; Saha et al., 1998). This phosphorylation is controlled by cyclin A/Cdk2, which allows for export of Cdc6 soon after replication has initiated (Petersen et al., 1999). Some evidence exists that human Cdc6 is degraded in Sphase, just as in yeast, a point that will need resolution (Coverley et al., 2000; Mendez and Stillman, 2000). Recently it was also reported that although exogenous Cdc6 is exported from the nucleus in S-phase, endogenous Cdc6 is not (Alexandrow and Hamlin, 2004). One further study demonstrates that although Cdc6 is released from chromatin and then degraded, it is constantly being resynthesized and immediately binds chromatin, replacing molecules which were displaced (Biermann et al., 2002). This observation may reconcile earlier observed differences, allowing Cdc6 to bind chromatin in S-phase in a tenuous manner so that rapid regulation can be achieved when needed.

Cdc6 is a AAA+ ATPase as is its homolog Orc1 (Neuwald et al., 1999), and is therefore also regulated in *cis*. The recently solved structure of an archaeal Cdc6 ortholog confirms the presence of a AAA+ ATPase domain, containing Walker A and B motifs, and well as several sensor regions thought to detect nucleotide binding status (Liu et al., 2000). Several studies have been carried out to characterize the importance of its ATP binding and hydrolysis activities. From these studies, it appears that the Walker A motif (nucleotide binding) may be important for Cdc6 binding to chromatin, and Walker B motif (nucleotide hydrolysis) is involved in MCM2-7 loading (Herbig et al., 1999; Perkins and Diffley, 1998; Weinreich et al., 1999). In addition to replication defects caused by mutation in the Walker A and B regions, certain mutations in the sensor regions are also detrimental to replication, often by failing to recruit MCM (Schepers and Diffley, 2001). These studies indicate the ATP binding and hydrolysis of Cdc6 are critical to its function in many different organisms.

In addition to its direct regulation, which serves to limit replication to once, and only once per cell cycle by controlling MCM loading, Cdc6 also has several other secondary roles. Interestingly, Cdc6 seems to regulate ORC by inhibiting its non-specific DNA binding (Harvey and Newport, 2003; Mizushima et al., 2000). By increasing sequence specificity of ORC, Cdc6 may be playing an indirect role in origin selection, especially in higher eukaryotes where consensus initiators have been elusive. This function may also help prevent inefficient fork firing by ensuring ORCs are directed to specific sites, presumably spaced evenly along the chromosome.

Cdc6 is not only regulated by the cell-cycle; it is also cleaved or degraded during apoptosis (Blanchard et al., 2002; Pelizon et al., 2002). It is not entirely clear why Cdc6 would be a target for apoptotic machinery; the cell no longer needs to worry about proper replication, as it is dying. However, this loss of Cdc6 may be part of the programmed cell death, halting replication initiation in preparation for DNA fragmentation. This intriguing finding illustrates the importance of Cdc6 in replication, and thus, the advantage of being able to tightly regulate its function in pre-RC formation.

3.4
MCM2-7

The MCM (mini chromosome maintenance) genes were originally identified in several independent screens as mutants having cell cycle defects, or mini-chromosome perpetuation defects [reviewed in (Dutta and Bell, 1997)]. These proteins were found to be the complex in *Xenopus* responsible for licensing, a regulatory activity that allows cells to replicate in S-phase, and not again until an intervening mitosis occurs (Chong et al., 1995; Kubota et al., 1997; Madine et al., 1995; Thommes et al., 1997). As mentioned earlier, MCM2-7 complex requires the chromatin loading of Cdt1 and Cdc6 (and thus, ORC) for its own chromatin loading. Although MCM2-7 binding depends on ORC, Cdc6, and Cdt1, once the complex is loaded onto chromatin, ORC and Cdc6 are no longer required for DNA replication. This suggests ORC and Cdc6 act to load MCM2-7, but not to maintain this chromatin association (Hua and Newport, 1998; Rowles et al., 1999).

The chromatin loading of MCM2-7 is regulated in a complex and redundant manner. Its additional role as the replication licensing factor illustrates its importance to replication as a whole. As such, its function has long been an intriguing mystery. MCM2-7 seemed to colocalize with DNA polymerase ε using ChIP (Aparicio et al., 1997; Zou and Stillman, 2000) and is critical for replication elongation in vivo (Labib et al., 2000). Early bioinformatics analysis suggested the MCMs may be involved in strand opening (Koonin, 1993). Biochemical analysis of the proteins confirmed this. The mammalian MCM4,6,7 complex was purified, and this complex was found to have a moderate helicase (DNA unwinding) activity (Ishimi, 1997; You et al., 1999). The fission yeast MCM4,6,7 complex also has helicase activity (Lee and Hurwitz, 2000). Interestingly, when MCM2 or MCM3,5 were present in the complex, helicase activity was lost (Lee and Hurwitz, 2000; Sato et al., 2000; You et al., 1999). An archaeal MCM protein has been identified, and also has a helicase activity (Chong et al., 2000; Kelman et al., 1999; Shechter et al., 2000). Electron microscopy studies of the MCM complex indicate that they form a heterohexamer ring structure (Adachi et al., 1997; Sato et al., 2000). This is supported by a recent report which proposes double head-to-head hexamers from a crystal structure of an archaeal MCM (Fletcher et al., 2003). This structure shares similar features with that of T-antigen, suggesting a common function to unwind double strand DNA (Li et al., 2003a) Despite the inhibitory nature of MCM2, 3, and 5, they, like MCM4,6,7, are essential in yeasts [for review see (Dutta and Bell, 1997; Kelly and Brown, 2000)]. This puzzling result seems to indicate that MCM4,6,7 serves as the catalytic helicase domain, while the other subunits act to modulate MCM activity. The exact mechanism of such regulation is still unknown.

Like certain ORC subunits and Cdc6, each member of the MCM2-7 complex has ATPase activity. This ATPase activity is critical for viability; mutants in the Walker A domains show S-phase defects and cell cycle arrest (Schwacha and Bell, 2001). Biochemical analysis of MCM helicase activity showed that the ATPase activity of these proteins is required for the helicase activity (Ishimi, 1997; Lee and Hurwitz, 2000; You et al., 1999). These studies indicated that the helicase activity of these proteins is most likely the in vivo function required for replication. Interestingly, Walker A mutations in MCM4,6,7 are much more toxic to yeast than similar mutations in MCM2,3,5. Although MCM2,3,5 are actually inhibitory for helicase activity (as described above), they are required for optimal ATPase activity of the complex (Schwacha and Bell, 2001). This further supports the role of the MCM4,6,7 complex as the catalytic domain, and MCM2,3,5 serving a regulatory function.

The chromatin association of MCM2-7 is regulated by several other methods to control replication. In budding yeast, MCM2-7 is exported from the nucleus beginning in S-phase (Labib et al., 1999; Nguyen et al., 2000). In higher eukaryotes, this behavior has not been described. However, *Xenopus* egg extract MCMs dissociate from the chromatin as S-phase progresses (Kubota et al., 1997; Thommes et al., 1997). In humans, MCMs also seem to dissociate during S-phase, and reassociate in late G2/M-early G1 (Mendez and Stillman, 2000). This dissociation could be a critical step to prevent re-replication. Once MCM complex is removed, it cannot reassociate until an intervening mitosis occurs (Seki and Diffley, 2000). This aspect of MCM regulation led it to be thought of as the critical factor necessary for chromatin licensing. Once it is removed from chromatin, it requires the activity of ORC, Cdc6 and Cdt1 to be reloaded. However, these proteins are unable to reload MCM in S-phase; they are held inactive by Cdk activity through various mechanisms described above. Therefore, in order to reload MCM, a period of low Cdk activity must be reached (in G1) to activate these proteins, allowing them to reload the MCM2-7 complex.

3.5
Geminin

Geminin is a small protein originally identified in a search for proteins degraded by mitotic *Xenopus* egg extracts (McGarry and Kirschner, 1998). Characterization of this protein indicated it was cell cycle regulated, with maximum expression occurring in S/G2, followed by degradation in mitosis. Contrary to other replication factors discussed here, geminin inhibits replication by preventing loading of MCM2-7 complex. It was subsequently determined that geminin binds Cdt1, thus inhibiting replication. This inhibition was rescued by the addition of excess Cdt1 (Tada et al., 2001; Wohlschlegel et al., 2000). In vitro evidence demonstrates that geminin binding to Cdt1 may disrupt an interaction between Cdt1 and MCM6 (Yanagi et al., 2002). The disruption of MCM6 binding may be preventing the recruitment of the MCM2-7 complex. Additionally, the same study suggested that Cdt1 can bind DNA, and this interaction was also disrupted by geminin inhibition. A more recent study suggests that a Cdt1 binding activity to MCM2 and to Cdc6 is also inhibited by geminin binding (Cook et al., 2004). Structural studies indicate that geminin forms a coiled-coil dimer which, together with an N-terminal flexible portion, executes a bipartite interaction with Cdt1 (Lee et al., 2004; Saxena et al., 2004). Initiation inhibition by geminin is now thought to be a part of the replication licensing system that prevents multiple rounds of replication in *Xenopus* egg extracts (Arias and Walter, 2004; Li and Blow, 2004). Although Cdt1 is the primary target of regulation, geminin provides an additional pathway to inactivate Cdt1, and thus preventing inappropriate replication. In mammalian systems, overexpression of Cdt1 or depletion of geminin is sufficient to cause massive rereplication and checkpoint activation (Melixetian et al., 2004; Zhu et al., 2004). Abrogating this checkpoint causes mitotic distress and eventual cell death, demonstrating the importance of limiting replication to a single event per cell cycle.

3.6
Summary

The pre-RC is composed of a variety of proteins that serve to recruit MCM2-7 to selected genomic regions, establishing a potential site for replication initiation (Fig. 1). The chromatin loading of MCM2-7 depends on the loading of Cdt1 and Cdc6, which in turn depend on the loading of ORC. This step-

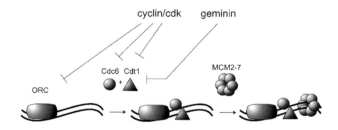

Regulation of the pre-RC

Fig. 1 Regulation of the pre-RC. The pre-RC functions to load MCM2-7 complex, the potential replicative helicase. ORC binds chromatin initially at origins. Its binding is required for the chromatin association of Cdt1 and Cdc6, which are themselves required to load the MCM2-7 complex. This loading occurs in late mitosis until late G1. Upon entry into S-phase, increasing cyclin/Cdk activity inhibits the loading of MCM2-7. This increased cyclin/Cdk activity promotes Orc1 and Cdc6 dissociation from chromatin, and degradation of Cdt1 (and perhaps also Orc1 and Cdc6.) Additionally, expression of geminin, a protein inhibitory of Cdt1, increases in S-phase, further preventing MCM2-7 loading

wise assembly of factors is regulated at each stage, mostly through changes in protein levels, chromatin affinities, or activity throughout the cell cycle. The activity of the pre-RC is limited to the time between late M and late G1/early S. This period is notable due to the absence of Cdk activity. As many pre-RC factors are inhibited by Cdk activity, they can only function to load MCM2-7 in the late M to late G1 periods. Once the cells pass into S-phase, cyclin/Cdk activity is high, and several pre-RC components dissociate from the chromatin, or are degraded, and thus, are no longer able to load MCM2-7. Additionally, the inhibitor protein geminin is synthesized in early S-phase, further preventing any MCM loading by directly inhibiting Cdt1. These results fit the original model of a "replication licensing" activity (Blow and Laskey, 1988). Early observation showed that once a given region of DNA has replicated, it cannot replicate again until an intervening mitosis occurs. It now appears that MCM2-7 loading is the licensing activity, as it is required for replication, and yet the loading cannot happen once replication has begun. This is due to the inhibition of the pre-RC components by cyclin/Cdk activity and geminin. These regulatory mechanisms ensure that licensing (MCM2-7 loading) can only occur once in each cell cycle before replication. Intriguingly, viral episomes subjected to similar once-per-cell cycle replication control, like Epstein-Barr-Virus derived episomes, also appear to use the cellular initiation factors for origin licensing (Chaudhuri et al., 2001; Dhar et al., 2001b; Schepers et al., 2001)

4
Pre-Initiation Complex

4.1
Mcm10

Mcm10 (also named DNA43) was originally identified in two independent screens; it was identified in a screen for DNA synthesis errors (Solomon et al., 1992) as well as in the mini-chromosome maintenance screen that identified the members of MCM2-7 complex (Maine et al., 1984). Mcm10 mutants showed defects in S-phase, specifically with origin firing and in fork elongation, and were shown to interact with the MCM2-7 complex (Aves et al., 1998; Merchant et al., 1997). The interaction with MCM7 in particular is critical for proper replication (Homesley et al., 2000). These results point to a vital role for yeast Mcm10 in replication.

Several recent studies have helped determine a likely function for Mcm10. In *Xenopus* egg extracts and fission yeast, Mcm10 is loaded onto chromatin in an MCM2-7 dependent manner, and is itself required to load Cdc45 in G1/early S [discussed below] (Gregan et al., 2003; Wohlschlegel et al., 2002b).

In budding yeast, the role of Mcm10 is not as clear. Earlier studies point to a role for Mcm10 in loading the MCM2-7 complex (Homesley et al., 2000), whereas a more recent study finds the opposite: MCM2-7 is required to load Mcm10 (Ricke and Bielinsky, 2004). Additionally, Mcm10 has been reported to load the pre-initiation kinase Dbf4/Cdc7, which is required to activate MCM2-7 [see below] (Lee et al., 2003). Mcm10 may also have a role in loading the polymerase complex as well. It was recently shown that Mcm10 stabilizes members of the Pol-α primase complex, and is important for chromatin loading of these subunits (Ricke and Bielinsky, 2004). Work in human cells has revealed a regulatory mechanism to control Mcm10 function. In G2, Mcm10 is phosphorylated, released from the chromatin, and subsequently degraded by the proteasome. It reappears and rebinds chromatin in late G1/early S (Izumi et al., 2001). Together, these data point to a role for Mcm10 in replication through a variety of potential mechanisms. While past work has revealed much about the importance of Mcm10 in replication, many of the details concerning these diverse functions remain to be worked out.

4.2
Cdc45

Cdc45 was identified in a yeast screen for cold sensitive cell cycle mutants (Moir et al., 1982). Cdc45 has been shown to interact genetically with members of the MCM2-7 complex, and mutants show replication defects (Hennessy et al., 1991; Hopwood and Dalton, 1996; Zou et al., 1997). Supporting a role in replication, Cdc45 has been found to bind chromatin in G1 and release in S-phase. Interestingly, this binding requires cyclin/Cdk activity (Zou and Stillman, 1998). Such a requirement implies a separation from earlier initiation stages; those steps leading up to loading of the MCM2-7 complex are negatively regulated by cyclin/Cdk phosphorylation. Pre-RCs can only be loaded in the absence of Cdk activity, at which point Cdc45 cannot be loaded. Only when cyclin/Cdks become active can Cdc45 be loaded and replication begin.

In addition to cyclin/Cdk regulation, Cdc45 also cooperates with Dbf4/Cdc7 kinase complex to bind chromatin (see below). These proteins seem to require each other for activating replication (Owens et al., 1997; Zou and Stillman, 2000). As mentioned earlier, Cdc45 loading requires Mcm10 loading (Gregan et al., 2003; Wohlschlegel et al., 2002b). The activating function of Cdc45 may be the chromatin loading of polymerase α, which was shown to depend on Cdc45 chromatin loading in *Xenopus* extracts (Mimura and Takisawa, 1998). In addition to its requirement in Pol-α loading, Cdc45 and RPA (single strand binding protein complex) depend on each other for chromatin binding (Zou and Stillman, 2000). This co-loading of Cdc45 and RPA may aid origin unwinding, allowing Pol-α access to template DNA (Walter and Newport, 2000). This is supported by data that show a Cdc45 requirement for helicase activity of MCM2-7 and in vivo origin unwinding (Masuda et al., 2003; Pacek and Walter, 2004).

Recent work has discovered a novel protein that interacts with Cdc45. This protein, Sld3, has been shown in budding yeast to bind chromatin with Cdc45 in G1 and S-phase. This binding occurs at origins, suggesting a replication initiation role for Sld3. Expression of a mutant Sld3 inhibits the interaction between Cdc45 and MCM2, and prevents loading of RPA (Kamimura et al., 2001). Studies in fission yeast suggest that Sld3 has a role in loading Cdc45 onto chromatin (Nakajima and Masukata, 2002; Yamada et al., 2004), and that Sld3 is also regulated by Dbf4/Cdc7. There is also evidence in these studies that Sld3 mediates the interaction between Cdc45 and MCM2-7, further linking pre-RC formation with pre-IC loading. Cdc45 appears to have many roles in promoting replication initiation. It will therefore be interesting to evaluate the extent of regulation of Cdc45 by the cell-cycle and the response to DNA damage.

4.3
Dbf4/Cdc7

Dbf4/Cdc7 forms a regulator/kinase pair very similar to cyclin/Cdks. Dbf4 association is required for the kinase activity of Cdc7, and Dbf4 itself is cell cycle regulated [for review see (Bell and Dutta, 2002; Masai and Arai, 2002)]. Cdc7/Dbf4 mutants show replication defects in yeast. Although early studies indicated a role in S-phase entry, more recent work shows the complex to be important all throughout S-phase (Bousset and Diffley, 1998; Donaldson et al., 1998). This is most likely due to a function for Dbf4/Cdc7 in activating both early and late firing origins locally. Dbf4/Cdc7 is thought to have several potential roles in origin firing. Dbf4/Cdc7 phosphorylates the MCM2-7 complex, both in vitro and in vivo (Kihara et al., 2000; Lei et al., 1997; Oshiro et al., 1999; Owens et al., 1997; Sato et al., 1997; Weinreich and Stillman, 1999). This phosphorylation is thought to activate the MCM2-7 complex, allowing replication to proceed (Lei et al., 1997). Consistent with this, mutations in MCM5 (Bob1) allows yeast to bypass the requirement of Dbf4/Cdc7 (Hardy et al., 1997). Additionally, Dbf4/Cdc7 is required for Cdc45 chromatin association, and subsequent loading of the replication machinery (Owens et al., 1997; Zou and Stillman, 2000). Although a specific mechanistic change resulting from phosphorylation is not known, Dbf4/Cdc7 seems to act as a late stage signaling step to promote loading of the replication machinery.

In order to control such a signal, Dbf4/Cdc7 is regulated in several ways. The predominant regulatory mechanism is the requirement of Dbf4 for Cdc7 kinase activity. This allows the cells tight control; Dbf4 is cell cycle regulated, and so the kinase is only active at the appropriate time in the cell cycle. Dbf4 is degraded by the APC in mitosis, and is resynthesized at G1/S, allowing kinase activity to rise at the G1/S transition, and fall at the end of the cell cycle (Brown and Kelly, 1999; Chapman and Johnston, 1989; Ferreira et al., 2000; Kumagai et al., 1999; Oshiro et al., 1999; Weinreich and Stillman, 1999). Even when Dbf4/Cdc7 is activated, its phosphorylation of MCM2-7 still requires Mcm10 (Lee et al., 2003). By controlling the chromatin loading of early factors, as well as the presence of activating regulators, the cell has many signaling pathways available to prevent DNA replication when conditions are not ideal.

Such unfavorable conditions would undoubtedly include DNA damage. When the template condition is poor, accurate duplication is a difficult prospect. However, recent studies have revealed a mechanism to prevent late origin firing at a local level. Dbf4 homologues in various organisms are hyperphosphorylated in response to hydroxyurea (HU) treatment (Brown and Kelly, 1999; Takeda et al., 1999; Weinreich and Stillman, 1999). The responsible kinases were the Chk2 yeast homologues Rad53/Cds1, which are implicated in DNA damage signaling. This checkpoint mediated phosphorylation inhibits the kinase activity of Dbf4/Cdc7 in vitro (Kihara et al., 2000). Similarly, single strand gaps in *Xenopus* egg extracts activate an ATR mediated checkpoint that also inhibits Dbf4/Cdc7 activity (Costanzo et al., 2003). The absence of kinase activity is presumably responsible for the dissociation of Cdc45 from the chromatin. This link between checkpoint pathways and replication initiation stops a potentially damaging replication cycle from occurring.

4.4
GINS

Recent work from two independent studies has identified a novel protein complex required for replication initiation. In one study, a combination of sequential multicopy suppressor screens (in an Sld5 mutant background) and co-immunoprecipitations identified a complex of four proteins; Sld5, Psf1 (partner of sld five), Psf2 and Psf3. This complex was termed GINS, short for the numbers 5,1,2,3 in Japanese [Go, Ichi, Nii, San] (Takayama et al., 2003). A similar complex was also purified from *Xenopus* egg extracts (Kubota et al., 2003). In a separate screen using an inducible degron system, novel replication deficient mutants were identified (Kanemaki et al., 2003). The identified proteins turned out to be the same four proteins constituting GINS.

All three studies show that GINS are critical for replication. The complex associates with chromatin, and travels with the replication fork (Kanemaki et al., 2003). Additionally, GINS chromatin association requires S-Cdk activity, similar to other pre-initiation complex members (Kanemaki et al., 2003; Kubota et al., 2003). However, licensing plays a role to restrict GINS activity, as it requires pre-RC loading for its own chromatin binding. This chromatin binding is also co-dependent with Cdc45, Dpb11, and Sld3; binding of any of these proteins depended on the presence of the others (Kubota et al., 2003; Takayama et al., 2003). Although its mechanistic role remains unclear, GINS has a critical function for replication initiation and elongation.

4.5
DPB11

Dpb11 (also known as TopBP1 in humans, Cut5/Rad4 in fission yeast, and Mus101 in *Drosophila* and *Xenopus*) was originally identified as a suppressor of mutations in DNA polymerase ε (Araki et al., 1995). It also physically interacts with Pol-ε by cross-linked coimmunoprecipitation (Masumoto et al., 2000). This protein is important for replication in yeasts, and has been implicated in polymerase recruitment to chromatin [for review, see (Bell and Dutta, 2002)]. Several studies describe genetic interactions between Dpb11 and Cdc45 (Kamimura et al., 1998; Reid et al., 1999), and recent studies show that Cdc45 loading requires Dpb11 (Dolan et al., 2004; Van Hatten et al., 2002).

In addition to its role in replication, Dpb11 seems to also be important for checkpoint signaling. Mutants in Dpb11 and its homologues show DNA damage sensitivity (Araki et al., 1995; Boyd et al., 1976; Saka and Yanagida, 1993). Dpb11 seems to be required for Rad53 (Chk2) activation in budding yeast (Wang and Elledge, 1999), and localizes to DNA damage foci (Yamane et al., 2002). Dpb11 is also important for ATR recruitment to damage sites, as well as subsequent Chk1 activation (Parrilla-Castellar and Karnitz, 2003). Interestingly, in this study using *Xenopus* egg extracts, and another using human cells (Yamane et al., 2003), Dpb11 was not required to activate Chk2, illustrating a difference between budding yeast checkpoint signaling. This role for Dpb11 suggests a link between replication and checkpoint that could be a mechanism for DNA damage sensing by the replication fork complex.

The regulation of Dpb11 is important for its function in fork loading and checkpoint activation. Like many other important genes, Dpb11 is regulated during the cell cycle. It was found to be under control of an E2F regulated promoter, allowing expression in late G1/S (Yoshida and Inoue, 2004). In addition to binding Pol-ε, Dpb11 interacts with Sld2, a protein required for replication (Kamimura et al., 1998). Sld2 is phosphorylated by Cdks in S-phase, which is also critical for replication (Masumoto et al., 2002). This cell cycle regulation of Dpb11, together with a phosphorylation requirement of its binding partner Sld2, adds to the many levels of control cells have over replication.

4.6
Summary

The pre-initiation complex builds upon the pre-replication complex to recruit additional proteins required for replication initiation. A variety of factors cooperate to bind chromatin and pre-RCs, activate the presumptive helicase activity of MCM2-7, and recruit the actual replication machinery (Fig. 2). In a marked difference from the pre-replication complex, the pre-initiation complex requires cyclin/Cdk activity for its function. Whether phosphorylation by Cdk is required for activity or chromatin binding, this activation is essential for loading the final replication machinery. At times where Cdk activity is low, the replication machinery cannot be loaded. This effectively limits fork firing to the time *after* pre-RCs have been loaded. Interestingly, the MCM2-7 complex can be thought of as a bridge between the two phases of replication initiation. It can only be loaded during periods of low Cdk activity, but requires Cdk activity for its function. This supports the role of MCM2-7 as the licensing factor that prevents re-replication. By limiting certain aspects of replication initiation to certain times in the cell cycle, a temporal order is maintained that allows replication only when certain conditions are met. In summary, the cell prepares for replication by loading MCM2-7 complex to

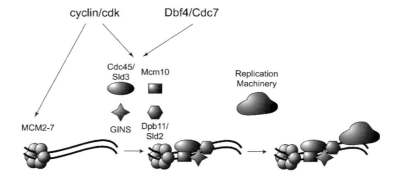

Regulation of the pre-IC

Fig. 2 Regulation of the pre-IC. Once the MCM2-7 complex is loaded, further factors are equired to ultimately recruit the replication machinery to the origins. These factors include Cdc45/Sld3, Mcm10, GINS, and Dpb11/Sld2. Chromatin association of these factors depends on chromatin bound MCM2-7, and on each other. These factors require cyclin/Cdk activation for their function, and some members also require activation via the cyclin/Cdk like complex Dbf4/Cdc7. MCM2-7 itself requires cyclin/Cdk activity for its function. These steps can only occur in S-phase, due to the requirement cyclin/Cdk activity. However, chromatin loading of MCM2-7 is also required, and this can only happen in M-G1. This system of regulation allows establishment of a temporal order to ensure that replication machinery is loaded to the correct sites once, and only once, per cell cycle

distinct sites in the genome. Once this is accomplished, the replication machinery is loaded, and replication begins. Each step in this process is subject to regulation, allowing the cell to halt replication under a wide variety of adverse conditions, including DNA damage.

5
S-phase Regulation and Cancer

Proper reproduction of the genome is critical for viability. Mutations and mistakes in this process can lead to a variety of detrimental effects. Such mutations in simpler organisms may mean a growth or metabolic disadvantage, and subsequent death. In higher, more complex organisms, this can lead to unregulated cell proliferation, a key cause of cancer. Since replication is a process with great potential for genomic damage, one might expect loss of regulation to be linked to tumorigenesis. Indeed, overexpression of Cdt1 and/or Cdc6 in fission yeast (Nishitani and Nurse, 1995; Nishitani et al., 2000), *Xenopus* egg extracts (Arias and Walter, 2004; Li and Blow, 2004), and human cells (Vaziri et al., 2003) causes rereplication. If not corrected, rereplication increases the chances of gene duplications that may lead to cancer. Supporting this idea, a mouse homolog of Cdt1 was identified by its oncogenic potential, and was able to form tumors in mice (Arentson et al., 2002). These results suggest that defects in the replication regulation machinery have the potential to lead to cancer.

But are these proteins implicated in clinical cancers? Work on cervical cancer has revealed that Cdc6 and MCM5 are often overexpressed in abnormal cervical cells (Williams et al., 1998). In fact, the authors demonstrate that Cdc6 and MCM5 overexpression seems to be a better marker for abnormal cells than Ki-67 and PCNA, two classical biomarkers for proliferation. This suggests a strong link between replication initiation factors and the proliferation seen in cancers. A recent study on patients with non-small-cell lung carcinomas demonstrated that Cdt1 and Cdc6 levels were higher in tumors than in non-tumor tissue (Karakaidos et al., 2004). In the majority of cases, these proteins were either both overexpressed or both normal. Interestingly, in cases showing Cdc6 and Cdt1 overexpression, p53 status affected the ploidy of the cells. When p53 was mutated, 73% of the cases showed aneuploidy, compared to only 28% of cases with wild-type p53. This result demonstrates that p53 can protect against the damaging effects of high levels of Cdt1 and Cdc6 and agrees with results in tumor cell cultures that show p53 can inhibit rereplication caused by Cdc6 and Cdt1 overexpression (Vaziri et al., 2003). Many replication factors seem to be highly expressed in cancer, and are under study for use as molecular markers, perhaps eventually for use in diagnosis and prognosis [for review see (Semple and Duncker, 2004)]. Intriguingly, geminin is also elevated in proliferating cancer cells (Karakaidos et al., 2004; Wohlschlegel et al., 2002a). As geminin is expressed only in the latter half of the cell cycle (from S until M), it also serves as a marker of proliferation.

In addition to genomic replication, replication of pathogenic organisms has a variety of implications for cancer. In particular, Epstein-Barr Virus (EBV) is extremely common in the human population, and is variably associated with nasopharyngeal carcinoma, several B-cell lymphomas (especially in immunocomprimised patients), and occasionally with gastric carcinoma [for review see (Young and Rickinson, 2004)]. As is often true of viruses, EBV relies on the host cell for its propagation. Specifically in this case, EBV utilizes the human ORC proteins to facilitate loading of cellular replication factors (Chaudhuri et al., 2001; Dhar et al., 2001b; Ritzi et al., 2003). Our lab has shown that decrease of ORC inhibits EBV replication, while allowing cells to survive. Addition of geminin can also inhibit virus replication (Dhar et al., 2001b). By inhibiting formation of a functional replication complex, viral episome replication can be prevented, thereby decreasing the chance of virus-mediated transformation. As the viral episomes are more sensitive to decreased replication factor levels, targeting these proteins may decrease viral load at concentrations that have minimal effects on the chromosomal replication of cells. Many other viruses are implicated in cancer, and some of these may also utilize host replication initiation factors. For example, Kaposi's Sarcoma-Associated herpesvirus (KSHV) was shown to utilize ORC for latent replication (Stedman et al., 2004). Therefore, studies on replication initiation factors can give insight not only into initiation and progression of cancers, but also aid in the diagnosis, prognosis, and treatment of cancer.

6
Conclusion

In order to copy its genome faithfully and reliably, an organism must have a mechanism to ensure exact alternation of genome replication and segregation. Rereplication without an intervening mitosis would yield extra copies of genes, causing a wide variety of problems for cells. Indeed, one of the hallmarks of human cancer is an abnormal genome content. Organisms ensure this exact alternation by a variety of ways. Immediately after mitosis, pre-RCs begin binding to chromatin-bound ORCs at distinct sites in the genome. Loading of the pre-RC components Cdt1 and Cdc6 proceeds in the absence of cyclin/Cdk activity, and functions to load the MCM2-7 complex to the chromatin. As cyclin/Cdk activity increases, various pre-RC components are removed from the chromatin, and further MCM2-7 loading is therefore prevented. In addition, geminin is activated, and inhibits any residual Cdt1, further preventing pre-RC activity in S-phase. The loss of the pre-RC does not affect MCM2-7 that has been loaded; this population sets the stage for downstream factor recruitment. The loading of the downstream factors requires MCM2-7, and begins with Mcm10. Once Mcm10 is loaded, the activity of cyclin/Cdk is required for the subsequent loading of Cdc45/Sld3, Dbf4/Cdc7, Dpb11/Sld2 and GINS. These proteins depend on each other for loading, and themselves are required to load the replicative polymerase machinery.

The alternating inhibition and activation of cyclin/Cdks sets up a timer to allow or inhibit these activities as the cell cycle progresses. Cells allow licensing (MCM2-7 loading) to occur in the absence of Cdk activity. When Cdk activity increases, MCM2-7 can no longer be loaded, but previously loaded MCM2-7 is now ready to recruit downstream factors to initiate replication. Once replication has begun at a given region, new forks cannot be fired, as MCM2-7 cannot be reloaded. Only after replication has been completed, and the cells have divided the newly synthesized genome, can MCM2-7 be reloaded.

By increasing the efficiency of replication, cells can move through this potentially damaging event with minimum time. This helps decrease the exposure of single strand DNA, and allows the cell to proceed through the cell cycle in a timely fashion. To increase efficiency, cells should replicate from evenly spaced origins. Achieving this goal requires a way to indicate where replication should start. It is evident that cells mark origins of replication, as specific regions seem to act as origins in each cell cycle. Origins are marked by consensus sequence patterns, or, in more complex organisms, by a more intricate combination of sequence, chromatin environment, and DNA structure.

While the field of replication initiation control has recently made great progress, there are many questions yet to be answered. Although the general licensing system has been described, more detailed mechanistic problems still abound. How does ORC recognize origins? How are origins defined in humans? How do the members of pre-RC interact to load MCM2-7? MCM2-7 itself is still somewhat mysterious, as its in vitro helicase activity is only apparent as a subcomplex; it is seemingly inhibited as the full six-subunit helicase. Interaction between pre-IC components and replication machinery has been demonstrated, but the exact function of each member in loading polymerase, primase, and others, is still under investigation. Further examination of this multifaceted event will not only yield a greater understanding of replication control, but will also help us determine the protections such control offers, as well as the consequences of deregulation.

Acknowledgements Work in AD's lab is supported by NIH grants CA60499 and CA89406.

References

Adachi Y, Usukura J, Yanagida M (1997) A globular complex formation by Nda1 and the other five members of the MCM protein family in fission yeast. Genes Cells 2:467–479

Aggarwal BD, Calvi BR (2004) Chromatin regulates origin activity in Drosophila follicle cells. Nature 430:372–376

Aladjem MI, Rodewald LW, Kolman JL, Wahl GM (1998) Genetic dissection of a mammalian replicator in the human beta-globin locus. Science 281:1005–1009

Alexandrow MG, Hamlin JL (2004) Cdc6 chromatin affinity is unaffected by serine-54 phosphorylation, S-phase progression, and overexpression of cyclin A. Mol Cell Biol 24:1614–1627

Altman AL, Fanning E (2004) Defined sequence modules and an architectural element cooperate to promote initiation at an ectopic mammalian chromosomal replication origin. Mol Cell Biol 24:4138–4150

Aparicio OM, Weinstein DM, Bell SP (1997) Components and dynamics of DNA replication complexes in S. cerevisiae: redistribution of MCM proteins and Cdc45p during S phase. Cell 91:59–69

Araki H, Leem SH, Phongdara A, Sugino A (1995) Dpb11, which interacts with DNA polymerase II(epsilon) in Saccharomyces cerevisiae, has a dual role in S-phase progression and at a cell cycle checkpoint. Proc Natl Acad Sci USA 92:11791–11795

Araki M, Wharton RP, Tang Z, Yu H, Asano M (2003) Degradation of origin recognition complex large subunit by the anaphase-promoting complex in Drosophila. Embo J 22:6115–6126

Arentson E, Faloon P, Seo J, Moon E, Studts JM, Fremont DH, Choi K (2002) Oncogenic potential of the DNA replication licensing protein CDT1. Oncogene 21:1150–1158

Arias EE, Walter JC (2005) Replication-dependent destruction of Cdt1 limits DNA replication to a single round per cell cycle in Xenopus egg extracts. Genes Dev 19:114–126

Aves SJ, Tongue N, Foster AJ, Hart EA (1998) The essential schizosaccharomyces pombe cdc23 DNA replication gene shares structural and functional homology with the Saccharomyces cerevisiae DNA43 (MCM10) gene. Curr Genet 34:164–171

Bell SP, Stillman B (1992) ATP-dependent recognition of eukaryotic origins of DNA replication by a multiprotein complex. Nature 357:128–134

Bell SP, Dutta A (2002) DNA replication in eukaryotic cells. Annu Rev Biochem 71:333–374

Biermann E, Baack M, Kreitz S, Knippers R (2002) Synthesis and turn-over of the replicative Cdc6 protein during the HeLa cell cycle. Eur J Biochem 269:1040–1046

Blanchard F, Rusiniak ME, Sharma K, Sun X, Todorov I, Castellano MM, Gutierrez C, Baumann H, Burhans WC (2002) Targeted destruction of DNA replication protein Cdc6 by cell death pathways in mammals and yeast. Mol Biol Cell 13:1536–1549

Blow JJ, Laskey RA (1988) A role for the nuclear envelope in controlling DNA replication within the cell cycle. Nature 332:546–548

Bousset K, Diffley JF (1998). The Cdc7 protein kinase is required for origin firing during S phase. Genes Dev 12:480–490

Boyd JB, Golino MD, Nguyen TD, Green MM (1976) Isolation and characterization of X-linked mutants of Drosophila melanogaster which are sensitive to mutagens. Genetics 84:485–506

Brown GW, Kelly TJ (1999) Cell cycle regulation of Dfp1, an activator of the Hsk1 protein kinase. Proc Natl Acad Sci USA 96:8443–8448

Burhans WC, Vassilev LT, Caddle MS, Heintz NH, DePamphilis ML (1990) Identification of an origin of bidirectional DNA replication in mammalian chromosomes. Cell 62:955–965

Carpenter PB, Mueller PR, Dunphy WG (1996) Role for a Xenopus Orc2-related protein in controlling DNA replication. Nature 379:357–360

Chapman JW, Johnston LH (1989) The yeast gene, DBF4, essential for entry into S phase is cell cycle regulated. Exp Cell Res 180:419–428

Chaudhuri B, Xu H, Todorov I, Dutta A, Yates JL (2001) Human DNA replication initiation factors, ORC and MCM, associate with oriP of Epstein-Barr virus. Proc Natl Acad Sci USA 98:10085–10089

Chesnokov I, Remus D, Botchan M (2001) Functional analysis of mutant and wild-type Drosophila origin recognition complex. Proc Natl Acad Sci USA 98:11997–12002

Chesnokov I, Gossen M, Remus D, Botchan M (1999) Assembly of functionally active Drosophila origin recognition complex from recombinant proteins. Genes Dev 13:1289–1296

Chong JP, Mahbubani HM, Khoo CY, Blow JJ (1995) Purification of an MCM-containing complex as a component of the DNA replication licensing system. Nature 375:418–421

Chong JP, Hayashi MK, Simon MN, Xu RM, Stillman B (2000) A double-hexamer archaeal minichromosome maintenance protein is an ATP-dependent DNA helicase. Proc Natl Acad Sci USA 97:1530–1535

Cocker JH, Piatti S, Santocanale C, Nasmyth K, Diffley JF (1996) An essential role for the Cdc6 protein in forming the pre-replicative complexes of budding yeast. Nature 379:180–182

Coleman TR, Carpenter PB, Dunphy WG (1996) The Xenopus Cdc6 protein is essential for the initiation of a single round of DNA replication in cell-free extracts. Cell 87:53–63

Cook JG, Chasse DA, Nevins JR (2004) The regulated association of Cdt1 with minichromosome maintenance proteins and Cdc6 in mammalian cells. J Biol Chem 279:9625–9633

Costanzo V, Shechter D, Lupardus PJ, Cimprich KA, Gottesman M, Gautier J (2003) An ATR- and Cdc7-dependent DNA damage checkpoint that inhibits initiation of DNA replication. Mol Cell 11:203–213

Coverley D, Pelizon C, Trewick S, Laskey RA (2000) Chromatin-bound Cdc6 persists in S and G2 phases in human cells, while soluble Cdc6 is destroyed in a cyclin A-cdk2 dependent process. J Cell Sci 113:1929–1938

Delmolino LM, Saha P, Dutta A (2001) Multiple mechanisms regulate subcellular localization of human CDC6. J Biol Chem 276:26947–26954

Dhar SK, Dutta A (2000) Identification and characterization of the human ORC6 homolog. J Biol Chem 275:34983–34988

Dhar SK, Delmolino L, Dutta A (2001a) Architecture of the human origin recognition complex. J Biol Chem 276:29067–29071

Dhar SK, Yoshida K, Machida Y, Khaira P, Chaudhuri B, Wohlschlegel JA, Leffak M, Yates J, Dutta A (2001b) Replication from oriP of Epstein-Barr virus requires human ORC and is inhibited by geminin. Cell 106:287–296

Dijkwel PA, Hamlin JL (1995) The Chinese hamster dihydrofolate reductase origin consists of multiple potential nascent-strand start sites. Mol Cell Biol 15:3023–3031

Dijkwel PA, Wang S, Hamlin JL (2002) Initiation sites are distributed at frequent intervals in the Chinese hamster dihydrofolate reductase origin of replication but are used with very different efficiencies. Mol Cell Biol 22:3053–3065

Dolan WP, Sherman DA, Forsburg SL (2004) Schizosaccharomyces pombe replication protein Cdc45/Sna41 requires Hsk1/Cdc7 and Rad4/Cut5 for chromatin binding. Chromosoma 113:145–156

Donaldson AD, Fangman WL, Brewer BJ (1998). Cdc7 is required throughout the yeast S phase to activate replication origins. Genes Dev 12:491–501

Donovan S, Harwood J, Drury LS, Diffley JF (1997) Cdc6p-dependent loading of Mcm proteins onto pre-replicative chromatin in budding yeast. Proc Natl Acad Sci USA 94:5611–5616

Drury LS, Perkins G, Diffley JF (1997) The Cdc4/34/53 pathway targets Cdc6p for proteolysis in budding yeast. Embo J 16:5966–5976

Dubey DD, Kim SM, Todorov IT, Huberman JA (1996) Large, complex modular structure of a fission yeast DNA replication origin. Curr Biol 6:467–473

Dutta A, Bell SP (1997) Initiation of DNA replication in eukaryotic cells. Annu Rev Cell Dev Biol 13:293–332

Elsasser S, Chi Y, Yang P, Campbell JL (1999) Phosphorylation controls timing of Cdc6p destruction: A biochemical analysis. Mol Biol Cell 10:3263–3277

Ferreira MF, Santocanale C, Drury LS, Diffley JF (2000) Dbf4p, an essential S phase-promoting factor, is targeted for degradation by the anaphase-promoting complex. Mol Cell Biol 20:242–248

Findeisen M, El-Denary M, Kapitza T, Graf R, Strausfeld U (1999) Cyclin A-dependent kinase activity affects chromatin binding of ORC, Cdc6, and MCM in egg extracts of Xenopus laevis. Eur J Biochem 264:415–426

Fletcher RJ, Bishop BE, Leon RP, Sclafani RA, Ogata CM, Chen XS (2003) The structure and function of MCM from archaeal M. Thermoautotrophicum. Nat Struct Biol 10:160–167

Fujita M, Ishimi Y, Nakamura H, Kiyono T, Tsurumi T (2002) Nuclear organization of DNA replication initiation proteins in mammalian cells. J Biol Chem 277:10354–10361

Fujita M, Yamada C, Goto H, Yokoyama N, Kuzushima K, Inagaki M, Tsurumi T (1999) Cell cycle regulation of human CDC6 protein. Intracellular localization, interaction with the human mcm complex, and CDC2 kinase-mediated hyperphosphorylation. J Biol Chem 274:25927–25932

Gavin KA, Hidaka M, Stillman B (1995) Conserved initiator proteins in eukaryotes. Science 270:1667–1671

Gilbert DM, Miyazawa H, DePamphilis ML (1995) Site-specific initiation of DNA replication in Xenopus egg extract requires nuclear structure. Mol Cell Biol 15:2942–2954

Gillespie PJ, Li A, Blow JJ (2001) Reconstitution of licensed replication origins on Xenopus sperm nuclei using purified proteins. BMC Biochem 2:15

Goren A, Cedar H (2003) Replicating by the clock. Nat Rev Mol Cell Biol 4:25–32

Gossen M, Pak DT, Hansen SK, Acharya JK, Botchan MR (1995) A Drosophila homolog of the yeast origin recognition complex. Science 270:1674–1677

Gregan J, Lindner K, Brimage L, Franklin R, Namdar M, Hart EA, Aves SJ, Kearsey SE (2003) Fission yeast Cdc23/Mcm10 functions after pre-replicative complex formation to promote Cdc45 chromatin binding. Mol Biol Cell 14:3876–3887

Hardy CF, Dryga O, Seematter S, Pahl PM, Sclafani RA (1997) mcm5/cdc46-bob1 bypasses the requirement for the S phase activator Cdc7p. Proc Natl Acad Sci USA 94:3151–3155

Hartwell LH (1976) Sequential function of gene products relative to DNA synthesis in the yeast cell cycle. J Mol Biol 104:803–817

Hartwell LH, Culotti J, Pringle JR, Reid BJ (1974) Genetic control of the cell division cycle in yeast. Science 183:46–51

Harvey KJ, Newport J (2003) Metazoan origin selection: origin recognition complex chromatin binding is regulated by CDC6 recruitment and ATP hydrolysis. J Biol Chem 278:48524–48528

Heintz NH, Hamlin JL (1982) An amplified chromosomal sequence that includes the gene for dihydrofolate reductase initiates replication within specific restriction fragments. Proc Natl Acad Sci USA 79:4083–4087

Heintz NH, Milbrandt JD, Greisen KS, Hamlin JL (1983) Cloning of the initiation region of a mammalian chromosomal replicon. Nature 302:439–441

Hennessy KM, Lee A, Chen E, Botstein D (1991) A group of interacting yeast DNA replication genes. Genes Dev 5:958–969

Herbig U, Marlar CA, Fanning E (1999) The Cdc6 nucleotide-binding site regulates its activity in DNA replication in human cells. Mol Biol Cell 10:2631–2645

Higa LA, Mihaylov IS, Banks DP, Zheng J, Zhang H (2003) Radiation-mediated proteolysis of CDT1 by CUL4-ROC1 and CSN complexes constitutes a new checkpoint. Nat Cell Biol 5:1008–1015

Hofmann JF, Beach D (1994) cdt1 is an essential target of the Cdc10/Sct1 transcription factor: requirement for DNA replication and inhibition of mitosis. Embo J 13:425–434

Homesley L, Lei M, Kawasaki Y, Sawyer S, Christensen T, Tye BK (2000) Mcm10 and the MCM2-7 complex interact to initiate DNA synthesis and to release replication factors from origins. Genes Dev 14:913–926

Hopwood B, Dalton S (1996) Cdc45p assembles into a complex with Cdc46p/Mcm5p, is required for minichromosome maintenance, and is essential for chromosomal DNA replication. Proc Natl Acad Sci USA 93:12309–12314

Hu J, McCall CM, Ohta T, Xiong Y (2004) Targeted ubiquitination of CDT1 by the DDB1-CUL4A-ROC1 ligase in response to DNA damage. Nat Cell Biol 6:1003–1009

Hua XH, Newport J (1998) Identification of a preinitiation step in DNA replication that is independent of origin recognition complex and cdc6, but dependent on cdk2. J Cell Biol 140:271–281

Hyrien O, Mechali M (1992) Plasmid replication in Xenopus eggs and egg extracts: a 2D gel electrophoretic analysis. Nucleic Acids Res 20:1463–1469

Hyrien O, Mechali M (1993) Chromosomal replication initiates and terminates at random sequences but at regular intervals in the ribosomal DNA of Xenopus early embryos. Embo J 12:4511–4520

Hyrien O, Maric C, Mechali M (1995) Transition in specification of embryonic metazoan DNA replication origins. Science 270:994–997

Ishimi Y (1997) A DNA helicase activity is associated with an MCM4, -6, and -7 protein complex. J Biol Chem 272:24508–24513

Izumi M, Yatagai F, Hanaoka F (2001) Cell cycle-dependent proteolysis and phosphorylation of human Mcm10. J Biol Chem 276:48526–48531

Jacob F, Brenner S, Cuzin F (1963) On the regulation of DNA replication in bacteria. Cold Spring Harbor Symp Quant Biol 28:329–348

Jallepalli PV, Brown GW, Muzi-Falconi M, Tien D, Kelly TJ (1997) Regulation of the replication initiator protein p65cdc18 by CDK phosphorylation. Genes Dev 11:2767–2779

Jeon Y, Bekiranov S, Karnani N, Kapranov P, Ghosh S, Macalpine DM, Lee C, Hwang DS, Gingeras TR, Dutta A (2005) Temporal profile of replication of human chromosomes. Proc Natl Acad Sci USA 102:6419–6424

Jiang W, Wells NJ, Hunter T (1999) Multistep regulation of DNA replication by Cdk phosphorylation of HsCdc6. Proc Natl Acad Sci USA 96:6193–6198

Kalejta RF, Li X, Mesner LD, Dijkwel PA, Lin HB, Hamlin JL (1998) Distal sequences, but not ori-beta/OBR-1, are essential for initiation of DNA replication in the Chinese hamster DHFR origin. Mol Cell 2:797–806

Kamimura Y, Masumoto H, Sugino A, Araki H (1998) Sld2, which interacts with Dpb11 in Saccharomyces cerevisiae, is required for chromosomal DNA replication. Mol Cell Biol 18:6102–6109

Kamimura Y, Tak YS, Sugino A, Araki H (2001) Sld3, which interacts with Cdc45 (Sld4), functions for chromosomal DNA replication in Saccharomyces cerevisiae. Embo J 20:2097–2107

Kanemaki M, Sanchez-Diaz A, Gambus A, Labib K (2003) Functional proteomic identification of DNA replication proteins by induced proteolysis in vivo. Nature 423:720–724

Karakaidos P, Taraviras S, Vassiliou LV, Zacharatos P, Kastrinakis NG, Kougiou D, Kouloukoussa M, Nishitani H, Papavassiliou AG, Lygerou Z, Gorgoulis VG (2004) Overexpression of the replication licensing regulators hCdt1 and hCdc6 characterizes a subset of non-small-cell lung carcinomas: synergistic effect with mutant p53 on tumor growth and chromosomal instability–evidence of E2F-1 transcriptional control over hCdt1. Am J Pathol 165:1351–1365

Kearsey SE, Montgomery S, Labib K, Lindner K (2000) Chromatin binding of the fission yeast replication factor mcm4 occurs during anaphase and requires ORC and cdc18. Embo J 19:1681–1690

Kelly TJ, Brown GW (2000) Regulation of chromosome replication. Annu Rev Biochem 69:829–880

Kelman Z, Lee JK, Hurwitz J (1999) The single minichromosome maintenance protein of Methanobacterium thermoautotrophicum DeltaH contains DNA helicase activity. Proc Natl Acad Sci USA 96:14783–14788

Kihara M, Nakai W, Asano S, Suzuki A, Kitada K, Kawasaki Y, Johnston LH, Sugino A (2000) Characterization of the yeast Cdc7p/Dbf4p complex purified from insect cells. Its protein kinase activity is regulated by Rad53p. J Biol Chem 275:35051–35062

Kitsberg D, Selig S, Keshet I, Cedar H (1993) Replication structure of the human beta-globin gene domain. Nature 366:588–590

Kobayashi T, Rein T, DePamphilis ML (1998) Identification of primary initiation sites for DNA replication in the hamster dihydrofolate reductase gene initiation zone. Mol Cell Biol 18:3266–3277

Kondo T, Kobayashi M, Tanaka J, Yokoyama A, Suzuki S, Kato N, Onozawa M, Chiba K, Hashino S, Imamura M et al. (2004) Rapid degradation of Cdt1 upon UV-induced DNA damage is mediated by SCF^{Skp2} complex. J Biol Chem 279:27315–27319

Kong D, DePamphilis ML (2001) Site-specific DNA binding of the Schizosaccharomyces pombe origin recognition complex is determined by the Orc4 subunit. Mol Cell Biol 21:8095–8103

Koonin EV (1993). A common set of conserved motifs in a vast variety of putative nucleic acid-dependent ATPases including MCM proteins involved in the initiation of eukaryotic DNA replication. Nucleic Acids Res 21:2541–2547

Kubota Y, Mimura S, Nishimoto S, Masuda T, Nojima H, Takisawa H (1997) Licensing of DNA replication by a multi-protein complex of MCM/P1 proteins in Xenopus eggs. Embo J 16:3320–3331

Kubota Y, Takase Y, Komori Y, Hashimoto Y, Arata T, Kamimura Y, Araki H, Takisawa H (2003) A novel ring-like complex of Xenopus proteins essential for the initiation of DNA replication. Genes Dev 17:1141–1152

Kumagai H, Sato N, Yamada M, Mahony D, Seghezzi W, Lees E, Arai K, Masai H (1999) A novel growth- and cell cycle-regulated protein, ASK, activates human Cdc7-related kinase and is essential for G1/S transition in mammalian cells. Mol Cell Biol 19:5083–5095

Labib K, Diffley JF, Kearsey SE (1999) G1-phase and B-type cyclins exclude the DNA-replication factor Mcm4 from the nucleus. Nat Cell Biol 1:415–422

Labib K, Tercero JA, Diffley JF (2000) Uninterrupted MCM2-7 function required for DNA replication fork progression. Science 288:1643–1647

Leatherwood J, Lopez-Girona A, Russell P (1996) Interaction of Cdc2 and Cdc18 with a fission yeast ORC2-like protein. Nature 379:360–363

Lee C, Hong B, Choi JM, Kim Y, Watanabe S, Ishimi Y, Enomoto T, Tada S, Cho Y (2004) Structural basis for inhibition of the replication licensing factor Cdt1 by geminin. Nature 430:913–917

Lee JK, Hurwitz J (2000) Isolation and characterization of various complexes of the minichromosome maintenance proteins of Schizosaccharomyces pombe. J Biol Chem 275:18871–18878

Lee JK, Seo YS, Hurwitz J (2003) The Cdc23 (Mcm10) protein is required for the phosphorylation of minichromosome maintenance complex by the Dfp1-Hsk1 kinase. Proc Natl Acad Sci USA 100:2334–2339

Lee JK, Moon KY, Jiang Y, Hurwitz J (2001) The Schizosaccharomyces pombe origin recognition complex interacts with multiple AT-rich regions of the replication origin DNA by means of the AT-hook domains of the spOrc4 protein. Proc Natl Acad Sci USA 98:13589–13594

Lei M, Kawasaki Y, Young MR, Kihara M, Sugino A, Tye BK (1997) Mcm2 is a target of regulation by Cdc7-Dbf4 during the initiation of DNA synthesis. Genes Dev 11:3365–3374

Li A, Blow JJ (2004) Cdt1 downregulation by proteolysis and geminin inhibition prevents DNA re-replication in Xenopus. Embo J 24:395–404

Li CJ, DePamphilis ML (2002) Mammalian Orc1 protein is selectively released from chromatin and ubiquitinated during the S-to-M transition in the cell division cycle. Mol Cell Biol 22:105–116

Li CJ, Vassilev A, DePamphilis ML (2004) Role for Cdk1 (Cdc2)/cyclin A in preventing the mammalian origin recognition complex's largest subunit (Orc1) from binding to chromatin during mitosis. Mol Cell Biol 24:5875–5886

Li D, Zhao R, Lilyestrom W, Gai D, Zhang R, DeCaprio JA, Fanning E, Jochimiak A, Szakonyi G, Chen XS (2003a) Structure of the replicative helicase of the oncoprotein SV40 large tumour antigen. Nature 423:512–518

Li X, Zhao Q, Liao R, Sun P, Wu X (2003b) The SCF^{Skp2} ubiquitin ligase complex interacts with the human replication licensing factor Cdt1 and regulates Cdt1 degradation. J Biol Chem 278:30854–30858

Liang C, Weinreich M, Stillman B (1995) ORC and Cdc6p interact and determine the frequency of initiation of DNA replication in the genome. Cell 81:667–676

Liu G, Malott M, Leffak M (2003) Multiple functional elements comprise a mammalian chromosomal replicator. Mol Cell Biol 23:1832–1842

Liu J, Smith CL, DeRyckere D, DeAngelis K, Martin GS, Berger JM (2000) Structure and function of Cdc6/Cdc18: implications for origin recognition and checkpoint control. Mol Cell 6:637–648

Macalpine DM, Rodriguez HK, Bell SP (2004) Coordination of replication and transcription along a Drosophila chromosome. Genes Dev 18:3094–3105

Madine MA, Khoo CY, Mills AD, Laskey RA (1995) MCM3 complex required for cell cycle regulation of DNA replication in vertebrate cells. Nature 375:421–424

Mahbubani HM, Paull T, Elder JK, Blow JJ (1992) DNA replication initiates at multiple sites on plasmid DNA in Xenopus egg extracts. Nucleic Acids Res 20:1457–1462

Maine GT, Sinha P, Tye BK (1984) Mutants of S. cerevisiae defective in the maintenance of minichromosomes. Genetics 106:365–385

Maiorano D, Moreau J, Mechali M (2000) XCDT1 is required for the assembly of pre-replicative complexes in Xenopus laevis. Nature 404:622–625

Marahrens Y, Stillman B (1992) A yeast chromosomal origin of DNA replication defined by multiple functional elements. Science 255:817–823

Masai H, Arai K (2002) Cdc7 kinase complex: a key regulator in the initiation of DNA replication. J Cell Physiol 190:287–296

Masuda HP, Ramos GB, de Almeida-Engler J, Cabral LM, Coqueiro VM, Macrini CM, Ferreira PC, Hemerly AS (2004) Genome based identification and analysis of the prereplicative complex of Arabidopsis thaliana. FEBS Lett 574:192–202

Masuda T, Mimura S, Takisawa H (2003) CDK- and Cdc45-dependent priming of the MCM complex on chromatin during S-phase in Xenopusenopus egg extracts: possible activation of MCM helicase by association with Cdc45. Genes Cells 8:145–161

Masumoto H, Sugino A, Araki H (2000) Dpb11 controls the association between DNA polymerases alpha and epsilon and the autonomously replicating sequence region of budding yeast. Mol Cell Biol 20:2809–2817

Masumoto H, Muramatsu S, Kamimura Y, Araki H (2002) S-Cdk-dependent phosphorylation of Sld2 essential for chromosomal DNA replication in budding yeast. Nature 415:651–655

McGarry TJ, Kirschner MW (1998) Geminin, an inhibitor of DNA replication, is degraded during mitosis. Cell 93:1043–1053

Melixetian M, Ballabeni A, Masiero L, Gasparini P, Zamponi R, Bartek J, Lukas J, Helin K (2004) Loss of Geminin induces rereplication in the presence of functional p53. J Cell Biol 165:473–482

Mendez J, Stillman B (2000) Chromatin association of human origin recognition complex, cdc6, and minichromosome maintenance proteins during the cell cycle: assembly of prereplication complexes in late mitosis. Mol Cell Biol 20:8602–8612

Mendez J, Zou-Yang XH, Kim SY, Hidaka M, Tansey WP, Stillman B (2002) Human origin recognition complex large subunit is degraded by ubiquitin-mediated proteolysis after initiation of DNA replication. Mol Cell 9:481–491

Merchant AM, Kawasaki Y, Chen Y, Lei M, Tye BK (1997) A lesion in the DNA replication initiation factor Mcm10 induces pausing of elongation forks through chromosomal replication origins in Saccharomyces cerevisiae. Mol Cell Biol 17:3261–3271

Mesner LD, Li X, Dijkwel PA, Hamlin JL (2003) The dihydrofolate reductase origin of replication does not contain any nonredundant genetic elements required for origin activity. Mol Cell Biol 23:804–814

Mimura S, Takisawa H (1998) Xenopus Cdc45-dependent loading of DNA polymerase alpha onto chromatin under the control of S-phase Cdk. Embo J 17:5699–5707

Mizushima T, Takahashi N, Stillman B (2000) Cdc6p modulates the structure and DNA binding activity of the origin recognition complex in vitro. Genes Dev 14:1631–1641

Moir D, Stewart SE, Osmond BC, Botstein D (1982) Cold-sensitive cell-division-cycle mutants of yeast: isolation, properties, and pseudoreversion studies. Genetics 100:547–563

Moon KY, Kong D, Lee JK, Raychaudhuri S, Hurwitz J (1999) Identification and reconstitution of the origin recognition complex from Schizosaccharomyces pombe. Proc Natl Acad Sci USA 96:12367–12372

Muller M, Lucchini R, Sogo JM (2000) Replication of yeast rDNA initiates downstream of transcriptionally active genes. Mol Cell 5:767–777

Muzi-Falconi M, Brown GW, Kelly TJ (1996) cdc18+ regulates initiation of DNA replication in Schizosaccharomyces pombe. Proc Natl Acad Sci USA 93:1566–1570

Muzi-Falconi M, Kelly TJ (1995) Orp1, a member of the Cdc18/Cdc6 family of S-phase regulators, is homologous to a component of the origin recognition complex. Proc Natl Acad Sci USA 92:12475–12479

Nakajima R, Masukata H (2002) SpSld3 is required for loading and maintenance of SpCdc45 on chromatin in DNA replication in fission yeast. Mol Biol Cell 13:1462–1472

Neuwald AF, Aravind L, Spouge JL, Koonin EV (1999) AAA+: A class of chaperone-like ATPases associated with the assembly, operation, and disassembly of protein complexes. Genome Res 9:27–43

Nguyen VQ, Co C, Li JJ (2001) Cyclin-dependent kinases prevent DNA re-replication through multiple mechanisms. Nature 411:1068–1073

Nguyen VQ, Co C, Irie K, Li JJ (2000) Clb/Cdc28 kinases promote nuclear export of the replication initiator proteins Mcm2-7. Curr Biol 10:195–205

Nishitani H, Nurse P (1995) p65cdc18 plays a major role controlling the initiation of DNA replication in fission yeast. Cell 83:397–405

Nishitani H, Lygerou Z, Nishimoto T, Nurse P (2000) The Cdt1 protein is required to license DNA for replication in fission yeast. Nature 404:625–628

Nishitani H, Taraviras S, Lygerou Z, Nishimoto T (2001) The human licensing factor for DNA replication Cdt1 accumulates in G1 and is destabilized after initiation of S-phase. J Biol Chem 276:44905–44911

Ohtani K, Tsujimoto A, Ikeda M, Nakamura M (1998) Regulation of cell growth-dependent expression of mammalian CDC6 gene by the cell cycle transcription factor E2F. Oncogene 17:1777–1785

Oshiro G, Owens JC, Shellman Y, Sclafani RA, Li JJ (1999) Cell cycle control of Cdc7p kinase activity through regulation of Dbf4p stability. Mol Cell Biol 19:4888–4896

Owens JC, Detweiler CS, Li JJ (1997) CDC45 is required in conjunction with CDC7/DBF4 to trigger the initiation of DNA replication. Proc Natl Acad Sci USA 94:12521–12526

Pacek M, Walter JC (2004) A requirement for MCM7 and Cdc45 in chromosome unwinding during eukaryotic DNA replication. Embo J 23:3667–3676

Paixao S, Colaluca IN, Cubells M, Peverali FA, Destro A, Giadrossi S, Giacca M, Falaschi A, Riva S, Biamonti G (2004) Modular structure of the human lamin B2 replicator. Mol Cell Biol 24:2958–2967

Parrilla-Castellar ER, Karnitz LM (2003) Cut5 is required for the binding of Atr and DNA polymerase alpha to genotoxin-damaged chromatin. J Biol Chem 278:45507–45511

Pelizon C, d'Adda di Fagagna F, Farrace L, Laskey RA (2002) Human replication protein Cdc6 is selectively cleaved by caspase 3 during apoptosis. EMBO Rep 3:780–784

Perkins G, Diffley JF (1998) Nucleotide-dependent prereplicative complex assembly by Cdc6p, a homolog of eukaryotic and prokaryotic clamp-loaders. Mol Cell 2:23–32

Petersen BO, Lukas J, Sorensen CS, Bartek J, Helin K (1999) Phosphorylation of mammalian CDC6 by cyclin A/CDK2 regulates its subcellular localization. Embo J 18:396–410

Pinto S, Quintana DG, Smith P, Mihalek RM, Hou ZH, Boynton S, Jones CJ, Hendricks M, Velinzon K, Wohlschlegel JA et al. (1999) latheo encodes a subunit of the origin recognition complex and disrupts neuronal proliferation and adult olfactory memory when mutant. Neuron 23:45–54

Quintana DG, Hou Z, Thome KC, Hendricks M, Saha P, Dutta A (1997) Identification of HsORC4, a member of the human origin of replication recognition complex. J Biol Chem 272:28247–28251

Quintana DG, Thome KC, Hou ZH, Ligon AH, Morton CC, Dutta A (1998) ORC5L, a new member of the human origin recognition complex, is deleted in uterine leiomyomas and malignant myeloid diseases. J Biol Chem 273:27137–27145

Raghuraman MK, Winzeler EA, Collingwood D, Hunt S, Wodicka L, Conway A, Lockhart DJ, Davis RW, Brewer BJ, Fangman WL (2001) Replication dynamics of the yeast genome. Science 294:115–121

Reid RJ, Fiorani P, Sugawara M, Bjornsti MA (1999) CDC45 and DPB11 are required for processive DNA replication and resistance to DNA topoisomerase I-mediated DNA damage. Proc Natl Acad Sci USA 96:11440–11445

Remus D, Beall EL, Botchan MR (2004) DNA topology, not DNA sequence, is a critical determinant for Drosophila ORC-DNA binding. Embo J 23:897–907

Ricke RM, Bielinsky AK (2004) Mcm10 regulates the stability and chromatin association of DNA polymerase-alpha. Mol Cell 16:173–185

Ritzi M, Tillack K, Gerhardt J, Ott E, Humme S, Kremmer E, Hammerschmidt W, Schepers A (2003) Complex protein-DNA dynamics at the latent origin of DNA replication of Epstein-Barr virus. J Cell Sci 116:3971–3984

Rowles A, Tada S, Blow JJ (1999) Changes in association of the Xenopus origin recognition complex with chromatin on licensing of replication origins. J Cell Sci 112:2011–2018

Rowley A, Cocker JH, Harwood J, Diffley JF (1995) Initiation complex assembly at budding yeast replication origins begins with the recognition of a bipartite sequence by limiting amounts of the initiator, ORC. Embo J 14:2631–2641

Saha P, Chen J, Thome KC, Lawlis SJ, Hou ZH, Hendricks M, Parvin JD, Dutta A (1998) Human CDC6/Cdc18 associates with Orc1 and cyclin-cdk and is selectively eliminated from the nucleus at the onset of S phase. Mol Cell Biol 18:2758–2767

Saha S, Shan Y, Mesner LD, Hamlin JL (2004) The promoter of the Chinese hamster ovary dihydrofolate reductase gene regulates the activity of the local origin and helps define its boundaries. Genes Dev 18:397–410

Saka Y, Yanagida M (1993) Fission yeast cut5+, required for S phase onset and M phase restraint, is identical to the radiation-damage repair gene rad4+. Cell 74:383–393

Sasaki T, Sawado T, Yamaguchi M, Shinomiya T (1999) Specification of regions of DNA replication initiation during embryogenesis in the 65-kilobase DNApolalpha-dE2F locus of Drosophila melanogaster. Mol Cell Biol 19:547–555

Sato M, Gotow T, You Z, Komamura-Kohno Y, Uchiyama Y, Yabuta N, Nojima H, Ishimi Y (2000) Electron microscopic observation and single-stranded DNA binding activity of the Mcm4,6,7 complex. J Mol Biol 300:421–431

Sato N, Arai K, Masai H (1997) Human and Xenopus cDNAs encoding budding yeast Cdc7-related kinases: in vitro phosphorylation of MCM subunits by a putative human homologue of Cdc7. Embo J 16:4340–4351

Saxena S, Yuan P, Dhar SK, Senga T, Takeda D, Robinson H, Kornbluth S, Swaminathan K, Dutta A (2004) A dimerized coiled-coil domain and an adjoining part of geminin interact with two sites on Cdt1 for replication inhibition. Mol Cell 15:245–258

Schepers A, Diffley JF (2001) Mutational analysis of conserved sequence motifs in the budding yeast Cdc6 protein. J Mol Biol 308:597–608

Schepers A, Ritzi M, Bousset K, Kremmer E, Yates JL, Harwood J, Diffley JF, Hammerschmidt W (2001) Human origin recognition complex binds to the region of the latent origin of DNA replication of Epstein-Barr virus. Embo J 20:4588–4602

Schubeler D, Scalzo D, Kooperberg C, van Steensel B, Delrow J, Groudine M (2002) Genome-wide DNA replication profile for Drosophila melanogaster: a link between transcription and replication timing. Nat Genet 32:438–442

Schwacha A, Bell SP (2001) Interactions between two catalytically distinct MCM subgroups are essential for coordinated ATP hydrolysis and DNA replication. Mol Cell 8:1093–1104

Seki T, Diffley JF (2000) Stepwise assembly of initiation proteins at budding yeast replication origins in vitro. Proc Natl Acad Sci USA 97:14115–14120

Semple JW, Duncker BP (2004) ORC-associated replication factors as biomarkers for cancer. Biotechnol Adv 22:621–631

Shechter DF, Ying CY, Gautier J (2000) The intrinsic DNA helicase activity of Methanobacterium thermoautotrophicum delta H minichromosome maintenance protein. J Biol Chem 275:15049–15059

Solomon NA, Wright MB, Chang S, Buckley AM, Dumas LB, Gaber RF (1992) Genetic and molecular analysis of DNA43 and DNA52: two new cell-cycle genes in Saccharomyces cerevisiae. Yeast 8:273–289

Stedman W, Deng Z, Lu F, Lieberman PM (2004) ORC, MCM, and histone hyperacetylation at the Kaposi's sarcoma-associated herpesvirus latent replication origin. J Virol 78:12566–12575

Stinchcomb DT, Struhl K, Davis RW (1979) Isolation and characterisation of a yeast chromosomal replicator. Nature 282:39–43

Stoeber K, Mills AD, Kubota Y, Krude T, Romanowski P, Marheineke K, Laskey RA, Williams GH (1998) Cdc6 protein causes premature entry into S phase in a mammalian cell-free system. Embo J 17:7219–7229

Struhl K, Stinchcomb DT, Scherer S, Davis RW (1979) High-frequency transformation of yeast: autonomous replication of hybrid DNA molecules. Proc Natl Acad Sci USA 76:1035–1039

Sugimoto N, Tatsumi Y, Tsurumi T, Matsukage A, Kiyono T, Nishitani H, Fujita M (2004) Cdt1 phosphorylation by cyclin A-dependent kinases negatively regulates its function without affecting geminin binding. J Biol Chem 279:19691–19697

Tada S, Li A, Maiorano D, Mechali M, Blow JJ (2001) Repression of origin assembly in metaphase depends on inhibition of RLF-B/Cdt1 by geminin. Nat Cell Biol 3:107–113

Takayama Y, Kamimura Y, Okawa M, Muramatsu S, Sugino A, Araki H (2003) GINS, a novel multiprotein complex required for chromosomal DNA replication in budding yeast. Genes Dev 17:1153–1165

Takeda DY, Parvin JD, Dutta A (2005) Degradation of Cdt1 during S-phase is SKP2 independent and is required for efficient progression of mammlian cells through S-phase. J Biol Chem (in press)

Takeda T, Ogino K, Matsui E, Cho MK, Kumagai H, Miyake T, Arai K, Masai H (1999) A fission yeast gene, him1(+)/dfp1(+) encoding a regulatory subunit for Hsk1 kinase, plays essential roles in S-phase initiation as well as in S-phase checkpoint control and recovery from DNA damage. Mol Cell Biol 19:5535–5547

Tanaka S, Diffley JF (2002) Interdependent nuclear accumulation of budding yeast Cdt1 and Mcm2-7 during G1 phase. Nat Cell Biol 4:198–207

Tanaka T, Knapp D, Nasmyth K (1997) Loading of an Mcm protein onto DNA replication origins is regulated by Cdc6p and CDKs. Cell 90:649–660

Tatsumi Y, Ohta S, Kimura H, Tsurimoto T, Obuse C (2003) The ORC1 cycle in human cells: I. cell cycle-regulated oscillation of human ORC1. J Biol Chem 278:41528–41534

Thommes P, Kubota Y, Takisawa H, Blow JJ (1997) The RLF-M component of the replication licensing system forms complexes containing all six MCM/P1 polypeptides. Embo J 16:3312–3319

Tugal T, Zou-Yang XH, Gavin K, Pappin D, Canas B, Kobayashi R, Hunt T, Stillman B (1998) The Orc4p and Orc5p subunits of the Xenopus and human origin recognition complex are related to Orc1p and Cdc6p. J Biol Chem 273:32421–32429

Van Hatten RA, Tutter AV, Holway AH, Khederian AM, Walter JC, Michael WM (2002) The Xenopus Xmus101 protein is required for the recruitment of Cdc45 to origins of DNA replication. J Cell Biol 159:541–547

Vashee S, Simancek P, Challberg MD, Kelly TJ (2001) Assembly of the human origin recognition complex. J Biol Chem 276:26666–26673

Vashee S, Cvetic C, Lu W, Simancek P, Kelly TJ, Walter JC (2003) Sequence-independent DNA binding replication initiation by the human origin recognition complex. Genes Dev 17:1894–1908

Vaughn JP, Dijkwel PA, Hamlin JL (1990) Replication initiates in a broad zone in the amplified CHO dihydrofolate reductase domain. Cell 61:1075–1087

Vaziri C, Saxena S, Jeon Y, Lee C, Murata K, Machida Y, Wagle N, Hwang DS, Dutta A (2003) A p53-dependent checkpoint pathway prevents rereplication. Mol Cell 11:997–1008

Walter J, Newport J (2000) Initiation of eukaryotic DNA replication: origin unwinding and sequential chromatin association of Cdc45, RPA, and DNA polymerase alpha. Mol Cell 5:617–627

Wang H, Elledge SJ (1999) DRC1, DNA replication and checkpoint protein 1, functions with DPB11 to control DNA replication and the S-phase checkpoint in Saccharomyces cerevisiae. Proc Natl Acad Sci USA 96:3824–3829

Wang L, Lin CM, Brooks S, Cimbora D, Groudine M, Aladjem MI (2004) The human beta-globin replication initiation region consists of two modular independent replicators. Mol Cell Biol 24:3373–3386

Weinreich M, Stillman B (1999) Cdc7p-Dbf4p kinase binds to chromatin during S phase and is regulated by both the APC and the RAD53 checkpoint pathway. Embo J 18:5334–5346

Weinreich, M, Liang C, Stillman B (1999) The Cdc6p nucleotide-binding motif is required for loading mcm proteins onto chromatin. Proc Natl Acad Sci USA 96:441–446

Williams GH, Romanowski P, Morris L, Madine M, Mills AD, Stoeber K, Marr J, Laskey RA, Coleman N (1998) Improved cervical smear assessment using antibodies against proteins that regulate DNA replication. Proc Natl Acad Sci USA 95:14932–14937

Williams RS, Shohet RV, Stillman B (1997) A human protein related to yeast Cdc6p. Proc Natl Acad Sci USA 94:142–147

Wilmes GM, Archambault V, Austin RJ, Jacobson MD, Bell SP, Cross FR (2004) Interaction of the S-phase cyclin Clb5 with an "RXL" docking sequence in the initiator protein Orc6 provides an origin-localized replication control switch. Genes Dev 18:981–991

Wohlschlegel JA, Kutok JL, Weng AP, Dutta A (2002a) Expression of geminin as a marker of cell proliferation in normal tissues and malignancies. Am J Pathol 161:267–273

Wohlschlegel JA, Dhar SK, Prokhorova TA, Dutta A, Walter JC (2002b) Xenopus Mcm10 binds to origins of DNA replication after Mcm2-7 and stimulates origin binding of Cdc45. Mol Cell 9:233–240

Wohlschlegel JA, Dwyer BT, Dhar SK, Cvetic C, Walter JC, Dutta A (2000) Inhibition of eukaryotic DNA replication by geminin binding to Cdt1. Science 290:2309–2312

Woodfine K, Fiegler H, Beare DM, Collins JE, McCann OT, Young BD, Debernardi S, Mott R, Dunham I, Carter NP (2004) Replication timing of the human genome. Hum Mol Genet 13:191–202

Wu JR, Gilbert DM (1996) A distinct G1 step required to specify the Chinese hamster DHFR replication origin. Science 271:1270–1272

Wuarin J, Buck V, Nurse P, Millar JB (2002) Stable association of mitotic cyclin B/Cdc2 to replication origins prevents endoreduplication. Cell 111:419–431

Yamada Y, Nakagawa T, Masukata H (2004) A novel intermediate in initiation complex assembly for fission yeast DNA replication. Mol Biol Cell 15:3740–3750

Yamane K, Wu X, Chen J (2002) A DNA damage-regulated BRCT-containing protein, TopBP1, is required for cell survival. Mol Cell Biol 22:555–566

Yamane K, Chen J, Kinsella TJ (2003) Both DNA topoisomerase II-binding protein 1 and BRCA1 regulate the G2-M cell cycle checkpoint. Cancer Res 63:3049–3053

Yan Z, DeGregori J, Shohet R, Leone G, Stillman B, Nevins JR, Williams RS (1998) Cdc6 is regulated by E2F and is essential for DNA replication in mammalian cells. Proc Natl Acad Sci USA 95:3603–3608

Yanagi K, Mizuno T, You Z, Hanaoka F (2002) Mouse geminin inhibits not only Cdt1-MCM6 interactions but also a novel intrinsic Cdt1 DNA binding activity. J Biol Chem 277:40871–40880

Yoshida K, Inoue I (2004) Expression of MCM10 and TopBP1 is regulated by cell proliferation and UV irradiation via the E2F transcription factor. Oncogene 23:6250–6260

You Z, Komamura Y, Ishimi Y (1999) Biochemical analysis of the intrinsic Mcm4-Mcm6-Mcm7 DNA helicase activity. Mol Cell Biol 19:8003–8015

Young LS, Rickinson AB (2004) Epstein-Barr virus:40 years on. Nat Rev Cancer 4:757–768

Zhong W, Feng H, Santiago FE, Kipreos ET (2003) CUL-4 ubiquitin ligase maintains genome stability by restraining DNA-replication licensing. Nature 423:885–889

Zhu W, Chen Y, Dutta A (2004) Rereplication by depletion of geminin is seen regardless of p53 status and activates a G2/M checkpoint. Mol Cell Biol 24:7140–7150

Zou L, Stillman B (1998) Formation of a preinitiation complex by S-phase cyclin CDK-dependent loading of Cdc45p onto chromatin. Science 280:593–596

Zou L, Stillman B (2000) Assembly of a complex containing Cdc45p, replication protein A, and Mcm2p at replication origins controlled by S-phase cyclin-dependent kinases and Cdc7p-Dbf4p kinase. Mol Cell Biol 20:3086–3096

Zou L, Mitchell J, Stillman B (1997) CDC45, a novel yeast gene that functions with the origin recognition complex and Mcm proteins in initiation of DNA replication. Mol Cell Biol 17:553–563

Checkpoint and Coordinated Cellular Responses to DNA Damage

Xiaohong H. Yang[1] · Lee Zou[1,2] (✉)

[1]MGH Cancer Center, Harvard Medical School, Charlestown, MA 02129, USA
zou.lee@mgh.harvard.edu

[2]Department of Pathology, Harvard Medical School, Charlestown, MA 02129, USA
zou.lee@mgh.harvard.edu

Abstract The DNA damage and replication checkpoints are signaling mechanisms that regulate and coordinate cellular responses to genotoxic conditions. The activation of checkpoints not only attenuates cell cycle progression, but also facilitates DNA repair and recovery of faulty replication forks, thereby preventing DNA lesions from being converted to inheritable mutations. It has become increasingly clear that the activation and signaling of the checkpoint are intimately linked to the cellular processes directly involved in chromosomal metabolism, such as DNA replication and DNA repair. Thus, the checkpoint pathway is not just a surveillance system that monitors genomic integrity and regulates cell proliferation, but also an integral part of the processes that work directly on chromosomes to maintain genomic stability. In this article, we discuss the current models of DNA damage and replication checkpoints, and highlight recent advances in the field.

1
Introduction

The survival of organisms relies on faithful duplication and segregation of their genomes. To ensure that the entire genome is accurately transmitted to the next generation of cells, duplication and segregation of the genome must be highly coordinated. The maintenance of genomic stability is a mounting task for cells because the DNA content in every cell is constantly challenged by both extrinsic insults and intrinsic stresses. For example, DNA lesions can arise from failures of the DNA replication machine or from various types of DNA-damaging molecules or radiations in the environment. If DNA damage is not accurately and quickly repaired, or if the coordinated genome duplication and segregation cannot be maintained, detrimental events such as loss of genetic information, deleterious chromosomal rearrangements, and mutations disrupting the control of cell proliferation might result. To maintain genomic stability in the presence of DNA damage or DNA replication interference, cells use a complex signaling pathway called the checkpoint to regulate and coordinate many cellular processes to remove DNA damage and to alleviate stresses on their genomes (Hartwell and Weinert 1989). It was initially discovered that activation of the checkpoint by DNA damage leads to cell cycle arrest, probably providing more time for DNA repair (Hartwell and Weinert 1989). Now it has become increasingly clear that the checkpoint also plays vital roles in the regulation of DNA replication, DNA repair, chromatin structure, and many other cellular processes important for genomic stability (Zhou and Elledge 2000; Osborn et al. 2002; Kastan and Bartek 2004). Most interestingly, many of the processes regulated by the checkpoint might also play roles in generating or transmitting DNA damage signals through the checkpoint pathway (Osborn et al. 2002). Thus, the checkpoint pathway is not only a surveillance system that monitors genomic integrity, but also an integral part of the processes that directly work on chromosomes to maintain their stability. Recent studies have revealed important mechanisms by which cells detect various types of DNA damage and activate the checkpoint pathway, as well as the mechanisms by which the checkpoint regulates key downstream processes such as DNA replication, DNA repair, and chromatin modulation. In this review, we will discuss an updated model of checkpoint signaling that begins to explain how different processes involved in the maintenance of genomic stability are integrated and coordinated by the checkpoint pathway.

2
Sensing DNA Damage and DNA Replication Stress

ATM (*ataxia telangiectasia mutated*) and ATR (*ATM- and Rad3-related*) are two large PI3K-like protein kinases that play central roles in the checkpoint-signaling pathway (Abraham 2001). In response to DNA damage, ATM and ATR phosphorylate Chk1 and Chk2, two downstream effector kinases, and numerous substrates involved in various cellular processes (e.g. p53, Brca1, Nbs1). The DNA damage specificities of ATM and ATR are distinct from each other. While ATM primarily responds to double-strand DNA breaks (DSBs), ATR is involved in the responses to DSBs as well as a broad spectrum of DNA damage caused by DNA replication interference. Although ATM is important for genomic stability, patients, mice, and cells lacking ATM are viable (Kastan and Bartek 2004), suggesting that ATM is not essential for normal cell proliferation in the absence of significant DSBs. On the other hand, ATR is indispensable for the proliferation of human and mouse cells (Brown and Baltimore 2000; Cortez et al. 2001). These findings indicate that ATR has a critical role even in normal cell cycles, and that the function of ATR might be regulated by certain DNA structures generated by intrinsic DNA metabolism.

2.1
Recruitment of ATR to DNA

What is the DNA structure sensed by the ATR kinase? Several important clues came from the studies of Mec1, the budding yeast homologue of ATR (see Table 1). In human cells, most ATR exists in a complex with its partner ATRIP (Cortez et al. 2001). Yeast Mec1 also forms a similar complex with its partner Ddc2 (Paciotti et al. 2000; Rouse and Jackson 2000; Wakayama et al. 2001). Importantly, the checkpoint functions of ATR and Mec1 are dependent upon ATRIP and Ddc2, respectively, indicating that ATR-ATRIP and Mec1-Ddc2 function as complexes in checkpoint signaling. In budding yeast, Mec1 can be activated in the *cdc13* mutant and by DSBs generated by the HO endonuclease (Lydall and Weinert 1995; Pellicioli et al. 2001). Interestingly, single-stranded DNA (ssDNA) is generated at telomeres in the *cdc13* mutant and at DSBs after they are recessed by exonucleases. Furthermore, Mec1 and Ddc2 have been shown to localize to telomeres in the *cdc13* mutant and to the HO-induced DSBs (Kondo et al. 2001; Melo et al. 2001; Rouse and Jackson 2002; Zou and Elledge 2003; Lisby et al. 2004). These findings indicated that ssDNA might be part of the DNA structure recognized by the DNA damage sensors that

Table 1 The proteins involved in checkpoint signaling in human and yeast cells

	Human	Budding Yeast
PI3K-like kinase	ATR	Mec1
	ATM	Tel1
ATR/Mec1 regulatory partner	ATRIP	Ddc2
Replication protein A	Rpa1-3	Rpa1-3
RFC-like complex	Rad17	Rad24
	Rfc2-5	Rfc2-5
PCNA-like complex	Rad9	Ddc1
	Rad1	Rad17
	Hus1	Mec3
MRN complex	Mre11	Mre11
	Rad50	Rad50
	Nbs1	Xrs2
Mediators and other signaling molecules	Claspin	Mrc1
	Brca1/53BP1/Mdc1	Rad9
	TopBP1	Dpb11
	hTim1	Tof1
	hTipin	Csm3
Effector kinase	Chk1	Chk1
	Chk2	Rad53

recruit the ATR-ATRIP kinase complex. Consistent with this idea, increased amounts of ssDNA was also observed at DNA replication forks halted by hydroxyurea (HU) treatment (Sogo et al. 2002), suggesting that ssDNA might also be important for the activation of Mec1 by replication fork stalling.

Studies using *Xenopus* egg extracts have also revealed important clues of how ATR is recruited to DNA. In *Xenopus* extracts, ATR associates with chromatin during S-phase in a replication-dependent manner (Hekmat-Nejad et al. 2000). Depletion of RPA, an ssDNA-binding protein complex essential for DNA replication, abolished the chromatin association of ATR (You et al. 2002), suggesting that RPA is either directly or indirectly required for the recruitment of ATR to chromatin. Similarly, in *Xenopus* extracts RPA is also needed for the recruitment of ATR to DNA lesions generated by etoposide (Costanzo et al. 2003), a DNA topoisomerase II inhibitor. This finding indicates that RPA itself or the DNA repair process involving RPA is required for ATR recruitment.

In human cells, RPA is required for the localization of ATR to DNA damage-induced nuclear foci and the efficient phosphorylation of Chk1 by ATR (Zou and Elledge 2003). In yeast, depletion of RPA in the cells arrested in G2 abolished the localization of Ddc2 to the HO-induced DSBs (Zou and Elledge 2003), suggesting that RPA is required for the recruitment of Ddc2 to DNA damage in vivo and its function is independent of its role in DNA replication. Indeed, *rfa1-t11*, a mutant of RPA that is proficient for DNA replication but partially defective for checkpoint responses (Umezu et al. 1998; Kim and Brill 2001; Pellicioli et al. 2001), exhibits diminished ability to recruit Ddc2 to DSBs and stalled replication forks in vivo (Zou and Elledge 2003; Lucca et al. 2004). All these findings suggest that RPA probably plays a rather direct role in the recruitment of ATR-ATRIP and Mec1-Ddc2. The direct role of RPA in the recruitment of ATR-ATRIP was eventually demonstrated by a series of in vitro biochemical experiments (Zou and Elledge 2003) (Fig. 1a, b). First, the purified human ATRIP protein is efficiently recruited to ssDNA only in the presence of RPA. Second, RPA renders ATRIP capable of distinguishing ssDNA from double-stranded DNA (dsDNA). Finally, the ATR-ATRIP complex, but not ATR alone, binds to ssDNA more efficiently in the presence of RPA. Together, these findings strongly suggest that RPA-coated ssDNA is sufficient for the recruitment of ATR-ATRIP to the sites of DNA damage. Consistently, both purified yeast Ddc2 and *Xenopus* ATRIP are efficiently recruited to ssDNA in an RPA-dependent manner (Zou and Elledge 2003; Kumagai et al. 2004). Notably, the recruitment of *Xenopus* ATRIP to the DNA structures formed by polyT-polyA oligomers, which can elicit ATR signaling in *Xenopus* extracts, is also dependent upon RPA (Kumagai et al. 2004).

Although RPA is required for the recruitment of ATR-ATRIP to DNA damage in vivo and RPA-coated ssDNA is sufficient for recruiting ATR-ATRIP in vitro, the possibility that ATR-ATRIP can interact with RPA-ssDNA through

Fig. 1 Models for checkpoint activation by DSBs and stalled replication forks. **a** Checkpoint activation by DSBs. ATM is autophosphorylated in response to DSBs. The MRN complex may recruit ATM to DSBs and activate ATM. ATM phosphorylates Nbs1, Brca1, and Smc1 at DSBs. When DSBs are recessed, the resulting ssDNA is coated by RPA. ATR-ATRIP is recruited to DSBs by RPA-ssDNA. RPA also stimulates the loading of 9-1-1 complexes by the Rad17 complex. **b** Checkpoint activation by stalled replication forks. Long stretches of ssDNA are generated at stalled replication forks. ATR-ATRIP, Rad17, and 9-1-1 complexes are recruited to RPA-ssDNA and junctions of double/single-stranded DNA at stalled forks. Although the checkpoint can be activated through different mechanisms, cell cycle arrest and inhibition of DNA replication are common results. The activated checkpoint may play important roles at DSBs and stalled replication forks to facilitate DNA repair and fork recovery

other proteins cannot be ruled out. Furthermore, it is also possible that the interaction between ATR-ATRIP and RPA-ssDNA is regulated by other proteins in vivo (see below). It was recently reported that ATR-ATRIP can associate with proteins such as Claspin, Msh2 and Mcm7 (Chini and Chen 2003; Wang and Qin 2003; Cortez et al. 2004). It is possible that the interactions of ATR-ATRIP with additional proteins on DNA also contribute to the localization of ATR-ATRIP to specific types of DNA damage.

2.2
DNA Damage Recognition by the RFC- and PCNA-like Checkpoint Complexes

Although ssDNA plays a crucial role in the recruitment of ATR-ATRIP, ssDNA alone is not sufficient to elicit the checkpoint responses, suggesting that additional DNA structures induced by DNA damage are also necessary for the activation of ATR-ATRIP. In addition to ATR-ATRIP itself, several other checkpoint proteins are also required for the initiation of checkpoint signaling. In human cells, these proteins include Rad17, Rad9, Rad1, and Hus1. Rad17 is a homologue of all five subunits of RFC, and its forms an RFC-like protein complex with the four small subunits of RFC (Lindsey-Boltz et al. 2001; Shiomi et al. 2002; Ellison and Stillman 2003; Zou et al. 2003). Rad9, Rad1, and Hus1, on the other hand, are all structurally related to PCNA (Venclovas and Thelen 2000), and they assemble into a hetero-trimeric ring-shaped complex (termed the 9-1-1 complex) resembling PCNA (Volkmer and Karnitz 1999). Ablation of Rad17 or the 9-1-1 complex in human or mouse cells resulted in severe chromosomal instability and failures in activating the ATR-mediate checkpoint (Weiss et al. 2002; Zou et al. 2002; Roos-Mattjus et al. 2003; Wang et al. 2003; Bao et al. 2004; Loegering et al. 2004). During DNA replication, RFC specifically recognizes the 3′ primer-template junctions on DNA and recruits PCNA onto DNA where it functions as the processivity factor for DNA polymerases. Interestingly, in response to DNA damage, the 9-1-1 complex is recruited onto chromatin in an Rad17-dependent manner (Zou et al. 2002), indicating that the RFC-like Rad17 complex might recog-

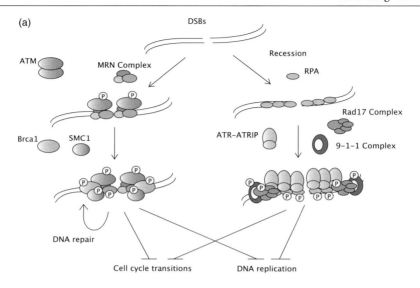

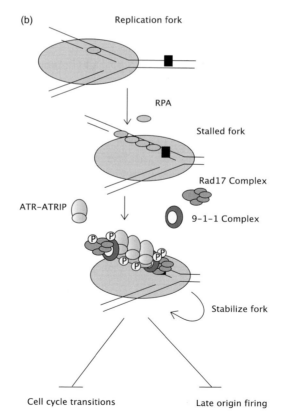

nize certain damage-induced DNA structure and recruit the 9-1-1 complex in a manner similar to the loading of PCNA by RFC.

What then is the DNA structure recognized by the Rad17 complex? In *Xenopus* extracts, the 9-1-1 complex associates with chromatin during S-phase like ATR does (You et al. 2002). However, unlike ATR, the chromatin association of 9-1-1 complex requires DNA polymerase α (You et al. 2002), indicating that the synthesis of RNA-DNA primer might be either directly or indirectly required for the function of the Rad17 complex. In budding yeast, both the HO-induced DSBs and the telomeres in *cdc13* mutant cells are recessed by 5'-to-3' exonucleases, resulting in 5' junctions of dsDNA and ssDNA (Lydall and Weinert 1995; Lee et al. 1998). Furthermore, the yeast PCNA-like checkpoint complex can be specifically recruited to the HO-induced DSBs (Kondo et al. 2001), suggesting that the RFC-like checkpoint complex might recognize the junctions of dsDNA/ssDNA. Interestingly, in yeast the efficient recruitment of the PCNA-like complex to DSBs also requires the function of RPA (Zou et al. 2003; Nakada et al. 2004). Consistent with these in vivo observations, purified Rad17 complexes recruit 9-1-1 complexes onto primed ssDNA or gapped DNA structures in an RPA-dependent manner (Ellison and Stillman 2003; Zou et al. 2003) (Fig. 1a, b). Unlike RFC, which only uses the 3' dsDNA/ssDNA junctions to load PCNA, the Rad17 complex can apparently use the 5' dsDNA/ssDNA junctions to recruit 9-1-1 complexes, providing a possible explanation of the DNA damage specificity of the Rad17 complex.

Together, the studies described above have revealed that the ATR-ATRIP, Rad17, and 9-1-1 complexes can independently recognize damage-induced DNA structures such as ssDNA and junctions of dsDNA/ssDNA. Moreover, RPA appears to play important roles in the damage recognition by both the ATR-ATRIP and the Rad17 complexes. Once recruited to the sites of DNA damage, the Rad17 and 9-1-1 complexes might facilitate the recognition and phosphorylation of ATR substrates by interacting with ATR-ATRIP and/or other proteins at the damage sites. Alternatively, the Rad17 and 9-1-1 complexes might stimulate the kinase activity of ATR on DNA. Although studies using *Xenopus* extracts suggested that ATR can be activated on DNA (Kumagai et al. 2004), how ATR is stimulated by DNA or its regulators on DNA remains to be elucidated.

2.3
Processing of DNA Lesions

Single-stranded DNA is a common DNA structure generated by DNA replication and many types of DNA repair processes. The discovery that ssDNA plays a crucial role in the activation of ATR provides a plausible explanation for why ATR can respond to different types of DNA damage. Furthermore, it clearly suggests that the processing of DNA lesions plays an important role in the activation of ATR. Many cellular processes involved in the maintenance of genomic stability, such as DNA replication, DNA repair, and chromatin modulation, have been implicated in the processing of DNA lesions. Thus, the activation of ATR appears to be an effective way to integrate the DNA damage signals generated by a variety of processes monitoring the integrity of the genome.

During S-phase of the cell cycle, the entire genome is duplicated by DNA replication forks. When replication forks encounter certain types of DNA damage or interference, they will activate the ATR-mediated checkpoint. Using yeast and *Xenopus* extracts, it has been shown that the activation of checkpoint by DNA damage generated by UV (ultraviolet light) or MMS (methyl methanesulfonate), a DNA-alkylating agent, is primarily through mechanisms dependent upon DNA replication (Lupardus et al. 2002; Stokes et al. 2002; Tercero et al. 2003). Furthermore, direct inhibition of DNA synthesis by aphidicolin, a DNA polymerase inhibitor, or HU, an inhibitor of dNTP synthesis, also elicits the ATR-mediated checkpoint. Thus, in addition to duplicating the genome, replication forks also function to scan the genome for DNA damage or other types of interference, and to translate these replication stresses into DNA structures that can be recognized by the checkpoint sensors. Several studies using human, *Xenopus*, and yeast systems have suggested that increased amounts of RPA-ssDNA are commonly induced at the replication forks encountering various types of DNA damage and interference (Tanaka and Nasmyth 1998; Michael et al. 2000; Lupardus et al. 2002; Zou and Elledge 2003). The accumulation of ssDNA at the stressed replication forks might be a result of the uncoupling of different fork components, such as helicases and DNA polymerases. Interestingly, the recruitment of ATRIP to RPA-ssDNA in vitro is dependent on the length of ssDNA (Zou and Elledge 2003), suggesting a quantitative mechanism for monitoring the stress that a replication fork encounters. Furthermore, elevated levels of DNA polymerase α were also observed on chromatin after DNA damage (Michael et al. 2000; Lupardus et al. 2002), suggesting that junctions of dsDNA/ssDNA might also accumulate at the stressed forks and they might facilitate the functions of Rad17 and 9-1-1 complexes.

DNA repair also plays important roles in the processing of DNA lesions and the activation of checkpoint. In yeast, the DSBs generated by the HO endonuclease are recessed by exonucleases in the 5′-to-3′ direction. The recession of DSBs requires Xrs2, the yeast homologue of the human Nbs1 protein, and Exo1 (exonuclease I) (Nakada et al. 2004). The generation of ssDNA at the ends of DSBs not only presents a signal for Mec1-Ddc2 activation, but also creates an important DNA intermediate for homologous recombination (Wang and Haber 2004). Furthermore, the recession of DSBs is controlled by the cyclin-dependent kinase Cdk1 (Ira et al. 2004). Hence, the processing of DSBs in yeast cells has presented a clear example how the activation of checkpoint by a particular type of DNA damage is coupled to a specific DNA repair pathway as well as the cell cycle regulatory apparatus. Certain types of

DNA damage might be sensed by both replication-dependent and replication-independent mechanisms. For example, in the cells arrested in G0, G1 or G2, UV-induced DNA damage can activate the checkpoint in a NER (nucleotide excision repair)-dependent manner (Neecke et al. 1999; O'Driscoll et al. 2003). Moreover, the endonuclease and helicase involved in NER are needed to process the DNA lesions into structures that can elicit checkpoint responses (Giannattasio et al. 2004). In addition to processing DNA lesions, many DNA repair proteins also physically interact with the checkpoint sensors. For example, the NER protein Rad14 associates with the 9-1-1 complex in yeast (Giannattasio et al. 2004), and the mismatch repair protein Msh2 binds to ATR in human cells (Wang and Qin 2003). The interactions among the repair and checkpoint proteins might contribute to the recruitment of checkpoint sensors to specific types of DNA damage.

Both DNA damage and the activation of checkpoint can lead to changes in the chromatin structure at the sites of DNA damage (see discussion below). It was recently reported that yeast histone H2A is phosphorylated by Mec1 and Tel1 (the yeast homologue of ATM) in the vicinity of the HO-induced DSBs (Shroff et al. 2004), and that the phosphorylated histone H2A functions to recruit the INO80 chromatin remodeling complex to the DSBs (Morrison et al. 2004; van Attikum et al. 2004). Interestingly, the INO80 complex is required for the extensive recession of DSBs (van Attikum et al. 2004), revealing a regulatory role of chromatin modulation in the processing of DNA lesions and the control of checkpoint signaling.

2.4
MRN Complex and Activation of ATM and ATR

Unlike ATR that responds to a variety of types of DNA damage, ATM is primarily involved in the responses to DSBs. The phosphorylation of ATM substrates can be observed in a very short time after DNA damage, suggesting that compared with ATR, the activation of ATM is less dependent on the processing of DNA lesions. In human cells, the Mre11-Rad50-Nbs1 protein complex (the MRN complex), a complex that exhibits both exo- and endonuclease activities in vitro (Paull and Gellert 1999), plays a crucial role in the activation of ATM (Carson et al. 2003; Uziel et al. 2003; Kitagawa et al. 2004). ATM fails to associate with chromatin and to phosphorylate many of its substrates in the cells lacking functional Nbs1 or Mre11 (Stewart et al. 1999; Carson et al. 2003; Uziel et al. 2003).

Recent studies by two laboratories have revealed important clues of how ATM is activated by DNA damage. A study by Kastan's laboratory reported that ATM is autophosphorylated on serine 1981 after DNA damage (Bakkenist and Kastan 2003) (Fig. 1a). Following this autophosphorylation, ATM dissociates from its multimeric form to become monomers (Bakkenist and Kastan 2003). The monomerization of ATM appears to be an important step for its activation. Intriguingly, the autophosphorylation of ATM can be induced by very low doses of DNA damage or treatments disrupting chromatin structures in the absence of detectable DSBs, leading to the hypothesis that ATM might be activated by changes of chromatin structures (Bakkenist and Kastan 2003). Although monomeric ATM is capable of phosphorylating non-DNA-bound substrates such as p53, the phosphorylation of other substrates at the sites of DSBs requires the MRN complex and Brca1 (Kitagawa et al. 2004). The mechanisms by which the autophosphorylation of ATM is regulated are yet to be elucidated. Proteins including Nbs1, 53BP1, Mdc1, PP5, PP2A, and p18 might be involved in the regulation of ATM autophosphorylation at various stages of signaling (Mochan et al. 2003; Uziel et al. 2003; Ali et al. 2004; Goodarzi et al. 2004; Park et al. 2005). Elevated kinase activity of ATM can be detected in vitro after DNA damage (Canman et al. 1998). A study by Paull's laboratory demonstrated that the MRN complex stimulates the phosphorylation of ATM substrates in vitro even in the absence of DNA (Lee and Paull 2004). Although this study did not reveal the contribution of DSBs in the activation of ATM, it clearly shows a direct role of the MRN complex in the stimulation of ATM. Together these recent findings suggest that the activation of ATM is a multistep process that involves the autophosphorylation of ATM, the interactions of ATM with other regulatory factors, and the localization of ATM to DSBs. The MRN complex is not only important for the localization of ATM to DSBs, but also critical for the activation of the kinase activity of ATM.

The MRN complex might also be involved in the activation of ATR (Carson et al. 2003; Pichierri and Rosselli 2004; Stiff et al. 2005). The budding and fission yeast mutants lacking the MRN complex display defective checkpoint responses after HU or MMS treatments (D'Amours and Jackson 2001; Chahwan et al. 2003). In human cells, like ATR, the MRN complex is implicated in the checkpoint signaling elicited by cross-linked DNA (Pichierri and Rosselli 2004). Very recently, it was shown that in the cells lacking functional Nbs1, ATR is unable to stably associate with RPA-ssDNA and to phosphorylate Chk1 after HU treatment (Stiff et al. 2005), raising the possibility that the MNR complex might regulate the interaction between ATR-ATRIP and RPA-ssDNA, and might contribute to the activation of ATR.

3
Transduction of DNA Damage Signals

After ATM and ATR are activated by DNA damage, they phosphorylate numerous substrates including two downstream checkpoint kinases, Chk1 and Chk2 (Liu et al. 2000; Matsuoka et al. 1998). Existing evidence suggests that Chk2 is primarily phosphorylated by ATM in response to DSBs (Matsuoka et al. 2000), whereas Chk1 is mainly phosphorylated by ATR after DNA damage interfering with DNA replication (Brown and Baltimore 2003; Cortez 2003). The phosphorylation of Chk1 and Chk2 by ATR and ATM are required for the activation of these kinases (Matsuoka et al. 2000; Zhao and Piwnica-Worms 2001). Once activated, Chk1 and Chk2 can then phosphorylate substrates such as p53, Brca1, and Cdc25A (Matsuoka et al. 1998; Hirao et al. 2000; Zhang et al. 2004), providing another layer of regulation on the cellular processes involved in the maintenance of genomic stability. Chk1 has been shown to play crucial roles in the regulation of cell cycle progression and the stability of DNA replication forks (Zhao et al. 2002; Zachos et al. 2003). Chk2, one the other hand, appears to play a less significant role in arresting the cell cycle but is important for the regulation of cell death (Hirao et al. 2002; Takai et al. 2002).

The phosphorylation of Chk1 and Chk2 by ATR and ATM is regulated by several other proteins in vivo. For example, in addition to the MRN complex (Buscemi et al. 2001), the phosphorylation of Chk2 requires 53BP1, Mdc1, and Brca1 (Foray et al. 2003; Peng and Chen 2003; Wang et al. 2002). Similarly, in addition to the Rad17 and 9-1-1 complexes, the phosphorylation of Chk1 requires Claspin, Mcm7, Brca1, CtIP, and TopBP1 (Kumagai and Dunphy 2000; Yarden et al. 2002; Yamane et al. 2003; Cortez et al. 2004; Lin et al. 2004; Tsao et al. 2004; Yu and Chen 2004). Because these groups of proteins function to mediate the DNA damage signals between ATM/ATR and Chk1/Chk2, they were appropriately termed "mediators". However, the exact biochemical activities of these proteins in checkpoint signaling are not known. The mediator proteins do share several features that might be important for their functions in signaling. First, most if not all mediators are present at the sites of DNA damage or stalled replication forks. Some mediators, such as Claspin and Mcm7, are probably components of DNA replication forks (Chini and Chen 2003; Lee et al. 2003), whereas others, such as Mdc1, 53BP1, and Brca1, are recruited to sites of DNA damage (Scully et al. 1997; Schultz et al. 2000; Goldberg et al. 2003; Lou et al. 2003; Peng and Chen 2003; Stewart et al. 2003; Xu and Stern 2003). Second, most mediator are themselves substrates of ATM or ATR after DNA damage (Tibbetts et al. 2000; Chini and Chen 2003; Stewart et al. 2003). Third, many mediators possess phospho-serine/threonine-binding motifs, such as the BRCT and the FHA motifs (Durocher et al. 1999; Manke et al. 2003; Rodriguez et al. 2003; Yu et al. 2003). It has been shown that certain mediators can associate with each other or with downstream kinase in a phosphorylation-dependent manner (Kumagai and Dunphy 2003; Yu and Chen 2004), suggesting that the damage-induced phosphorylation of these mediators and their phospho-Ser/Thr-binding motifs might be involved in organizing the protein complex at the sites of DNA damage and/or recruiting downstream kinases. Consistent with the idea that mediators are important for the recruitment of downstream kinases to ATM/ATR, certain mediators in budding and fission yeast can be partially bypassed by fusing the downstream kinase with Ddc2 or its fission yeast counterpart Rad26 (Lee et al. 2004; Tanaka and Russell 2004).

Although the recruitment of downstream kinase appears to be an important function of the mediators, some mediators might also affect the localization or activity of ATM or ATR. For example, the autophosphorylation of ATM is diminished by the depletion of 53BP1 and Mdc1 (Mochan et al. 2003). Furthermore, several mediators, including Brca1 and Claspin, have been shown to possess affinities to DNA (Paull et al. 2001; Sar et al. 2004). Although the DNA-binding activities of these mediators might contribute to their localizations and their functions, they do not seem to provide the DNA-structure specificity necessary for the initial recognition of DNA lesions. It is possible that these mediators function to amplify the DNA damage signals generated by the basal sensors. Alternatively, these mediators might enhance the functions of the damage sensors to detect specific types of DNA damage.

It should be noted that the mediators are not simply proteins built for transmitting checkpoint signals. Many mediators themselves are important for the processes directly involved in the maintenance of genomic stability. For example, Brca1 plays important roles in both homologous recombination (HR) and non-homologous end joining [NHEJ] (Moynahan et al. 1999; Zhong et al. 2002a,b; Zhang et al. 2004). Mrc1, the yeast homologue of Claspin, might be important for coupling replication forks and DNA synthesis upon replication stress (Katou et al. 2003). Thus, besides their roles in checkpoint signaling, the phosphorylation of mediators by ATM and ATR are probably important for other cellular processes occurring at the sites of DNA damage and replication interference. This is another example of how checkpoint signaling is intimately coupled to various processes functioning on chromatin. Unlike the basal checkpoint sensors conserved between yeast and humans, the mediators are clearly more numerous and complex in human cells compared to yeast. This may provide additional means for regulating checkpoint signaling in human cells. Besides Chk1 and Chk2, mediators also transduce damage signals to other checkpoint effectors. It is possible that different subsets of mediators are involved in transducing different DNA damage signals generated by different DNA lesions. Alternatively, it is possible that different mediators can transmit the same damage signal to different downstream effectors.

4
Regulation of Downstream Cellular Processes

The activation of checkpoint signaling eventually leads to regulation of many cellular processes that can help cells to remove or overcome the stresses on their genomes and to maintain genomic stability. Furthermore, the checkpoint seems to regulate different cellular processes in a coordinated manner, giving the cells the greatest probability to survive under stressed conditions. It is important to remember that many processes regulated by the checkpoint can crosstalk to each other, and they might also affect how checkpoint signals flow in cells. It should also be noted that although checkpoint activation can result in responses at the cellular level (e.g. arrest of the cell cycle), many other checkpoint-mediated responses might take place more locally at the sites of DNA damage. Instead of being a on-or-off switch, the checkpoint control on many cellular processes might be quantitative. In many cases, the checkpoint regulates multiple effectors involved in the same cellular process important for the stability of genome, ensuring that these processes are under fine-tuned controls.

4.1
Regulation of the Cell Cycle

The activation of checkpoint can lead to cell cycle arrest at either G1/S or G2/M transitions, depending on the nature of DNA damage and where cells are in the cell cycle when they encounter DNA damage. Furthermore, activated checkpoint can also slow down the progression of S-phase by inhibiting the firing of late replication origins and perhaps by slowing down the elongating replication forks (Santocanale and Diffley 1998; Merrick et al. 2004). The arrest of cell cycle presumably provides more time for DNA repair processes to remove DNA lesions. Different repair mechanisms may be used by cells arrested at different stages of the cell cycle. The branches of the checkpoint pathway that lead to cell cycle arrest in G1, S, and G2 are often referred to as the G1, intra-S, and G2/M checkpoints, respectively. Both ATM and ATR were shown to be important for the cell cycle arrest, but their roles seem to be different in cells at different cell-cycle stages and in response to different types of DNA damage (Beamish et al. 1996; Cortez et al. 2001; Brown and Baltimore 2003). The downstream kinase Chk1 is also important for the G1, intra-S, and G2/M checkpoints (Zhao et al. 2002; Zachos et al. 2003). The function of Chk2 in cell cycle arrest is less clear. Studies using knockout mice suggested that Chk2 might play a role in the G1 checkpoint and the maintenance of G2/M checkpoint arrest (Hirao et al. 2002; Jack et al. 2002; Takai et al. 2002).

During the G1 phase of the cell cycle, one of the critical targets of the checkpoint is p53. p53 can be phosphorylated by ATM/ATR after DNA damage (Khanna et al. 1998; Delia et al. 2000), leading to the stabilization of p53 protein and the consequent expression of the Cdk inhibitor p21. In addition, Cdc25A, a phosphatase that activates Cdk1 and Cdk2, is phosphorylated by Chk1 and Chk2 in G1 after DNA damage (Falck et al. 2001). Once phosphorylated by Chk1 and other kinases, Cdc25A is rapidly degraded by the ubiquitin-mediated proteolysis, leading to reduced Cdk activity and G1 arrest (Busino et al. 2003; Jin et al. 2003; Sorensen et al. 2003). Upon DNA damage, Cdc25A may also be phosphorylated and degraded during S-phase. The reduction of Cdk activity during S-phase will result in less efficient replication origin firing. Studies using *Xenopus* extracts have also suggested that the activation of the ATM- and ATR-mediated checkpoint pathways can lead to inhibition of Cdk2 and Cdc7, two protein kinases essential for the initiation of DNA replication (Costanzo et al. 2003). During G2, in addition to Cdc25A, another Cdk-activating phosphatase Cdc25C is also phosphorylated by Chk2 and perhaps Chk1 (Matsuoka et al. 1998). The phosphorylation of Cdc25C leads to its export from the nucleus (Lopez-Girona et al. 1999) and perhaps its inactivation (Lopez-Girona et al. 2001), providing additional mechanisms to prevent Cdk1 activation and entry into mitosis.

4.2
Regulation of DNA Replication Forks

ATR and many other checkpoint proteins in the ATR pathway are essential for cell proliferation, suggesting that this pathway plays a critical role even in normal cell cycles. The first glance on the essential function of the ATR pathway came from the studies using budding yeast. In yeast, although Mec1 is an essential protein, cells lacking Mec1 can survive when dNTP levels are elevated, suggesting that Mec1 might have an critical role coping with stress on DNA replication (Desany et al. 1998; Zhao et al. 1998). Indeed, Mec1 was later shown to be crucial for the stability of replication forks in the presence of DNA damage and for the recovery form replication blocks (Desany et al. 1998; Lopes et al. 2001; Tercero and Diffley 2001). Furthermore, certain mutant alleles of Mec1 display high genomic instability even in the absence of exogenously introduced replication stress (Cha and Kleckner 2002). Consistently, vertebrate ATR and Chk1 are also important for the stability of replication forks (Brown and Baltimore 2003; Zachos et al. 2003). Cells lacking ATR or Chk1 fail to properly recover from replication blocks and exhibit elevated chromosomal fragility even in the absence of replication-blocking agents (Brown and Baltimore 2000; Liu et al. 2000; Casper et al. 2002). Recent studies using yeast have suggested that the progression of some DNA replication forks might be hindered by certain endogenous DNA or DNA-protein structures (Ivessa et al. 2003). Therefore, it is plausible that the ATR pathway plays a vital role in facilitating the elongation of DNA replication through various chromosomal regions during normal S-phase. If this is indeed the case, both the activation of ATR and the essential function of ATR are tightly coupled to the progression of replication forks.

How the ATR checkpoint pathway stabilizes replication forks is not understood. Several proteins at replication forks, including RPA, Claspin, and Mcm2, are phosphorylated by ATR/ATM upon DNA damage (Kumagai and Dunphy 2003; Block et al. 2004; Cortez et al. 2004; Yoo et al. 2004a,b). Nevertheless, how the phosphorylation of these proteins affects the stability of replication forks is unclear. Several other proteins that might be involved in the repair/recombination processes at stalled replication forks, such as Mus81 and BLM, are also phosphorylated after DNA damage (Boddy et al. 2000; Davies et al. 2004). Whether and how the phosphorylation of these proteins is implicated in the recovery of replication forks need to be further examined. A complete understanding of the function of ATR at replication forks awaits extensive biochemical and cell biological studies.

4.3
Regulation of DNA Repair

As DNA replication, DNA repair processes are also intimately linked to checkpoint signaling. In response to ionizing irradiation (IR), Nbs1 and Brca1 are both phosphorylated by ATM (Cortez et al. 1999; Lim et al. 2000; Wu et al. 2000; Zhao et al. 2000). Cells lacking ATM, Nbs1, or Brca1 are highly sensitive to IR, indicating that these proteins are directly or indirectly involved in the repair of DSBs. Biochemical and cell biological studies have implicated both the MRN complex and Brca1 in HR and NHEJ, two major pathways of DSB repair (Moynahan et al. 1999; Chen et al. 2001; Tauchi et al. 2002; Zhong et al. 2002a,b). Moreover, a recent study has provided direct evidence that ATM is required for the repair of a subset of DSBs (Riballo et al. 2004). Thus, the checkpoint signaling through the ATM-Nbs1-Brca1 pathway is likely important for the repair of DSBs. Furthermore, SMC1, a component of the cohesion complex, is phosphorylated by ATM in a Nbs1- and Brca1-dependent manner (Kim et al. 2002; Yazdi et al. 2002; Kitagawa et al. 2004). It has been recently reported that the cohesion complex is specifically recruited to DSBs in yeast (Unal et al. 2004), suggesting the possibility that SMC1 might play a rather direct role in the repair of DSBs. Cells expressing the SMC1 mutant lacking the ATM phosphorylation site are highly sensitive to IR, indicating that SMC1 is a critical target of the ATM-Nbs1-Brca1 signaling pathway (Kitagawa et al. 2004). Consistent with the idea that the repair of DSBs is regulated by the checkpoint, the budding yeast Rad55, a protein involved in HR, is also phosphorylated in a checkpoint-dependent manner after DNA damage (Bashkirov et al. 2000).

In addition to DSBs, the repair of DNA lesions generated by inter-strand cross-linkers (such as mitomycin C (MMC)) might also involve the MRN complex. In response to treatments with DNA cross-linker, Nbs1 is phosphorylated by ATR (Pichierri and Rosselli 2004). Furthermore, FANC-D2 (Fanconi anemia D2), a protein that might be involved in the repair of cross-linked DNA and DSBs, is phosphorylated by ATM after IR (Taniguchi et al. 2002) and is mono-ubiquitinated in a ATR- and Nbs1-dependent manner after MMC treatment (Andreassen et al. 2004; Pichierri and Rosselli 2004).

The above examples give us only a glimpse of the importance of checkpoint for the control of DNA repair. Checkpoint proteins such as ATR and the 9-1-1 complex or their yeast homologues have been shown to associate with proteins involved in nucleotide excision repair (XPA/Rad14) (Giannattasio et al. 2005), mismatch repair (Msh2) (Wang and Qin 2003), and base excision repair (MutY) (Chang and Lu 2005). It is conceivable that many repair proteins can be phosphorylated and regulated by the checkpoint kinases. The checkpoint-signaling pathway might also play a role in coordinating different types of repair. A better understanding of the regulatory roles of the checkpoint in DNA repair requires the identification of the checkpoint kinase substrates involved in DNA repair, and the elucidation of the functional effects of the phosphorylation events.

4.4
Regulation of Telomeres

Although in many ways telomeres resemble DSBs, they do not normally elicit checkpoint responses. This is perhaps because telomeres are usually protected by the telomere-binding proteins and/or secondary DNA structures such as the T-loop (Griffith et al. 1999). Interestingly, many checkpoint proteins have important roles in the maintenance of normal telomere structures. AT cells lacking ATM display shortened telomeres (Vaziri et al. 1997). In budding yeast, RPA, Mec1, and Tel1 (the yeast ATM homologue) are recruited to telomeres at different stages of the cell cycle (Takata et al. 2004). The budding yeast mutant lacking both Mec1 and Tel1 exhibits severe telomere defects as the telomerase-null cells, indicating that telomerase cannot function at all in the absence of Mec1 and Tel1 (Chan and Blackburn 2003). Telomere defects were also found in certain budding and fission yeast RPA mutants (Smith et al. 2000; Ono et al. 2003). The 9-1-1 and MRN complexes are also detected at telomeres in budding yeast, fission yeast, and human cells (Zhu et al. 2000; Nakamura et al. 2002; Katou et al. 2003). The fission yeast mutants lacking the 9-1-1 or MRN complex also display shortened telomeres (Nakamura et al. 2002; Chahwan et al. 2003). It is known that heterochromatin structures are formed at telomeres, resulting in slow progression of DNA replication forks in these regions. The single-stranded regions at the end of telomeres might be transiently exposed during late S-phase when telomeres undergo replication. Furthermore, a transient telomerase activity can be detected in normal human fibroblast during S-phase (Masutomi et al. 2003). Together, these findings raise the possibility that checkpoint signaling might locally and transiently exist at the telomeres in dividing cells. The functions of the checkpoint proteins might be important for the organization of the telomere structures, the regulation of DNA replication in telomere regions, or the function of telomerase.

5
Interplay between Checkpoint Signaling and Chromatin

In eukaryotic cells, chromatin structures play crucial roles in the regulation of many cellular processes such as transcription, DNA replication, DNA repair, and chromosome segregation. Recent studies have clearly revealed that chromatin structures are also important for the signaling and function of the checkpoint pathway, unveiling a new dimension of the intertwined relationship between the checkpoint and the cellular processes involved in the maintenance of genomic stability.

In human cells, histone H2AX, a variant of the histone H2A, is rapidly phosphorylated by ATM or DNA-PK in response to DSBs (Rogakou et al. 1999; Burma et al. 2001). Importantly, the phosphorylation of H2AX only occurs in the regions proximal to the DNA breaks (Rogakou et al. 1999), indicating that a special chromatin structure is generated at the sites of DNA damage. Although histone H2AX is not essential for checkpoint signaling (Celeste et al. 2002), many proteins involved in checkpoint responses can be recruited to the sites of DNA damage through their interactions with the phosphorylated histone H2AX (γ-H2AX). These proteins include Nbs1, Mdc1, and 53BP1 (Kobayashi et al. 2002; Stewart et al. 2003; Ward et al. 2003). Recently, it was shown in budding yeast that the cohesin complex (Unal et al. 2004), the INO80 chromatin-remodeling complex (Morrison et al. 2004; van Attikum et al. 2004), the NuA4 histone acetyl transferase (HAT) complex (Downs et al. 2004), and perhaps the 19S proteasome (Krogan et al. 2004) are also recruited to DSBs in a manner dependent upon γ-H2AX. Many of the protein complexes recruited by the γ-H2AX were shown to be important for the efficient repair of DSBs. Indeed, cells lacking histone H2AX exhibit defects in repair of DSBs (Bassing et al. 2002). Furthermore, the H2AX-null cells display defects in checkpoint response when treated with low does of IR, indicating that the recruitment of the checkpoint proteins through γ-H2AX might be important for the amplification of checkpoint signals (Fernandez-Capetillo et al. 2002). Together, these studies suggest that γ-H2AX helps to enrich repair and checkpoint proteins at the sites of DNA damage, and its role might be important in the presence of physiological levels of DNA damage or at the sites of a specific subset of DNA lesions.

In addition to the phosphorylation of histone H2AX, other histone modifications are also detected in the vicinity of DSBs. For example, histone H4 is shown to be acetylated at DSBs (Bird et al. 2002), and the acetylation of H4 is important for the repair of DNA breaks. Furthermore, the methylation of histone H3 on Lys79 is required for the recruitment of 53BP1 to damage-induced nuclear foci and the damage-induced phosphorylation of budding yeast Rad9, the homologue of 53BP1 (Giannatasio et al. 2005; Huyen et al. 2004). The ubiquitination of histone H2B on Lys123, which might facilitate the methylation of histone H3 on Lys79, is important for checkpoint signaling in budding yeast (Giannattasio et al. 2005). The methylation of histone H4 on Lys20 is also required for the damage-induced foci formation and phosphorylation of Crb2, the fission yeast homologue of 53BP1 (Sanders et al. 2004). The fission yeast mutant lacking the methylation site in histone H4 is sensitive to some types of DNA damage and defective in certain checkpoint responses. Although the methylations of H3 and H4 are not induced by DNA damage, the changes in chromatin structure after DNA damage might expose these histone modifications to present a damage-specific "histone code" to proteins involved in DNA repair or checkpoint signaling

6
Perspectives

The DNA damage and DNA replication checkpoints were originally defined as the signaling pathways that arrest cell cycle progression in response to DNA damage or replication blocks. The initial definition and the simple linear structure of the checkpoint have now been significantly extended by the recent findings of this pathway. First, it is now clear that the checkpoint not only controls cell cycle arrest but also many other cellular processes crucial for the stability of genome, such as DNA replication, DNA repair, and chromatin modulation. Second, the functions of checkpoint are important not only for responding to extrinsic DNA damage but also for coping with the intrinsic stresses generated during normal metabolism. Third, in addition to the responses at the cellular level, checkpoint signaling might also quantitatively and locally regulate many cellular processes at specific sites on chromosomes. Fourth, although checkpoint signaling affects the many cellular processes occurring on chromatin, its activation and signal transduction are tightly coupled to and are dependent upon these processes. Fifth, it has become clear that chromatin structures play crucial roles in the activation and signaling of checkpoint signals. Finally, the identification of numerous mediator proteins in human cells has suggested that the structure of the checkpoint pathway is far from linear, and that much of the regulation of checkpoint signaling has not been appreciated. There are many important questions about the checkpoint pathway waiting to be addressed. For example, although many proteins involved in checkpoint activation have been identified, it is still unclear how these proteins function together to bring about ATR and ATM activation in a damage-specific manner. It is also not known how exactly the checkpoint proteins interact with various cellular machines such as the DNA replication and DNA repair machines. Furthermore, we are just beginning to understand the contributions of the mediators and chromatin structures. The elucidation of how checkpoint signals are quantitatively and spatially controlled, how checkpoint signaling is regulated by different types of DNA damage, and how different downstream cellular processes are coordinated will require much more extensive cell biological and biochemical studies. Finally, the involvement of the checkpoint pathway in the regulation of various cell type-specific, tissue-specific, or developmentally controlled cellular events remains to be thoroughly examined. The next decade is promised to be an exciting period for the research of the DNA damage and DNA replication checkpoint.

Acknowledgements The authors would like to apologize to those whose work has not been cited due to space limitations. L. Zou is supported in part by a Smith Family New Investigator Award from the Medical Foundation.

References

Abraham RT (2001) Cell cycle checkpoint signaling through the ATM and ATR kinases. Genes Dev 15:2177–2196

Ali A, Zhang J, Bao S, Liu I, Otterness D, Dean NM, Abraham RT, Wang XF (2004) Requirement of protein phosphatase 5 in DNA-damage-induced ATM activation. Genes Dev 18:249–254

Andreassen PR, D'Andrea AD, Taniguchi T (2004) ATR couples FANCD2 monoubiquitination to the DNA-damage response. Genes Dev 18:1958–1963

Bakkenist CJ, Kastan MB (2003) DNA damage activates ATM through intermolecular autophosphorylation and dimer dissociation. Nature 421:499–506

Bao S, Lu T, Wang X, Zheng H, Wang LE, Wei Q, Hittelman WN, Li L (2004) Disruption of the Rad9/Rad1/Hus1 (9-1-1) complex leads to checkpoint signaling and replication defects. Oncogene 23:5586–5593

Bashkirov VI, King JS, Bashkirova EV, Schmuckli-Maurer J, Heyer WD (2000) DNA repair protein Rad55 is a terminal substrate of the DNA damage checkpoints. Mol Cell Biol 20:4393–4404

Bassing CH, Chua KF, Sekiguchi J, Suh H, Whitlow SR, Fleming JC, Monroe BC, Ciccone DN, Yan C, Vlasakova K, Livingston DM, Ferguson DO, Scully R, Alt FW (2002) Increased ionizing radiation sensitivity and genomic instability in the absence of histone H2AX. Proc Natl Acad Sci USA 99:8173–8178

Beamish H, Williams R, Chen P, Lavin MF (1996) Defect in multiple cell cycle checkpoints in ataxia-telangiectasia postirradiation. J Biol Chem 271:20486–20493

Bird AW, Yu DY, Pray-Grant MG, Qiu Q, Harmon KE, Megee PC, Grant PA, Smith MM, Christman MF (2002) Acetylation of histone H4 by Esa1 is required for DNA double-strand break repair. Nature 419:411–415

Block WD, Yu Y, Lees-Miller SP (2004) Phosphatidyl inositol 3-kinase-like serine/threonine protein kinases (PIKKs) are required for DNA damage-induced phosphorylation of the 32 kDa subunit of replication protein A at threonine 21. Nucleic Acids Res 32:997–1005

Boddy MN, Lopez-Girona A, Shanahan P, Interthal H, Heyer WD, Russell P (2000) Damage tolerance protein Mus81 associates with the FHA1 domain of checkpoint kinase Cds1. Mol Cell Biol 20:8758–8766

Brown EJ, Baltimore D (2000) ATR disruption leads to chromosomal fragmentation and early embryonic lethality. Genes Dev 14:397–402

Brown EJ, Baltimore D (2003) Essential and dispensable roles of ATR in cell cycle arrest and genome maintenance. Genes Dev 17:615–628

Burma S, Chen BP, Murphy M, Kurimasa A, Chen DJ (2001) ATM phosphorylates histone H2AX in response to DNA double-strand breaks. J Biol Chem 276:42462–42467

Buscemi G, Savio C, Zannini L, Micciche F, Masnada D, Nakanishi M, Tauchi H, Komatsu K, Mizutani S, Khanna K, Chen P, Concannon P, Chessa L, Delia D (2001) Chk2 activation dependence on Nbs1 after DNA damage. Mol Cell Biol 21:5214–5222

Busino L, Donzelli M, Chiesa M, Guardavaccaro D, Ganoth D, Dorrello NV, Hershko A, Pagano M, Draetta GF (2003) Degradation of Cdc25A by beta-TrCP during S phase and in response to DNA damage. Nature 426:87–91

Canman CE, Lim DS, Cimprich KA, Taya Y, Tamai K, Sakaguchi K, Appella E, Kastan MB, Siliciano JD (1998) Activation of the ATM kinase by ionizing radiation and phosphorylation of p53. Science 281:1677–1679

Carson CT, Schwartz RA, Stracker TH, Lilley CE, Lee DV, Weitzman MD (2003) The Mre11 complex is required for ATM activation and the G2/M checkpoint. Embo J 22:6610–6620

Casper AM, Nghiem P, Arlt MF, Glover TW (2002) ATR regulates fragile site stability. Cell 111:779–789

Celeste A, Petersen S, Romanienko PJ, Fernandez-Capetillo O, Chen HT, Sedelnikova OA, Reina-San-Martin B, Coppola V, Meffre E, Difilippantonio MJ, Redon C, Pilch DR, Olaru A, Eckhaus M, Camerini-Otero RD, Tessarollo L, Livak F, Manova K, Bonner WM, Nussenzweig MC, Nussenzweig A (2002) Genomic instability in mice lacking histone H2AX. Science 296:922–927

Cha RS, Kleckner N (2002) ATR homolog Mec1 promotes fork progression, thus averting breaks in replication slow zones. Science 297:602–606

Chahwan C, Nakamura TM, Sivakumar S, Russell P, Rhind N (2003) The fission yeast Rad32 (Mre11)-Rad50-Nbs1 complex is required for the S-phase DNA damage checkpoint. Mol Cell Biol 23:6564–6573

Chan SW, Blackburn EH (2003) Telomerase and ATM/Tel1p protect telomeres from non-homologous end joining. Mol Cell 11:1379–1387

Chang DY, Lu AL (2005) Interaction of checkpoint proteins Hus1/Rad1/Rad9 with DNA base excision repair enzyme MutY homolog in fission yeast, Schizosaccharomyces pombe. J Biol Chem 280:408–417

Chen L, Trujillo K, Ramos W, Sung P, Tomkinson AE (2001) Promotion of Dnl4-catalyzed DNA end-joining by the Rad50/Mre11/Xrs2 and Hdf1/Hdf2 complexes. Mol Cell 8:1105–1115

Chini CC, Chen J (2003) Human claspin is required for replication checkpoint control. J Biol Chem 278:30057–30062

Cortez D (2003) Caffeine inhibits checkpoint responses without inhibiting the ataxia-telangiectasia-mutated (ATM) and ATM- and Rad3-related (ATR) protein kinases. J Biol Chem 278:37139–37145

Cortez D, Wang Y, Qin J, Elledge SJ (1999) Requirement of ATM-dependent phosphorylation of brca1 in the DNA damage response to double-strand breaks. Science 286:1162–1166

Cortez D, Guntuku S, Qin J, Elledge SJ (2001) ATR and ATRIP: partners in checkpoint signaling. Science 294:1713–1716

Cortez D, Glick G, Elledge SJ (2004) Minichromosome maintenance proteins are direct targets of the ATM and ATR checkpoint kinases. Proc Natl Acad Sci USA 101:10078–10083

Costanzo V, Shechter D, Lupardus PJ, Cimprich KA, Gottesman M, Gautier J (2003) An ATR- and Cdc7-dependent DNA damage checkpoint that inhibits initiation of DNA replication. Mol Cell 11:203-213

D'Amours D, Jackson SP (2001) The yeast Xrs2 complex functions in S phase checkpoint regulation. Genes Dev 15:2238-2249

Davies SL, North PS, Dart A, Lakin ND, Hickson ID (2004) Phosphorylation of the Bloom's syndrome helicase and its role in recovery from S-phase arrest. Mol Cell Biol 24:1279-1291

Delia D, Mizutani S, Panigone S, Tagliabue E, Fontanella E, Asada M, Yamada T, Taya Y, Prudente S, Saviozzi S, Frati L, Pierotti MA, Chessa L (2000) ATM protein and p53-serine 15 phosphorylation in ataxia-telangiectasia (AT) patients and at heterozygotes. Br J Cancer 82:1938-1945

Desany BA, Alcasabas AA, Bachant JB, Elledge SJ (1998) Recovery from DNA replicational stress is the essential function of the S-phase checkpoint pathway. Genes Dev 12:2956-2970

Downs JA, Allard S, Jobin-Robitaille O, Javaheri A, Auger A, Bouchard N, Kron SJ, Jackson SP, Cote J (2004) Binding of chromatin-modifying activities to phosphorylated histone H2A at DNA damage sites. Mol Cell 16:979-990

Durocher D, Henckel J, Fersht AR, Jackson SP (1999) The FHA domain is a modular phosphopeptide recognition motif. Mol Cell 4:387-394

Ellison V, Stillman B (2003) Biochemical characterization of DNA damage checkpoint complexes: clamp loader and clamp complexes with specificity for 5' recessed DNA. PLoS Biol 1:E33

Falck J, Mailand N, Syljuasen RG, Bartek J, Lukas J (2001) The ATM-Chk2-Cdc25A checkpoint pathway guards against radioresistant DNA synthesis. Nature 410:842-847

Fernandez-Capetillo O, Chen HT, Celeste A, Ward I, Romanienko PJ, Morales JC, Naka K, Xia Z, Camerini-Otero RD, Motoyama N, Carpenter PB, Bonner WM, Chen J, Nussenzweig A (2002) DNA damage-induced G2-M checkpoint activation by histone H2AX and 53BP1. Nat Cell Biol 4:993-997

Foray N, Marot D, Gabriel A, Randrianarison V, Carr AM, Perricaudet M, Ashworth A, Jeggo P (2003) A subset of ATM- and ATR-dependent phosphorylation events requires the BRCA1 protein. Embo J 22:2860-2871

Ghaemmaghami S, Huh WK, Bower K, Howson RW, Belle A, Dephoure N, O'Shea EK, Weissman JS (2003) Global analysis of protein expression in yeast. Nature 425:737-741

Giannattasio M, Lazzaro F, Longhese MP, Plevani P, Muzi-Falconi M (2004) Physical and functional interactions between nucleotide excision repair and DNA damage checkpoint. Embo J 23:429-438

Giannattasio M, Lazzaro F, Plevani P, Muzi-Falconi M (2005) The DNA damage checkpoint response requires histone H2B ubiquitination by Rad6-Bre1 and H3 methylation by Dot1. J Biol Chem 280:9879-9886

Goldberg M, Stucki M, Falck J, D'Amours D, Rahman D, Pappin D, Bartek J, Jackson SP (2003) MDC1 is required for the intra-S-phase DNA damage checkpoint. Nature 421:952-956

Goodarzi AA, Jonnalagadda JC, Douglas P, Young D, Ye R, Moorhead GB, Lees-Miller SP, Khanna KK (2004) Autophosphorylation of ataxia-telangiectasia mutated is regulated by protein phosphatase 2A. Embo J 23:4451-4461

Griffith JD, Comeau L, Rosenfield S, Stansel RM, Bianchi A, Moss H, de Lange T (1999) Mammalian telomeres end in a large duplex loop. Cell 97:503-514

Hartwell LH, Weinert TA (1989) Checkpoints: controls that ensure the order of cell cycle events. Science 246:629-634

Hekmat-Nejad M, You Z, Yee MC, Newport JW, Cimprich KA (2000) *Xenopus* ATR is a replication-dependent chromatin-binding protein required for the DNA replication checkpoint. Curr Biol 10:1565–1573

Hirao A, Cheung A, Duncan G, Girard PM, Elia AJ, Wakeham A, Okada H, Sarkissian T, Wong JA, Sakai T, De Stanchina E, Bristow RG, Suda T, Lowe SW, Jeggo PA, Elledge SJ, Mak TW (2002) Chk2 is a tumor suppressor that regulates apoptosis in both an ataxia telangiectasia mutated (ATM)-dependent and an ATM-independent manner. Mol Cell Biol 22:6521–6532

Hirao A, Kong YY, Matsuoka S, Wakeham A, Ruland J, Yoshida H, Liu D, Elledge SJ, Mak TW (2000) DNA damage-induced activation of p53 by the checkpoint kinase Chk2. Science 287:1824–1827

Huyen Y, Zgheib O, Ditullio RA, Jr, Gorgoulis VG, Zacharatos P, Petty TJ, Sheston EA, Mellert HS, Stavridi ES, Halazonetis TD (2004) Methylated lysine 79 of histone H3 targets 53BP1 to DNA double-strand breaks. Nature 432:406–411

Ira G, Pellicioli A, Balijja A, Wang X, Fiorani S, Carotenuto W, Liberi G, Bressan D, Wan L, Hollingsworth NM, Haber JE, Foiani M (2004) DNA end resection, homologous recombination and DNA damage checkpoint activation require Cdk1. Nature 431:1011–1017

Ivessa AS, Lenzmeier BA, Bessler JB, Goudsouzian LK, Schnakenberg SL, Zakian VA (2003) The Saccharomyces cerevisiae helicase Rrm3p facilitates replication past nonhistone protein-DNA complexes. Mol Cell 12:1525–1536

Jack MT, Woo RA, Hirao A, Cheung A, Mak TW, Lee PW (2002) Chk2 is dispensable for p53-mediated G1 arrest but is required for a latent p53-mediated apoptotic response. Proc Natl Acad Sci USA 99:9825–9829

Jin J, Shirogane T, Xu L, Nalepa G, Qin J, Elledge SJ, Harper JW (2003) SCF$^{\beta-TRCP}$ links Chk1 signaling to degradation of the Cdc25A protein phosphatase. Genes Dev 17:3062–3074

Kastan MB, Bartek J (2004) Cell-cycle checkpoints and cancer. Nature 432:316–323

Katou Y, Kanoh Y, Bando M, Noguchi H, Tanaka H, Ashikari T, Sugimoto K, Shirahige K (2003) S-phase checkpoint proteins Tof1 and Mrc1 form a stable replication-pausing complex. Nature 424:1078–1083

Khanna KK, Keating KE, Kozlov S, Scott S, Gatei M, Hobson K, Taya Y, Gabrielli B, Chan D, Lees-Miller SP, Lavin MF (1998) ATM associates with and phosphorylates p53: mapping the region of interaction. Nat Genet 20:398–400

Kim HS, Brill SJ (2001) Rfc4 interacts with Rpa1 and is required for both DNA replication and DNA damage checkpoints in *Saccharomyces cerevisiae*. Mol Cell Biol 21:3725–3737

Kim ST, Xu B, Kastan MB (2002) Involvement of the cohesin protein, Smc1, in Atm-dependent and independent responses to DNA damage. Genes Dev 16:560–570

Kitagawa R, Bakkenist CJ, McKinnon PJ, Kastan MB (2004) Phosphorylation of SMC1 is a critical downstream event in the ATM-NBS1-BRCA1 pathway. Genes Dev 18:1423–1438

Kobayashi J, Tauchi H, Sakamoto S, Nakamura A, Morishima K, Matsuura S, Kobayashi T, Tamai K, Tanimoto K, Komatsu K (2002) NBS1 localizes to gamma-H2AX foci through interaction with the FHA/BRCT domain. Curr Biol 12:1846–1851

Kondo T, Wakayama T, Naiki T, Matsumoto K, Sugimoto K (2001) Recruitment of Mec1 and Ddc1 checkpoint proteins to double-strand breaks through distinct mechanisms. Science 294:867–870

Krogan NJ, Lam MH, Fillingham J, Keogh MC, Gebbia M, Li J, Datta N, Cagney G, Buratowski S, Emili A, Greenblatt JF (2004) Proteasome involvement in the repair of DNA double-strand breaks. Mol Cell 16:1027–1034

Kumagai A, Dunphy WG (2000) Claspin, a novel protein required for the activation of Chk1 during a DNA replication checkpoint response in *Xenopus* egg extracts. Mol Cell 6:839–849

Kumagai A, Dunphy WG (2003) Repeated phosphopeptide motifs in Claspin mediate the regulated binding of Chk1. Nat Cell Biol 5:161–165

Kumagai A, Kim SM, Dunphy WG (2004) Claspin and the activated form of ATR-ATRIP collaborate in the activation of Chk1. J Biol Chem 279:49599–49608

Lee J, Kumagai A, Dunphy WG (2003) Claspin, a Chk1-regulatory protein, monitors DNA replication on chromatin independently of RPA, ATR, and Rad17. Mol Cell 11:329–340

Lee JH, Paull TT (2004) Direct activation of the ATM protein kinase by the Mre11/Rad50/Nbs1 complex. Science 304:93–96

Lee SE, Moore JK, Holmes A, Umezu K, Kolodner RD, Haber JE (1998) Saccharomyces Ku70, mre11/rad50 and RPA proteins regulate adaptation to G2/M arrest after DNA damage. Cell 94:399–409

Lee SJ, Duong JK, Stern DF (2004) A Ddc2-Rad53 fusion protein can bypass the requirements for RAD9 and MRC1 in Rad53 activation. Mol Biol Cell 15:5443–5455

Lim DS, Kim ST, Xu B, Maser RS, Lin J, Petrini JH, Kastan MB (2000) ATM phosphorylates p95/Nbs1 in an S-phase checkpoint pathway. Nature 404:613–617

Lin SY, Li K, Stewart GS, Elledge SJ (2004) Human Claspin works with BRCA1 to both positively and negatively regulate cell proliferation. Proc Natl Acad Sci USA 101:6484–6489

Lindsey-Boltz LA, Bermudez VP, Hurwitz J, Sancar A (2001) Purification and characterization of human DNA damage checkpoint Rad complexes. Proc Natl Acad Sci USA 98:11236–1141

Lisby M, Barlow JH, Burgess RC, Rothstein R (2004) Choreography of the DNA damage response: spatiotemporal relationships among checkpoint and repair proteins. Cell 118:699–713

Liu Q, Guntuku S, Cui XS, Matsuoka S, Cortez D, Tamai K, Luo G, Carattini-Rivera S, DeMayo F, Bradley A, Donehower LA, Elledge SJ (2000) Chk1 is an essential kinase that is regulated by Atr and required for the G2/M DNA damage checkpoint. Genes Dev 14:1448–1459

Loegering D, Arlander SJ, Hackbarth J, Vroman BT, Roos-Mattjus P, Hopkins KM, Lieberman HB, Karnitz LM, Kaufmann SH (2004) Rad9 protects cells from topoisomerase poison-induced cell death. J Biol Chem 279:18641–18647

Lopes M, Cotta-Ramusino C, Pellicioli A, Liberi G, Plevani P, Muzi-Falconi M, Newlon CS, Foiani M (2001) The DNA replication checkpoint response stabilizes stalled replication forks. Nature 412:557–661

Lopez-Girona A, Furnari B, Mondesert O, Russell P (1999) Nuclear localization of Cdc25 is regulated by DNA damage and a 14-3-3 protein. Nature 397:172–175

Lopez-Girona A, Kanoh J, Russell P (2001) Nuclear exclusion of Cdc25 is not required for the DNA damage checkpoint in fission yeast. Curr Biol 11:50–54

Lou Z, Minter-Dykhouse K, Wu X, Chen J (2003) MDC1 is coupled to activated CHK2 in mammalian DNA damage response pathways. Nature 421:957–961

Lucca C, Vanoli F, Cotta-Ramusino C, Pellicioli A, Liberi G, Haber J, Foiani M (2004) Checkpoint-mediated control of replisome-fork association and signalling in response to replication pausing. Oncogene 23:1206–1213

Lupardus PJ, Byun T, Yee MC, Hekmat-Nejad M, Cimprich KA (2002) A requirement for replication in activation of the ATR-dependent DNA damage checkpoint. Genes Dev 16:2327–2332

Lydall D, Weinert T (1995) Yeast checkpoint genes in DNA damage processing: implications for repair and arrest. Science 270:1488–1491

Manke IA, Lowery DM, Nguyen A, Yaffe MB (2003) BRCT repeats as phosphopeptide-binding modules involved in protein targeting. Science 302:636–639

Masutomi K, Yu EY, Khurts S, Ben-Porath I, Currier JL, Metz GB, Brooks MW, Kaneko S, Murakami S, DeCaprio JA, Weinberg RA, Stewart SA, Hahn WC (2003) Telomerase maintains telomere structure in normal human cells. Cell 114:241–253

Matsuoka S, Huang M, Elledge SJ (1998) Linkage of ATM to cell cycle regulation by the Chk2 protein kinase. Science 282:1893–1897

Matsuoka S, Rotman G, Ogawa A, Shiloh Y, Tamai K, Elledge SJ (2000) Ataxia telangiectasia-mutated phosphorylates Chk2 in vivo and in vitro. Proc Natl Acad Sci USA 97:10389–10394

Melo JA, Cohen J, Toczyski DP (2001) Two checkpoint complexes are independently recruited to sites of DNA damage in vivo. Genes Dev 15:2809–28021

Merrick CJ, Jackson D, Diffley JF (2004) Visualization of altered replication dynamics after DNA damage in human cells. J Biol Chem 279:20067–20075

Michael WM, Ott R, Fanning E, Newport J (2000) Activation of the DNA replication checkpoint through RNA synthesis by primase. Science 289:2133–2137

Mochan TA, Venere M, DiTullio RA, Jr, Halazonetis TD (2003) 53BP1 and NFBD1/MDC1-Nbs1 function in parallel interacting pathways activating ataxia-telangiectasia mutated (ATM) in response to DNA damage. Cancer Res 63:8586–8591

Morrison AJ, Highland J, Krogan NJ, Arbel-Eden A, Greenblatt JF, Haber JE, Shen X (2004) INO80 and gamma-H2AX interaction links ATP-dependent chromatin remodeling to DNA damage repair. Cell 119:767–775

Moynahan ME, Chiu JW, Koller BH, Jasin M (1999) Brca1 controls homology-directed DNA repair. Mol Cell 4:511–518

Nakada D, Hirano Y, Sugimoto K (2004) Requirement of the Mre11 complex and exonuclease 1 for activation of the Mec1 signaling pathway. Mol Cell Biol 24:10016–10025

Nakamura TM, Moser BA, Russell P (2002) Telomere binding of checkpoint sensor and DNA repair proteins contributes to maintenance of functional fission yeast telomeres. Genetics 161:1437–1452

Neecke H, Lucchini G, Longhese MP (1999) Cell cycle progression in the presence of irreparable DNA damage is controlled by a Mec1- and Rad53-dependent checkpoint in budding yeast. Embo J 18:4485–4497

O'Driscoll M, Ruiz-Perez VL, Woods CG, Jeggo PA, Goodship JA (2003) A splicing mutation affecting expression of ataxia-telangiectasia and Rad3-related protein (ATR) results in Seckel syndrome. Nat Genet 33:497–501

Ono Y, Tomita K, Matsuura A, Nakagawa T, Masukata H, Uritani M, Ushimaru T, Ueno M (2003) A novel allele of fission yeast rad11 that causes defects in DNA repair and telomere length regulation. Nucleic Acids Res 31:7141–7149

Osborn AJ, Elledge SJ, Zou L (2002) Checking on the fork: the DNA-replication stress-response pathway. Trends Cell Biol 12:509–516

Paciotti V, Clerici M, Lucchini G, Longhese MP (2000) The checkpoint protein Ddc2, functionally related to S. pombe Rad26, interacts with Mec1 and is regulated by Mec1-dependent phosphorylation in budding yeast. Genes Dev 14:2046–2059

Park BJ, Kang JW, Lee SW, Choi SJ, Shin YK, Ahn YH, Choi YH, Choi D, Lee KS, Kim S (2005) The haploinsufficient tumor suppressor p18 upregulates p53 via interactions with ATM/ATR. Cell 120:209–221

Paull TT, Gellert M (1999) Nbs1 potentiates ATP-driven DNA unwinding and endonuclease cleavage by the Mre11/Rad50 complex. Genes Dev 13:1276–1288

Paull TT, Cortez D, Bowers B, Elledge SJ, Gellert M (2001) Direct DNA binding by Brca1. Proc Natl Acad Sci USA 98:6086–6091

Pellicioli A, Lee SE, Lucca C, Foiani M, Haber JE (2001) Regulation of *Saccharomyces* Rad53 checkpoint kinase during adaptation from DNA damage-induced G2/M arrest. Mol Cell 7:293–300

Peng A, Chen PL (2003) NFBD1, like 53BP1, is an early and redundant transducer mediating Chk2 phosphorylation in response to DNA damage. J Biol Chem 278:8873–8876

Pichierri P, Rosselli F (2004) The DNA crosslink-induced S-phase checkpoint depends on ATR-CHK1 and ATR-NBS1-FANCD2 pathways. Embo J 23:1178–1187

Riballo E, Kuhne M, Rief N, Doherty A, Smith GC, Recio MJ, Reis C, Dahm K, Fricke A, Krempler A, Parker AR, Jackson SP, Gennery A, Jeggo PA, Lobrich M (2004) A pathway of double-strand break rejoining dependent upon ATM, Artemis, and proteins locating to gamma-H2AX foci. Mol Cell 16:715–724

Rodriguez M, Yu X, Chen J, Songyang Z (2003) Phosphopeptide binding specificities of BRCA1 COOH-terminal (BRCT) domains. J Biol Chem 278:52914–52918

Rogakou EP, Boon C, Redon C, Bonner WM (1999) Megabase chromatin domains involved in DNA double-strand breaks in vivo. J Cell Biol 146:905–916

Roos-Mattjus P, Hopkins KM, Oestreich AJ, Vroman BT, Johnson KL, Naylor S, Lieberman HB, Karnitz LM (2003) Phosphorylation of human Rad9 is required for genotoxin-activated checkpoint signaling. J Biol Chem 278:24428–24433

Rouse J, Jackson SP (2000) LCD1: an essential gene involved in checkpoint control and regulation of the MEC1 signalling pathway in *Saccharomyces cerevisiae*. Embo J 19:5801–58012

Rouse J, Jackson SP (2002) Lcd1p recruits Mec1p to DNA lesions in vitro and in vivo. Mol Cell 9:857–869

Sanders SL, Portoso M, Mata J, Bahler J, Allshire RC, Kouzarides T (2004) Methylation of histone H4 lysine 20 controls recruitment of Crb2 to sites of DNA damage. Cell 119:603–614

Santocanale C, Diffley JF (1998) A Mec1- and Rad53-dependent checkpoint controls late-firing origins of DNA replication. Nature 395:615–618

Sar F, Lindsey-Boltz LA, Subramanian D, Croteau DL, Hutsell SQ, Griffith JD, Sancar A (2004) Human claspin is a ring-shaped DNA-binding protein with high affinity to branched DNA structures. J Biol Chem 279:39289–39295

Schultz LB, Chehab NH, Malikzay A, Halazonetis TD (2000) p53 binding protein 1 (53BP1) is an early participant in the cellular response to DNA double-strand breaks. J Cell Biol 151:1381–1390

Scully R, Chen J, Ochs RL, Keegan K, Hoekstra M, Feunteun J, Livingston DM (1997) Dynamic changes of BRCA1 subnuclear location and phosphorylation state are initiated by DNA damage. Cell 90:425–435

Shiomi Y, Shinozaki A, Nakada D, Sugimoto K, Usukura J, Obuse C, Tsurimoto T (2002) Clamp and clamp loader structures of the human checkpoint protein complexes, Rad9-1-1 and Rad17-RFC. Genes Cells 7:861–868

Shroff R, Arbel-Eden A, Pilch D, Ira G, Bonner WM, Petrini JH, Haber JE, Lichten M (2004) Distribution and dynamics of chromatin modification induced by a defined DNA double-strand break. Curr Biol 14:1703–1711

Smith J, Zou H, Rothstein R (2000) Characterization of genetic interactions with RFA1: the role of RPA in DNA replication and telomere maintenance. Biochimie 82:71–78

Sogo JM, Lopes M, Foiani M (2002) Fork reversal and ssDNA accumulation at stalled replication forks owing to checkpoint defects. Science 297:599–602

Sorensen CS, Syljuasen RG, Falck J, Schroeder T, Ronnstrand L, Khanna KK, Zhou BB, Bartek J, Lukas J (2003) Chk1 regulates the S phase checkpoint by coupling the physiological turnover and ionizing radiation-induced accelerated proteolysis of Cdc25A. Cancer Cell 3:247–258

Stewart GS, Maser RS, Stankovic T, Bressan DA, Kaplan MI, Jaspers NG, Raams A, Byrd PJ, Petrini JH, Taylor AM (1999) The DNA double-strand break repair gene hMRE11 is mutated in individuals with an ataxia-telangiectasia-like disorder. Cell 99:577–587

Stewart GS, Wang B, Bignell CR, Taylor AM, Elledge SJ (2003) MDC1 is a mediator of the mammalian DNA damage checkpoint. Nature 421:961–966

Stiff T, Reis C, Alderton GK, Woodbine L, O'Driscoll M, Jeggo PA (2005) Nbs1 is required for ATR-dependent phosphorylation events. Embo J 24:199–208

Stokes MP, Van Hatten R, Lindsay HD, Michael WM (2002) DNA replication is required for the checkpoint response to damaged DNA in *Xenopus* egg extracts. J Cell Biol 158:863–872

Takai H, Naka K, Okada Y, Watanabe M, Harada N, Saito S, Anderson CW, Appella E, Nakanishi M, Suzuki H, Nagashima K, Sawa H, Ikeda K, Motoyama N (2002) Chk2-deficient mice exhibit radioresistance and defective p53-mediated transcription. Embo J 21:5195–5205

Takata H, Kanoh Y, Gunge N, Shirahige K, Matsuura A (2004) Reciprocal association of the budding yeast ATM-related proteins Tel1 and Mec1 with telomeres in vivo. Mol Cell 14:515–522

Tanaka K, Russell P (2004) Cds1 phosphorylation by Rad3-Rad26 kinase is mediated by forkhead-associated domain interaction with Mrc1. J Biol Chem 279:32079–32086

Tanaka T, Nasmyth K (1998) Association of RPA with chromosomal replication origins requires an Mcm protein, and is regulated by Rad53, and cyclin- and Dbf4-dependent kinases. Embo J 17:5182–5191

Taniguchi T, Garcia-Higuera I, Xu B, Andreassen PR, Gregory RC, Kim ST, Lane WS, Kastan MB, D'Andrea AD (2002) Convergence of the fanconi anemia and ataxia telangiectasia signaling pathways. Cell 109:459–472

Tauchi H, Kobayashi J, Morishima K, van Gent DC, Shiraishi T, Verkaik NS, vanHeems D, Ito E, Nakamura A, Sonoda E, Takata M, Takeda S, Matsuura S, Komatsu K (2002) Nbs1 is essential for DNA repair by homologous recombination in higher vertebrate cells. Nature 420:93–98

Tercero JA, Diffley JF (2001) Regulation of DNA replication fork progression through damaged DNA by the Mec1/Rad53 checkpoint. Nature 412:553–557

Tercero JA, Longhese MP, Diffley JF (2003) A central role for DNA replication forks in checkpoint activation and response. Mol Cell 11:1323–1336

Tibbetts RS, Cortez D, Brumbaugh KM, Scully R, Livingston D, Elledge SJ, Abraham RT (2000) Functional interactions between BRCA1 and the checkpoint kinase ATR during genotoxic stress. Genes Dev 14:2989–3002

Tsao CC, Geisen C, Abraham RT (2004) Interaction between human MCM7 and Rad17 proteins is required for replication checkpoint signaling. Embo J 23:4660–4669

Umezu K, Sugawara N, Chen C, Haber JE, Kolodner RD (1998) Genetic analysis of yeast RPA1 reveals its multiple functions in DNA metabolism. Genetics 148:989–1005

Unal E, Arbel-Eden A, Sattler U, Shroff R, Lichten M, Haber JE, Koshland D (2004) DNA damage response pathway uses histone modification to assemble a double-strand break-specific cohesin domain. Mol Cell 16:991–1002

Uziel T, Lerenthal Y, Moyal L, Andegeko Y, Mittelman L, Shiloh Y (2003) Requirement of the MRN complex for ATM activation by DNA damage. Embo J 22:5612–5621 van Attikum H, Fritsch O, Hohn B, Gasser SM (2004) Recruitment of the INO80 complex by H2A phosphorylation links ATP-dependent chromatin remodeling with DNA double-strand break repair. Cell 119:777–788

Vaziri H, West MD, Allsopp RC, Davison TS, Wu YS, Arrowsmith CH, Poirier GG, Benchimol S (1997) ATM-dependent telomere loss in aging human diploid fibroblasts and DNA damage lead to the post-translational activation of p53 protein involving poly(ADP-ribose) polymerase. Embo J 16:6018–6033

Venclovas C, Thelen MP (2000) Structure-based predictions of Rad1, Rad9, Hus1 and Rad17 participation in sliding clamp and clamp-loading complexes. Nucleic Acids Res 28:2481–2493

Volkmer E, Karnitz LM (1999) Human homologs of Schizosaccharomyces pombe rad1, hus1, and rad9 form a DNA damage-responsive protein complex. J Biol Chem 274:567–570

Wakayama T, Kondo T, Ando S, Matsumoto K, Sugimoto K (2001) Pie1, a protein interacting with Mec1, controls cell growth and checkpoint responses in Saccharomyces cerevisiae. Mol Cell Biol 21:755–764

Wang B, Matsuoka S, Carpenter PB, Elledge SJ (2002) 53BP1, a mediator of the DNA damage checkpoint. Science 298:1435–1438

Wang X, Haber JE (2004) Role of saccharomyces single-stranded DNA-binding protein RPA in the strand invasion step of double-strand break repair. PLoS Biol 2:E21

Wang X, Zou L, Zheng H, Wei Q, Elledge SJ, Li L (2003) Genomic instability and endoreduplication triggered by RAD17 deletion. Genes Dev 17:965–970

Wang Y, Qin J (2003) MSH2 and ATR form a signaling module and regulate two branches of the damage response to DNA methylation. Proc Natl Acad Sci USA 100:15387–15392

Ward IM, Minn K, Jorda G, Chen (2003) Accumulation of checkpoint protein 53BP1 at DNA breaks involves its binding to phosphorylated histone H2AX. J Biol Chem 278:19579–19582

Weiss S, Matsuoka S, Elledge SJ, Leder P (2002) Hus1 acts upstream of chk1 in a mammalian DNA damage response pathway. Curr Biol 12:73–77

Wu X, Ranganathan V, Weisman DS, Heine WF, Ciccone DN, O'Neill TB, Crick KE, Pierce KA, Lane WS, Rathbun G, Livingston DM, Weaver DT (2000) ATM phosphorylation of Nijmegen breakage syndrome protein is required in a DNA damage response. Nature 405:477–482

Xu X, Stern DF (2003) NFBD1/KIAA0170 is a chromatin-associated protein involved in DNA damage signaling pathways. J Biol Chem 278:8795–8803

Yamane K, Chen J, Kinsella TJ (2003) Both DNA topoisomerase II-binding protein 1 and BRCA1 regulate the G2-M cell cycle checkpoint. Cancer Res 63:3049–3053

Yarden RI, Pardo-Reoyo S, Sgagias M, Cowan KH, Brody LC (2002) BRCA1 regulates the G2/M checkpoint by activating Chk1 kinase upon DNA damage. Nat Genet 30:285–289

Yazdi PT, Wang Y, Zhao S, Patel N, Lee EY, Qin J (2002) SMC1 is a downstream effector in the ATM/NBS1 branch of the human S-phase checkpoint. Genes Dev 16:571–582

Yoo HY, Kumagai A, Shevchenko A, Shevchenko A, Dunphy WG (2004a) Adaptation of a DNA replication checkpoint response depends upon inactivation of Claspin by the Polo-like kinase. Cell 117:575–588

Yoo HY, Shevchenko A, Shevchenko A, Dunphy WG (2004b) Mcm2 is a direct substrate of ATM and ATR during DNA damage and DNA replication checkpoint responses. J Biol Chem 279:53353–53364

You Z, Kong L, Newport J (2002) The role of single-stranded DNA and polymerase alpha in establishing the ATR, Hus1 DNA replication checkpoint. J Biol Chem 277:27088–27093

Yu X, Chen J (2004) DNA damage-induced cell cycle checkpoint control requires CtIP, a phosphorylation-dependent binding partner of BRCA1 C-terminal domains. Mol Cell Biol 24:9478-9486

Yu X, Chini CC, He M, Mer G, Chen J (2003) The BRCT domain is a phospho-protein binding domain. Science 302:639-642

Zachos G, Rainey MD, Gillespie DA (2003) Chk1-deficient tumour cells are viable but exhibit multiple checkpoint and survival defects. Embo J 22:713-723

Zhang J, Willers H, Feng Z, Ghosh JC, Kim S, Weaver DT, Chung JH, Powell SN, Xia F (2004) Chk2 phosphorylation of BRCA1 regulates DNA double-strand break repair. Mol Cell Biol 24:708-718

Zhao H, Piwnica-Worms H (2001) ATR-mediated checkpoint pathways regulate phosphorylation and activation of human Chk1. Mol Cell Biol 21:4129-4139

Zhao H, Watkins JL, Piwnica-Worms H (2002) Disruption of the checkpoint kinase 1/cell division cycle 25A pathway abrogates ionizing radiation-induced S and G2 checkpoints. Proc Natl Acad Sci USA 99:14795-14800

Zhao S, Weng YC, Yuan SS, Lin YT, Hsu HC, Lin SC, Gerbino E, Song MH, Zdzienicka MZ, Gatti RA, Shay JW, Ziv Y, Shiloh Y, Lee EY (2000) Functional link between ataxia-telangiectasia and Nijmegen breakage syndrome gene products. Nature 405:473-477

Zhao X, Muller EG, Rothstein R (1998) A suppressor of two essential checkpoint genes identifies a novel protein that negatively affects dNTP pools. Mol Cell 2:329-340

Zhong Q, Boyer TG, Chen PL, Lee WH (2002a) Deficient nonhomologous end-joining activity in cell-free extracts from Brca1-null fibroblasts. Cancer Res 62:3966-3970

Zhong Q, Chen CF, Chen PL, Lee WH (2002b) BRCA1 facilitates microhomology-mediated end joining of DNA double strand breaks. J Biol Chem 277:28641-28647

Zhou BB, Elledge SJ (2000) The DNA damage response: putting checkpoints in perspective. Nature 408:433-439

Zhu XD, Kuster B, Mann M, Petrini JH, de Lange T (2000) Cell-cycle-regulated association of RAD50/MRE11/NBS1 with TRF2 and human telomeres. Nat Genet 25:347-352

Zou L, Elledge SJ (2003) Sensing DNA damage through ATRIP recognition of RPA-ssDNA complexes. Science 300:1542-1548

Zou L, Cortez D, Elledge SJ (2002) Regulation of ATR substrate selection by Rad17-dependent loading of Rad9 complexes onto chromatin. Genes Dev 16:198-208

Zou L, Liu D, Elledge SJ (2003) Replication protein A-mediated recruitment and activation of Rad17 complexes. Proc Natl Acad Sci USA 100:13827-13832

… Results Probl Cell Differ (42)
P. Kaldis: Cell Cycle Regulation
DOI 10.1007/b138827/Published online: 23 September 2005
© Springer-Verlag Berlin Heidelberg 2005

Protein Kinases Involved in Mitotic Spindle Checkpoint Regulation

Ingrid Hoffmann

Cell Cycle Control and Carcinogenesis (F045), German Cancer Research Center (DKFZ), Im Neuenheimer Feld 242, 69120 Heidelberg, Germany
Ingrid.Hoffmann@dkfz.de

Abstract A number of checkpoint controls function to preserve the genome by restraining cell cycle progression until prerequisite events have been properly completed. Chromosome attachment to the mitotic spindle is monitored by the spindle assembly checkpoint. Sister chromatid separation in anaphase is initiated only once all chromosomes have been attached to both poles of the spindle. Premature separation of sister chromatids leads to the loss or gain of chromosomes in daughter cells (aneuploidy), a prevalent form of genetic instability of human cancer. The spindle assembly checkpoint ensures that cells with misaligned chromosomes do not exit mitosis and divide to form aneuploid cells. A number of protein kinases and checkpoint phosphoproteins are required for the function of the spindle assembly checkpoint. This review discusses the recent progress in understanding the role of protein kinases of the mitotic checkpoint complex in the surveillance pathway of the checkpoint.

1
Introduction

During cell division, accurate transmission of the genome is essential for survival. Entry into mitosis is controlled by checkpoints that monitor DNA damage and replication, whereas exit from mitosis is controlled by checkpoints that monitor assembly and position of the mitotic spindle. The mitotic spindle checkpoint is activated by the lack of microtubule occupancy and tension at the kinetochores and leads to cell cycle arrest in prometaphase. It is a tightly conserved signal transduction pathway that prevents sister chromatid separation until all chromosomes achieve bipolar attachment to the mitotic spindle. The presence of even a single misaligned or unattached chromosome is sufficient to activate the checkpoint. In response to defects in the mitotic apparatus, it blocks the activity of the anaphase-promoting complex or cyclosome (APC/C), a large multisubunit ubiquitin ligase required for chromosome segregation. After all sister chromatids have achieved biorientation, the APC/C in association with one of its substrate-binding cofactors, Cdc20, tags the anaphase-inhibiting protein securin with polyubiquitin chains, leading to its degradation by the proteasome (Peters 2002; Harper et al. 2002). Sister chromatids are held together by cohesin and cleavage of cohesin will result in loss of sister chromatid cohesion and the onset of sister chromatid separation (Nasmyth 2002). Degradation of securin activates separase, a protease which cleaves the Scc1 subunit of cohesin.

2
The Spindle Assembly Checkpoint

The molecular components of the spindle assembly checkpoint were identified initially in *Saccharomyces cerevisiae* (Gorbsky 2001; Shah and Cleveland 2000). They include Mad1-3 (mitotic arrest deficiency) (Li and Murray 1991), Bub1-3 (Hoyt et al. 1991; Roberts et al. 1994), and Mps1 (Weiss and Winey 1996) (Table 1). Homologues of these checkpoint proteins were later found in other organisms, including mammals. Checkpoint proteins accumulate at unattached kinetochores in prometaphase, but disappear from kinetochores later in mitosis or meiosis upon microtubule attachment and tension. In higher eukaryotes the checkpoint control proteins comprise Mad1, Mad2, Bub3 and the protein kinases Bub1, BubR1 (Mad3 in budding yeast), and Mps1. In addition to these basic checkpoint components, other proteins such as CENP-E (a member of the kinesin superfamily) (Abrieu et al. 2000; Yao et al. 2000), Rod, ZW10 (Chan et al. 2000), Aurora B (Biggins and Murray 2001; Kallio et al. 2002; Ditchfield et al. 2003) and mitogen-activated protein kinase (MAPK) (Shapiro et al. 1998; Zecevic et al. 1998) play a role in the spindle checkpoint (Table 1).

Table 1 Proteins involved in mitotic spindle checkpoint regulation

Protein	Proposed function in the spindle assembly checkpoint
Mad1	Coiled-coil protein, binds to Mad2 and recruits Mad2 to kinetochores phosphorylated by Mps1 and Bub1 upon checkpoint activation
Mad2	Binds to Mad1, binds and inhibits APC/C^{Cdc20}
BubR1 (Mad3)	Protein kinase, binds to Bub3 and APC/C^{Cdc20}, binds to the mitotic motor protein CENP-E
Bub1	Protein kinase, binds to and recruits Bub3, Mad1 and Mad2
Bub3	Contains WD-40 repeats, binds to Bub1 and BubR1
Mps1	Protein kinase, essential for establishment and maintenance of the spindle checkpoint
Aurora B	Protein kinase, binds to INCENP, other substrates: CENP-A, Rec8, vimentin, desmin, the kinesin MCAK, histone H3
Aurora C	Protein kinase, binds to INCENP and Aurora B
MAPK	Protein kinase
Rod	Identified in Drosophila, binds to Zw10
Zw10	Identified in Drosophila, binds to Rod

Subcellular localization studies have placed all these checkpoint proteins at the kinetochores. Ablation or suppression of function of any of these proteins substantially compromises mitotic checkpoint control (Lew and Burke 2003). These checkpoint control proteins form a complex intracellular network, the mitotic checkpoint complex (MCC), to block the action of APC/C^{Cdc20}.

Mad2 interacts with other components of the spindle checkpoint and plays a key role in the signaling pathway of the checkpoint. In interphase, it binds to Mad1 and is preferentially found on the nuclear periphery (Chen et al. 1999). This localization is strictly dependent on Mad1 since in a fission yeast strain lacking Mad1, Mad2 is no longer found on the nuclear periphery (Ikui et al. 2002). Upon the onset of mitosis, Mad2 translocates into the nucleus and is guided to unattached kinetochores by Mad1 (Chen et al. 1999). Recruiting Mad2 to kinetochores is the only known function of Mad1 to date. From early mitosis on, Mad2 is found in a complex with its target, Cdc20. Human Mad2 is modified through phosphorylation on multiple serine residues in vivo in a cell cycle-dependent manner. Only unphosphorylated Mad2 interacts with Mad1 or the APC/C in vivo (Wassmann et al. 2003). Injection of anti-Mad2 antibodies drives prophase cells into a premature anaphase and overrides the arrest induced by microtubule depolymerization, indicating that the checkpoint activation in situations with unattached kinetochores requires Mad2 (Gorbsky et al. 1998). A Mad2 mutant containing serine to aspartic acid mutations mimicking the C-terminal phosphorylation events fails to interact with Mad1 or the APC/C and acts as a dominant-negative antagonist of wild-type Mad2 (Wassmann et al. 2003). Although yeast strains lacking Mad2 are viable, deletion of Mad2 in mouse causes cell lethality. Mad2-/- mouse cells do not arrest in response to spindle damage, show widespread chromosome missegregation, and undergo apoptosis during initiation of gastrulation (Dobles et al. 2000).

Upon checkpoint activation, both BubR1-Bub3 and Mad2 are capable of blocking the activity of APC/C through their direct binding to Cdc20 (Yu 2002; Bharadwaj and Yu 2004). Binding of spindle microtubules to kinetochores, disrupts the interaction between Mad1 and Mad2 and ultimately disables the arrest (Fig. 1a).

3
Regulation of the Spindle Checkpoint by Protein Kinases

3.1
Bub1

Bub1 is a protein kinase and an essential checkpoint component that resides at kinetochores during mitosis. Bub1 was first described in a genetic screen searching for budding yeast mutants that were sensitive to the spin-

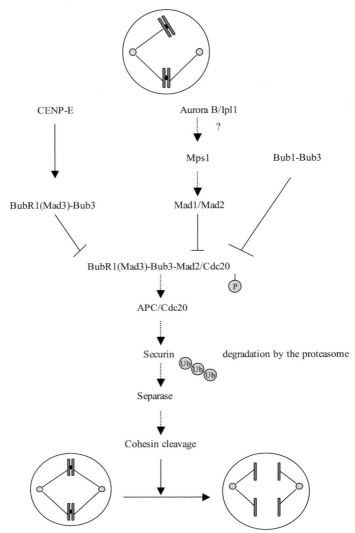

Fig. 1 Functions of protein kinases in the spindle checkpoint. Attachment of chromosomes to the mitotic spindle is monitored by the spindle checkpoint. Sensing mechanisms may involve Aurora B/Ipl1 and CENP-E. Upon checkpoint activation both BubR1-Bub3 and Mad2 interact with Cdc20 and lead to an inhibition of APC. Inactivation of the checkpoint occurs upon bipolar attachment to the mitotic spindle. APC is activated and ubiquitinates securin. Degradation of securin activates a protease called separase which cleaves cohesin, resulting in a loss of sister chromatid cohesion, leading to the the onset of anaphase

dle poison benomyl (Hoyt et al. 1991). Bub1 binds to Bub3 throughout the cell cycle and phosphorylates Bub3 in vitro (Roberts et al. 1994). Overexpression of a dominant allele of Bub1 in yeast causes a mitotic delay without spindle damage that is dependent on the functions of Bub2, Bub3, and Mad1-3 (Farr and Hoyt 1998). In yeast, the Bub1–Bub3 complex interacts with Mad1 when the spindle checkpoint is activated (Brady and Hardwick 2000). In vertebrates, Bub1 is required for the kinetochore localization of Mad1 and Mad2 (Sharp-Baker and Chen 2001; Johnson et al. 2004; Vigneron et al. 2004). This activity of Bub1 seems to be independent of its kinase activity since a kinase-inactive mutant of Bub1 is fully capable of recruiting Mad1 and Mad2 to the kinetochores in *Xenopus* egg extracts (Sharp-Baker and Chen 2001). Immunodepletion of Bub1 abolishes the spindle checkpoint and the kinetochore binding of the checkpoint proteins Mad1-3, Bub3, BubR1 and the kinetochore motor protein CENP-E (Sharp-Baker and Chen 2001; Johnson et al. 2004). Recently, it was shown that mammalian Bub1 also has a downstream function in the spindle checkpoint since it directly phosphorylates Cdc20 (Tang et al. 2004). HeLa cells depleted for Bub1 by RNA interference (RNAi) are defective in checkpoint signaling. Bub1 directly phosphorylates Cdc20 in vitro and in vivo and inhibits the ubiquitin ligase activity of APC/C^{Cdc20} catalytically (Tang et al. 2004). Six Ser/Thr residues in Cdc20 were phosphorylated by Bub1 in vitro. Ectopic expression of a Cdc20 protein where all six residues were mutated to alanine is refractory to Bub1-mediated phosphorylation and inhibition. Overexpression of this Cdc20 mutant protein impairs the function of the spindle checkpoint. Bub1 function seems to be regulated by several upstream kinases. Bub1 becomes hyperphosphorylated and its kinase activity is induced specifically at unattached chromosomes (Chen 2004). MAPK contributes to this phosphorylation, as inhibiting MAPK or altering MAPK consensus sites in Bub1 abolishes the phosphorylation and activation on chromosomes. The activation of Bub1 seems to be important in maintaining the checkpoint towards late prometaphase when the cell contains only a few kinetochores or a single unattached kinetochore. It has been shown that the MAPK downstream target Rsk activates cytosolic Bub1 during frog oocyte maturation (Schwab et al. 2001), but Rsk does not seem to be involved in Bub1 phosphorylation at kinetochores, because immunodepletion of Rsk does not have an effect on the phosphorylation. Fission yeast Bub1 is phosphorylated during mitosis and the protein is a substrate for Cdc2 (Yamaguchi et al. 2003). Mutation at four putative Cdc2 sites abolishes the checkpoint function (Yamaguchi et al. 2003). Both Cdc2 and MAPK have similar consensus phosphorylation sites. In egg extracts, inhibition of MAPK abolishes Bub1 phosphorylation without an effect on Cdc2 activity, indicating that Cdc2 is not involved in Bub1 phosphorylation in the frog. Finally, Bub1 has a noncheckpoint function at the kinetochores and preserves cohesion through the MEI-S332/shugoshin family of proteins (Salic et al. 2004; Tang et al. 2004).

3.2
BubR1

BubR1 was isolated as a Mad3/Bub1-related protein kinase on the basis of its similarities with the N-terminal domain of the yeast checkpoint protein Mad3 (Taylor et al. 1998). Thus, BubR1 is thought to be the homologue of Mad3; in higher eukaryotes BubR1 directly binds to CENP-E (Chan et al. 1998). It is a mitosis-specific kinase and is inactive during interphase (Chan et al. 1999). Microinjection of antibodies against BubR1 into HeLa cells abrogated mitotic arrest after nocodazole-induced spindle disassembly (Chan et al. 1998; Chan et al. 1999). In *Xenopus* egg extracts, immunodepletion of BubR1 also prevented mitotic arrest in response to spindle damage (Chen 2002). BubR1 accumulates and becomes hyperphosphorylated at unattached kinetochores. Immunodepletion of BubR1 greatly reduces kinetochore binding of Bub1, Bub3, Mad1, Mad2, and CENP-E. These defects can be rescued by wild-type, kinase-dead, or a truncated BubR1 protein that lacks its kinase domain, indicating that the kinase activity of BubR1 is not essential for the spindle checkpoint in egg extracts (Chen 2002). Whether phosphorylation of BubR1 leads to an activation of the kinase is not known. BubR1 accumulates to a higher level and becomes hyperphosphorylated at unattached kinetochores compared with that at metaphase kinetochores. This phosphorylation requires Mad1 or its downstream effector, but not Mad2 (Chen 2002). Expression of a kinase-inactive mutant of BubR1 abolished mitotic arrest induced by microtubule disassembly (Chan et al. 1999). RNAi-mediated depletion of BubR1 causes severe chromosome misalignment and results in the loss of kinetochore-microtubule attachment (Chan et al. 1999). Attachment in these cells can be restored by inhibition of Aurora kinase, which is known to stabilize kinetochore-microtubule interactions. BubR1 similar to Mad2 also associates and can phosphorylate Cdc20 in vitro leading to inactivation of the APC/C (Sudakin et al. 2001; Tang et al. 2001; Fang 2002). Both Mad2 and BubR1 can indirectly bind to Cdc20 in vitro and either independently or cooperatively inhibit polyubiquitination of APC/C^{Cdc20} substrates. Quantitative analysis indicates that BubR1 binds to Cdc20 with a higher affinity and is more potent than Mad2 in inhibiting the activation of APC by Cdc20 (Fang 2002). The two pathways seem to act synergistically since inactivation of either Mad2 or BubR1 by microinjection of inhibitors shows that the activity of both proteins is required for the metaphase delay at 23 °C (Shannon et al. 2002). But why does the cell need two different inhibitors of APC in the same checkpoint pathway? It is possible that binding of Cdc20 to BubR1 recruits Cdc20 to kinetochores, where BubR1 promotes the formation of the Mad2–Cdc20 complex, which subsequently diffuses away from kinetochores and inhibits APC throughout the cell. Alternatively, BubR1 and Mad2 might inhibit APC in response to different checkpoint signals. In *Xenopus* egg extracts, CENP-E dependent activation of BubR1 kinase activity at kinetochores is necessary for establishing the mitotic checkpoint (Mao et al. 2003). Although BubR1 binds to CENP-E, its kinase activity seems to be required for Mad2, but not CENP-E, recruitment to kinetochores (Mao et al. 2003). Graded reduction of BubR1 expression in mouse embryonic fibroblasts causes increased aneuploidy and senescence (Baker et al. 2004). Male and female mutant mice have defects in meiotic chromosome segregation and are infertile (Baker et al. 2004). BubR1-knockout mice die during early development (beyond day 8.5 in utero) as a result of an increased prevalence of apoptosis (Baker et al. 2004). Downregulation of BubR1 by RNAi is associated with the formation of polyploidy. This seems to be the result of a prolonged mitotic arrest which leads to a decrease in BubR1 levels and a concomitant increase in polyploid cells (Shin et al. 2003). These data suggest that BubR1 is not only a sensor that monitors the mitotic checkpoint but that it is also involved in apoptotic signaling and chromosome instability.

3.3
Aurora B

The Ipl1/Aurora family of protein kinases plays multiple roles in mitosis, including chromosome segregation and cell division (Meraldi et al. 2004b). In budding yeast, Ipl1 ensures accurate chromosome segregation by resolving syntelic orientations, possible by monitoring tension at centromeres and destabilizing inappropriately bound microtubules (Tanaka et al. 2002). In an Ipl1 mutant, the checkpoint remains functional when triggered by disruption of the spindle by nocodazole (Biggins and Murray 2001). Higher eukaryotes express three Aurora kinases, Aurora A, Aurora B, and Aurora C. Aurora A and Aurora B have very distinct localizations and functions. Aurora A is involved in regulation of microtubule nucleation at centrosomes. Aurora B is found at the inner centromeric regions of chromosomes from prophase through the metaphase–anaphase transition as part of a "chromosomal passenger protein complex," where it appears to promote correct bipolar microtubule-kinetochore attachments. Thus, Aurora B seems to be involved in several mitotic processes, including chromosome condensation through phosphorylation of histone H3, chromosome alignment, kinetochore disjunction, the spindle assembly checkpoint, and cytokinesis. After anaphase onset, Aurora B relocalizes to the central microtubules of the anaphase spindle and then to the midbody during the completion of cytokinesis (Meraldi et al. 2004b). Aurora B requires the association to the inner centromere protein and a founding member of the chromosome passenger proteins, INCENP, for its localization (Adams et al. 2000). For over a decade INCENP has been implicated in the regulation of cytokinesis as it localizes to the cortex in late anaphase/telophase, before cleavage furrow ingression and before the recruitment of myosin (Cooke et al. 1987). Aurora B also affects the localization of a taxin family member, TACC1, to the midbody, leading to abnormal cell division and multinucleated cells (Delaval et al. 2004). In addition, survivin, a conserved inhibitor of apoptosis also seems to be required for the localization of Aurora B (Romano et al. 2003). The role of Aurora B kinase activity has been adressed by ectopically expressing a kinase-negative version, Aurora B^{K109R}, by RNAi, and by small molecule inhibitors. One report described that cells expressing Aurora B^{K109R} completed mitosis, but failed to undergo cytokinesis, suggesting that Aurora B activity is not required for chromosome segregation (Tatsuka et al. 1998). However, another study revealed that the kinase-negative mutant prevents chromosome alignment owing to the failure of kinetochore-microtubule interactions (Murata-Hori and Wang 2002). Ablation of Aurora B function by RNAi results in a dramatic increase in polyploidy as assayed by the presence of binucleate and multinucleated cells (Ditchfield et al. 2003; Scrittori et al. 2005). In addition, treatment with short interfering RNA (siRNA) affected the prometaphase–metaphase transition, yielding a significant increase in the percentage of mitotic cells at prometaphase in transfected cells in comparison with control cells (Scrittori et al. 2005). Microinjection of anti-Aurora B antibodies blocked chromosome alignment and segregation, leading to abrogation of the spindle checkpoint (Gorbsky 2001). However, in these approaches it is difficult to distinguish between effects due to the lack of protein itself, where Aurora B-containing complexes and subcomplexes do not form, and those simply due to lack of kinase acitivity, where substrate phosphorylation is the initial defect. The use of the small molecule inhibitor ZM447439 (AstraZeneca) which inhibits Aurora B kinase activity allowed to study of the phenotype which is solely due to the lack of kinase activity. When ZM447439 was added to mammalian cell cultures, cells entered mitosis and formed a mitotic spindle, but phosphorylation of histone H3 was reduced, the spindle was disorganized, and cytokinesis was blocked (Ditchfield et al. 2003). The use of ZM447439 in *Xenopus* egg cycling extracts made it possible to study the effect of the inhibitor in a system on the basic cell cycle machinery in the absence of functional checkpoints. Checkpoint pathways including the spindle checkpoint do not opperate in *Xenopus* early embryonic cell cycles. Gadea and Ruderman (2005) found that ZM447439 had striking effects on chromosome morphology since chromosome condensation began to schedule but then failed to progress properly owing to premature decondensation during mid-mitosis. ZM447439 strongly interfered with mitotic spindle assembly by inhibiting the formation of microtubules that are nucleated/stabilized by chromatin. Another inhibitor, hesperadin, causes defects in mitosis and cytokinesis and inhibits Aurora B in vitro (Hauf et al. 2003). The use of each of these inhibitors phenocopies Aurora B RNAi. A possible substrate of Aurora B is the Kin1 kinesin MCAK (Andrews et al. 2004). Aurora B inhibits the microtubule depolymerizing activity of MCAK in vitro. Moreover, disruption of Aurora B function by expression of a kinase-dead mutant or RNAi prevented centromeric targeting of MCAK. PP1 is known to antagonize Aurora B activity and inhibit its kinase activity (Francisco and Chan 1994; Sassoon et al. 1999). Another known substrate of

Aurora B is the GTPase activating protein, MgcRacGAP, which colocalizes with the kinase at the midbody. Aurora B phosphorylates MgcRacGAP at serine residues and that this modification induces latent GAP activity towards RhoA in vitro (Minoshima et al. 2003), thus functionally converting the protein.

Little is known about the localization and function of the third mammalian Aurora kinase family member, Aurora C. Aurora C was reported to express in testis and some human cancer cell lines with the highest level detected at G2/M (Kimura et al. 1999). Recent studies indicate that Aurora C is a chromosome passenger protein similar to Aurora B (Li et al. 2004; Sasai et al. 2004). Aurora C is tightly bound to mitotic chromosomes where it interacts with Aurora B and INCENP. Elevated expression of Aurora C in cancer cells might play a critical role in perturbing the chromosomal passenger complexes. In the future, it would seem important to analyze in greater detail the Aurora C pathway in mammalian cells.

3.4
Mps1

The Mps1 (monopolar spindle 1) protein kinase was first described in *S. cerevisiae*, where it is implicated in the duplication of the spindle pole body (Winey et al. 1991). Mps1 was identified initially as a centrosomal protein required for the assembly of bipolar spindles, but it was later shown to play a role in the spindle checkpoint as well. In yeast, overexpression of Mps1p or a mutant form of Bub1p, Bub-5p, causes mitotic arrest in the apparent absence of defects in spindle assembly (Hardwick et al. 1996; Farr and Hoyt 1998). The mitotic arrest of each is dependent on all other checkpoint proteins and both Mps1p and Bub1-5p induced arrests are interdependent. These results also suggest that the activation of Mps1 is an early event in checkpoint signaling and may lead to the recruitment of other checkpoint proteins to kinetochores. Using *Xenopus* egg extracts, it was shown that the *Xenopus* homologue of yeast Mps1p is a kinetochore-associated protein kinase, whose activity is necessary to establish and maintain the spindle checkpoint (Abrieu et al. 2001). Since high levels of Mad2 overcome checkpoint loss in Mps1-depleted extracts, Mps1 acts upstream of Mad2-mediated inhibition of APC/C. Mps1 is essential for the checkpoint because it is required for recruitment and retention of active CENP-E at kinetochores, which is in turn required for association of both Mad1 and Mad 2 (Abrieu et al. 2001). The human homologue of yeast Mps1p, hMps1, is a cell cycle-regulated kinase and both its protein levels and its kinase activity peak during progression through the M phase (Stucke et al. 2002). Microinjection of anti-hMps1 antibodies and silencing of the kinase by siRNA interferes with the spindle assembly checkpoint (Stucke et al. 2002). hMps1 is hyperphosphorylated in mitosis and is dephosphorylated when cells exit mitosis (Liu et al. 2003). Similar to the *Xenopus* homologue, human Mps1 is required for the recruitment of Mad1 and Mad2 to kinetochores (Liu et al. 2003) and it associates with APC/C in both interphase and mitosis (Liu et al. 2003). Thus, it is possible that Mps1 phosphorylates the APC/C during mitosis and that these modifications may be part of the mechanism by which the checkpoint inhibits APC/C. A recent study of *Drosophila* Mps1 revealed that Mps1 is required for the spindle checkpoint by demonstrating that embryos harbor a transposon insertion mutation. In this mutation, called Mps1[1], the single fly Mps1 orthologue is disrupted; therefore, Mps1[1] embryos do not undergo a mitotic arrest in response to the microtubule poison colcemid (Fischer et al. 2004). In addition the authors showed that the metaphase-to-anaphase transition is accelerated in Mps1[1] embryos similar to what was observed for other spindle checkpoint proteins in vertebrate cells (Meraldi et al. 2004a). Finally, as with any protein kinase, the identification of substrates is critical, particularly those substrates that reveal their specific functions in the various stages of mitosis. Thus, it is possible that Mps1 phosphorylates the APC/C during mitosis and that these modifications may be part of the mechanism by which the checkpoint inhibits the APC/C.

3.5
Mitogen-activated protein kinase

MAPKs are serine/threonine-specific protein kinases that are activated in response to extracellular signals and play important roles as effectors for diverse cellular functions, including growth, differentiation, movement, and secretion (L'Allemain 1994). In vertebrate cells, a small fraction of MAPK is activated and enriched at kinetochores during mitosis, and the level of active MAPK decreases at and after metaphase (Shapiro et al. 1998; Zecevic et al. 1998). MAPK activity is important for the spindle checkpoint both in egg extracts and in somatic cells. It has also been shown that MAPK interacts with the kinetochore motor protein CENP-E and that MAPK phosphorylates CENP-E in vitro to create a tension-sensitive epitope that is recognized by a monoclonal antibody (Zecevic et al. 1998). In *Xenopus*, MAPK also contributes to Cdc20 phosphorylation, and this phosphorylation is required for the checkpoint proteins to bind and inhibit Cdc20 (Chung and Chen 2003). Cdc20 mutants that are phosphorylation-deficient are able to activate the APC in *X. laevis* egg extracts (Chung and Chen 2003). Thus, dephosphorylation of Cdc20 at multiple MAPK sites may provide a mechanism to disassemble the existing spindle checkpoint complex.

4
The Spindle Checkpoint and Cancer

The survival of a cell depends on the accuracy of mitosis and errors in the mechanisms controlling mitosis can lead to genomic instability. Cancer cells are highly aneuploid which is caused by chromosomal instability. Spindle checkpoint disruption appears to be one of the mechanisms leading to aneuploidy in human cancers. Thus, defects in the spindle checkpoint might promote aneuploidy and tumorigenesis (Bharadwaj and Yu 2004). A high percentage of solid tumors fail to arrest in response to microtubule poisons such as nocodazole, suggesting an impaired spindle checkpoint. A number of genes required for chromosome segregation are mutated or increased in their expression levels in human cancers.

Members of the Aurora kinase family are required for multiple aspects of mitosis. The Aurora kinases are frequently overexpressed in human tumors. In addition, the three Aurora genes map to chromosomal loci that are frequently altered in human tumors: Aurora A to chromosome 20q13, Aurora B to chromosome 17p13, and Aurora C to 19q13. Both Aurora A and Aurora B are overexpressed in primary breast and colon tumor samples (Sen et al. 2002; Warner et al. 2003), and their levels are significantly elevated in tumor samples compared with normal tissues (Keen and Taylor 2004). Interestingly, a systematic analysis of the expression levels of Aurora kinases reveals that Aurora A and Aurora B levels seem to rise or decline in parallel in tumor samples. Aurora C did not seem to be overexpressed in this study, and its expression did not correlate with Aurora A and Aurora B expression (Keen and Taylor 2004). As Aurora A and Aurora B kinases are overexpressed in many tumors it is likely that they are useful targets of anticancer drugs. Recently several Aurora kinase inhibitors were described that inhibit their enzymatic activity by occupying the ATP binding site: Hesperadin (Hauf et al. 2003), ZM447439 (Ditchfield et al. 2003), and VX-680 (Harrington et al. 2004). All three inhibitors prevent cell division and phosphorylation of histone H3 on serine 10 (Ditchfield et al. 2003; Hauf et al. 2003; Harrington et al. 2004). Treatment of cells with hesperadin shows that many of the chromosomes are oriented in a syntelic manner (Hauf et al. 2003). Despite the presence of misaligned chromosomes, cells treated with ZM447439 or hesperadin undergo anaphase and exit mitosis prematurely, indicating that the spindle checkpoint is impaired (Ditchfield et al. 2003; Hauf et al. 2003). In the presence of ZM447439 or hesperadin the checkpoint proteins Bub1 and BubR1 do not properly localize to kinetochores (Ditchfield et al. 2003; Hauf et al. 2003). VX-680 blocks cell cycle progression, inhibits proliferation and induces apoptosis in a diverse range of human tumor types, while noncycling cells remain unaffected (Harrington et al. 2004). VX-60 caused a marked reduction in tumor size in a human AML (HL-60) xenograft model (Harrington et al. 2004).

Bub1 and BubR1 have been shown to be mutated in a subset of colon cancers with chromosomal instability (Cahill et al. 1998; Lengauer and Wang 2004). Truncations and missense mutations in the Bub1b gene which encodes BubR1 were identified in families with mosaic variegated aneuploidy, a rare recessive condition characterized amongst others by childhood cancer (Hanks et al. 2004). This important study supports the idea that there is a genetic basis for aneuploidy in cancers. In addition, deletions of Bub1 and BubR1 were detected in lung cancer (Sato et al. 2000). The adenomatous polyposis coli (Apc) gene also plays a an important role in regulating genomic stability. A functional interaction exists between BubR1 and Apc genes in vivo and BubR1 deficiency confers the suceptibility of Apc Min/+ mice to develop colonic tumors and in development and progression of colorectal cancer (Hanks et al. 2004). Mutant mice that express low levels of BubR1 protein develop progressive aneuploidy but form tumors only after being challenged with a carcinogen (Dai et al. 2004).

Given the importance of Aurora B, Bub1, and BubR1 kinases in regulating the spindle checkpoint it will be intriguing to generate selective inhibitors of their activity. Combinations with microtubule-targeted drugs as paclitaxel (Taxol) and vinca alkaloids will be interesting to evaluate.

5
Conclusions

The initiation of chromosome segregation at anaphase is linked by the spindle assembly checkpoint to the completion of chromosome-microtubule attachment during metaphase. Several components of the spindle checkpoint (Mad2, BubR1, and Bub3) form a complex that prevents entry into anaphase by binding to Cdc20, an essential activator of the anaphase-promoting complex (APC/C). The kinases Bub1 and Aurora B display multiple functions in mitosis. Strategies that allow the separation of these functions will be invaluable for future studies on the role of these proteins in mitosis. The nature of the direct molecular interactions between checkpoint proteins are poorly understood. Therefore, unraveling the molecular mechanisms of this checkpoint, how the proteins involved in checkpoint regulation collaborate to inhibit the APC/C, will be required to develop novel and effective strategies to combat cancer.

Acknowledgements I thank my coworkers Onur Cizmecioglu and Daniel Spengler for discussions and critical reading of the manuscript.

References

Abrieu A, Kahana JA, Wood KW, Cleveland DW (2000) CENP-E as an essential component of the mitotic checkpoint in vitro. Cell 102:817–826
Abrieu A, Magnaghi-Jaulin L, Kahana JA, Peter M, Castro A, Vigneron S, Lorca T, Cleveland DW, Labbe JC (2001) Mps1 is a kinetochore-associated kinase essential for the vertebrate mitotic checkpoint. Cell 106:83–93
Adams RR, Wheatley SP, Gouldsworthy AM, Kandels-Lewis SE, Carmena M, Smythe C, Gerloff DL, Earnshaw WC (2000) INCENP binds the Aurora-related kinase AIRK2 and

is required to target it to chromosomes, the central spindle and cleavage furrow. Curr Biol 10:1075–1078

Andrews PD, Ovechkina Y, Morrice N, Wagenbach M, Duncan K, Wordeman L, Swedlow JR (2004) Aurora B regulates MCAK at the mitotic centromere. Dev Cell 6:253–268

Baker DJ, Jeganathan KB, Cameron JD, Thompson M, Juneja S, Kopecka A, Kumar R, Jenkins RB, de Groen PC, Roche P, van Deursen JM (2004) BubR1 insufficiency causes early onset of aging-associated phenotypes and infertility in mice. Nat Genet 36:744–749

Bharadwaj R, Yu H (2004) The spindle checkpoint, aneuploidy, and cancer. Oncogene 23:2016–2027

Biggins S, Murray AW (2001) The budding yeast protein kinase Ipl1/Aurora allows the absence of tension to activate the spindle checkpoint. Genes Dev 15:3118–3129

Brady DM, Hardwick KG (2000) Complex formation between Mad1p, Bub1p and Bub3p is crucial for spindle checkpoint function. Curr Biol 10:675–678

Cahill DP, Lengauer C, Yu J, Riggins GJ, Willson JK, Markowitz SD, Kinzler KW, Vogelstein B (1998) Mutations of mitotic checkpoint genes in human cancers. Nature 392:300–303

Chan GK, Schaar BT, Yen TJ (1998) Characterization of the kinetochore binding domain of CENP-E reveals interactions with the kinetochore proteins CENP-F and hBUBR1. J Cell Biol 143:49–63

Chan GK, Jablonski SA, Sudakin V, Hittle JC, Yen TJ (1999) Human BUBR1 is a mitotic checkpoint kinase that monitors CENP-E functions at kinetochores and binds the cyclosome/APC. J Cell Biol 146:941–954

Chan GK, Jablonski SA, Starr DA, Goldberg ML, Yen TJ (2000) Human Zw10 and ROD are mitotic checkpoint proteins that bind to kinetochores. Nat Cell Biol 2:944–947

Chen RH (2002) BubR1 is essential for kinetochore localization of other spindle checkpoint proteins and its phosphorylation requires Mad1. J Cell Biol 158:487–496

Chen RH (2004) Phosphorylation and activation of Bub1 on unattached chromosomes facilitate the spindle checkpoint. EMBO J 23:3113–3121

Chen RH, Brady DM, Smith D, Murray AW, Hardwick KG (1999) The spindle checkpoint of budding yeast depends on a tight complex between the Mad1 and Mad2 proteins. Mol Biol Cell 10:2607–2618

Chung E, Chen RH (2003) Phosphorylation of Cdc20 is required for its inhibition by the spindle checkpoint. Nat Cell Biol 5:748–753

Cooke CA, Heck MM, Earnshaw WC (1987) The inner centromere protein (INCENP) antigens: movement from inner centromere to midbody during mitosis. J Cell Biol 105:2053–2067

Dai W, Wang Q, Liu T, Swamy M, Fang Y, Xie S, Mahmood R, Yang YM, Xu M, Rao CV (2004) Slippage of mitotic arrest and enhanced tumor development in mice with BubR1 haploinsufficiency. Cancer Res 64:440–445

Delaval B, Ferrand A, Conte N, Larroque C, Hernandez-Verdun D, Prigent C, Birnbaum D (2004) Aurora B-TACC1 protein complex in cytokinesis. Oncogene 23:4516–4522

Ditchfield C, Johnson VL, Tighe A, Ellston R, Haworth C, Johnson T, Mortlock A, Keen N, Taylor SS (2003) Aurora B couples chromosome alignment with anaphase by targeting BubR1, Mad2, and Cenp-E to kinetochores. J Cell Biol 161:267–280

Dobles M, Liberal V, Scott ML, Benezra R, Sorger PK (2000) Chromosome missegregation and apoptosis in mice lacking the mitotic checkpoint protein Mad2. Cell 101:635–645

Fang G (2002) Checkpoint protein BubR1 acts synergistically with Mad2 to inhibit anaphase-promoting complex. Mol Biol Cell 13:755–766

Farr KA, Hoyt MA (1998) Bub1p kinase activates the Saccharomyces cerevisiae spindle assembly checkpoint. Mol Cell Biol 18:2738–2747

Fischer MG, Heeger S, Hacker U, Lehner CF (2004) The mitotic arrest in response to hypoxia and of polar bodies during early embryogenesis requires Drosophila Mps1. Curr Biol 14:2019–2024

Francisco L, Chan CS (1994) Regulation of yeast chromosome segregation by Ipl1 protein kinase and type 1 protein phosphatase. Cell Mol Biol Res 40:207–213

Gadea BB, Ruderman JV (2005) Aurora kinase inhibitor ZM447439 blocks chromosome-induced spindle assembly, the completion of chromosome condensation, and the establishment of the spindle integrity checkpoint in *Xenopus* egg extracts. Mol Biol Cell 16:1305–1318

Gorbsky GJ (2001) The mitotic spindle checkpoint. Curr Biol 11:R1001–1004

Gorbsky GJ, Chen RH, Murray AW (1998) Microinjection of antibody to Mad2 protein into mammalian cells in mitosis induces premature anaphase. J Cell Biol 141:1193–1205

Hanks S, Coleman K, Reid S, Plaja A, Firth H, Fitzpatrick D, Kidd A, Mehes K, Nash R, Robin N, Shannon N, Tolmie J, Swansbury J, Irrthum A, Douglas J, Rahman N (2004) Constitutional aneuploidy and cancer predisposition caused by biallelic mutations in BUB1B. Nat Genet 36:1159–1161

Hardwick KG, Weiss E, Luca FC, Winey M, Murray AW (1996) Activation of the budding yeast spindle assembly checkpoint without mitotic spindle disruption. Science 273:953–956

Harper JW, Burton JL, Solomon MJ (2002) The anaphase-promoting complex: it's not just for mitosis any more. Genes Dev 16:2179–2206

Harrington EA, Bebbington D, Moore J, Rasmussen RK, Ajose-Adeogun AO, Nakayama T, Graham JA, Demur C, Hercend T, Diu-Hercend A, Su M, Golec JM, Miller KM (2004) VX-680, a potent and selective small-molecule inhibitor of the Aurora kinases, suppresses tumor growth in vivo. Nat Med 10:262–267

Hauf S, Cole RW, LaTerra S, Zimmer C, Schnapp G, Walter R, Heckel A, van Meel J, Rieder CL, Peters JM (2003) The small molecule Hesperadin reveals a role for Aurora B in correcting kinetochore-microtubule attachment and in maintaining the spindle assembly checkpoint. J Cell Biol 161:281–294

Hoyt MA, Totis L, Roberts BT (1991) S. cerevisiae genes required for cell cycle arrest in response to loss of microtubule function. Cell 66:507–517

Ikui AE, Furuya K, Yanagida M, Matsumoto T (2002) Control of localization of a spindle checkpoint protein, Mad2, in fission yeast. J Cell Sci 115:1603–1610

Johnson VL, Scott MI, Holt SV, Hussein D, Taylor SS (2004) Bub1 is required for kinetochore localization of BubR1, Cenp-E, Cenp-F and Mad2, and chromosome congression. J Cell Sci 117:1577–1589

Kallio MJ, McCleland ML, Stukenberg PT, Gorbsky GJ (2002) Inhibition of aurora B kinase blocks chromosome segregation, overrides the spindle checkpoint, and perturbs microtubule dynamics in mitosis. Curr Biol 12:900–905

Keen N, Taylor S (2004) Aurora-kinase inhibitors as anticancer agents. Nat Rev Cancer 4:927–936

Kimura M, Matsuda Y, Yoshioka T, Okano Y (1999) Cell cycle-dependent expression and centrosome localization of a third human aurora/Ipl1-related protein kinase, AIK3. J Biol Chem 274:7334–7340

L'Allemain G (1994) Deciphering the MAP kinase pathway. Prog Growth Factor Res 5:291–334

Lengauer C, Wang Z (2004) From spindle checkpoint to cancer. Nat Genet 36:1144–1145

Lew DJ, Burke DJ (2003) The spindle assembly and spindle position checkpoints. Annu Rev Genet 37:251–282

Li R, Murray AW (1991) Feedback control of mitosis in budding yeast. Cell 66:519–531

Li X, Sakashita G, Matsuzaki H, Sugimoto K, Kimura K, Hanaoka F, Taniguchi H, Furukawa K, Urano T (2004) Direct association with inner centromere protein (INCENP) activates the novel chromosomal passenger protein, Aurora-C. J Biol Chem 279:47201–47211

Liu ST, Chan GK, Hittle JC, Fujii G, Lees E, Yen TJ (2003) Human MPS1 kinase is required for mitotic arrest induced by the loss of CENP-E from kinetochores. Mol Biol Cell 14:1638–1651

Mao Y, Abrieu A, Cleveland DW (2003) Activating and silencing the mitotic checkpoint through CENP-E-dependent activation/inactivation of BubR1. Cell 114:87–98

Meraldi P, Draviam VM, Sorger PK (2004a) Timing and checkpoints in the regulation of mitotic progression. Dev Cell 7:45–60

Meraldi P, Honda R, Nigg EA (2004b) Aurora kinases link chromosome segregation and cell division to cancer susceptibility. Curr Opin Genet Dev 14:29–36

Minoshima Y, Kawashima T, Hirose K, Tonozuka Y, Kawajiri A, Bao YC, Deng X, Tatsuka M, Narumiya S, May WS Jr, Nosaka T, Semba K, Inoue T, Satoh T, Inagaki M, Kitamura T (2003) Phosphorylation by aurora B converts MgcRacGAP to a RhoGAP during cytokinesis. Dev Cell 4:549–560

Murata-Hori M, Wang YL (2002) The kinase activity of aurora B is required for kinetochore-microtubule interactions during mitosis. Curr Biol 12:894–899

Nasmyth K (2002) Segregating sister genomes: the molecular biology of chromosome separation. Science 297:559–565

Peters JM (2002) The anaphase-promoting complex: proteolysis in mitosis and beyond. Mol Cell 9:931–943

Roberts BT, Farr KA, Hoyt MA (1994) The Saccharomyces cerevisiae checkpoint gene BUB1 encodes a novel protein kinase. Mol Cell Biol 14:8282–8291

Romano A, Guse A, Krascenicova I, Schnabel H, Schnabel R, Glotzer M (2003) CSC-1: a subunit of the Aurora B kinase complex that binds to the survivin-like protein BIR-1 and the incenp-like protein ICP-1. J Cell Biol 161:229–236

Salic A, Waters JC, Mitchison TJ (2004) Vertebrate shugoshin links sister centromere cohesion and kinetochore microtubule stability in mitosis. Cell 118:567–578

Sasai K, Katayama H, Stenoien DL, Fujii S, Honda R, Kimura M, Okano Y, Tatsuka M, Suzuki F, Nigg EA, Earnshaw WC, Brinkley WR, Sen S (2004) Aurora-C kinase is a novel chromosomal passenger protein that can complement Aurora-B kinase function in mitotic cells. Cell Motil Cytoskel 59:249–263

Sassoon I, Severin FF, Andrews PD, Taba MR, Kaplan KB, Ashford AJ, Stark MJ, Sorger PK, Hyman AA (1999) Regulation of Saccharomyces cerevisiae kinetochores by the type 1 phosphatase Glc7p. Genes Dev 13:545–555

Sato M, Sekido Y, Horio Y, Takahashi M, Saito H, Minna JD, Shimokata K, Hasegawa Y (2000) Infrequent mutation of the hBUB1 and hBUBR1 genes in human lung cancer. Jpn J Cancer Res 91:504–509

Schwab MS, Roberts BT, Gross SD, Tunquist BJ, Taieb FE, Lewellyn AL, Maller JL (2001) Bub1 is activated by the protein kinase p90^{Rsk} during Xenopus oocyte maturation. Curr Biol 11:141–150

Scrittori L, Skoufias DA, Hans F, Gerson V, Sassone-Corsi P, Dimitrov S, Margolis RL (2005) A small C-terminal sequence of Aurora B is responsible for localization and function. Mol Biol Cell 16:292–305

Sen S, Zhou H, Zhang RD, Yoon DS, Vakar-Lopez F, Ito S, Jiang F, Johnston D, Grossman HB, Ruifrok AC, Katz RL, Brinkley W, Czerniak B (2002) Amplification/overexpression of a mitotic kinase gene in human bladder cancer. J Natl Cancer Inst 94:1320–1329

Shah JV, Cleveland DW (2000) Waiting for anaphase: Mad2 and the spindle assembly checkpoint. Cell 103:997–1000

Shannon KB, Canman JC, Salmon ED (2002) Mad2 and BubR1 function in a single checkpoint pathway that responds to a loss of tension. Mol Biol Cell 13:3706–3719

Shapiro PS, Vaisberg E, Hunt AJ, Tolwinski NS, Whalen AM, McIntosh JR, Ahn NG (1998) Activation of the MKK/ERK pathway during somatic cell mitosis: direct interactions of active ERK with kinetochores and regulation of the mitotic 3F3/2 phosphoantigen. J Cell Biol 142:1533–1545

Sharp-Baker H, Chen RH (2001) Spindle checkpoint protein Bub1 is required for kinetochore localization of Mad1, Mad2, Bub3, and CENP-E, independently of its kinase activity. J Cell Biol 153:1239–1250

Shin HJ, Baek KH, Jeon AH, Park MT, Lee SJ, Kang CM, Lee HS, Yoo SH, Chung DH, Sung YC, McKeon F, Lee CW (2003) Dual roles of human BubR1, a mitotic checkpoint kinase, in the monitoring of chromosomal instability. Cancer Cell 4:483–497

Stucke VM, Sillje HH, Arnaud L, Nigg EA (2002) Human Mps1 kinase is required for the spindle assembly checkpoint but not for centrosome duplication. EMBO J 21:1723–1732

Sudakin V, Chan GK, Yen TJ (2001) Checkpoint inhibition of the APC/C in HeLa cells is mediated by a complex of BUBR1, BUB3, CDC20, and MAD2. J Cell Biol 154:925–936

Tanaka TU, Rachidi N, Janke C, Pereira G, Galova M, Schiebel E, Stark MJ, Nasmyth K (2002) Evidence that the Ipl1-Sli15 (Aurora kinase-INCENP) complex promotes chromosome bi-orientation by altering kinetochore-spindle pole connections. Cell 108:317–329

Tang Z, Bharadwaj R, Li B, Yu H (2001) Mad2-Independent inhibition of APC^{Cdc20} by the mitotic checkpoint protein BubR1. Dev Cell 1:227–237

Tang Z, Shu H, Oncel D, Chen S, Yu H (2004) Phosphorylation of Cdc20 by Bub1 provides a catalytic mechanism for APC/C inhibition by the spindle checkpoint. Mol Cell 16:387–397

Tatsuka M, Katayama H, Ota T, Tanaka T, Odashima S, Suzuki F, Terada Y (1998) Multinuclearity and increased ploidy caused by overexpression of the aurora- and Ipl1-like midbody-associated protein mitotic kinase in human cancer cells. Cancer Res 58:4811–4816

Taylor SS, Ha E, McKeon F (1998) The human homologue of Bub3 is required for kinetochore localization of Bub1 and a Mad3/Bub1-related protein kinase. J Cell Biol 142:1–11

Vigneron S, Prieto S, Bernis C, Labbe JC, Castro A, Lorca T (2004) Kinetochore localization of spindle checkpoint proteins: who controls whom? Mol Biol Cell 15:4584–4596

Warner SL, Bearss DJ, Han H, Von Hoff DD (2003) Targeting Aurora-2 kinase in cancer. Mol Cancer Ther 2:589–595

Wassmann K, Liberal V, Benezra R (2003) Mad2 phosphorylation regulates its association with Mad1 and the APC/C. EMBO J 22:797–806

Weiss E, Winey M (1996) The Saccharomyces cerevisiae spindle pole body duplication gene MPS1 is part of a mitotic checkpoint. J Cell Biol 132:111–123

Winey M, Goetsch L, Baum P, Byers B (1991) MPS1 and MPS2: novel yeast genes defining distinct steps of spindle pole body duplication. J Cell Biol 114:745–754

Yamaguchi S, Decottignies A, Nurse P (2003) Function of Cdc2p-dependent Bub1p phosphorylation and Bub1p kinase activity in the mitotic and meiotic spindle checkpoint. EMBO J 22:1075–1087

Yao X, Abrieu A, Zheng Y, Sullivan KF, Cleveland DW (2000) CENP-E forms a link between attachment of spindle microtubules to kinetochores and the mitotic checkpoint. Nat Cell Biol 2:484–491

Yu H (2002) Regulation of APC-Cdc20 by the spindle checkpoint. Curr Opin Cell Biol 14:706–714

Zecevic M, Catling AD, Eblen ST, Renzi L, Hittle JC, Yen TJ, Gorbsky GJ, Weber MJ (1998) Active MAP kinase in mitosis: localization at kinetochores and association with the motor protein CENP-E. J Cell Biol 142:1547–1558

The Centrosome Cycle

Christopher P. Mattison · Mark Winey (✉)

MCD Biology, University of Colorado-Boulder, CB347, Boulder, CO 80309-0347, USA
Mark.Winey@Colorado.edu

Abstract Centrosomes are dynamic organelles involved in many aspects of cell function and growth. Centrosomes act as microtubule organizing centers, and provide a site for concerted regulation of cell cycle progression. While there is diversity in microtubule organizing center structure among eukaryotes, many centrosome components, such as centrin, are conserved. Experimental analysis has provided an outline to describe centrosome duplication, and numerous centrosome components have been identified. Even so, more work is needed to provide a detailed understanding of the interactions between centrosome components and their roles in centrosome function and duplication. Precise duplication of centrosomes once during each cell cycle ensures proper mitotic spindle formation and chromosome segregation. Defects in centrosome duplication or function are linked to human diseases including cancer. Here we provide a multifaceted look at centrosomes with a detailed summary of the centrosome cycle.

1
Introduction

1.1
History

The centrosome is a unique organelle, first described and fully appreciated for its importance in the cell cycle by Theodor Boveri in early 1900 (for a review of Boveri's work, see Moritz and Sauer 1996). Working with roundworm and purple sea urchin embryos, he carefully documented the assembly of the mitotic spindle and thoughtfully described the centrosome in detail. He noted the presence of two centrioles engulfed by a dense sphere of material and also provided a detailed outline of the centrosome cycle. His experimentation indicated that centrosomes duplicate once per cell cycle and act as equivalent force generating anchors for the mitotic spindle. In his thorough observations, he also noted the occurrence of multipolar and monopolar spindles during the cell cycle as spontaneously occurring abnormalities. Notably, in the analysis of his own work, he insightfully conceptualized the idea that centrosome malfunction and the resulting chromosome inequality might lead to malignant tumors, as documented in his book *The Origin of Malignant Tumors* (1914) (for a review, see Manchester 1995). The past 15 years has seen a surge in interest and activity in centrosome study and an increase in understanding of this organelle. In this review, we consider the progress made in understanding centrosome complexity, function, link to disease, and duplication. As you will see, much of the recent progress made in understanding centrosome duplication and function serves to reinforce Boveri's conclusions and highlight the importance of his pioneering contributions.

The past 15 years has seen a surge in interest and activity in centrosome study and an increase in the understanding of this complex organelle. In this review, we first provide a brief look at centrosome related organelles and survey the growing list of centrosome functions. We also describe centrosome structure, complexity and link to disease. Next, we delve into the main body of this text and provide a comprehensive review of the components and regulators involved in centrosome duplication.

1.2
Microtubule Organizing Centers

The centrosome is one example of a broad class of structures called microtubule organizing centers (MTOCs). MTOCs from different organisms are morphologically distinct, but serve the same function to nucleate microtubules. These structures also contain many conserved components such as centrin and tubulins. There are centrosome related organelles such as basal bodies in ciliates, asters in plants, and spindle pole bodies (SPBs) in yeast. Each of these organelles is used to arrange microtubules into distinct functional arrays. Basal bodies are also found in specialized ciliated and flagellated cells in the kidney, lung and sperm in mammals. *Chlamydomonas* and sperm basal bodies are converted into centrosomes. This observation serves to reinforce the equivalent nature of these organelles. The yeast spindle pole body (SPB) serves as the centrosome, but differs in that it is a membrane bound organelle and is duplicated during G1 rather than S-phase as in mammalian cells. There are also differences among the centrioles of metazoans. For example, *D. melanogaster* centrioles contain doublet microtubule pairs while in *C. elegans* they are singlets. Studies of different systems and organelles have complemented one another in many instances and provided the building blocks for our current understanding of centrosome function and duplication.

1.3
Centrosome Functions

Centrosomes are unique organelles that function to organize cellular microtubules, and they are important for mitotic spindle positioning, cell division, and cell cycle progression (Heald et al. 1997; Hinchcliffe et al. 2001; Khodjakov and Rieder 2001; Piel et al. 2001). However, bipolar mitotic spindles can form in the absence of centrosomes (Khodjakov et al. 2000). More recently, centrosomes have been shown to serve as a signaling platform for many cell cycle decisions (Kramer et al. 2004; Doxsey 2005). Centrosomes also play a role in organization of interphase microtubules to maintain proper cell shape and influence nuclear translocation and cell migration (Schliwa et al. 1999; Abal et al. 2002; Malone et al. 2003).

1.4
Centrosome Dysfunction and Cancer/Disease

Proper duplication of centrosomes is essential for bipolar spindle formation and equal segregation of chromosomal DNA to daughter cells. Centrosome defects can lead to genetic imbalance, and centrosome abnormalities have been shown to be a marker for cancer [see reviews by Nigg (2002); Sluder and Nordberg (2004)]. For example, studies involving a panel of cells from both normal breast and breast tumor tissue have shown a direct relationship between centrosome abnormalities and chromosome instability/aneuploidy (Lingle et al. 2002). Similarly, a survey of prostate and other cancers has shown that centrosome defects can be found at the earliest detectable stages of cancer and increase with chromosome instability, and more importantly with tumor grade (Pihan et al. 2003). An interesting observation from these analyses is the lack of correlation between p53 mutation and centrosome defects or chromosome instability (Lingle et al. 2002; Pihan et al. 2003). However, p53 mutation is correlated with increased microtubule nucleation by centrosomes, and p53 status can dramatically affect centrosome number, both by affecting centrosome duplication and cytokinesis (Fukasawa et al. 1996, 1997; Lingle et al. 2002; Tarapore and Fukasawa 2002). In addition to p53, the retinoblastoma protein (RB) and breast cancer 1 (BRCA1) tumor suppressor proteins localize to centrosomes with cell cycle specific timing, suggesting an important link between centrosomes, proper cell cycle regulation, and genetic stability (Thomas et al. 1996; Hsu and White 1998). Thus there is a clear link between centrosome defects and cancer. It should be noted that additional mechanisms are probably involved in mediating compensation for prolonged cell division from a progenitor cell(s) derived from multipolar spindles and cancer progression [see reviews by Nigg (2002); Sluder and Nordberg (2004)]. Finally, there are at least eight ciliated cell types in humans, and diseases including Bardet-Biedl syndrome and polycystic kidney disease, and several neuronal disorders have been linked to mutations in basal body, centriolar, and pericentriolar material (PCM) proteins (Afzelius 2004; Snell et al. 2004; Bodano et al. 2005) providing additional links to centriole/basal body dysfunction and disease.

1.5
Centrosome Structure

Electron microscopy studies have provided informative details of centrosome structure and offered an outline of the centrosome cycle (Chretien et al. 1997; Kuriyama and Borisy 1981; Vorobjev and Nadezhdina 1987; see Figure 1). The centrosome of vertebrates contains two orthogonally spaced, loosely connected, cylindrically shaped centrioles at its core. The centrioles are composed of nine-triplet microtubule bundles symmetrically organized around a circular hub. It is estimated that centrioles are $\sim 1\,\mu M^3$ in size. At the base or proximal end of each centriole is a cartwheel structure, and this end anchors the minus ends of the centriole microtubules. On the mother centriole, this end also serves as the initiation platform for daughter centriole construction. The distal end contains the plus end of centriole microtubules and is the site for the assembly of distal appendages. Distal appendages are fin-like structures giving centrioles a rocket-like shape, but are not found in all centrioles. They function in microtubule nucleation, membrane attachment of primary cilia, and serve as a maturation marker. Surrounding the paired centrioles is the PCM, a dense cloud of structured matter (On et al. 2004). Interactions between PCM components and centrioles are important for proper centrosome function and duplication. The PCM is composed of proteins that anchor the gamma-tubulin ring complex (γ-TuRC) and other components involved in microtubule nucleation. Lastly, some PCM components are shared with centrioles and are also important for accurate centrosome duplication.

2
The Centrosome Cycle

2.1
Introduction

This chapter is focused on the centrosome cycle, the changes that occur within the centrosome throughout the cell cycle, and the molecules that regulate these processes. For additional material reviewing centrosomes, see Ou and Rattner (2004). The typical mammalian G1 cell contains a single centrosome with two centrioles, an older mother and younger daughter, at its core. As the cell progresses from G1 through S-phase, the centriole pair within the centrosome lose their orthogonal position, although centriole disorientation has been observed to occur as early as telophase. During S-phase, centriole duplication ensues with each centriole serving as the template for a daughter centriole. In general centriole duplication is initiated first in the mother centriole (White et al. 2000), although untemplated centriole duplication has

The Centrosome Cycle

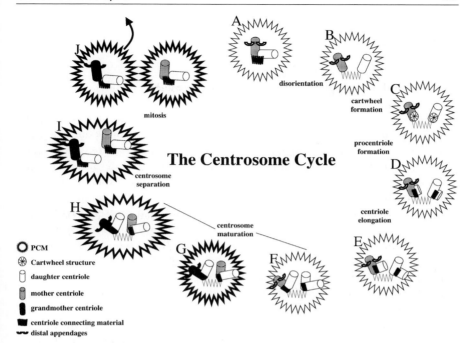

Fig. 1 The centrosome cycle. **a** The centrosome in an early G1 cell contains a mother and daughter centriole pair in an orthogonal orientation. **b** In late G1 or at the G1/S transition, the centriole pair lose their tight association and disorient. **c** In S-phase, cartwheel structures form at the proximal end of both centrioles. **d** Procentrioles form and **e** continue to elongate in S-phase. **f–h** Centrosome maturation begins in late S-phase and continues throughout G2. Maturation includes the recruitment of additional PCM components, increased microtubule nucleation, and the addition of distal appendages to the oldest centriole. **i** At the G2/M transition centrosomes separate, **j** move apart to form the mitotic spindle poles, and mitosis ensues. Following mitosis, centrosomes lose much of the additional PCM and return to a G1 state

been observed (Khodjakov et al. 2002). Centrioles continue to elongate, and it is not until after G2 that two complete centrosomes have formed. These centrosomes are not equivalent, as one contains a grandmother/daughter pair of centrioles and the other a mother/daughter pair. During G2 and the G2/M transition, proteins are added to the two parental centrioles within a centrosome in a maturation process that causes morphological and functional distinctions between the centriole pairs. In addition, the amount of PCM increases. Concomitantly, duplicated centrosomes separate and migrate to opposite sides of the nucleus in preparation for mitotic spindle assembly. After mitosis, centrosomes return to a G1 state in which they have a reduced/altered microtubule nucleation capacity. Our description will be based on the canonical mammalian centrosome cycle (Fig. 1), but where helpful we incorporate information from other systems to provide a more complete picture.

2.2
Centrosome Duplication

As mentioned, EM studies of centrosome duplication have provided an excellent guide for dissecting the centrosome duplication cycle into discrete steps (Kuriyama and Borisy 1981; Vorobjev and Nadezhdina 1987; Chretien et al. 1997) and are summarized in Hinchcliffe and Sluder (2001), Meraldi and Nigg (2002) and Delattre and Gonczy (2004). Experiments using labeled tubulin show that only the newly forming centriole incorporates tubulin during duplication of centrioles, indicating a conservative mechanism of duplication (Kochansky and Borisy 1990). Once centrosome separation occurs, centrioles are partitioned such that each cell receives either the grandmother/daughter or mother/daughter centriole pair (semi-conservative distribution), and therefore each cell has a unique centrosome assembly (Kochanski and Borisy 1991). Importantly, using cell fusion assays, it has been shown that a G1 cell duplicates its centrosome when fused to S or G2 phase cells, but G2 centrosomes cannot duplicate in G1/S cytoplasm (Wong and Stearns 2003). This indicates there are factors inherent to the centrosome that prevent reduplication during a normal cell cycle.

2.2.1
Cyclin-Dependent Kinase 2, Cdk2

Mammalian centrosome duplication occurs once each cell cycle and normally begins in late G1 with centriole disorientation. Centriole duplication proceeds through S-phase, and the duplicated centrosomes mature during G2 and M. Regulated expression of different cyclins in association with specific members of the conserved serine/threonine cyclin-dependent kinase (Cdk) family is important for the timing and progression of cell cycle events, including centrosome duplication (Hinchcliffe and Sluder 2002). Several studies have implicated cyclin-dependent kinase 2 (Cdk2) as an important regulator of initiation and progression of the centrosome cycle. Cells arrested in S-phase are permissive for extra rounds of centrosome duplication, and this observation has provided an assay to demonstrate the requirement of factors in centrosome duplication (Kuriyama et al. 1986; Balczon et al. 1995). Cdk2 activity has a role in DNA replication (reviewed in Woo and Poon 2003) and also participates in centrosome re-duplication in S-phase arrested cells (Matsumoto et al. 1999; Meraldi et al. 1999). Similarly, Cdk2/cyclin E activity, which drives the G1/S transition, contributes to centrosome duplication in *X. laevis* extracts (Hinchcliffe et al. 1999; Lacey et al. 1999; Matsumoto et al. 1999). Cdk2 also associates with cyclin A, and centrosomes can reduplicate in CHO cells under conditions thought to be specific for Cdk2/cyclin A activity (Meraldi et al. 1999) or in HeLa cells overexpressing cyclin A (Balczon 2001). While the role of Cdk2 is not completely understood, these and other studies clearly demonstrate a role for Cdk2 coupled to cyclin A or E in centrosome duplication (Tarapore et al. 2002).

Proper regulation of centrosome duplication requires the function of several transcription factors. For example, phosphorylation of the Rb protein and release of bound E2F transcription factors is required for centrosome duplication (Meraldi et al. 1999). Release of E2Fs presumably leads to the upregulation of cyclins, other cell cycle regulators, and centrosome structural components. In addition, the status of the p53 protein/transcription factor can affect centrosome number. Careful studies of p53-/- mouse embryonic fibroblasts (MEFs) have shown that centrosome abnormalities in the absence of p53 can arise from cytokinesis defects, but the major route for p53-dependent centrosome number defects arises from inappropriate initiation of centrosome duplication and/or prevention of reduplication (Fukasawa et al. 1996; Fukasawa et al. 1997; Tarapore and Fukasawa 2002). The p53 protein is known to control transcription of the Cdk inhibitor p21, and this indirectly influences centrosome duplication; however, p53 localizes to centrosomes and can also directly affect centrosome duplication (Tarapore et al. 2001a,b; Tarapore and Fukasawa 2002). Cdk2/cyclin E or A phosphorylates p53 at serine 315 (Wang and Prives 1995), and while mutations at this site do not affect p53 transcriptional activity (Crook et al. 1994; Fuchs et al. 1995), they can affect localization of p53 to centrosomes. The p53-S315A mutation prevents localization to unduplicated centrosomes, suggesting a direct link to regulation of centrosome duplication (Tarapore et al. 2001b; Tarapore and Fukasawa 2002).

2.2.2
Cdk2 Substrates

While there is clearly a role for Cdk2-cyclin A/E in centrosome duplication, the mechanism by which Cdk2 controls centrosome duplication is not completely understood. There are currently four known Cdk2 substrates relevant to centrosome duplication, including p53 (described above), Nucleophosmin, CP110, and Mps1. Nucleophosmin (NPM/B23/numatrin) is a nucleolar protein involved in ribosome biogenesis and also regulates the stability and transcriptional activity of p53 (Colombo et al. 2002). NPM localizes to unduplicated centrosomes and is phosphorylated on Thr199 by Cdk2/cyclin E late in G1 (Okuda et al. 2000; Tokuyama et al. 2001). This phosphorylation dissociates NPM from unduplicated centrosomes and is required to allow duplication to proceed (Okuda et al. 2000; Tokuyama et al. 2001; Tarapore et al. 2002). Conversely, there is evidence that phosphorylation of NPM on two threonine residues, amino acid 234 and 237, by Cdk1/cyclin B is important for its recruitment to centrosomes after nuclear envelope breakdown

(Cha et al. 2004). This may impart some change in microtubule dynamics at the centrosome for spindle assembly, or it may provide a regulatory circuit to reset each centrosome prior to duplication for the next cell cycle. These observations suggest that phosphorylation of NPM by different kinases is expression of a non-phosphorylatable important for multiple stages of centrosome duplication.

Another Cdk2 substrate is CP110, which was isolated in a screen for proteins that bound to a dominant negative Cdk2 allele. CP110 is a substrate for Cdk2/cyclin A or E and Cdk1/cyclin B (Chen et al. 2002). CP110 is an integral coiled-coil centrosome component whose protein level peaks during S-phase. Treatment of cells with CP110 RNA interference (RNAi) or CP110 allele leads to premature centrosome separation. In addition, CP110 RNAi treatment prevents centrosome reduplication in S-phase arrested cells (Chen et al. 2002). Taken together, these data suggest that CP110 is a structural component that may be important for centriole disorientation, or play a role in the timing of centrosome separation.

Lastly, the Mps1 protein kinase has been demonstrated to be a Cdk2 substrate whose stability during S-phase is regulated by Cdk2 phosphorylation (Fisk and Winey 2001). The mouse and human Mps1 proteins (mMps1 and hMps1) localize to centrosomes throughout the cell cycle and their overexpression can drive centrosome re-duplication (Fisk and Winey 2001; Fisk et al. 2003; Liu et al. 2003; Quintyne et al. 2005). Conversely, expression of kinase inactive Mps1 or treatment with MPS1 RNAi prevents the normal duplication of centrosomes during S-phase (Fisk and Winey 2001; Fisk et al. 2003). The threshold level of hMps1 activity required for centrosome duplication seems to be very low, and a severe decrease in hMps1 level is required to reveal its role in centrosome duplication (Fisk et al. 2003). Experiments that do not sufficiently reduce hMps1 activity have no effect on centrosome duplication (Stucke et al. 2002). In addition, hMps1 autophosphorylation complicates the analysis of immunofluorescence microscopy localization data, due to the varying efficacy with which numerous antibodies recognize Mps1 (Stucke et al. 2004). Consequently, while there are data implicating mammalian Mps1 in centrosome duplication, conclusions about its role must be tempered while waiting for continued analysis. In some other systems, it seems Mps1 is not involved in centrosome duplication. For example, no role for the *S. pombe* and *Drosophila* Mps1 proteins in SPB/centrosome duplication has been detected (He et al. 1998; Bettencourt-Dias et al. 2004; Fischer et al. 2004). In *S. cerevisiae* however, there are clearly multiple roles for Mps1 in SPB duplication (Jaspersen and Winey 2004). Further, analogous to mMps1, *S. cerevisiae* Mps1 is also a Cdc28 (Cdk homolog) substrate, and this interaction stabilizes the Mps1 protein (Fisk and Winey 2001; Jaspersen et al. 2004). However, even in *S. cerevisiae* the role of Mps1 in centrosome duplication is only partly understood, and there are only three identified Mps1 substrates relevant to centrosome duplication/function (reviewed in Jaspersen and Winey 2004).

The only potential centrosome related substrate described to date for hMps1 is the transforming acidic coiled-coil protein-2 (TACC2). TACC2 interacts with hMps1 in mitotic lysates, and its centrosome localization is disrupted by expression of kinase inactive hMps1 (Dou et al. 2004). While this interaction may be important for spindle stability, it is not clear what role, if any, it plays in centrosome duplication. It is important to note that Mps1 proteins also play a highly conserved role in the mitotic spindle checkpoint (Abrieu et al. 2001; Martin-Lluesma et al. 2002; Poss et al. 2002; Stucke et al. 2002, 2004; Liu et al. 2003; Fischer et al. 2004; Fisk and Winey 2004). Stucke et al. (2004) also observed that microtubules can increase hMps1 kinase activity *in vitro*, suggesting a possible regulatory mechanism *in vivo*.

While these studies strongly suggest an essential role for Cdk2 in centrosome duplication, it seems likely that there is redundancy among Cdk/cyclin complexes and that other Cdk/cyclin complexes can function to drive centrosome duplication. In support of this, Cdk2 is dispensable for normal mouse development (Berthet et al. 2003; Ortega et al. 2003). In addition, mice lacking both cyclin E1 and E2 have problems with DNA replication in some specialized cells required for complete gestation, but appear to have normal early embryonic development (Geng et al. 2003). Nonetheless, there is sufficient data to indicate that Cdk2 and its associated cyclins are important for the regulated duplication of centrosomes. Undoubtedly there are additional Cdk substrates important for centrosome duplication, and further analysis is required to identify them. Moreover, the interactions between Cdk2 and other kinases such as Mps1 must be better understood to provide a more complete picture of how centrosome duplication is regulated. Lastly, the differences in the ability of Cdk2 cyclin A or E associated activity to regulate centrosome duplication in the various model systems may provide clues to the regulation of the process or its coupling to the cell cycle.

2.2.3
Centriole Disorientation

In G1, a cell has a single centrosome containing a mother/daughter pair of centrioles. Each of these centrioles can provide a template site for the initiation of a new centriole. The first step in centrosome duplication is disorientation (also called splitting or loss of orthogonal positioning) of the centrioles just prior to initiation of daughter centriole formation (Fig. 1). Centriole disorientation can occur as early as anaphase/telophase in the previous cell cycle, but is most often thought to occur late in G1 or at the G1/S transition.

Several proteins have been implicated in disorientation, but it is not clear what specific cues are required. First, Cdk2 activity is important for centrosome duplication, and in human cells Cdk2 paired with cyclin A or E overexpression can induce disorientation of parental centrioles (Meraldi and

Nigg 2001). This may reflect an early role for Cdk2 prior to centriole duplication. Secondly, a highly conserved component of centrioles is the centrin protein (discussed below), and human centrin2 is phosphorylated by protein kinase A (PKA). Further, elevated PKA expression can cause interphase centrioles to separate, suggesting that this phosphorylation event may be an important cue for initiation of centriole disorientation (Lutz et al. 2001). Thirdly, a protein linkage is thought to connect paired centrioles within a centrosome, and this linkage may need to be modified or broken to allow centriole disorientation. Thus, it is likely that proteolysis may be important for this early step in centriole duplication. In support of this, embryos from *Drosophila* Fizzy mutants (a homolog of Cdc20 involved in proteolysis during the metaphase to anaphase transition) have delayed centriole disorientation (Vidwans et al. 1999). Further, mammalian components of the Skp1-Cullin-F-box (SCF) complex (important for entry into S-phase) localize to centrosomes, and inhibition of SCF components in *Xenopus* extracts blocks centriole disorientation (Freed et al. 1999). While these observations establish a role for proteolysis in centriole disorientation, the targets are not yet known.

In summary, centriole disorientation is the first identifiable step in the process of centriole duplication, and occurs at or prior to the G1 to S-phase transition. It is thought that proteolysis or other modification of proteins linking paired centrioles in their orthogonal orientation is required for this step. Further research will hopefully clarify this step and provide a more complete list of the proteins involved and their regulation.

2.2.4
Daughter Centriole Formation

We favor the idea that centriole assembly/elongation resembles the self-assembly of viral capsids. That is, once centriole duplication has been initiated with a template, it is propagated by the sequential addition of proteins independently of the mother centriole. An initial step thought to occur is the generation of a cartwheel structure (Fig. 1). The coiled-coil Bld10 protein has been localized to the cartwheel structure by EM, and it is the only identified cartwheel component (Matsuura et al. 2004). Bld10 was isolated as a flagella-less mutant from *Chlamydomonas rheinhardtii* and characterization of Bld10 mutants shows an absence of any basal body structures, indicating that it is required for the earliest steps of basal body assembly (Matsuura et al. 2004). Following cartwheel formation, procentrioles form and centriole duplication continues in S-phase. Not all the genes described in the following sections have been shown to function solely in centriole elongation. Some of them may function prior to elongation or at multiple steps in the pathway, but further investigation is required to determine this.

2.2.5
Centrin and Calcium Binding Proteins

Centrin is a highly conserved component of the centriole and is essential for centriole duplication. Centrins were first identified as components of the basal body in *Chlamydomonas* and are small (~ 20 kDa) calcium-binding proteins belonging to the EF-hand super-family of proteins (Huang et al. 1988; Salisbury 1995; Schiebel and Bornens 1995). There are at least four mammalian centrins, including three human (CETN1-3) and a fourth identified in mice (CETN4), that localize to centrosomes (Lee and Huang 1993; Errabolu et al. 1994; Middendorp et al. 1997; Gavet et al. 2003). Human CETN2 is concentrated in the distal lumen of centrioles and is essential for centrosome duplication, as treatment with Centrin2 RNAi prevents centriole duplication and leads to a progressive loss of centrosomes, abnormal spindle pole morphology, cytokinesis defects, and an increase in cell ploidy (Paoletti et al. 1996; Salisbury et al. 2002). Experiments with human CETN3 in *Xenopus* embryos suggest it may also be required for centriole duplication (Middendorp et al. 1997). Centrins in other systems are also involved in centriole duplication. A single *Chlamydomonas* centrin has been identified and shown to be essential for basal body formation (Koblenz et al. 2003). Similarly, fission and budding yeast CDC31/centrin is required for SPB duplication (Winey et al. 1991; Paoletti et al. 2003).

Another centrin-related calcium binding protein, calmodulin, is required for the G1-S transition in *X. laevis* embryo extracts (Matsumoto and Maller 2002). Interestingly, a calmodulin binding partner, the calcium-calmodulin dependent protein kinase II (CaMKII), is found at centrosomes (Ohta et al. 1990), and addition of CaMKII inhibitors prevents centrosome duplication in extracts (Matsumoto and Maller 2002). These results reinforce the idea that calcium and calcium binding proteins are essential to centrosome duplication.

2.2.6
Polo Kinases

Polo kinases (PLKs) were originally identified in *Drosophila* (polo) and yeast (Cdc5) and were shown to be required for mitosis (Sunkel and Glover 1988; Llamazares et al. 1991; Kitada et al. 1993; Lowery et al. 2005). Polo kinase family members contain a serine/threonine amino-terminal kinase domain and conserved ~ 30 amino acid long "Polo" boxes at the carboxy terminus (Dai 2005; Lowery et al. 2005). The Polo boxes are crucial to substrate recognition and proper localization of Polo kinases. Mammalian PLKs function in many steps of the cell cycle, including G1/S and G2/M transitions, mitotic exit, and DNA damage. There are four mammalian PLKs and human PLK2 (hPlk2) is important for centrosome duplication (Warnke et al. 2004). hPlk2 kinase activity increases during S-phase coincident with centriole duplication, and hPlk2 localizes to centrosomes *in vivo* and *in vitro*. In addition, overexpression of hPlk2 in Hydroxyurea (HU) arrested cells leads to centrosome reduplication. Conversely, treatment with hPLK2 RNAi or overexpression of kinase inactive hPLK2 prevents centriole duplication in HU arrested cells (Warnke et al., 2004). Similarly, treatment with hPLK1 RNAi in S-phase arrested cells prevents centrosome reduplication (Liu and Erikson 2002), suggesting that it also has a role. It is unclear how PLKs function to control centrosome duplication, and the relevant substrates have not been identified. hPlk1 phosphorylates NPM on serine 4, and mutation of this residue leads to pleiotropic effects, including centrosome abnormalities (Zhang et al. 2004). Thus, NPM phosphorylation by hPlk1 may be part of a mechanism that licenses centrosomes for duplication in the next cell cycle.

2.2.7
C. elegans Proteins

Several approaches using screens for embryonic lethality in *C. elegans* have uncovered at least five genes involved in centriole duplication. For example, the *C. elegans* zygote defective-1 (ZYG-1) gene was isolated in a screen for cell division mutants. ZYG-1 mutants have a centrosome duplication defect and do not progress past the two-cell stage (O'Connell et al. 1998). In maternal mutants daughter centrioles do not form, while paternal ZYG-1 mutants only form a single centriole during spermatogenesis, indicating dual paternal and maternal requirements (O'Connell et al. 2001). ZYG-1 is a unique kinase that localizes to centrosomes throughout the cell cycle (Dammermann et al. 2004), but the substrate(s) of ZYG-1 remain undiscovered.

The spindle assembly-4 (SAS-4) mutant was isolated in a screen using GFP-tubulin fusions and looking for mitotic spindle defects in early *C. elegans* embryos. SAS-4 is a coiled-coil protein that localizes to centriole walls throughout the cell cycle, and RNAi experiments indicate that it is required for centriole duplication/elongation (Kirkham et al. 2003; Leidel and Gonczy 2003). While there is no clear SAS-4 homolog, the human centrosomal protein 4.1-associated protein (CPAP) involved in microtubule stability may be a related molecule (Hung et al. 2000, 2004). Marked mating experiments and FRAP analysis of the SAS-4 protein indicate that like tubulin it is a stable component of newly forming centrioles (Kirkham et al. 2003; Leidel and Gonczy 2003). Partial reduction of SAS-4 not only leads to defective new centrioles, but those new centrioles that are formed contain fewer PCM components such as γ-tubulin. This asymmetric spindle pole phenotype suggests an additional role in microtubule nucleation (Kirkham et al. 2003). While originally thought to be unique to SAS-4 reduction, partial reduction of other centrosome components such as ZYG-1 also leads to defective centriole duplication and asymmetric spindle morphology (Dammermann et al. 2004; Leidel et al.

2005). Thus, there is a concentration dependent correlation between these components and proper centriole duplication.

The SAS-5 gene, encoding a coiled-coil protein, was isolated as a cell division mutant, and further investigation using a GFP fusion to the PCM component TAC-1 [a TACC homolog (Le Bot et al. 2003)] revealed a lack of centrosome duplication (Delattre et al. 2004). Treatment with SAS-5 RNAi in cells expressing GFP-SAS-4 or GFP-ZYG-1 indicated that centriole duplication was defective. Similar to ZYG-1, both paternal and maternal SAS-5 is required for normal embryonic centriole duplication. SAS-5 localizes primarily to both mother/daughter centrioles in a ZYG1-dependent manner but is not a stable centriole component as FRAP analysis and mixed mating experiments indicate it rapidly exchanges with a cytoplasmic pool. SAS-5 RNAi treatment prevents centriole duplication and partial SAS-5 reduction leads to a phenotype similar to that of SAS-4 reduction as new centrioles are only partially formed, suggesting that SAS-5 is also a dose-dependent regulator of centriole duplication. Finally, because the residual centriolar SAS-5 left after RNAi treatment cannot support centriole duplication, it may be that cytoplasmic shuttling of SAS-5 is required for centriole duplication (Delattre et al. 2004).

Both SAS-5 and SAS-6 were isolated using genome wide RNAi screens for cell division mutants in *C. elegans* (Dammermann et al. 2004; Leidel et al. 2005). Using treatment with SAS-6 RNAi and capitalizing on the stable incorporation of SAS-4 (fused to GFP) into newly forming centrioles, it has been demonstrated that SAS-6 is required for SAS-4 incorporation during centriole duplication (Dammermann et al. 2004; Leidel et al. 2005). Dammermann et al. (2004) also use SAS-4 incorporation into newly forming centrioles to confirm the role of ZYG-1, SAS-5, and SPD-2 (discussed below) in centriole duplication. FRAP analysis indicates that SAS-6 is recruited to centrioles at the onset of duplication and remains stably incorporated throughout the cell cycle (Leidel et al. 2005). ZYG-1 is required for SAS-6 centriole localization, and SAS-5 and SAS-6 physically associate and are co-dependent for centriole localization (Leidel et al. 2005). The 56 kDa SAS-6 protein is evolutionarily conserved containing a central coiled-coil domain and an amino-terminal present in SAS-6 (PISA) motif (Dammermann et al. 2004; Leidel et al. 2005). The PISA motif is conserved in similar human and *Drosophila* proteins, and the human homolog was previously identified in a proteomic analysis of human centrosomes (Andersen et al. 2003). The human SAS-6 homolog (HsSAS-6) also localizes to centrioles, and its overexpression in U2OS cells results in the formation of extra centrioles (Dammermann et al. 2004; Leidel et al. 2005). Further, treatment with HsSAS-6 RNAi prevents both centrosome reduplication in S-phase arrested cells and normal centrosome duplication, indicating a conserved role for SAS-6 in the centrosome duplication cycle (Leidel et al. 2005).

Some genes required for centriole duplication may have more than one role, such as the SAS-4 gene that affects centriole duplication and centrosome maturation. Another example of this is the *C. elegans* spindle deffective-2 (SPD-2) gene. SPD-2 was isolated in a screen for cell division mutants and was characterized as a gene required for anteroposterior axis polarization and sperm aster formation (O'Connell et al. 1998, 2000). Subsequently, it was shown to be required for both centriole duplication and centrosome maturation/PCM assembly (Kemp et al. 2004; Pelletier et al. 2004). The coiled-coil SPD-2 protein is conserved, sharing a 200 residue "SPD-2 domain" with human and *Drosophila* proteins, and, similar to SAS-6, was previously isolated from human centrosomes (Andersen et al. 2003). SPD-2 localizes to centrioles throughout the cell cycle, but its centriole localization can be disrupted from pre-existing centrioles, suggesting it is not a core centriole component (Kemp et al. 2004; Pelletier et al. 2004). SPD-2 may be complexed with another PCM component, SPD-5, as their PCM localization is co-dependent (Kemp et al. 2004; Pelletier et al. 2004). Loss of SPD-2 also disrupts PCM recruitment of AIR-1, PLK-1, ZYG-9, and γ-tubulin (O'Connell et al. 2000; Kemp et al. 2004), strongly suggesting that SPD-2 has a role in linking PCM components to centrioles. Finally, treatment with SPD-5 RNAi does not prevent centriole duplication, suggesting that SPD-2 has a specific centriole duplication function in addition to its role in PCM recruitment (Kemp et al. 2004; Pelletier et al. 2004). One way to explain the dual role of SPD-2 centers on its interaction with γ-tubulin. γ-tubulin is required for both centriole duplication and microtubule recruitment (discussed below), and it seems that SPD-2 may be needed for γ-tubulin recruitment to both structures (Kemp et al. 2004).

These *C. elegans* studies have identified several proteins important for centriole duplication. While the *C. elegans* genome lacks some important highly conserved centriolar molecules such as centrin, at least three of these molecules are conserved in other systems, including SAS-4, SAS-6 and SPD-2. While further study of their human counterparts is needed to determine the functional significance regarding centrosome duplication, analysis of HsSAS-6 indicates functional conservation as well. Therefore, although *C. elegans* centrioles are structurally different, the application of this new knowledge has benefited other systems.

2.2.8
Tubulins

Several members of the tubulin superfamily are components of centrioles and are important for centriole duplication and function. The α- and β-tubulins are essential centriole components and are incorporated into the centriole microtubule blades. α/β-tubulin dimers form the triplet microtubule blades in mammals that extend the length of the centriole. The longest 13 protofilament A-tube is the innermost, while the shorter 10 and 11 protofilament B- and C-tubes extend radially outward. Tubulin modifications such as glutamylation, a marker for centriole microtubules, are thought to stabilize these microtubule bundles (Bobinnec et al. 1998).

γ-Tubulin is considered an essential PCM component required for maturation and proper microtubule nucleation, but a role for γ-tubulin in centriole assembly has been demonstrated by studies in *C. elegans* and ciliates. For example, studies of *C. elegans* embryos depleted of γ-tubulin by RNAi treatment show severe defects in centriole duplication (Dammermann et al. 2004). This type of effect has also been seen in ciliates, where inhibition of γ-tubulin expression leads to defects in basal body duplication (Ruiz et al. 1999; Shang et al. 2002). In contrast, RNAi mediated depletion of γ-tubulin in cultured *Drosophila* cells did not prevent centriole duplication (Raynaud-Messina et al. 2004). Similar studies have not been performed in mammalian cells; however, the ciliate studies may indicate dual roles for γ-tubulin in microtubule recruitment and centriole duplication.

Other members of the tubulin family, δ-tubulin and ε-tubulin, have also been implicated in centriole duplication. In *Chlamydomonas*, mutation of the UNI3 (δ-tubulin) gene leads to defects in spindle positioning, cell division, and basal body centriole formation (Dutcher and Trabuco 1998). The defective centrioles contain doublet rather than triplet microtubules, suggesting a δ-tubulin requirement in centriole microtubule bundling or maintenance (Dutcher and Trabuco 1998). Human δ-tubulin was cloned based upon its homology to the *Chlamydomonas* δ-tubulin, and similar sequences have been found in mice and rats (Chang and Stearns 2000). Human δ-tubulin is also found associated with centrioles and is most intense in the space between centrioles of a centrosome (Chang and Stearns 2000), suggesting it may have a role in centriole cohesion in addition to those functions suggested by the analysis of *Chlamydomonas* δ-tubulin.

ε-Tubulin serves as a maturation marker to distinguish the mother centriole, as well as functioning in centriole duplication. Human ε-tubulin localizes to the mature centrosome, co-localizing with γ-tub in the PCM, prior to and during duplication, but during maturation is also recruited to the daughter centrosome (Chang and Stearns 2000). EM analysis shows that human ε-tubulin localizes specifically to the sub- distal appendages of the grandmother centriole in the older centrosome, and is recruited to the mother centriole of the new centrosome only after S-phase. *X. laevis* ε-tubulin depleted extracts fail to duplicate centrioles, although centrioles do separate, suggesting ε-tubulin functions at a later step, possibly in centriole elongation (Chang et al. 2003). Similarly, aster formation in *Xenopus* extracts also requires ε-tubulin, presumably to recruit or anchor microtubule-organizing components such as ninein, cenexin, and γ-tubulin. Additional support for ε-tubulin in centriole duplication comes from *Chlamydomonas*. The *Chlamydomonas* BLD2 gene encodes an ε-tubulin that localizes around the basal bodies in a diffuse manner. Further, mutation of this gene leads to short defective centrioles with singlet rather than triplet microtubules, suggesting that it is important for proper centriole duplication (Dutcher et al. 2002).

Centrioles are complex organelles, and their composition is not entirely known. Several proteins have been identified as centriole components, and while their significance is not well understood, it appears they contribute to centriole microtubule stability. The tektins are a family of at least three isoforms of ~ 50 kDa alpha helical proteins first discovered in sea urchin sperm flagella (Linck 1976). They are conserved as tektin antibodies have been used to detect 50 kDa tektin-like proteins in HeLa and CHO cell centrosomes (Steffen et al. 1994). In addition, the mouse Tekt1 protein localizes to centrioles in spermatids and may be important for basal body assembly (Larsson et al. 2000). Tektins are found as equimolar components in basal bodies and centrioles, form filamentous polymers, and remain as part of the cartwheel structure after tubulin depolymerization (Linck et al. 1985). They are thought to be important for the structural integrity or morphology of centriole microtubule bundles/junctions (Stephens et al. 1989; Stephens and Lemieux 1998).

Two proteins, Sp77 and Sp83, were isolated biochemically from sea urchin sperm flagella as part of an insoluble protofilament ribbon from axonemal microtubules along with tektins and α/β-tubulin (Hinchcliffe and Linck 1998). While both of these proteins localize to sea urchin basal bodies, immunofluorescence with polyclonal antibodies to Sp83 indicates that it, but not Sp77, localizes to CHO cell centrosomes. Further, Sp83 antibodies recognized two central dots in mitotic spindle poles and this signal is independent of microtubules, suggesting that Sp83 is a core component of centrioles. Their localization to centrioles and flagellar microtubules suggests that Sp83 and Sp77 may be important for stability of microtubule bundles (Hinchcliffe and Linck 1998).

EM analysis has clearly demonstrated that α/β-tubulin are a major centriole component. Other tubulin superfamily members, γ-tubulin, δ-tubulin, and ε-tubulin, are also centriole components. The tubulins and other proteins such as the tektins and Sp83 are organized into highly ordered structures, such as the microtubule blades, within the centriole. One major question that is not understood is how these and other proteins assemble properly and are organized to form centrioles. Future research identifying additional centriole components and characterizing the interactions between these proteins will hopefully answer this question.

2.3
Centrosome Maturation

Centrosome maturation occurs during late S/G2 and early mitosis and continues as cells enter mitosis. It is characterized by a significant increase in centrosome size and microtubule nucleation through the recruitment of additional PCM proteins (Palazzo et al. 2000). The PCM is a dynamic structure that exchanges components and changes in size, depending on cell cycle stage (Ou et al. 2004). Two common themes shared with many PCM components are 1) the ability to recruit, stabilize, and organize microtubules for interphase centrosomes and/or mitotic spindle poles, and 2) as markers for centrosome maturation. In some cases, they are also important for centrosome duplication.

2.3.1
Aurora-A and Polo kinases

At least two kinases, Aurora-A and Polo, are involved in regulating centrosome maturation. Aurora kinases are a conserved family of serine/threonine kinases that regulate mitotic progression, and humans have three aurora isoforms, A, B, and C. Aurora-A localizes to centrosomes and is involved in centrosome maturation, separation, and bipolar spindle assembly (for a more complete review of all Aurora-A mitotic functions, see Marumoto et al. 2005). Aurora-A also mediates centrosome maturation in *C. elegans* and *D. melanogaster*. For example, *Drosophila* Aurora-A mutants do not recruit centrosomin or γ-tub to centrosomes (Berdnik and Knoblich 2002), and Aurora-A RNAi treatment in *C. elegans* prevents increased microtubule nucleation, recruitment of γ-tubulin, and recruitment of other PCM components (Hannak et al. 2001). FRAP studies in *Drosophila* of Aurora-A centrosome localization indicate a transient association (Berdnik and Knoblich 2002), and localization in *Xenopus* is mediated mainly by the non-catalytic domain and is influenced by microtubules (Giet and Prigent 2001). Two-hybrid experiments indicate an interaction between human Aurora-A with TACC-1, the microtubule-associated protein minispindles (MSPS), and the XMAP-215 homolog colonic and hepatic tumor overexpressed gene (ch-TOG) (Conte et al. 2003). A hint as to the functional relevance of these interactions is provided from experiments in *Drosophila* and *C. elegans*. Maturation of centrosomes by Aurora-A in *Drosophila* seems to be mediated at least in part by interaction with the *Drosophila*, D-TACC. D-TACC is an Aurora-A substrate whose centrosome localization is dependent on Aurora-A, and D-TACC also forms a complex with MSPS and XMAP215 (Giet et al. 2002). Similarly, TAC-1 centrosome association in *C. elegans* is dependent on Aurora-A (AIR-1), and TAC-1 interacts with the ZYG-9 protein, an XMAP215 family member (Bellanger and Gonczy 2003). Members of the XMAP215 family of proteins regulate microtubules, and are known to function both in microtubule growth and destruction (Popov and Karsenti 2003). This may provide a mechanism for regulation of microtubule dynamics during centrosome maturation. Human Aurora-A is also required for the recruitment of cyclin B1 to centrosomes, an important step for mitotic progression that may also be part of centrosome maturation (Hirota et al. 2003).

In addition to its role in centrosome duplication, PLK1 is also important for centrosome separation and maturation. For example, cells injected with antibodies to human Plk1 have unseparated centrosomes that are severely reduced in size and lack γ-tubulin (Lane and Nigg 1996). Similarly, Polo mutants or treatment with Polo RNAi in *Drosophila* fail to recruit normal amounts of γ-tubulin at mitotic spindle poles (Donaldson et al. 2001; Bettencourt-Dias et al. 2004). There are at least six identified PLK1 substrates potentially important for microtubule regulation during centrosome maturation. The abnormal spindles (Asp) protein is a Polo substrate that is important for microtubule nucleation (Avides et al. 2001). Ninein-like protein (Nlp) was isolated as a two-hybrid interacting protein with *X. laevis* kinase-dead polo (PLX1) (Casenghi et al. 2003). Nlp is a coiled-coil protein with homology to ninein that localizes to centrosomes during interphase and is displaced during mitosis. Nlp is phosphorylated by PLX1, binds to γ-tubulin, and is important for microtubule nucleation. *In vivo* studies with a hyperactive allele of PLX1 indicate that PLX1 phosphorylation of Nlp leads to its dissociation from centrosomes and increased γ-tubulin signal. Conversely, expression of kinase inactive PLX1 prevents γ-tubulin enrichment at centrosomes, suggesting that inhibitory regulation of Nlp by PLX1 is required for proper centrosome maturation (Casenghi et al. 2003). CEP170 is another PLK1 substrate and is a marker for centrosome maturation (Guarguaglini et al. 2004). CEP170 is found at the sub-distal appendages of mature interphase centrosomes and also associates with spindle microtubules during mitosis (Guarguaglini et al. 2004). Polo also phosphorylates the translationally controlled tumor protein (TCTP), thought to stabilize microtubules (Yarm 2002). DMAP-85 functions to stabilize microtubules and is phosphorylated by Polo *in vitro*, suggesting that it may be regulated by Polo *in vivo* (Cambiazo et al. 2000). Finally, the Op18 protein destabilizes microtubules, and phosphorylation by Polo is thought to inhibit Op18 function (Budde et al. 2001).

2.3.2
γ-Tubulin Ring Complex

Polo kinase is also important for recruitment of γ-tubulin to the centrosome during maturation. γ-Tubulin is an important centrosome component essential for the increase in microtubule nucleation that occurs prior to mitosis (for a review, see Oakley and Akkari 1999; Gunawardane et al. 2000). For example, human γ-tubulin localizes to the centrosome, and its level increases during M phase (Khodjakov and Rieder 1999). γ-Tubulin is a highly conserved member of the tubulin superfamily and is a central component of the γ-TuRC (Oakley and Oakley 1989; Stearns et al. 1991; Zheng et al. 1991; Stearns and Kirschner 1994). EM of seven purified γ-TuRC proteins from *X. laevis* shows they form an open ring structure that can nucleate microtubules *in vitro* (Moritz et al. 1995; Zheng et al. 1995). The *S. cerevisiae* γ-TuRC is composed of only three proteins including Spc97, Spc98, and γ-tubulin, making it the smallest γ-TuRC (Knop and Schiebel 1997). There are at least six human γ-TuRC proteins, including γ-tubulin, GCP2 (Spc97), GCP3 (Spc98), GCP4, GCP5 and GCP6 (Murphy et al. 1998; Fava et al. 1999; Murphy et al. 2001). Recruitment of γ-TuRC to the centrosome is mediated by binding of PCM proteins to γ-TuRC components. In yeast, two proteins Spc110 and Spc72 attach the γ-TuRC to the spindle pole body (Knop et al. 1997; Knop and Schiebel 1998). In humans, the pericentrin protein, a homolog of Spc110, functions to anchor γ-TuRC to the centrosome (Doxsey et al. 1994; Li et al. 2001; Flory and Davis 2003; Zimmerman et al. 2004). Pericentrin is not the only protein thought to recruit the γ-TuRC to centrosomes. Another Spc110 homolog, the A-kinase anchoring protein-450 (AKAP450 or CG-NAP), shares the centrosome targeting PACT domain with pericentrin and binds γ-TuRC components at the centrosome (Takahashi et al. 1999, 2002; Gillingham and Munro 2000). The *Drosophila* CP309 protein shares homology with AKAP450 and pericentrin, binds γ-TuRC, and provides a similar function (Kawaguchi and Zheng 2004).

PCM components function to anchor the γ-TuRC and allow increased microtubule recruitment at centrosomes, which is essential for proper mitotic spindle formation. The interaction of stably associated PCM components as well as signaling molecules, such as Aurora-A kinase, is important for this process. At least one aspect of centrosomes that requires additional characterization is the interaction between centriole and PCM components. These interactions provide cues for centriole duplication as well as for microtubule recruitment and maturation. Continued identification and analysis of proteins at the centriole-PCM interface should provide important information into how these interactions govern centrosome function.

2.3.3
Maturation and Appendages

The PCM contains many proteins, some of which serve as maturation markers, and others that function in microtubule recruitment in addition to the γ-TuRC. The small GTPase Ran is one of these, and can promote microtubule nucleation at centrosomes (see Zheng 2004 for a review of Ran functions). There are a few proteins that have been characterized due to their specific association with mature centrosomes. For example, the 96 kDa cenexin protein was isolated by an immunofluorescence screening assay and is recruited only to the mother centriole at the G2/M transition (Lange and Gull 1995). Another marker for mature centrosomes is the presence of sub-distal appendages on the older centriole of a pair. Sub-distal appendages are thought to be a site for microtubule anchoring. The sub-distal appendages of mother centrioles contain several components including ε-tubulin (Chang et al. 2003). In addition, the adenomatous polyposis coli (APC) and end-binding protein 1 (EB1) proteins, important for microtubule organization, localize to the mother centriole (Louie et al. 2004). EB1 is found at the distal end of centrioles, co-localizing with ε-tubulin, and EB1 RNAi treatment depletion reduces microtubule recruitment (Louie et al. 2004). Thus, several structures and protein complexes are important for microtubule nucleation at centrosomes.

2.4
Centrosome Separation

2.4.1
NIMA-Related Kinase-2 (NEK2)

Once duplicated, the tether between grandmother and mother centrioles must be severed for the duplicated centriole pairs/centrosomes to separate prior to mitosis. The conserved NIMA-related kinase 2 (NEK2) is a homolog of the *Aspergillus* never in mitosis A (NIMA) protein involved in mitotic progression (Lu and Hunter 1995; Fry 2002; O'Connell et al. 2003). NEK2 is a serine/threonine kinase that can autophosphorylate, and whose activity peaks in S and G2 (Fry et al. 1995, 1998b, 1999). NEK2 is expressed as two alternative spliced variants, A and B, both of which localize to centrosomes throughout the cell cycle (Fry 2002). NEK2B, which lacks a carboxy terminal coiled-coil domain, is important for centrosome formation in *Xenopus* embryos (Fry et al. 2000; Uto and Sagata 2000; Twomey et al. 2004). Studies of human NEK2A indicate it has a role in centrosome separation. Cell lines expressing inducible human NEK2A show NEK2A at interphase centrosomes but not at mitotic spindle poles, supporting previous results (Faragher and Fry 2003; Fry et al. 1998b). In addition, NEK2A overexpression stimulates separation of interphase centrioles, and causes at least a partial loss of the centrosomal NEK2-associated protein 1 (C-Nap1) protein from centrioles (Faragher and Fry 2003). Conversely, overexpression of kinase inactive NEK2A prevents centrosome separation and leads to mono-polar spindles (Faragher and Fry 2003). Recent studies indicate the majority of NEK2A localizes to small cytoplasmic particles in addition to its centrosome localization (Hames et al. 2005). There is rapid exchange of cytoplasmic NEK2A with centrosome associated NEK2A however, the significance of this observation is not fully understood (Hames et al. 2005).

2.4.2
NEK2 Substrates

NEK2A has at least four substrates, NPM (described above), C-Nap1, also called CEP250 (Mack et al. 1998), the serine/threonine protein phosphatase 1 (PP1) and itself through autophosphorylation (Fry et al. 1998b, 1999). NEK2A has been shown to phosphorylate NPM and regulate its association with centrosomes during mitosis (Yao et al. 2004); however, the dynamics and functional significance of NPM phosphorylation by NEK2A is not completely understood. Alternatively, NEK2A regulation of C-Nap1 has been implicated in centriole cohesion. C-Nap1 is a large coiled-coil protein identified in a two-hybrid screen as a NEK2A interacting protein (Fry et al. 1998a). A role for C-Nap1 in centriole cohesion is supported by the observation that it localizes to the proximal end of both interphase centrioles, and expression of C-Nap1 fragments or interference of C-Nap1 by antibody injection leads to unregulated separation of parental centrioles, a phenotype identical to NEK2A overexpression (Fry et al. 1998a; Mayor et al. 2000, 2002). In addition, phosphorylation by NEK2A is potentially responsible for the partial reduction of C-Nap1 at separated centrosomes and more complete loss at mitotic spindle poles (Mayor et al. 2000, 2002; Faragher and Fry 2003). It should be noted that while microtubules have also been shown to play a role in centrosome cohesion, their role is not required for NEK2A mediated separation (Mayor et al. 2000, 2002; Meraldi and Nigg 2001).

PP1 regulates a number of cellular processes (Ceulemans and Bollen 2004) and interactions between NEK2A and PP1 are important for centriole cohesion. These proteins act antagonistically, as PP1 can dephosphorylate both C-Nap1 and NEK2A, reducing its kinase activity, while NEK2A phosphorylation reduces PP1 activity (Helps et al. 2000). Accordingly, PP1 inhibitors stimulate centrosome separation while PP1 overexpression can suppress the effects of NEK2A overexpression (Meraldi and Nigg 2001). Finally, evidence of an *in vivo* complex containing NEK2A, PP1, and C-Nap1 suggests a tight balance of enzymatic activity and serves to reinforce a model wherein this complex regulates centrosome separation (Helps et al. 2000; Meraldi and Nigg 2001; Fry 2002).

2.4.3
Motor Proteins

The microtubule network is involved in centrosome separation, and is required to direct the movement of motor proteins such as kinesin (discussed below) and dynein. Dynein, a multi-subunit protein complex, is a minus-end directed motor that is involved in many cellular processes, including centrosome separation. In support of this, injection of antibodies to human dynein prevents centrosome separation and leads to monopolar spindles (Vaisberg et al. 1993). In addition, mutation or RNAi treatment of *C. elegans* dynein subunits prevents centrosome separation (Gonczy et al. 1999; Yoder and Han 2001; Schmidt et al. 2005). Similarly, the dynein intermediate chain in *Dictyostelium* is required for proper centrosome separation (Ma et al. 1999). Dynein function is also required for proper positioning of centrosomes during interphase, and to coalesce aberrantly generated multiple centrosomes into a single centrosome cluster to prevent multipolar spindles from forming during mitosis (Koonce et al. 1999; Burakov et al. 2003; Quintyne et al. 2005). Finally, dynein function is also required to recruit PCM components, such as pericentrin and γ-tubulin, to the centrosome (Purohit et al. 1999; Young et al. 2000).

2.4.4
Aurora-A and Kinesin

A role for Aurora-A in centrosome separation via its interaction with the plus-ended motor protein kinesin has also been suggested. Aurora-A specific RNAi treatment or injection of anti-Aurora-A antibodies in HeLa cells leads to abnormal centrosome separation (Marumoto et al. 2003). Similarly, expression of kinase inactive Aurora-A (Eg2) in *X. laevis* extracts prevents bipolar spindle assembly (Roghi et al. 1998), and Eg2 phosphorylates the kinesin related protein Eg5 (Giet et al. 1999). Microinjection of human Eg5 (HsEg5) antibodies disrupts the microtubule network and prevents centrosome separation leading to mono-polar spindles in HeLa cells (Blangy et al. 1995; Whitehead and Rattner 1998). In addition, some *Drosophila* Aurora-A mutants fail to separate centrosomes, which results in mono-polar spindles (Glover et al. 1995). However, in *C. elegans* embryos treated with Aurora-A RNAi centrosomes do separate initially, but then collapse and fail to form a bipolar spindle, suggesting an additional role in spindle maintenance (Hannak et al. 2001). Further evidence for a role in spindle maintenance comes from studies using *Xenopus* egg extracts. Previously formed bipolar spindles are destabilized by the addition of kinase inactive Eg2 or inhibitory Eg2 antibodies (Giet and Prigent 2000). Thus, Aurora-A and its interaction with the kinesin Eg5, and potentially other motor proteins, contributes to the proper separation of centrosomes and the maintenance of spindle pole separation.

Several other kinases and phosphatases have also been implicated in regulating centrosome cohesion. For example, inhibition of the PCM localized Rho-dependent kinase p160ROCK leads to stable G1 centriole separation and movement of the mother centriole to the site of abscission (Chevrier et al. 2002). The latter observation may indicate a role for p160ROCK in centriole positioning during exit from mitosis. The human CDC14A (hCdc14A) phosphatase localizes to interphase centrosomes and may play a role in regulating centrosome separation (Kaiser et al. 2002; Mailand et al. 2002). Overexpression of active hCdc14A led to premature separation of duplicated centrosomes, while reduction by RNAi treatment had the opposite effect (Mailand et al. 2002). While hCdc14A has been demonstrated to act on some Cdk substrates *in vitro* (Kaiser et al. 2002), the specific substrate(s) targeted by hCdc14A with respect to its role in centrosome separation are not known. There is a *Xenopus* CDC14A homolog (Xcdc14β) that also localizes to centrosomes (Kaiser et al. 2004), although a connection to centrosome duplication has not been demonstrated for this molecule. Finally, PLK1 RNAi treatment indicates that PLK1 is required for centrosome separation and formation of bipolar spindle (van Vugt et al. 2004).

2.5
Licensing of Centrosome Duplication

To ensure genomic integrity there must be a mechanism that limits centrosome duplication to once per cell cycle. It is thought there is a "licensing" mechanism that prevents centrosomes from re-duplication until the next cell cycle. The Wong and Stearns (2003) cell fusion experiments indicated that G2 centrosomes cannot duplicate in a permissive S-phase environment, suggesting that there are centrosome intrinsic modifications that prevent reduplication. Cdk1/cyclin B is a key regulator of mitosis and is thought to be involved in licensing of centrosome duplication. B-type mitotic cyclins prevent the re-replication of chromosomal DNA, and the inactivation of B-type mitotic cyclins is essential for DNA replication in the next cell cycle in yeast (Dahmann et al. 1995; Noton and Diffley 2000). Similarly, yeast cells lacking the mitotic cyclin Clb1 reduplicate their centrosomes when arrested in S-phase, while those expressing Clb1 cannot reduplicate centrosomes (Haase et al. 2001). Human Cdk1 centrosome localization increases in early mitosis (Bailly et al. 1989; Pockwinse et al. 1997), and active Cdk1/cyclin B is first detected at centrosomes during prophase (Bailly et al. 1992; Jackman et al. 2003). These observations may suggest that modification of Cdk1/cyclin B substrate(s) or regulation of Cdk1/cyclin B complexes at the centrosome is important for licensing the next round of centrosome duplication.

One possible centrosome licensing mechanism involves Cdk1/cyclin B phosphorylation of NPM. NPM is known to reassociate with centrosomes during mitosis (Okuda et al. 2000), and there is evidence that phosphorylation of NPM on two threonine residues (234 and 237) by Cdk1/cyclin B is important for its recruitment to centrosomes after nuclear envelope breakdown (Cha et al. 2004). These observations suggest a model whereby centrosome re-association of NPM after phosphorylation by Cdk1/cyclin B may provide a regulatory circuit to prevent centrosome re-duplication prior to the next cell cycle. The presence of NPM at the centrosome would prevent centrosome duplication until Cdk2/cyclin E/A phosphorylation removes it from the centrosome in the following S-phase (Okuda 2002).

2.6
Post-Mitosis Return to G1

After mitosis centrosomes return to a G1 state, decreasing in size and microtubule nucleation capacity as cells prepare for duplication in the next cell cycle. How this reorganization and diminution of microtubule nucleation capacity back to G1/interphase levels occurs is not well understood. Using LLCPK1 cells expressing a photoactivatable GFP-α-tubulin, Rusan and colleagues (2005) followed microtubule behavior from metaphase to interphase and have shed some light on this matter (Rusan and Wadsworth 2005). They observed a large increase in release of microtubules from centrosomes during anaphase reaching a maximum in telophase. Dissociated microtubules and centrosome fragments move in a polar manner away from the chromosomal region toward the cell periphery. A constitutively active Cdk1/cyclin B suppressed centrosome fragmentation and microtubule release, suggesting an inhibitory role for Cdk1/cyclin B in regulating the deconstruction of mitotic centrosomes.

Lastly, an interesting aspect of centrioles is their migration after telophase. Using GFP labeled centrin as a centriole marker Piel et al. (2000) observed movement of the mother and daughter centrioles within the cell. The mother centriole migrated to the intercellular bridge during cell narrowing but moved back to the cell center before abscission (Piel et al. 2000). What role is served by centriole migration after mitosis is not known, but it may be important for cell abscission (Piel et al. 2001).

3
Conclusion

Centrosomes are complex organelles critical for microtubule organization and are composed of paired centrioles surrounded by an organized matrix of PCM proteins. While the tubular structure of centrioles has been well characterized by EM studies, it is not clear how this relates to their function. Nevertheless, their structure holds the blueprints for their duplication, as they are self-templating structures, and guides the organization of PCM. The PCM components that engulf centrioles work in concert with centrioles and are essential for the intricate microtubule patterns formed during cell growth. The network of interactions between centriole and PCM components must be further characterized to fully understand their role in centrosome duplication and the organization of microtubule networks guided by the centrosome.

The precise duplication of centrosomes is essential for proper cell growth and division. Centrosome duplication proceeds through distinct steps that coincide with different cell cycle stages (Fig. 1). Many signaling and structural proteins have been identified that are involved in the duplication of centrosomes, although the list is sure to grow. Several proteins critical to cell cycle regulation localize to centrosomes, and their localization is thought to be important for the coordinated relationship between cell cycle progression and the centrosome cycle.

Continued study of centrosomes is essential as centrosome/basal body abnormalities are connected to a growing number of diseases. For instance, basal body defects affect many types of ciliated cells leading to respiratory and kidney disease. Basal body defects can also lead problems with internal body symmetry and fertility. Finally, centrosome defects can lead to problems in migration of neuronal cells resulting in nervous disorders and chromosome segregation errors resulting in cancer. Recent advances have identified many new centrosome proteins, but continued study is needed in order to provide a complete map of the relationships between these proteins and how this relates to human disease.

Acknowledgements We thank Shelly Jones, Suzanne Naone, Alex Stemm-wolf, and Yvette Bren-Mattison for critical reading of this manuscript. This work was supported by National Institutes of Health Grant GM 51312 (to M.W.) and by a Fellowship award from the Department of Defense (Award number DAMD17-03-1-0404) (to C.P.M). The US Army Medical Research Acquisition Activity (820 Chandler Street, Fort Detrick, MD, 21702-5104) is the awarding and administering acquisition office for this award, and the content contained within this manuscript does not necessarily reflect the position or policy of the US Government and no official endorsement should be inferred.

References

Abal M, Piel M, Bouckson-Castaing V, Mogensen M, Sibarita JB, Bornens M (2002) Microtubule release from the centrosome in migrating cells. J Cell Biol 159:731–737

Abrieu A, Magnaghi-Jaulin L, Kahana JA, Peter M, Castro A, Vigneron S, Lorca T, Cleveland DW, Labbe JC (2001) Mps1 is a kinetochore-associated kinase essential for the vertebrate mitotic checkpoint. Cell 106:83–93

Afzelius BA (2004) Cilia-related diseases. J Pathol 204:470–477

Andersen JS, Wilkinson CJ, Mayor T, Mortensen P, Nigg EA, Mann M (2003) Proteomic characterization of the human centrosome by protein correlation profiling. Nature 426:570–574

Avides MdC, Tavares AAM, Glover DM (2001) Polo kinase and Asp are needed to promote the mitotic organizing activity of centrosomes. Nat Cell Biol 3:421–423

Bailly E, Doree M, Nurse P, Bornens M (1989) p34^{cdc2} is located in both nucleus and cytoplasm; part is centrosomally associated at G2/M and enters vesicles at anaphase. EMBO J 8:3985–3995

Bailly E, Pines J, Hunter T, Bornens M (1992) Cytoplasmic accumulation of cyclin B1 in human cells: association with a detergent-resistant compartment and with the centrosome. J Cell Sci 101:529–545

Balczon R, Bao L, Zimmer W, Brown K, Zinkowski R, Brinkley B (1995) Dissociation of centrosome replication events from cycles of DNA synthesis and mitotic division in hydroxyurea-arrested chinese hamster ovary cells. J Cell Biol. 130:105–115

Balczon RC (2001) Overexpression of cyclin A in human HeLa cells induces detachment of kinetochores and spindle pole/centrosome overproduction. Chromosoma 110:381–392

Bellanger JM, Gonczy P (2003) TAC-1 and ZYG-9 form a complex that promotes microtubule assembly in C. elegans embryos. Curr Biol 13:1488–1498

Berdnik D, Knoblich JA (2002) *Drosophila* Aurora-A is required for centrosome maturation and actin-dependent asymmetric protein localization during mitosis. Curr Biol 12:640–647

Berthet C, Aleem E, Coppola V, Tessarollo L, Kaldis P (2003) Cdk2 knockout mice are viable. Curr Biol 13:1775–1785

Bettencourt-Dias M, Giet R, Sinka R, Mazumdar A, Lock WG, Balloux F, Zafiropoulos PJ, Yamaguchi S, Winter S, Carthew RW, Cooper M, Jones D, Frenz L, Glover DM (2004) Genome-wide survey of protein kinases required for cell cycle progression. Nature 432:980–987

Blangy A, Lane HA, d'Herin P, Harper M, Kress M, Nigg EA (1995) Phosphorylation by p34^{cdc2} regulates spindle association of human Eg5, a kinesin-related motor essential for bipolar spindle formation *in vivo*. Cell 83:1159–1169

Bobinnec Y, Khodjakov A, Mir LM, Rieder CL, Edde B, Bornens M (1998) Centriole disassembly *in vivo* and its effect on centrosome structure and function in vertebrate cells. J Cell Biol 143:1575–1589

Bodano J, Teslovish T, Katsanis N (2005) The centrosome in human genetic disease. Nat Rev Gen 6:194–205

Budde PP, Kumagai A, Dunphy WG, Heald R (2001) Regulation of Op18 during spindle assembly in *Xenopus* egg extracts. J Cell Biol 153:149–158

Burakov A, Nadezhdina E, Slepchenko B, Rodionov V (2003) Centrosome positioning in interphase cells. J Cell Biol 162:963–969

Cambiazo V, Logarinho E, Pottstock H, Sunkel CE (2000) Microtubule binding of the *drosophila* DMAP-85 protein is regulated by phosphorylation *in vitro*. FEBS Lett 483:37–42

Casenghi M, Meraldi P, Weinhart U, Duncan PI, Korner R, Nigg EA (2003) Polo-like kinase 1 regulates Nlp, a centrosome protein involved in microtubule nucleation. Dev Cell 5:113–125

Ceulemans H, Bollen M (2004) Functional diversity of protein phosphatase-1, a cellular economizer and reset button. Physiol Rev 84:1–39

Cha H, Hancock C, Dangi S, Maiguel D, Carrier F, Shapiro P (2004) Phosphorylation regulates nucleophosmin targeting to the centrosome during mitosis as detected by cross-reactive phosphorylation-specific MKK1/MKK2 antibodies. Biochem J 378:857–865

Chang P, Stearns T (2000) Delta-tubulin and epsilon-tubulin: two new human centrosomal tubulins reveal new aspects of centrosome structure and function. Nat Cell Biol 2:30–35

Chang P, Giddings TH Jr, Winey M, Stearns T (2003) Epsilon-tubulin is required for centriole duplication and microtubule organization. Nat Cell Biol 5:71–76

Chen Z, Indjeian VB, McManus M, Wang L, Dynlacht BD (2002) CP110, a cell cycle-dependent CDK substrate, regulates centrosome duplication in human cells. Dev Cell 3:339–350

Chevrier V, Piel M, Collomb N, Saoudi Y, Frank R, Paintrand M, Narumiya S, Bornens M, Job D (2002) The Rho-associated protein kinase p160ROCK is required for centrosome positioning. J Cell Biol 157:807–817

Chretien D, Buendia B, Fuller SD, Karsenti E (1997) Reconstruction of the centrosome cycle from cryoelectron micrographs. J Struct Biol 120:117–133

Colombo E, Marine JC, Danovi D, Falini B, Pelicci PG (2002) Nucleophosmin regulates the stability and transcriptional activity of p53. Nat Cell Biol 4:529–533

Conte N, Delaval B, Ginestier C, Ferrand A, Isnardon D, Larroque C, Prigent C, Seraphin B, Jacquemier J, Birnbaum D (2003) TACC1-chTOG-Aurora A protein complex in breast cancer. Oncogene 22:8102–8116

Crook T, Marston NJ, Sara EA, Vousden KH (1994) Transcriptional activation by p53 correlates with suppression of growth but not transformation. Cell 79:817–827

Dahmann C, Diffley JFX, Futcher B (1995) S-phase-promoting kinases prevent re-replication by inhibiting the transition of replication origins to a pre-replicative state. Curr Biol 5:1257–1269

Dai W (2005) Polo-like kinases, an introduction. Oncogene 24:214–216

Dammermann A, Muller-Reichert T, Pelletier L, Habermann B, Desai A, Oegema K (2004) Centriole assembly requires both centriolar and pericentriolar material proteins. Dev Cell 7:815–829

Delattre M, Gonczy P (2004) The arithmetic of centrosome biogenesis. J Cell Sci 117:1619–1630

Delattre M, Leidel S, Wani K, Baumer K, Bamat J, Schnabel H, Feichtinger R, Schnabel R, Gonczy P (2004) Centriolar SAS-5 is required for centrosome duplication in *C. elegans*. Nat Cell Biol 6:656–664

Donaldson MM, Tavares AAM, Ohkura H, Deak P, Glover DM (2001) Metaphase arrest with centromere separation in polo mutants of *Drosophila*. J Cell Biol 153:663–675

Dou Z, Ding X, Zereshki A, Zhang Y, Zhang J, Wang F, Sun J, Huang H, Yao X (2004) TTK kinase is essential for the centrosomal localization of TACC2. FEBS Lett 572:51–56

Doxsey S, Zimmermann W, Mikule K (2005) Centrosome control of the cell cycle. Trends in Cell Biology (in press)

Doxsey S, Stein P, Evans L, Calarco P, Kirschner M (1994) Pericentrin, a highly conserved centrosome protein involved in microtubule organization. Cell 76:639–650

Dutcher SK, Trabuco EC (1998) The UNI3 gene is required for assembly of basal bodies of *Chlamydomonas* and encodes delta-tubulin, a new member of the tubulin superfamily. Mol Biol Cell 9:1293–1308

Dutcher SK, Morrissette NS, Preble AM, Rackley C, Stanga J (2002) Epsilon-tubulin is an essential component of the centriole. Mol Biol Cell 13:3859–3869

Errabolu R, Sanders MA, Salisbury JL (1994) Cloning of a cDNA encoding human centrin, an EF-hand protein of centrosomes and mitotic spindle poles. J Cell Sci 107:9–16

Faragher AJ, Fry AM (2003) Nek2A kinase stimulates centrosome disjunction and is required for formation of bipolar mitotic spindles. Mol Biol Cell 14:2876–2889

Fava F, Raynaud-Messina B, Leung-Tack J, Mazzolini L, Li M, Guillemot JC, Cachot D, Tollon Y, Ferrara P, Wright M (1999) Human 76p: a new member of the gamma-tubulin-associated protein family. J Cell Biol 147:857–868

Fischer MG, Heeger S, Hacker U, Lehner CF (2004) The mitotic arrest in response to hypoxia and of polar bodies during early embryogenesis requires *Drosophila* Mps1. Curr Biol 14:2019–2024

Fisk HA, Winey M (2001) The mouse Mps1p-like kinase regulates centrosome duplication. Cell 106:95–104

Fisk HA, Winey M (2004) Spindle regulation: Mps1 flies into new areas. Curr Biol 14:1058–1060

Fisk HA, Mattison CP, Winey M (2003) Human Mps1 protein kinase is required for centrosome duplication and normal mitotic progression. Proc Natl Acad Sci USA 100:14875–14880

Flory MR, Davis TN (2003) The centrosomal proteins pericentrin and kendrin are encoded by alternatively spliced products of one gene. Genomics 82:401–405

Freed E, Lacey KR, Huie P, Lyapina SA, Deshaies RJ, Stearns T, Jackson PK (1999) Components of an SCF ubiquitin ligase localize to the centrosome and regulate the centrosome duplication cycle. Genes Dev 13:2242–2257

Fry AM (2002) The Nek2 protein kinase: a novel regulator of centrosome structure. Oncogene 21:6184–6194

Fry AM, Schultz SJ, Bartek J, Nigg EA (1995) Substrate specificity and cell cycle regulation of the Nek2 protein kinase, a potential human homolog of the mitotic regulator NIMA of *Aspergillus nidulans*. J Biol Chem 270:12899–12905

Fry AM, Mayor T, Meraldi P, Stierhof YD, Tanaka K, Nigg EA (1998a) C-Nap1, a novel centrosomal coiled-coil protein and candidate substrate of the cell cycle-regulated protein kinase Nek2. J Cell Biol 141:1563–1574

Fry AM, Meraldi P, Nigg EA (1998b) A centrosomal function for the human Nek2 protein kinase, a member of the NIMA family of cell cycle regulators. EMBO J 17:470–481

Fry AM, Arnaud L, Nigg EA (1999) Activity of the human centrosomal kinase, Nek2, depends on an unusual leucine zipper dimerization motif. J Biol Chem 274:16304–163110

Fry AM, Descombes P, Twomey C, Bacchieri R, Nigg EA (2000) The NIMA-related kinase X-Nek2B is required for efficient assembly of the zygotic centrosome in *Xenopus laevis*. J Cell Sci 113:1973–1984

Fuchs B, Hecker D, Scheidtmann KH (1995) Phosphorylation studies on rat p53 using the baculovirus expression system. Manipulation of the phosphorylation state with okadaic acid and influence on DNA binding. Eur J Biochem 228:625–639

Fukasawa K, Choi T, Kuriyama R, Rulong S, Vande Woude GF (1996) Abnormal centrosome amplification in the absence of p53. Science 271:1744–1747

Fukasawa K, Wiener F, Vande Woude GF, Mai S (1997) Genomic instability and apoptosis are frequent in p53 deficient young mice. Oncogene 15:1295–1302

Gavet O, Alvarez C, Gaspar P, Bornens M (2003) Centrin4p, a novel mammalian centrin specifically expressed in ciliated cells. Mol Biol Cell 14:1818–1834

Geng Y, Yu Q, Sicinska E, Das M, Schneider JE, Bhattacharya S, Rideout WM, Bronson RT, Gardner H, Sicinski P (2003) Cyclin E ablation in the mouse. Cell 114:431–443

Giet R, Prigent C (2000) The *Xenopus laevis* aurora/Ip11p-related kinase pEg2 participates in the stability of the bipolar mitotic spindle. Exp Cell Res 258:145–151

Giet R, Prigent C (2001) The non-catalytic domain of the *Xenopus laevis* Aurora-A kinase localises the protein to the centrosome. J Cell Sci 114:2095–2104

Giet R, Uzbekov R, Cubizolles F, Le Guellec K, Prigent C (1999) The *Xenopus laevis* aurora-related protein kinase pEg2 associates with and phosphorylates the kinesin-related protein XlEg5. J Biol Chem 274:15005–15013

Giet R, McLean D, Descamps S, Lee MJ, Raff JW, Prigent C, Glover DM (2002) *Drosophila* Aurora A kinase is required to localize D-TACC to centrosomes and to regulate astral microtubules. J Cell Biol 156:437–451

Gillingham AK, Munro S (2000) The PACT domain, a conserved centrosomal targeting motif in the coiled-coil proteins AKAP450 and pericentrin. EMBO Rep 1:524–529

Glover D, Leibowitz M, McLean D, Parry H (1995) Mutations in aurora prevent centrosome separation leading to the formation of monopolar spindles. Cell 81:95–105

Gonczy P, Pichler S, Kirkham M, Hyman AA (1999) Cytoplasmic dynein is required for distinct aspects of MTOC positioning, including centrosome separation, in the one cell stage *Caenorhabditis elegans* embryo. J Cell Biol 147:135–150

Guarguaglini G, Duncan PI, Stierhof YD, Holmstrom T, Duensing S, Nigg EA (2004) The FHA-domain protein Cep170 interacts with Plk1 and serves as a marker for mature centrioles. Mol Biol Cell 16:1095–1107

Gunawardane RN, Lizarraga SB, Wiese C, Wilde A, Zheng Y (2000) gamma-Tubulin complexes and their role in microtubule nucleation. Curr Top Dev Biol 49:55–73

Haase SB, Winey M, Reed SI (2001) Multi-step control of spindle pole body duplication by cyclin-dependent kinase. Nat Cell Biol 3:38–42

Hames RS, Crookes RE, Straatman KR, Merdes A, Hayes MJ, Faragher AJ, Fry AM (2005) Dynamic recruitment of Nek2 kinase to the centrosome involves microtubules, PCM-1 and localized proteasomal degradation. Mol Biol Cell 16:1711–1724

Hannak E, Kirkham M, Hyman AA, Oegema K (2001) Aurora-A kinase is required for centrosome maturation in *Caenorhabditis elegans*. J Cell Biol 155:1109–1116

He X, Jones MH, Winey M, Sazer S (1998) Mph1, a member of the Mps1-like family of dual specificity protein kinases, is required for the spindle checkpoint in *S. pombe*. J Cell Sci 111:1635–1647

Heald R, Tournebize R, Habermann A, Karsenti E, Hyman A (1997) Spindle assembly in *Xenopus* egg extracts: respective roles of centrosomes and microtubule self-organization. J Cell Biol 138:615–628

Helps NR, Luo X, Barker HM, Cohen PT (2000) NIMA-related kinase 2 (Nek2), a cell-cycle-regulated protein kinase localized to centrosomes, is complexed to protein phosphatase 1. Biochem J 349:509–518

Hinchcliffe EH, Li C, Thompson EA, Maller JL, Sluder G (1999) Requirement of Cdk2-cyclin E activity for repeated centrosome reproduction in *Xenopus* egg extracts. Science 283:851–854

Hinchcliffe EH, Linck RW (1998) Two proteins isolated from sea urchin sperm flagella: structural components common to the stable microtubules of axonemes and centrioles. J Cell Sci 111:585–595

Hinchcliffe EH, Sluder G (2001) "It takes two to tango": understanding how centrosome duplication is regulated throughout the cell cycle. Genes Dev 15:1167–1181

Hinchcliffe EH, Sluder G (2002) Two for two: Cdk2 and its role in centrosome doubling. Oncogene 21:6154–6160

Hinchcliffe EH, Miller FJ, Cham M, Khodjakov A, Sluder G (2001) Requirement of a centrosomal activity for cell cycle progression through G1 into S-phase. Science 291:1547–1550

Hirota T, Kunitoku N, Sasayama T, Marumoto T, Zhang D, Nitta M, Hatakeyama K, Saya H (2003) Aurora-A and an interacting activator, the LIM protein Ajuba, are required for mitotic commitment in human cells. Cell 114:585–598

Hsu LC, White RL (1998) BRCA1 is associated with the centrosome during mitosis. Proc Natl Acad Sci USA 95:12983–12988

Huang B, Mengersen A, Lee V (1988) Molecular cloning of cDNA for caltractin, a basal body-associated Ca^{2+}-binding protein: homology in its protein sequence with calmodulin and the yeast CDC31 gene product. J Cell Biol 107:133–140

Hung LY, Tang CJ, Tang TK (2000) Protein 4.1 R-135 interacts with a novel centrosomal protein (CPAP) which is associated with the gamma-tubulin complex. Mol Cell Biol 20:7813–7825

Hung LY, Chen HL, Chang CW, Li BR, Tang TK (2004) Identification of a novel microtubule-destabilizing motif in CPAP that binds to tubulin heterodimers and inhibits microtubule assembly. Mol Biol Cell 15:2697–2706

Jackman M, Lindon C, Nigg EA, Pines J (2003) Active cyclin B1-Cdk1 first appears on centrosomes in prophase. Nat Cell Biol 5:143–148

Jaspersen SL, Winey M (2004) The budding yeast spindle pole body: structure, duplication, and function. Annu Rev Cell Dev Biol 20:1–28

Jaspersen SL, Huneycutt BJ, Giddings TH, Jr, Resing KA, Ahn NG, Winey M (2004) Cdc28/Cdk1 regulates spindle pole body duplication through phosphorylation of Spc42 and Mps1. Dev Cell 7:263–274

Kaiser BK, Zimmerman ZA, Charbonneau H, Jackson PK (2002) Disruption of centrosome structure, chromosome segregation, and cytokinesis by misexpression of human Cdc14A phosphatase. Mol Biol Cell 13:2289–2300

Kaiser BK, Nachury MV, Gardner BE, Jackson PK (2004) *Xenopus* Cdc14 alpha/beta are localized to the nucleolus and centrosome and are required for embryonic cell division. BMC Cell Biol 5:27

Kawaguchi S, Zheng Y (2004) Characterization of a *Drosophila* centrosome protein CP309 that shares homology with Kendrin and CG-NAP. Mol Biol Cell 15:37–45

Kemp CA, Kopish KR, Zipperlen P, Ahringer J, O'Connell KF (2004) Centrosome maturation and duplication in *C. elegans* require the coiled-coil protein SPD-2. Dev Cell 6:511–523

Khodjakov A, Rieder CL (1999) The sudden recruitment of gamma-tubulin to the centrosome at the onset of mitosis and its dynamic exchange throughout the cell cycle, do not require microtubules. J Cell Biol 146:585–596

Khodjakov A, Rieder CL (2001) Centrosomes enhance the fidelity of cytokinesis in vertebrates and are required for cell cycle progression. J Cell Biol 153:237–242

Khodjakov A, Cole RW, Oakley BR, Rieder CL (2000) Centrosome-independent mitotic spindle formation in vertebrates. Curr Biol 10:59–67

Khodjakov A, Rieder CL, Sluder G, Cassels G, Sibon O, Wang CL (2002) De novo formation of centrosomes in vertebrate cells arrested during S-phase. J Cell Biol 158:1171–1181

Kirkham M, Muller-Reichert T, Oegema K, Grill S, Hyman AA (2003) SAS-4 is a *C. elegans* centriolar protein that controls centrosome size. Cell 112:575–587

Kitada K, Johnson AL, Johnston LH, Sugino A (1993) A multicopy suppressor gene of the *Saccharomyces cerevisiae* G1 cell cycle mutant gene dbf4 encodes a protein kinase and is identified as CDC5. Mol Cell Biol 13:4445–4457

Knop M, Schiebel E (1997) Spc98p and Spc97p of the yeast γ-tubulin complex mediate binding to the spindle pole body via their interaction with Spc110p. EMBO J 18:6985–6995

Knop M, Schiebel E (1998) Receptors determine the cellular localization of a gamma-tubulin complex and thereby the site of microtubule formation. EMBO J 17:3952–3967

Knop M, Pereira G, Geissler S, Grein K, Schiebel E (1997) The spindle pole body component Spc97p interacts with the gamma-tubulin of *Saccharomyces cerevisiae* and functions in microtubule organization and spindle pole body duplication. EMBO J 16:1550–1564

Koblenz B, Schoppmeier J, Grunow A, Lechtreck KF (2003) Centrin deficiency in *Chlamydomonas* causes defects in basal body replication, segregation and maturation. J Cell Sci 116:2635–2646

Kochanski R, Borisy G (1991) Mode of centriole duplication and distribution. J Cell Biol 110:1599–1605

Koonce MP, Kohler J, Neujahr R, Schwartz JM, Tikhonenko I, Gerisch G (1999) Dynein motor regulation stabilizes interphase microtubule arrays and determines centrosome position. EMBO J 18:6786–6792

Kramer A, Lukas J, Bartek J (2004) Checking out the centrosome. Cell Cycle 3:1390–1393

Kuriyama R, Borisy G (1981) Microtubule-nucleating activity of centrosomes in Chinese hamster ovary cells is independent of the centriole cycle but coupled to the mitotic cycle. J Cell Biol 91:822–826

Kuriyama R, Dasgupta S, Borisy GG (1986) Independence of centriole formation and initiation of DNA synthesis in Chinese hamster ovary cells. Cell Motil Cytoskeleton 6:355–362

Lacey K, Jackson P, Stearns T (1999) Cyclin-dependent kinase control of centrosome duplication. Proc Natl Acad Sci USA 96:2817–2822

Lane HA, Nigg EA (1996) Antibody microinjection reveals an essential role for human polo-like kinase 1 (Plk1) in the functional maturation of mitotic centrosomes. J Cell Biol 135:1701–1713

Lange B, Gull K (1995) A molecular marker for centriole maturation in the mammalian cell cycle. J Cell Biol 130:919–927

Larsson M, Norrander J, Graslund S, Brundell E, Linck R, Stahl S, Hoog C (2000) The spatial and temporal expression of Tekt1, a mouse tektin C homologue, during spermatogenesis suggest that it is involved in the development of the sperm tail basal body and axoneme. Eur J Cell Biol 79:718–725

Le Bot N, Tsai MC, Andrews RK, Ahringer J (2003) TAC-1, a regulator of microtubule length in the *C. elegans* embryo. Curr Biol 13:1499–1505

Lee VD, Huang B (1993) Molecular cloning and centrosomal localization of human caltractin. Proc Natl Acad Sci USA 90:11039–11043

Leidel S, Delattre M, Cerutti L, Baumer K, Gonczy P (2005) SAS-6 defines a protein family required for centrosome duplication in *C. elegans* and in human cells. Nat Cell Biol 7:115–125

Leidel S, Gonczy P (2003) SAS-4 is essential for centrosome duplication in *C elegans* and is recruited to daughter centrioles once per cell cycle. Dev Cell 4:431–439

Li Q, Hansen D, Killilea A, Joshi HC, Palazzo RE, Balczon R (2001) Kendrin/pericentrin-B, a centrosome protein with homology to pericentrin that complexes with PCM-1. J Cell Sci 114:797–809

Linck RW (1976) Flagellar doublet microtubules: fractionation of minor components and alpha-tubulin from specific regions of the A-tubule. J Cell Sci 20:405–439

Linck RW, Amos LA, Amos WB (1985) Localization of tektin filaments in microtubules of sea urchin sperm flagella by immunoelectron microscopy. J Cell Biol 100:126–135

Lingle WL, Barrett SL, Negron VC, D'Assoro AB, Boeneman K, Liu W, Whitehead CM, Reynolds C, Salisbury JL (2002) Centrosome amplification drives chromosomal instability in breast tumor development. Proc Natl Acad Sci USA 99:1978–1983

Liu ST, Chan GK, Hittle JC, Fujii G, Lees E, Yen TJ (2003) Human MPS1 kinase is required for mitotic arrest induced by the loss of CENP-E from kinetochores. Mol Biol Cell 14:1638–1651

Liu X, Erikson RL (2002) Activation of Cdc2/cyclin B and inhibition of centrosome amplification in cells depleted of Plk1 by siRNA. Proc Natl Acad Sci USA 99:8672–8676

Llamazares S, Moreira A, Tavares A, Girdham C, Spruce BA, Gonzalez C, Karess RE, Glover DM, Sunkel CE (1991) polo encodes a protein kinase homolog required for mitosis in *Drosophila*. Genes Dev 5:2153–2165

Louie RK, Bahmanyar S, Siemers KA, Votin V, Chang P, Stearns T, Nelson WJ, Barth AI (2004) Adenomatous polyposis coli and EB1 localize in close proximity of the mother centriole and EB1 is a functional component of centrosomes. J Cell Sci 117:1117–1128

Lowery DM, Lim D, Yaffe MB (2005) Structure and function of Polo-like kinases. Oncogene 24:248–259

Lu KP, Hunter T (1995) The NIMA kinase: a mitotic regulator in *Aspergillus nidulans* and vertebrate cells. Prog Cell Cycle Res 1:187–205

Lutz W, Lingle WL, McCormick D, Greenwood TM, Salisbury JL (2001) Phosphorylation of centrin during the cell cycle and its role in centriole separation preceding centrosome duplication. J Biol Chem 276:20774–20780

Ma S, Trivinos-Lagos L, Graf R, Chisholm RL (1999) Dynein intermediate chain mediated dynein-dynactin interaction is required for interphase microtubule organization and centrosome replication and separation in *Dictyostelium*. J Cell Biol 147:1261–1273

Mack GJ, Rees J, Sandblom O, Balczon R, Fritzler MJ, Rattner JB (1998) Autoantibodies to a group of centrosomal proteins in human autoimmune sera reactive with the centrosome. Arthr Rheum 41:551–558

Mailand N, Lukas C, Kaiser BK, Jackson PK, Bartek J, Lukas J (2002) Deregulated human Cdc14A phosphatase disrupts centrosome separation and chromosome segregation. Nat Cell Biol 4:317–322

Malone CJ, Misner L, Le Bot N, Tsai MC, Campbell JM, Ahringer J, White JG (2003) The *C. elegans* hook protein, ZYG-12, mediates the essential attachment between the centrosome and nucleus. Cell 115:825–836

Manchester KL (1995) Theodor Boveri and the origin of malignant tumours. Trends Cell Biol 5:384–387

Martin-Lluesma S, Stucke VM, Nigg EA (2002) Role of Hec1 in spindle checkpoint signaling and kinetochore recruitment of Mad1/Mad2. Science 297:2267–2270

Marumoto T, Honda S, Hara T, Nitta M, Hirota T, Kohmura E, Saya H (2003) Aurora-A kinase maintains the fidelity of early and late mitotic events in HeLa cells. J Biol Chem 278:51786–51795

Marumoto T, Zhang D, Saya H (2005) Aurora-A – a guardian of poles. Nat Rev Cancer 5:42–50

Matsumoto Y, Maller JL (2002) Calcium, calmodulin, and CaMKII requirement for initiation of centrosome duplication in *Xenopus* egg extracts. Science 295:499–502

Matsumoto Y, Hayashi K, Nishida E (1999) Cyclin-dependent kinase 2 (Cdk2) is required for centrosome duplication in mammalian cells. Curr Biol 9:429–432

Matsuura K, Lefebvre PA, Kamiya R, Hirono M (2004) Bld10p, a novel protein essential for basal body assembly in *Chlamydomonas*: localization to the cartwheel, the first ninefold symmetrical structure appearing during assembly. J Cell Biol 165:663–671

Mayor T, Hacker U, Stierhof YD, Nigg EA (2002) The mechanism regulating the dissociation of the centrosomal protein C-Nap1 from mitotic spindle poles. J Cell Sci 115:3275–3284

Mayor T, Stierhof Y-D, Tanaka K, Fry AM, Nigg EA (2000) The centrosomal protein C-Nap1 is required for cell cycle-regulated centrosome cohesion. J Cell Biol 151:837–846

Meraldi P, Lukas J, Fry AM, Bartek J, Nigg EA (1999) Centrosome duplication in mammalian somatic cells requires E2F and Cdk2-cyclin A. Nat Cell Biol 1:88–93

Meraldi P, Nigg EA (2001) Centrosome cohesion is regulated by a balance of kinase and phosphatase activities. J Cell Sci 114:3749–3757

Meraldi P, Nigg EA (2002) The centrosome cycle. FEBS Lett 521:9–13

Middendorp S, Paoletti A, Schiebel E, Bornens M (1997) Identification of a new mammalian centrin gene, more closely related to *Saccharomyces cerevisiae* CDC31 gene. Proc Natl Acad Sci USA 94:9141–9146

Moritz KB, Sauer HW (1996) Boveri's contributions to developmental biology – a challenge for today. Int J Dev Biol 40:27–47

Moritz M, Braufeld M, Sedat J, Alberts B, Agard D (1995) Microtubule nucleation by gamma-tubulin-containing rings in the centrosome. Nature 378:638–640

Murphy SM, Urbani L, Stearns T (1998) The mammalian gamma-tubulin complex contains homologues of the yeast spindle pole body components spc97p and spc98p. J Cell Biol 141:663–674

Murphy SM, Preble AM, Patel UK, O'Connell KL, Dias DP, Moritz M, Agard D, Stults JT, Stearns T (2001) GCP5 and GCP6: two new members of the human gamma-tubulin complex. Mol Biol Cell 12:3340–3352

Nigg EA (2002) Centrosome aberrations: cause or consequence of cancer progression? Nat Rev Cancer 2:815–825

Noton E, Diffley JF (2000) CDK inactivation is the only essential function of the APC/C and the mitotic exit network proteins for origin resetting during mitosis. Mol Cell 5:85–95

O'Connell KF, Caron C, Kopish KR, Hurd DD, Kemphues KJ, Li Y, White JG (2001) The C. elegans zyg-1 gene encodes a regulator of centrosome duplication with distinct maternal and paternal roles in the embryo. Cell 105:547–558

O'Connell KF, Leys CM, White JG (1998) A genetic screen for temperature-sensitive cell-division mutants of *Caenorhabditis elegans*. Genetics 149:1303–1321

O'Connell KF, Maxwell KN, White JG (2000) The spd-2 gene is required for polarization of the anteroposterior axis and formation of the sperm asters in the *Caenorhabditis elegans* zygote. Dev Biol 222:55–70

O'Connell MJ, Krien MJ, Hunter T (2003) Never say never. The NIMA-related protein kinases in mitotic control. Trends Cell Biol 13:221–228

Oakley BR, Akkari YN (1999) Gamma-tubulin at ten: progress and prospects. Cell Struct Funct 24:365–372

Oakley CE, Oakley BR (1989) Identification of gamma-tubulin, a new member of the tubulin superfamily encoded by mipA gene of *Aspergillus nidulans*. Nature 338:662–664

Ohta Y, Ohba T, Miyamoto E (1990) Ca^{2+}/calmodulin-dependent protein kinase II: localization in the interphase nucleus and the mitotic apparatus of mammalian cells. Proc Natl Acad Sci USA 87:5341–5345

Okuda M (2002) The role of nucleophosmin in centrosome duplication. Oncogene 21:6170–6174

Okuda M, Horn HF, Tarapore P, Tokuyama Y, Smulian AG, Chan PK, Knudsen ES, Hofmann IA, Snyder JD, Bove KE, Fukasawa K (2000) Nucleophosmin/B23 is a target of CDK2/cyclin E in centrosome duplication. Cell 103:127–140

Ortega S, Prieto I, Odajima J, Martin A, Dubus P, Sotillo R, Barbero JL, Malumbres M, Barbacid M (2003) Cyclin-dependent kinase 2 is essential for meiosis but not for mitotic cell division in mice. Nat Genet 35:25–31

Ou Y, Rattner JB (2004) The centrosome in higher organisms: structure, composition, and duplication. Int Rev Cytol 238:119–182

Ou Y, Zhang M, Rattner JB (2004) The centrosome: the centriole-PCM coalition. Cell Motil Cytoskeleton 57:1–7

Palazzo RE, Vogel JM, Schnackenberg BJ, Hull DR, Wu X (2000) Centrosome maturation. Curr Top Dev Biol 49:449–470

Paoletti A, Moudjou M, Paintrand M, Salisbury JL, Bornens M (1996) Most of centrin in animal cells is not centrosome-associated and centrosomal centrin is confined to the distal lumen of centrioles. J Cell Sci 109:3089–3102

Paoletti A, Bordes N, Haddad R, Schwartz CL, Chang F, Bornens M (2003) Fission yeast cdc31p is a component of the half-bridge and controls SPB duplication. Mol Biol Cell 14:2793–2808

Pelletier L, Ozlu N, Hannak E, Cowan C, Habermann B, Ruer M, Muller-Reichert T, Hyman AA (2004) The *Caenorhabditis elegans* centrosomal protein SPD-2 is required for both pericentriolar material recruitment and centriole duplication. Curr Biol 14:863–873

Piel M, Meyer P, Khodjakov A, Rieder CL, Bornens M (2000) The respective contributions of the mother and daughter centrioles to centrosome activity and behavior in vertebrate cells. J Cell Biol 149:317–330

Piel M, Nordberg J, Euteneuer U, Bornens M (2001) Centrosome-dependent exit of cytokinesis in animal cells. Science 291:1550–1553

Pihan GA, Wallace J, Zhou Y, Doxsey SJ (2003) Centrosome abnormalities and chromosome instability occur together in pre-invasive carcinomas. Cancer Res 63:1398–1404

Pockwinse SM, Krockmalnic G, Doxsey SJ, Nickerson J, Lian JB, van Wijnen AJ, Stein JL, Stein GS, Penman S (1997) Cell cycle independent interaction of CDC2 with the centrosome, which is associated with the nuclear matrix-intermediate filament scaffold. Proc Natl Acad Sci USA 94:3022–3027

Popov AV, Karsenti E (2003) Stu2p and XMAP215: turncoat microtubule-associated proteins? Trends Cell Biol 13:547–550

Poss KD, Nechiporuk A, Hillam AM, Johnson SL, Keating MT (2002) Mps1 defines a proximal blastemal proliferative compartment essential for zebrafish fin regeneration. Development 129:5141–5149

Purohit A, Tynan SH, Vallee R, Doxsey SJ (1999) Direct interaction of pericentrin with cytoplasmic dynein light intermediate chain contributes to mitotic spindle organization. J Cell Biol 147:481–491

Quintyne NJ, Reing JE, Hoffelder DR, Gollin SM, Saunders WS (2005) Spindle multipolarity is prevented by centrosomal clustering. Science 307:127–129

Raynaud-Messina B, Mazzolini L, Moisand A, Cirinesi AM, Wright M (2004) Elongation of centriolar microtubule triplets contributes to the formation of the mitotic spindle in gamma-tubulin-depleted cells. J Cell Sci 117:5497–5507

Roghi C, Giet R, Uzbekov R, Morin N, Chartrain I, Le Guellec R, Couturier A, Doree M, Philippe M, Prigent C (1998) The *Xenopus* protein kinase pEg2 associates with the centrosome in a cell cycle-dependent manner, binds to the spindle microtubules and is involved in bipolar mitotic spindle assembly. J Cell Sci 111:557–572

Ruiz F, Beisson J, Rossier J, Dupuis-Williams P (1999) Basal body duplication in *Paramecium* requires gamma-tubulin. Curr Biol 9:43–46

Rusan NM, Wadsworth P (2005) Centrosome fragments and microtubules are transported asymmetrically away from division plane in anaphase. J Cell Biol 168:21–28

Salisbury JL (1995) Centrin, centrosomes, and mitotic spindle poles. Curr Opin Cell Biol 7:39–45

Salisbury JL, Suino KM, Busby R, Springett M (2002) Centrin-2 is required for centriole duplication in mammalian cells. Curr Biol 12:1287–1292

Schiebel E, Bornens M (1995) In search of a function for centrins. Trends Cell Biol 5:197–201

Schliwa M, Euteneuer U, Graf R, Ueda M (1999) Centrosomes, microtubules and cell migration. Biochem Soc Symp 65:223–231

Schmidt DJ, Rose DJ, Saxton WM, Strome S (2005) Functional analysis of cytoplasmic dynein heavy chain in *Caenorhabditis elegans* with fast-acting temperature-sensitive mutations. Mol Biol Cell 16:1200–1212

Shang Y, Li B, Gorovsky MA (2002) *Tetrahymena thermophila* contains a conventional gamma-tubulin that is differentially required for the maintenance of different microtubule-organizing centers. J Cell Biol 158:1195–1206

Sluder G, Nordberg JJ (2004) The good, the bad and the ugly: the practical consequences of centrosome amplification. Curr Opin Cell Biol 16:49–54

Snell WJ, Pan J, Wang Q (2004) Cilia and flagella revealed: from flagellar assembly in *Chlamydomonas* to human obesity disorders. Cell 117:693–697

Stearns T, Kirschner M (1994) *In vitro* reconstitution of centrosome assembly and function: The central role of gamma-tubulin. Cell 76:623–637

Stearns T, Evans L, Kirschner M (1991) Gamma-tubulin is a highly conserved component of the centrosome. Cell 65:825–836

Steffen W, Fajer EA, Linck RW (1994) Centrosomal components immunologically related to tektins from ciliary and flagellar microtubules. J Cell Sci 107:2095–2105

Stephens RE, Lemieux NA (1998) Tektins as structural determinants in basal bodies. Cell Motil Cytoskeleton 40:379–392

Stephens RE, Oleszko-Szuts S, Linck RW (1989) Retention of ciliary ninefold structure after removal of microtubules. J Cell Sci 92:391–402

Stucke VM, Sillje HH, Arnaud L, Nigg EA (2002) Human Mps1 kinase is required for the spindle assembly checkpoint but not for centrosome duplication. EMBO J 21:1723–1732

Stucke VM, Baumann C, Nigg EA (2004) Kinetochore localization and microtubule interaction of the human spindle checkpoint kinase Mps1. Chromosoma 113:1–15

Sunkel CE, Glover DM (1988) polo, a mitotic mutant of *Drosophila* displaying abnormal spindle poles. J Cell Sci 89:25–38

Takahashi M, Shibata H, Shimakawa M, Miyamoto M, Mukai H, Ono Y (1999) Characterization of a novel giant scaffolding protein, CG-NAP, that anchors multiple signaling enzymes to centrosome and the Golgi apparatus. J Biol Chem 274:17267–17274

Takahashi M, Yamagiwa A, Nishimura T, Mukai H, Ono Y (2002) Centrosomal proteins CG-NAP and kendrin provide microtubule nucleation sites by anchoring gamma-tubulin ring complex. Mol Biol Cell 13:3235–3245

Tarapore P, Fukasawa K (2002) Loss of p53 and centrosome hyperamplification. Oncogene 21:6234–6240

Tarapore P, Horn HF, Tokuyama Y, Fukasawa K (2001a) Direct regulation of the centrosome duplication cycle by the p53-p21$^{Waf1/Cip1}$ pathway. Oncogene 20:3173–3184

Tarapore P, Tokuyama Y, Horn HF, Fukasawa K (2001b) Difference in the centrosome duplication regulatory activity among p53 'hot spot' mutants: potential role of Ser 315 phosphorylation-dependent centrosome binding of p53. Oncogene 20:6851–6863

Tarapore P, Okuda M, Fukasawa K (2002) A mammalian *in vitro* centriole duplication system: evidence for involvement of CDK2/cyclin E and nucleophosmin/B23 in centrosome duplication. Cell Cycle 1:75–81

Thomas RC, Edwards MJ, Marks R (1996) Translocation of the retinoblastoma gene during mitosis. Exp Cell Res 223:227–232

Tokuyama Y, Horn HF, Kawamura K, Tarapore P, Fukasawa K (2001) Specific phosphorylation of nucleophosmin on Thr^{199} by cyclin-dependent kinase 2-cyclin E and its role in centrosome duplication. J Biol Chem 276:21529–21537

Twomey C, Wattam SL, Pillai MR, Rapley J, Baxter JE, Fry AM (2004) Nek2B stimulates zygotic centrosome assembly in *Xenopus laevis* in a kinase-independent manner. Dev Biol 265:384–398

Uto K, Sagata N (2000) Nek2B, a novel maternal form of Nek2 kinase is essential for the assembly or maintenance of centrosomes in early *Xenopus* embryos. EMBO J 19:1816–1826

Vaisberg EA, Koonce MP, McIntosh JR (1993) Cytoplasmic dynein plays a role in mammalian mitotic spindle formation. J Cell Biol 123:849–858 van Vugt MA, van de Weerdt BC, Vader G, Janssen H, Calafat J, Klompmaker R, Wolthuis RM, Medema RH (2004) Polo-like kinase-1 is required for bipolar spindle formation but is dispensable for anaphase promoting complex/Cdc20 activation and initiation of cytokinesis. J Biol Chem 279:36841–36854

Vidwans SJ, Wong ML, O'Farrell PH (1999) Mitotic regulators govern progress through steps in the centrosome duplication cycle. J Cell Biol 147:1371–1378

Vorobjev I, Nadezhdina E (1987) The centrosome and its role in the organization of microtubles. Int Rev Cytol 106:227–293

Wang Y, Prives C (1995) Increased and altered DNA binding of human p53 by S and G2/M but not G1 cyclin-dependent kinases. Nature 376:88–91

Warnke S, Kemmler S, Hames RS, Tsai HL, Hoffmann-Rohrer U, Fry AM, Hoffmann I (2004) Polo-like kinase-2 is required for centriole duplication in mammalian cells. Curr Biol 14:1200–1207

White RA, Pan Z, Salisbury JL (2000) GFP-centrin as a marker for centriole dynamics in living cells. Microsc Res Tech 49:451–457

Whitehead CM, Rattner JB (1998) Expanding the role of HsEg5 within the mitotic and post-mitotic phases of the cell cycle. J Cell Sci 111:2551–2561

Winey M, Goetsch L, Baum P, Byers B (1991) MPS1 and MPS2: Novel yeast genes defining distinct steps of spindle pole body duplication. J Cell Biol 114:745–754

Wong C, Stearns T (2003) Centrosome number is controlled by a centrosome-intrinsic block to reduplication. Nat Cell Biol 5:539–544

Woo RA, Poon RY (2003) Cyclin-dependent kinases and S-phase control in mammalian cells. Cell Cycle 2:316–324

Yao J, Fu C, Ding X, Guo Z, Zenreski A, Chen Y, Ahmed K, Liao J, Dou Z, Yao X (2004) Nek2A kinase regulates the localization of numatrin to centrosome in mitosis. FEBS Lett 575:112–118

Yarm FR (2002) Plk phosphorylation regulates the microtubule-stabilizing protein TCTP. Mol Cell Biol 22:6209–6221

Yoder JH, Han M (2001) Cytoplasmic dynein light intermediate chain is required for discrete aspects of mitosis in *Caenorhabditis elegans*. Mol Biol Cell 12:2921–2933

Young A, Dictenberg JB, Purohit A, Tuft R, Doxsey SJ (2000) Cytoplasmic dynein-mediated assembly of pericentrin and γ-tubulin onto centrosomes. Mol Biol Cell 11:2047–2056

Zhang H, Shi X, Paddon H, Hampong M, Dai W, Pelech S (2004) B23/nucleophosmin serine 4 phosphorylation mediates mitotic functions of polo-like kinase 1. J Biol Chem 279:35726–35734

Zheng Y (2004) G protein control of microtubule assembly. Annu Rev Cell Dev Biol 20:867–894

Zheng Y, Jung K, Oakley B (1991) Gamma-tubulin is present in *Drosophila melanogaster* and *Homo sapiens* and is associated with the centrosome. Cell 65:817–823

Zheng Y, Wong M, Alberts B, Mitchison T (1995) Nucleation of microtubule assembly by a gamma-tubulin-containing ring complex. Nature 378:578–583

Zimmerman WC, Sillibourne J, Rosa J, Doxsey SJ (2004) Mitosis-specific anchoring of gamma tubulin complexes by pericentrin controls spindle organization and mitotic entry. Mol Biol Cell 15:3642–3657

The ubiquitin-proteasome pathway in cell cycle control

Steven I. Reed

Department of Molecular Biology, MB-7, The Scripps Research Institute, 10550 North Torrey Pines Road, La Jolla, CA 92037, USA
sreed@scripps.edu

Abstract Ubiquitin-mediated proteolysis is one of the key mechanisms underlying cell cycle control. The removal of barriers posed by accumulation of negative regulators, as well as the clearance of proteins when they are no longer needed or deleterious, are carried out via the ubiquitin-proteasome system. Ubiquitin conjugating enzymes and protein-ubiquitin ligases collaborate to mark proteins destined for degradation by the proteasome by covalent attachment of multi-ubiquitin chains. Most regulated proteolysis during the cell cycle can be attributed to two families of protein-ubiquitin ligases. The anaphase promoting complex/cyclosome (APC/C) is activated during mitosis and G1 where it is responsible for eliminating proteins that impede mitotic progression and that would have deleterious consequences if allowed to accumulate during G1. SCF (Skp1/Culin/F-box protein) protein-ubiquitin ligases ubiquitylate proteins that are marked by phosphorylation at specific sequences known as phosphodegrons. Targeting of proteins for destruction by phosphorylation provides a mechanism for linking cell cycle regulation to internal and external signaling pathways via regulated protein kinase activities.

1
Introduction

The importance of ubiquitin-mediated proteolysis for cell cycle progression and control is now well beyond dispute and can be illustrated in a variety of ways. One of the most telling comes from the assignment of function to the initial collection of cell division cycle (cdc) genes identified by Hartwell and colleagues (Hartwell et al. 1974). Of the original 35 genes described in the Hartwell screen based on a division-defective phenotypic endpoint, seven, or fully 20%, either encode components of protein-ubiquitin ligases or ubiquitin conjugating enzymes. For comparison, only five genes in this set encode protein kinases or phosphatases. Even though this exercise cannot be construed as being of high quantitative significance, it does underscore the importance of ubiquitin-mediated proteolysis and at least suggests its rough equivalence with regulatory phosphorylation as an underlying mechanism in cell cycle progression. We now know that ubiquitin-mediated protein turnover and protein phosphorylation are not separable processes, but are indeed highly interconnected, collaborating to form a network of pathways and regulatory loops that control the cell cycle. The purpose of this review is to provide a current view the role of ubiquitylation, ubiquitin-mediated proteolysis, and proteasomes in cell cycle control in both yeasts and metazoans.

2
The ubiquitin-proteasome pathway

The process of ubiquitin-mediated proteolysis begins with the covalent attachment of multi-ubiquitin chains to targeted proteins (Fig. 1) (Hershko 1983). Ubiquitin is a small (76 amino acid) highly conserved protein (Hershko 1983). A cascade of enzyme-catalyzed reactions first activates monomeric ubiquitin and then effects the processive attachment of ubiquitin monomers first to lysines on the targeted protein and then to lysines of previously attached ubiquitins. The activation of ubiquitin by the formation of a high energy thioester bond with the C-terminal carboxylate of ubiquitin is carried out by a ubiquitin-activating enzyme or E1. The transfer of activated ubiquitin to the lysines of target proteins to form an isopeptide bond between the C-terminal carboxylate of ubiquitin and a lysine epsilon amino group is carried out through a collaboration between ubiquitin conjugating enzymes (E2) and protein-ubiquitin ligases (E3). The processive addition of ubiquitin monomers to the lysines of already attached ubiquitins leads to the decoration of proteins with long ubiquitin chains. Although several different lysines on ubiquitin can serve as acceptor sites for ubiquitin addition, the predominant linkage for protein turnover is through lysine 48 (Chen and Pickart 1990). Once chains of significant length have been produced, the

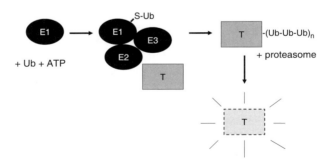

Fig. 1 The ubiquitin proteasome pathway. Ubiquitin is initially activated by formation of a thioester bond with E1 at the expense of a molecule of ATP. Activated ubiquitin is then transferred via a complex of E2 (ubiquitin conjugating enzymer) and E3 (protein-ubiquitin ligase) to the target molecule (T), Processive addition of ubiquitin to previously conjugated ubiquitins forms multi-ubiquitylated chains that are recognized by the 26S proteasome, leading to degradation of T

protein is recognized by a complex protease known as the proteasome, ultimately leading to its processing into small peptides and the recycling of ubiquitin. The active (26S) proteasome is composed of a barrel shaped catalytic core containing multiple protease activities on the inside surface (the 20S particle) and a regulatory (19S) particle at either end (Pickart and Cohen 2004). The regulatory particle is responsible for recognizing ubiquitylated targets, removing the polyubiquitin chains for recycling of ubiquitin, unfolding the protein to be degraded and opening up a pore in the 20S particle so that unfolded substrate proteins can enter and contact the protease active sites. The 19S cap contains a receptor for polyubiquitin chains. However, efficient substrate recognition of at least some polyubiquitinated targets requires "adapter" proteins that themselves bind ubiquitylated targets and dock to the 19S regulatory cap (Elsasser et al. 2004; Verma et al. 2004). The best characterized of these is Rad23, which contains two ubiquitin-binding Uba domains and a ubiquitin-like (Ubl) domain that interacts with the proteasome. One final level of regulation concerns activities known as E4, which lengthen ubiquitin chains on already polyubiquitylated proteasome substrates, presumably to prevent substrate escape due to deubiquitylating activities (Koegl et al. 1999; Hatakeyama et al. 2001).

3
Protein-ubiquitin ligases in the cell cycle core machinery

One of the earliest observations relevant to the molecular basis for cell cycle regulation was periodic synthesis and destruction of major proteins in sea urchin cleavage embryos (Evans et al. 1983). Although the function of these proteins, termed cyclins, was not known at the time, their accumulation during interphase and turnover during mitosis was suggestive of a critical cell cycle role. We now know that the cyclins observed in these sea urchin embryo studies are positive regulatory subunits of the cyclin-dependent kinase (Cdk), Cdk1 (Draetta et al. 1989; Meijer et al. 1989), which controls the mitotic state for all eukaryotes: activation of Cdk1 establishes the mitotic state, whereas inactivation determines mitotic exit into interphase. The inactivation of Cdk1, potentiating mitotic exit is mediated for the most part by the ubiquitin-mediated proteolysis of mitotic cyclins (Murray et al. 1989; Ghiara et al. 1991; Surana et al. 1993). Investigation into the basis for mitotic cyclin degradation has led to the discovery and characterization of a complex protein-ubiquitin ligase known as the anaphase promoting complex/cyclosome or APC/C (King et al. 1995; Sudakin et al. 1995). Composed of at least 13 core subunits (Zachariae et al. 1998b) and two alternative regulatory subunits (Visintin et al. 1997) (Fig. 2), the APC/C targets not only mititotic cyclins (A and B) but many other proteins that need to be degraded during mitosis

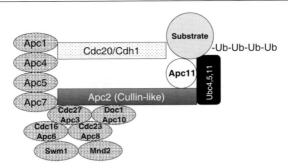

Fig. 2 The anaphase promoting complex/cyclosome (APC/C)

and/or the subsequent G1 interval (Table 1). Typically, the APC/C recognizes targets whose destruction at the population level is mandated at the times when the APC/C is active, from the metaphase-anaphase transition to the G1–S phase transition.

While the initial observations leading to the discovery of mitosis-specific protein degradation came from studies on the early embryonic cell cycles of

Table 1 Cell cycle targets of the anaphase promoting complex

Substrate	Organism	Specificity factor	Cell cycle function
Securin (Pds1)	S. cerevisiae (metazoan)	Cdc20	Anaphase inhibitor
Clb2	S. cerevisiae	Cdc20/Cdh1	B-type cyclin (mitosis)
Clb5	S. cerevisiae	Cdc20	B-type cyclin (S phase)
Cyclin B	Metazoan	Cdc20/Cdh1	Mitosis
Cyclin A	Metazoan	Cdc20/Cdh1	S phase, mitosis
Cdc20	S. cerevisiae, metazoan	Cdh1	Mitosis
Cdc5/Plk	S. cerevisiae, metazoan	Cdh1	Mitosis
Aurora A	vertebrate	Cdh1	Mitosis
Dbf4	S. cerevisiae	Cdc20	Replication
Ase1	S. cerevisiae	Cdh1	Mitosis
Nek2A	Vertebrate	Cdh1	Centrosome development
Cdc6	Vertebrate	Cdh1	Replication
Geminin	Metazoan	Cdh1	Replication licensing
Cin8/Kip1	S. cerevisiae	Cdh1	Mitotic spindle motor
Xkid	Vertebrate	Cdh1	Mitotic spindle motor
Hsl1	S. cerevisiae	Cdc20/Cdh1	Mitosis
Cdc25A	Vertebrates	Cdh1	S phase, mitosis
Skp2	Vertebrates	Cdh1	SCF cofactor

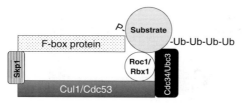

Fig. 3 SCF protein-ubiqutin ligases

marine invertebrates (Evans et al. 1983; Swenson et al. 1986), first insights into the role of proteolysis at the G1–S phase transition were a byproduct of genetic analysis of the yeast cell cycle. As mentioned above, a number of the first cell division cycle (cdc) mutants to be characterized were ultimately found to define components of protein-ubiquitin ligases and associated proteins. Of these, *cdc4* and *cdc34* conferred arrest at the G1/S boundary (Hartwell et al. 1974). In a subsequent round of cdc mutant isolation, mutations in a third gene, *cdc53*, were found to confer a similar phenotype. We now know that Cdc53 defines part of the catalytic core of a class of protein-ubiquitin ligases known as SCF (Skp1-Cullin-F-box protein) (Willems et al. 1996) (Fig. 3), whereas Cdc4 constitutes one of at least several SCF substrate specificity factors (Feldman et al. 1997; Skowyra et al. 1997). Cdc34 is the associated ubiquitin conjugating enzyme that works in conjunction with SCF to transfer ubiquitin to target proteins (Goebl et al. 1988). The cell cycle arrest phenotype conferred by mutations in the genes encoding these proteins results from an inability to degrade a Cdk inhibitor, Sic1 (Nugroho and Mendenhall 1994; Schwob et al. 1994). Since Cdk1 activity is required for initiation of DNA replication in yeast, failure to degrade Sic1 leads to arrest at the G1-S phase boundary. Subsequently, numerous other cell cycle targets of SCF ubiquitin ligases have been identified both in yeasts and metazoans. However, unlike the APC/C, SCF activities are also simultaneously targeted to non-cell-cycle-related proteins (although roles for the APC/C have recently been described in post-mitotic cells). This is possible because the SCF system is largely activated at the substrate level by substrate phosphorylation, allowing simultaneous targeting of individual marked proteins within diverse populations and with distinct functions. Nevertheless, as will be described below, SCF ligases constitute a core component of the cell cycle machinery. It is also interesting to note that the APC/C and SCF systems do not operate in isolation from each other. Recent evidence suggests that they are mutually toggled presumably to enforce coordination of cell cycle events (see below).

3.1
APC/C protein-ubiquitin ligases

As with SCF, a number of components of the APC/C were represented in the original collection of yeast cdc mutants assembled by Hartwell and colleagues (Hartwell et al. 1974). Of the genes defined, four (*CDC16*, *CDC23*, *CDC26* and *CDC27*) encode subunits of the ligase itself (Zachariae and Nasmyth 1996; Zachariae et al. 1996, 1998b), whereas one (*CDC20*) (Visintin et al. 1997) encodes an essential positive regulatory factor of the APC/C. Conditional mutations in all of these essential genes confer a mitotic arrest phenotype characterized by unseparated sister chromatids. However, the functions of these genes and their products was revealed only when it was discovered that cyclin B was degraded via the ubiquitin-proteasome pathway (Glotzer et al. 1991) and the activity responsible for ubiquitylating cyclin B was purified from mitotic clam and frog oocytes, respectively (King et al. 1995; Sudakin et al. 1995). The large (20S) ubiquitin ligase from frog oocytes was found to contain homologs of the yeast Cdc16 and Cdc27 proteins providing a rationale for mitotic arrest phenotypes associated with *cdc16* and *cdc27* mutants and leading a direct demonstration that *cdc16*, *cdc23* and *cdc27* mutants were defective in mitotic cyclin degradation in yeast (Zachariae and Nasmyth 1996). Analysis of purified complexes from both metazoans and yeast has revealed that the APC/C consists of 13 core polypeptides (Zachariae et al. 1998b; Grossberger et al. 1999) (Fig. 2). Three of these subunits (Cdc16, Cdc23 and Cdc27) contain a repeating motif known as TPR (for tetratricopeptide repeat) that is involved in protein-protein interactions important for assembling the APC/C macromolecular complex (Sikorski et al. 1990; Lamb et al. 1994). It is not clear why so many subunits are required for activity, since most protein-ubiquitin ligases are much smaller. Cryoelectron microscopy indicates that the APC/C has a hollow asymmetric structure (Gieffers et al. 2001). However, until the substrate binding and catalytic sites have been identified within this structure, its significance remains obscure. This should be possible, since two the APC/C subunits, Apc2 and Apc11, share significant homology with subunits of the catalytic core of SCF, Cul1/Cdc53 and Roc1/Rbx1, respectively (Ohta et al. 1999; Seol et al. 1999). In this context, Apc11 and Roc1/Rbx1 contain "ring finger" motifs that are a characteristic of the catalytic site of a major class of protein-ubiquitin ligases (Lorick et al. 1999).

The APC/C core is inactive without one of two structurally related positive regulatory cofactors, known as Cdc20 and Cdh1, respectively (Schwab et al. 1997; Visintin et al. 1997). One function of these regulatory factors is to recruit substrates (Hilioti et al. 2001; Pfleger et al. 2001; Schwab et al. 2001). It has been shown that Cdc20 preferentially recognizes a sequence known as the D-box with a consensus R-X-X-L-X-X-X-X-N/D/E (Glotzer et al. 1991; King et al. 1996), whereas Cdh1 recognizes both the D-box sequence as well as a second sequence known as the KEN-box (Pfleger and Kirschner 2000) and a third known as an A-box (Littlepage and Ruderman 2002), found specifically in the mitotic kinase Aurora A. However, the mechanisms of substrate recognition and targeting by different forms of APC/C have yet to be completely elucidated. For some targets, e.g. cyclin B, a single D-box is sufficient to mediate ubiquitylation and destruction, whereas for others, e.g. cyclin A, both a D-box and a KEN-box are required (Geley et al. 2001). Furthermore, a recent report suggests that the APC/C itself contributes to substrate recognition and binding, independent of Cdc20 and Cdh1, in that a D-box-containing affinity matrix retained APC/C without Cdc20 (Yamano et al. 2004). This result suggests that Cdc20 and Cdh1 may provide an activation function in addition to substrate recruitment.

Whereas Cdc20 and Cdh1 share a high degree of structural homology and presumably provide analogous positive regulatory functions at the enzymatic level, their biological functions are quite distinct. APC^{Cdc20} is primarily responsible for mediating the metaphase-anaphase transition and early phases of mitotic exit (Schwab et al. 1997; Visintin et al. 1997; Fang et al. 1999). APC^{Cdh1} completes mitotic exit and restricts mitotic proteins to low levels during the subsequent G1 phase (Schwab et al. 1997; Visintin et al. 1997; Fang et al. 1999). This division of labor is orchestrated in part by a complex regulatory relationship between Cdc20 and Cdh1. Whereas Cdc20 accumulates based on periodic transcription late in the cell cycle, largely accounting for the active window of APC^{Cdc20} (Prinz et al. 1998), APC^{Cdh1} is expressed constitutively throughout the cell cycle (Weinstein 1997; Prinz et al. 1998; Zhu et al. 2000). However, APC^{Cdh1} is negatively regulated by Cdk phosphorylation of Cdh1 (Zachariae et al. 1998a; Lukas et al. 1999; Sorensen et al. 2001). It is the APC^{Cdc20}-mediated ubiquitylation and degradation of S-phase and mitotic cyclins that allows dephosphorylation and activation of Cdh1 during mitotic exit. On the other hand, the inhibition of Cdh1 at the G1–S phase transition allows the accumulation of S phase cyclins, which in turn promotes the biosynthesis of Cdc20 via transcription. Cdh1 inhibition at the G1–S phase transition is initially mediated in mammalian cells by E2F-driven accumulation of the Cdh1/Cdc20 inhibitor, Emi1 (Hsu et al. 2002) and autoubiquitylation and degradation of a ubiquitin conjugating enzyme, Ubc10, that serves as a cofactor with APC^{Cdh1} (Rape and Kirschner 2004). The resultant accumulation of cyclin A and activation of Cdk2 provides additional inhibition of Cdh1 via phosphorylation. Accumulation of Cdc20 per se is not sufficient for activation of APC^{Cdc20}, as Cdc20 is also regulated posttranslationally. Like Cdh1, Emi1 also inhibits Cdc20 (Reimann et al. 2001a,b). Phosphorylation-dependent degradation of Emi1 upon mitotic entrance (see below) potentiates Cdc20 activation (Margottin-Goguet et al. 2003). In addition, the spindle assembly checkpoint, which will be discussed in greater detail below, maintains Cdc20 in an inactive state until bipolar attachment of replicated chromosomes to a functional mitotic spindle is accomplished, thus triggering anaphase (Lew and Burke 2003). Although phosphorylation of the APC/C by cyclin B-Cdk1 has been suggested to be critical for mitotic activation, the precise targets and mechanism(s) have remained elusive (Sudakin et al. 1995; Kraft et al. 2003). Finally, genetic experiments in budding yeast have revealed that *CDC20* is essential, whereas *CDH1* is dispensable (Visintin et al. 1997). This is because of a redundant pathway for downregulating Cdk activity in late mitosis and G1, allowing mitotic exit in the absence of complete cyclin proteolysis (Visintin et al. 1998; Shirayama et al. 1999). Indeed, it has been possible to dispense with the APC/C entirely in yeast if the primary target of APC^{Cdc20}, the anaphase inhibitor Pds1, is eliminated and Cdk activity is down regulated by non-proteolytic mechanisms (Thornton and Toczyski 2003).

3.2
APC/C substrates and biology

An increasing number of proteins has been shown to be targeted by the APC/C (Table 1). However, only for a few has this targeting been demonstrated to be absolutely essential. Yeast Pds1 (known generically as securin in other organisms) is perhaps the most critical target of the APC/C (Cohen-Fix et al. 1996; Yamamoto et al. 1996b; Zou et al. 1999; Zur and Brandeis 2001) (Fig. 4). Prior to mitotic activation of the APC/C, Pds1/securin is bound to the protease known as separase (Esp1 in yeast) (Ciosk et al. 1998). Although Pds1 has positive regulatory roles with regard to Esp1 (e.g. nuclear localization in yeast and chaparonin functions in mammalian cells) (Jensen et al. 2001; Hornig et al. 2002; Waizenegger et al. 2002), its primary function is to inhibit Esp1 protease activity (Ciosk et al. 1998; Waizenegger et al. 2002). Esp1 is the substrate level trigger of anaphase, mediated via the endoproteolytic cleavage of the Scc1 component of cohesin, a protein complex that binds sister chromatids together subsequent to DNA replication (Uhlmann et al. 1999; Waizenegger et al. 2002). Release from cohesion allows spindle-generated forces to separate sister chromatids and initiate anaphase. Esp1/separase also then targets other proteins that regulate anaphase spindle functions (Jensen et al. 2001; Stegmeier et al. 2002; Sullivan et al. 2001). Interestingly, deletion of *PDS1* in yeast is not lethal at moderate temperatures, although *pds1* nullizygous cells grow poorly (Yamamoto et al. 1996a). The explanation is most likely due to the balanced loss of both positive and negative Esp1 regulatory functions, as well as parallel secondary pathways that can restrict proteolytic targeting of Scc1 to an appropriate time frame (Alexandru et al. 2001). The other critical targets of the APC/C are cyclins (King et al. 1995; Sudakin et al. 1995; Zachariae and Nasmyth 1996). As mentioned above, in order to exit from mitosis, Cdk activities need to be down-regulated. The primary mechanism whereby this requirement is met is via the APC/C-dependent

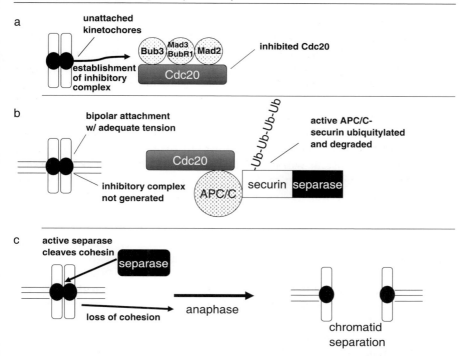

Fig. 4 The spindle assembly checkpoint. Unattached kinetochores establish an inhibitory complex consisting of Bub3, Mad3/BubR1 and Mad2, that binds to the APC/C cofactor Cdc20. Bipolar attachment of chromosomes with adequate tension leads to loss of the inhibitory complex. Free Cdc20 activates the APC/C, which leads to ubiquitylation and degradation of the separase inhibitor securin. Active separase cleaves cohesin, leading to loss of cohesion, sister chromatid separation and anaphase

degradation of mitotic cyclins. Initially, these cyclins are targeted by APC^{Cdc20} but they are also recognized by APC^{Cdh1}, which presumably is responsible for completing mitotic exit and restricting mitotic cyclin expression during G1 (Yeong et al. 2000; Wasch and Cross 2002). One key issue that remains unresolved with regard to the targeting of mitotic cyclins by the APC/C is the differential kinetics of ubiquitylation of cyclin A relative to cyclin B (Pines and Hunter 1990; Hunt et al. 1992; den Elzen and Pines 2001). Although both are targeted by APC^{Cdc20}, cyclin A is always ubiquitylated and degraded earlier in mitosis than cyclin B. This relationship is intrinsic to mitotic cyclins and the APC/C system, since it is observed in amphibian oocytes and early embryonic cell cycles as well as in vertebrate somatic cells. In addition to securin and cyclins, the APC/C has been shown to target many proteins for destruction as cells exit mitosis (Table 1). Although the turnover of these proteins is not necessary for mitotic exit or survival, it is presumed that their clearance is required for optimal cellular function.

3.3
APC/C and meiosis

Although meiosis constitutes a modified cell cycle, its unusual chromosome segregation characteristics would suggest that degradation of securin only be required at the second meiotic division, where sister chromatid segregation occurs. This is consistent with what has been reported for meiosis in *Xenopus* oocytes, where depletion of APC/C does not appear to affect progress through the first division (Peter et al. 2001; Taieb et al. 2001). However, in mouse oocytes, worms, and yeast, APC function is required for the first meiotic division (Salah and Nasmyth 2000; Davis et al. 2002; Terret et al. 2003). This is rationalized if one assumes that loss of sister chromatid cohesion is required to resolve meiotic recombination-generated cross-overs between chromosome arms prior to the first (reductional) division. Presumably, APC function is also required to allow gametes to exit from meiosis after the second (equational) division has been completed.

3.4
SCF protein-ubiquitin ligases

SCF protein-ubiquitin ligases constitute the second class of ubiquitinating enzymes that are central to cell cycle regulation in both lower and higher eukaryotes. The core of the SCF ligase consists of three polypeptides: Cul1/Cdc53, Rbx1/Roc1 and (Feldman et al. 1997; Lisztwan et al. 1998; Skowyra et al. 1999). Catalytic activity of the complex resides in a dimer composed of Cul1/Cdc53 and Rbx1/Roc1 (Ohta et al. 1999; Seol et al. 1999), the latter being a ring-finger protein, characteristic of many protein-ubiquitin ligases (Lorick et al. 1999). Structural studies also suggest that Cul1/Cdc53 serves as a scaffold for binding the substrate-specificity component of the SCF complex (Zheng et al. 2002). This consists of Skp1, an adapter protein that binds directly to Cul1/Cdc53, and one of several F-box-containing proteins. The 42-48 amino acid F-box motif constitutes a Skp1-binding domain (Bai et al. 1996). Although genomic analysis has revealed the existence of a large number of F-box proteins in both lower and higher eukaryotes, to date only a few have been confirmed as components of SCF protein ubiquitin ligases, although the number is likely to increase significantly. On the other hand, Skp1 has been shown to participate in complexes other than SCF, presumably recruiting some F-box proteins for roles distinct from ubiquitin ligation (Connelly and Hieter 1996; Russell et al. 1999). The F-box proteins involved in SCF function generally contain an F-box motif near their amino termini and one of several protein-protein interaction motifs carboxy terminal to the F-box. In addition, some members of the F-box protein family contain dimerization motifs amino terminal to the F-box, although the mechanistic implications of dimerization remain to be elucidated (Kominami et al. 1998; Suzuki et al. 2000). Interestingly, there are three classes of related F-box proteins in both yeast and mammalian cells that are involved in cell cycle control. Yeast Cdc4, originally identified via the Hartwell cdc mutant screen (Hartwell et al. 1974), has mammalian and *Drosophila* homologs, known as hCdc4/Fbw7 (Koepp et al. 2001; Strohmaier et al. 2001) and Archipelago (Ago) (Moberg et al. 2001), respectively. Cdc4 and its homologs contain eight tandem WD40 repeats that form a beta-propeller structure (Orlicky et al. 2003). This constitutes the substrate recruitment domain. hCdc4 also contains a dimerization domain upstream of the F-box (O. Sangfelt, F. van Drogen and S.I. Reed, unpublished data). In fission yeast (*S. pombe*), Cdc4 is expressed as two separate genes that encode closely related proteins, Pop1 and Pop2 (Kominami et al. 1998; Wolf et al. 1999). The active form has been shown to consist of a heterodimer, although the reason for this is not clear, since each monomer contains an F-box and a substrate interaction domain. It is conceivable that dimerization is required to configure a substrate binding domain properly with respect to the catalytic site of the SCF core. The essentiality of dimerization in the function of other Cdc4/Fbw7 homologs remains to be determined.

The second class of SCF cofactor F-box proteins involved in cell-cyclin control is defined by vertebrate β-TrCP (Fuchs et al. 1999; Kroll et al. 1999; Latres et al. 1999; Shirane et al. 1999; Suzuki et al. 1999; Tan et al. 1999; Winston et al. 1999), *Drosophila* Slimb (Bocca et al. 2001), and yeast Met30 (Kaiser et al. 1998; Patton et al. 1998). These F-box proteins share a similar topology with Cdc4/Fbw7 in that they contain WD40 repeats (seven for β-TrCP and five for Met30), an F-box and an amino terminal dimerization domain (Suzuki et al. 2000). As with Cdc4/Fbw7, the role and importance of dimerization has not been established.

The third class of cell-cycle relevant F-box protein is Skp2 (Lisztwan et al. 1998; Lyapina et al. 1998) in vertebrates and Grr1 (Li and Johnston 1997; Skowyra et al. 1997; Kishi et al. 1998) in yeast. Structurally, these SCF specificity factors are different from Cdc4/Fbw7 and β-TrCP in that the substrate interacting domain contains a motif known as a leucine-rich repeat (Kobe and Deisenhofer 1994, 1995) instead of WD40 repeats. Structural determination of Skp2 reveals that the 12 leucine-rich repeats of this molecule form a concave surface where substrate binding is likely to occur (Schulman et al. 2000). Although yeast Grr1 can be modeled to the Skp2 structure, there is little primary structure homology between the proteins, and it not clear whether they are functionally homologous in any sense.

3.5
SCF substrates and biology

Whereas APC/C activity is regulated primarily at the level of its cofactors Cdc20 and Cdh1, the primary mode of regulation of SCF activity appears to be substrate activation via phosphorylation (Skowyra et al. 1997), although regulation of F-box protein levels has also been reported (see below). SCF ubiquitin ligases mediate ubiquitylation and turnover of a large number of proteins involved in cell cycle control (Table 2). For both Cdc4/Fbw7 and β-TrCP, a specific phosphorylated consensus sequence on the substrate, designated a phosphodegron, has been described. The optimal phosphodegron sequence for Cdc4/Fbw7 is I/L-I/L/P-pT-P where basic residues are disfavored at the next four positions carboxy terminal to the proline at position +1 from the phosphothreonine (Nash et al. 2001). The crystal structure of yeast Cdc4 bound to a peptide corresponding to an ideal phosphodegron has revealed interactions between the negatively charged phosphate and several arginine residues of the surface created by the WD40-generated β-propeller structure (Orlicky et al. 2003). Mutation of any of the key arginines is sufficient to functionally inactivate Cdc4 (Koepp et al. 2001; Orlicky et al. 2003). It has been shown that substrates of SCF$^{Cdc4/Fbw7}$ can either contain a sin-

Table 2 Cell cycle targets of SCF ubiquitin ligases

Substrate	Organism	F-box protein	Cell cycle function
Sic1/Rum1	S. cerevisiae, S. pombe	Cdc4/ Pop1/2	G1–S transition inhibitor
Far1	S. cerevisiae	Cdc4	G1–S transition inhibitor
Cdc6/Cdc18	S. cerevisiae, S. pombe	Cdc4/ Pop1/2	DNA replication
Cln1,2	S. cerevisiae	Grr1	G1 cyclin
Gic1,2	S. cerevisiae	Grr1	Budding
Swe1	S. cerevisiae	Met30	Mitosis inhibitor
Met4	S. cerevisiae	Met30	G1-S transition inhibitor
Wee1	Vertebrates	β-TrCP	Mitosis inhibitor
Cyclin E	Metazoans	Cdc4/Fbw7/Ago	G1-S cyclin
p27^{Kip1}	Mammals	Skp2	G1-S transition inhibitor
p21^{Cip1}	Mammals	Skp2	G1-S transition inhibitor
p130	Mammals	Skp2	G1-S transition inhibitor
Ctd1	Mammals	Skp2	DNA replication
Orc1	Mammals	Skp2	DNA replication
Emi1	Vertebrates	β-TrCP	APC/C inhibitor
Cdc25A	Mammals	β-TrCP	S phase, mitosis
c-Myc	Mammals	Cdc4/Fbw7/Ago	S phase

gle high efficiency phosphodegron that closely matches the derived consensus or multiple low-efficiency sites, most of which are poor matches (Nash et al. 2001). It has been proposed that having a requirement for phosphorylation of multiple inefficient phosphodegron sequences within a protein constitutes a means of delaying ubiquitylation and turnover of a regulatory protein until kinase levels are sufficiently high to assure appropriate timing for a cell cycle transition (Nash et al. 2001). This model is largely based on the yeast Cdk inhibitor Sic1, which requires minimally six phosphorylation events creating six poor phosphodegron sequences in order to interact effectively with SCF^{Cdc4} (Nash et al. 2001). Achieving this level of phosphorylation requires robust accumulation of the G1 cyclins Cln1 and Cln2 and concomitant full activation of Cdk1, indicating that cells are ready to enter S phase. Ubiquitylation and degradation of Sic1 then leads to activation of S phase Cdk activities, allowing DNA replication to proceed. A similar inhibitor, Rum1, is targeted by the homologous ligase, SCF$^{Pop1/Pop2}$ in fission yeast (Kominami and Toda 1997; Wolf et al. 1999). Although Sic1 appears to be the most critical target for cell cycle progression in budding yeast, deletion of *SIC1* has revealed another key cell cycle target of SCF^{Cdc4}, as *cdc4 sic1* double mutants arrest in mitosis (Goh and Surana 1999). However, the critical M phase target remains to be identified. Other yeast cell cycle targets of SCF^{Cdc4} are listed in Table 2.

In mammalian and other metazoan cells, the best known target of SCF$^{Cdc4/Fbw7}$ is the G1 cyclin, cyclin E (Koepp et al. 2001; Moberg et al. 2001; Strohmaier et al. 2001). Unlike Sic1, cyclin E contains a phosphodegron sequence that conforms precisely to the optimized consensus (Nash et al. 2001), although a second less efficient phosphodegron can interact with SCF$^{Cdc4/Fbw7}$ when the primary phosphodegron is mutated (Strohmaier et al. 2001). Activation of the primary cyclin E phosphodegron requires autophosphorylation by Cdk2 at a site carboxyerminal to the phosphodegron, which then primes phosphorylation of the phosphodegron itself by the kinase GSK3β (Welcker et al. 2003). Since, unlike Sic1, cyclin E is a positive regulator of cell cycle progression, the inability to degrade cyclin E does not block cell cycle progression. However inability to degrade cyclin E on schedule is associated with chromosome instability and polyploidy (Spruck et al. 1999). It appears that persistence cyclin E during mitosis causes defects in mitosis itself (Rajagopalan et al. 2004) as well as pre-replication complex assembly during mitotic exit (Ekholm-Reed et al. 2004), possibly accounting for these phenotypes. Since genomic instability is a driving force behind human malignancy, it is not surprising that *CDC4/FBW7* has been found to be mutated in a broad spectrum of cancers (Moberg et al. 2001; Strohmaier et al. 2001; Spruck et al. 2002; Calhoun et al. 2003; Mao et al. 2004; Rajagopalan et al. 2004). However, although loss of Cdc4/Fbw7 function confers phonotypes very analogous to those associated with mutational stabilization/deregulation of cyclin E, it is difficult to attribute the tumorigenicity of Cdc4 loss solely to defects in cyclin E turnover, since several other targets of SCF$^{Cdc4/Fbw7}$ are associated with genomic instability and malignancy. Of these, c-Myc is this most noteworthy (Welcker et al. 2004a,b; Yada et al. 2004). c-Myc is a transcription factor associated with entry into S phase and apoptosis, although the critical targets for these responses have not yet been clearly established (Nilsson and Cleveland 2003). Overexpression of c-Myc has been associated with genomic instability (Mai and Mushinski 2003). Loss of Cdc4/Fbw7 function has been shown to result in overexpression and deregulation of c-Myc (Welcker et al. 2004b; Yada et al. 2004), most likely contributing to the overall associated genomic instability. Other possible targets of SCF$^{Cdc4/Fbw7}$ associated with malignancy are cytoplasmic signaling domains of Notch proteins (Gupta-Rossi et al. 2001; Oberg et al. 2001; Wu et al. 2001; Tetzlaff et al. 2004; Tsunematsu et al. 2004). Interestingly, complete loss of Cdc4/Fbw7 may not be necessary to promote tumorigenesis. In a mouse model, haploinsufficiency of Cdc4/Fbw7 was observed in a high percentage of tumors isolated from p53 heterozygous mice (Mao et al. 2004). These data suggest that in some cells, Cdc4/Fbw7 is likely to be rate-limiting for turnover of important SCF targets.

Although β-TrCP has a similar substrate binding motif composed of WD40 repeats, its preferred phosphodegron is quite different. β-TrCP recognizes the sequence D-pS-G-(X)$_n$-pS, where (X)$_n$ can be two or several amino acids and both serines need to phosphorylated (Yaron et al. 1997, 1998; Hattori et al. 1999; Orian et al. 2000). As with, Cdc4/Fbw7 and its phosphodegron, the phosphates of the β-TrCP phosphodegron interact with arginines on the WD40-repeat surface of β-TrCP (Wu et al. 2003). Although several proteins have been shown to be substrates of SCF$^{\beta-TrCP}$, those most relevant to cell cycle control are Emi1 (Guardavaccaro et al. 2003; Margottin-Goguet et al. 2003) and Wee1 (Watanabe et al. 2004). Emi1 is an inhibitor of the APC/C (Reimann et al. 2001a,b; Hsu et al. 2002). It binds to both Cdh1 and Cdc20, preventing them from interacting with the APC/C core. Emi1 accomplishes two important cell cycle functions. Its E2F-driven accumulation at the G1/S boundary downregulates Cdh1 activity, allowing levels of APC/C targets required for S phase and mitosis to begin to rise (Hsu et al. 2002). Notably, stabilization of cyclin A allows further inhibition of Cdh1 via phosphorylation and progression through S phase. Emi1 then couples APC/C activation to mitotic kinase activation. The primary kinase responsible for phosphodegron phosphorylation of Emi1 is Plk (polo-like kinase) (Hansen et al. 2004; Moshe et al. 2004). However, cyclin B-Cdk1 strongly stimulates this reaction, directly linking Emi1 destruction to mitotic entry (Hansen et al. 2004; Moshe et al. 2004). Loss of Emi1 then potentiates activation of APC^{Cdc20} and initiation of anaphase. A second mitotic inhibitor, Wee1, has also been shown to be targeted for turnover by SCF$^{\beta-TrCP}$ (Watanabe et al. 2004). Wee1, a kinase, prevents activation cyclin B-Cdk1 by phosphorylating Cdk1 on tyrosine 15, thereby preventing premature entrance into mitosis. Phosphorylation by

Plk and Cdk1, respectively, creates an unconventional phosphodegron that nevertheless is sufficient to promote β-TrCP binding and ubiquitylation by SCF$^{\beta\text{-TrCP}}$ (Watanabe et al. 2004). Presumably Cdk1-dependent degradation of Wee1, by creating a positive feedback loop, promotes irreversible progression through mitosis. Although it is debatable whether yeast Met30 constitutes a true homolog of vertebrate β-TrCP (it contains only five WD40 repeats instead of seven), it is interesting that SCF^{Met30} binds and ubiquitylates Swe1, the budding yeast Wee1 homologue (Kaiser et al. 1998). However, SCF^{Met30} has other cell cycle roles; *met30* mutants arrest primarily in G1 without buds (Kaiser et al. 1998; Patton et al. 2000). This phenotype can be linked to hyperactivation of the transcription factor Met4 (Patton et al. 2000), which is reversibly downregulated independently of degradation by SCF^{Met30} mediated ubiquitylation (Kaiser et al. 2000). However, the mechanism whereby Met4 hyperactivation confers G1 arrest has remained elusive. Since Met4 is activated in response to low levels of the biosynthetic 1-carbon donor, S-adenosyl methionine (SAM) (Thomas and Surdin-Kerjan 1997), required for synthesis of dTTP, the concomitant G1 arrest may constitute a mechanism to protect cells from initiating DNA replication under limiting deoxynucleotide pool conditions.

The leucine-rich-repeat containing F-box proteins constitute the third class of SCF cofactors with important cell cycle functions. In mammalian cells, the primary targets of SCF^{Skp2} are Cdk inhibitors p27 (Carrano et al. 1999; Sutterluty et al. 1999; Tsvetkov et al. 1999), p21 (Bornstein et al. 2003) and p130 (Tedesco et al. 2002; Bhattacharya et al. 2003). SCF^{Skp2} has also been shown to target the pre-replication complex assembly factor Cdt1 at the G1-S boundary (Li et al. 2003; Liu et al. 2004). Presumably this constitutes part of the mechanism that limits DNA replication to one round per cell cycle. Although Skp2 interaction and ubiquitylation of all of these targets are driven by substrate phosphorylation, no Skp2 phosphodegron sequence has emerged. Since no Skp2-substrate structure has been forthcoming, it is not clear how substrate recognition occurs, or whether all substrates bind to the same site within the leucine-rich repeats. Binding of Skp2 to substrates and their subsequent ubiquitylation, however, require a small cofactor, the protein Cks1 (Ganoth et al. 2001; Spruck et al. 2001), which itself binds near the carboxy terminus of Skp2 (Wang et al. 2003, 2004). The mechanism whereby Cks1 facilitates substrate interactions has not been elucidated, although several have been proposed, ranging from conferring a permissible conformation on Skp2 (Xu et al. 2003) to serving as a substrate-binding adapter (Ganoth et al. 2001). It has been suggested that an anion binding pocket demonstrated on one surface of Cks1 directly binds substrate phosphates (Ganoth et al. 2001). In addition, the ability of Cks1 to bind Cdks with high affinity appears to help dock substrates, most if not all of which are delivered as cyclin/Cdk complexes (Sitry et al. 2002). Cells from mice nullizygous for either *SKP2* or *CKS1* accumulate high levels of Cdk inhibitors and grow slowly, consistent with turnover of Cdk inhibitors being the limiting function of SCF^{Skp2} (Nakayama et al. 2000; Spruck et al. 2001). In budding yeast, the F-box protein Grr1 participates in an SCF ubiquitin ligase that presumably has structural similarities to SCF^{Skp2}. However, whether Grr1 and Skp2 constitute true structural homologs is not clear, since as with Skp2, no phosphodegron consensus has emerged. Indeed, a reconstituted recombinant SCF^{Grr1} system capable of ubiquitylating phosphorylated Cln1 (the yeast G1 cyclin) has been described, which unlike SCF^{Skp2}, does not require a cofactor for robust activity (Skowyra et al. 1999). This suggests that Grr1 and Skp2 are not structural homologues in the true sense. Modeling and mutational analysis of Grr1 suggests that, as with the WD-40 repeat-containing F-box proteins, positively charged residues on the concave surface formed by leucine-rich repeats are important for binding phosphorylated substrates (Hsiung et al. 2001). Finally, in addition to G1 cyclins, SCF^{Grr1} has been shown to ubiquitylate Gic2, an effector of the Cdc42 GTPase involved in establishment of cell polarity and budding, thus targeting it for turnover at the G1-S phase boundary immediately subsequent to bud emergence (Jaquenoud et al. 1998).

3.6
Regulation of SCF activity

Although, as stated above, SCF action is largely regulated by substrate phosphorylation, regulation of F-box protein components can also modulate substrate targeting. A case in point is Skp2. During G1, Skp2 is maintained at a low steady state level by ubiquitin-mediated proteolysis mediated by APC^{Cdh1} (Bashir et al. 2004; Wei et al. 2004). Presumably, this allows SCF^{Skp2} targets, e.g. p21, p27, p130 and Ctd1 to accumulate during G1, where their functions are required. At the G1-S phase boundary, when APC^{Cdh1} becomes inactive, Skp2 accumulates and these proteins are degraded, consistent with their biological functions. Thus, there appears to be a reciprocal relationship between SCF^{Skp2}, which targets proteins that normally need to accumulate in G1 but be restricted from other intervals of the cell cycle and APC^{Cdh1}, which targets proteins that need to be restricted from G1 but need to accumulate subsequent to G1. This cycle is enforced by direct targeting of Skp2 by APC^{Cdh1} during G1 (Bashir et al. 2004; Wei et al. 2004), and inhibition of Cdh1 mediated indirectly by targeting of Cdk inhibitors by SCF^{Skp2} subsequent to G1 (Carrano et al. 1999; Sutterluty et al. 1999; Tsvetkov et al. 1999; Tedesco et al. 2002; Bhattacharya et al. 2003; Bornstein et al. 2003). Activation of Cdks, notably cyclin A/Cdk2, via degradation of inhibitors promotes phosphorylation-dependent inactivation of Cdh1 (Lukas et al. 1999; Sorensen et al. 2001).

Although Cdc4/Fbw7 does not appear to undergo cell cycle regulation in mammalian cells, it is upregulated in response to genotoxic stress

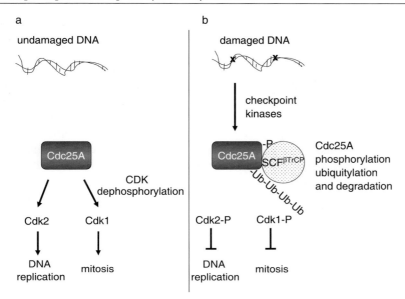

Fig. 5 DNA damage checkpoints and Cdc25A. In the absence of damaged DNA, active Cdc25A dephosphorylates Cdk2 and Cdk1, promoting DNA replication and mitosis. In the presence of DNA damage, checkpoint kinases activate the Cdc25A phosphodegron, leading to ubiquitylation by SCF$^{\beta-TrCP}$ and degradation by the proteasome. Cdk2 and Cdk1 become phosphorylated at negative regulatory sites, leading to inhibition of DNA replication and mitosis

from the phosphodegron (Donzelli et al. 2004). It has been suggested that phosphorylation by Chk1 and/or Chk2 primes Cdc25A for phosphorylation of its phosphodegron by an unidentified kinase, leading to ubiquitylation by SCF$^{\beta-TrCP}$ (Donzelli et al. 2004). Another checkpoint target regulated by ubiquitin-mediated proteolysis is prereplication complex component Cdt1. Normally, Cdt1 is degraded by SCF^{Skp2} at the G1/S boundary, presumably to prevent rereplication once origins have been fired (Li et al. 2003; Liu et al. 2004). However, UV-mediated DNA damage in G1 leads to phosphorylation and degradation of Cdt1 by SCF^{Skp2} prior to origin firing, thus blocking initiation of DNA replication (Kondo et al. 2004).

It should be noted that one of the key checkpoint proteins in vertebrates, p53, is largely regulated at the level of ubiquitin-mediated proteolysis (Momand et al. 2000). p53, a transcription factor that is activated in response to variety of cellular stresses, evokes a number of responses ranging from cell cycle arrest to programmed cell death, depending on the cell type and the situation (Oren 1999; Prives and Hall 1999; Stewart and Pietenpol 2001; Wahl and Carr 2001). Accumulation of p53 in response to DNA damage is mediated to a large extent by reducing its intrinsically high rate of turnover. Ubiquitylation of p53 is carried out by a single-polypeptide ubiquitin ligase, Mdm2

(Honda et al. 1997; Fang et al. 2000; Honda and Yasuda 2000), which binds directly to p53 at its amino terminal transactivation domain(Momand et al. 1992; Chen et al. 1993; Oliner et al. 1993; Kussie et al. 1996). Checkpoint activation, though a variety of phosphorylation-based mechanisms (Banin et al. 1998; Canman et al. 1998; Maya et al. 2001), interferes with Mdm2-mediated ubiquitylation of p53, leading to p53 accumulation and transcription of p53 target genes. One of the best characterized of these is *CDKN1A*, encoding the Cdk inhibitor p21^{Cip1} (Harper et al. 1993; Dulic et al. 1994; el-Deiry et al. 1994), which links p53 stabilization/activation to Cdk inhibition and cell cycle arrest in G1. Another transcriptional target of p53 is the Mdm2 ubiquitin ligase itself (Barak et al. 1993). Thus in addition to stimulating effector pathways that establish checkpoint responses, activated p53 also sets up a negative feedback loop that ultimately leads to its downregulation (Wu et al. 1993).

5
Atypical roles of proteasomes and ubiquitylation

In recent years, it has become apparent that proteasomes and ubiquitylation play roles in cell cycle control in contexts removed from protein degradation. For example, mutations in the ATPase subunits of the 19S regulatory component of the proteasome confer a mitotic arrest in yeast (Ghislain et al. 1993), whereas mutations in the 20S catalytic core do not. This suggests that non-proteolytic functions of the proteasome may be required for progression through mitosis. Indeed, it has been shown that 19S but not 20S proteasome function is required for efficient transcription of the *CDC20* gene accounting for mitotic arrest in 19S mutants (Morris et al. 2003). However, this phenomenon appears to be part of a larger requirement of 19S proteasome function for efficient transcription of many genes in yeast, of which *CDC20* appears to be the most limiting for proliferation (Ferdous et al. 2001, 2002; Yu et al. 2005). This interpretation that 20S function is not required for these transcriptional functions, however, comes with the caveat that it is difficult to completely inactivate proteolytic functions of the proteasome mutationally due to the presence of multiple catalytic sites. What these putative non-proteolytic functions of the proteasome might be remains to be determined, although it has been proposed that the ATPases of the 19S regulatory particle might be used to remodel chromatin in the context of transcriptional elongation.

Ubiquitylation can be used to mark proteins for fates other than proteolysis. The DNA replication processivity factor PCNA can be monoubiquitylated, polyubiquitylated or sumoylated (SUMO is a ubiquitin-like protein that can be conjugated to lysines via an isopeptide bond in a similar manner to ubiquitin) at the same residue (Hoege et al. 2002; Stelter and Ulrich 2003; Haracska et al. 2004). Whereas sumoylation targets PCNA to normal replicative complexes, polyubiquitylation is associated with error-free repair synthesis (Stelter and Ulrich 2003). Monoubiquitylation, however, targets PCNA to error prone DNA polymerases for translesion DNA synthesis (Stelter and Ulrich 2003).

As stated above, SCF^{Met30} is responsible for down-regulating the transcription factor Met4, thus preventing cell cycle arrest. However, in this context Met4 is not targeted for destruction. Under repressive conditions (high intracellular SAM levels) a short K48 linked polyubiquitin chain is attached to a single lysine (Kaiser et al. 2000; Flick et al. 2004). This modification does not target Met4 to the proteasome but neutralizes its transactivation functions (Kaiser et al. 2000; Flick et al. 2004). Thus in this context, ubiquitylation serves as a reversible regulatory switch, similar to phosphorylation. What limits the extent of Met4 ubiquitylation to a single short chain and why ubiquitylated Met4 is not recognized by the proteasome as a substrate are questions that remain unanswered.

6
Deubiquitylating enzymes

Ubiquitylation is a dynamic process, as there is evidence that the net level of protein ubiquitylation is based on both ubiquitin conjugation as well as the activities of specialized proteases that remove ubiquitin and ubiquitin chains. Genomic analysis in yeast suggests the existence of 17 genes encoding deubiquitylating enzymes or DUBs (Amerik et al. 2000), and many more presumably in mammalian cells. However, there is no strong growth or division phenotype in yeast associated with deletion of any of these (Amerik et al. 2000). This suggests either that there is a significant degree of redundancy in any cell cycle regulatory pathway depending on DUBs, or that active deubiquitylation is not important for regulating important cell cycle proteins. More extensive genetic analysis will be required to address this issue.

7
Conclusions

Research over the past 20 years has revealed that ubiquitin-mediated proteolysis is one of the principal regulatory motifs of cell cycle control. If there is any pervasive theme that provides a rationale for this, it is that regulated proteolysis of negative regulators of cell cycle transitions provides the advantage of irreversibility. For the cell division process in particular, the unidirectional progression through the transitions that make up the cell cycle are critical to maintain the integrity of the cell, particularly of its genetic contents. Negative regulatory barriers in the form of inhibitory proteins are inserted at key transitions to assure that all requirements for proceeding further are met. The two notable transitions that might be considered "points of no return" are the G1–S and metaphase-anaphase transitions. From the perspective of genomic integrity, it is easy to see why entry into and completion of S phase or initiation and completion of anaphase must be decisive and irreversible. Reversibility of either transition would almost certainly lead to loss of genomic integrity. The classical example of an S phase barrier is the Sic1 Cdk inhibitor of yeast. Sic1 accumulates during G1 and maintains S phase cyclin/Cdk complexes in an inhibited state (Schwob et al. 1994). Although targeted for ubiquitylation and destruction by SCF^{Cdc4}, Sic1 contains no consensus high-affinity phosphodegrons (Nash et al. 2001). Instead, multiple low affinity phosphodegrons need to be mobilized by phoshorylation in order for Sic1 to interact with Cdc4 (Nash et al. 2001). The need for multiple phosphorylation events couples Sic1 degradation to accumulation of high levels of G1 cyclins (Clns), and accumulation of these cyclins and generation of high levels of Cln–Cdk1 activity is linked to signaling pathways that insure that the cell is prepared for S phase. Thus Sic1 not only provides a barrier to S phase but also an intrinsic buffering system that responds to the amplitude of the forward signaling machinery. The irreversible degradation of Sic1 by SCF^{Cdc4} and activation of S phase Cdks assures that sufficient Cdk activity will be available for completion of a round of DNA replication and that this activity will remain sufficiently high until mitosis to prevent rereplication. In an analogous fashion, securin poses a barrier to anaphase (Cohen-Fix et al. 1996; Zou et al. 1999). Ubiquitin-dependent proteolysis of securin by APC^{Cdc20} leads to irreversible activation of the protease separase enforcing anaphase (Visintin et al. 1997; Ciosk et al. 1998; Waizenegger et al. 2002). Securin can only be ubiquitylated when negative regulatory signals impacting on the cofactor Cdc20 or on securin itself are relieved, which occurs only when bipolar chromosome attachment to a properly constituted spindle is achieved (Lew and Burke 2003). Thus, securin provides a barrier to anaphase, allowing all the essential prerequisites to be met, and its irreversible destruction by APC^{Cdc20} ensures that at the moment when preconditions are met, anaphase occurs without delay. Underscoring the importance of both of these proteolysis-based regulatory schemes for genome integrity and survival, deletion of either *SIC1* (Lengronne and Schwob 2002) or *PDS1* (encoding securin) (Yamamoto et al. 1996a) in yeast confers extreme genomic instability. Such levels of genomic instability would certainly impact the long-term survivability of a yeast population and lead to early embryonic lethality in a metazoan.

References

Agarwal R, Tang Z, Yu H, Cohen-Fix O (2003) Two distinct pathways for inhibiting pds1 ubiquitination in response to DNA damage. J Biol Chem 278:45027-44033

Alexandru G, Uhlmann F, Mechtler K, Poupart MA, Nasmyth K (2001) Phosphorylation of the cohesin subunit Scc1 by Polo/Cdc5 kinase regulates sister chromatid separation in yeast. Cell 105:459-472

Amerik AY, Li SJ, Hochstrasser M (2000) Analysis of the deubiquitinating enzymes of the yeast *Saccharomyces cerevisiae*. Biol Chem 381:981-992

Bai C, Sen P, Hofmann K, Ma L, Goebl M, Harper JW, Elledge SJ (1996) SKP1 connects cell cycle regulators to the ubiquitin proteolysis machinery through a novel motif, the F-box. Cell 86:263-274

Banin S, Moyal L, Shieh S, Taya Y, Anderson CW, Chessa L, Smorodinsky NI, Prives C, Reiss Y, Shiloh Y, Ziv Y (1998) Enhanced phosphorylation of p53 by ATM in response to DNA damage. Science 281:1674-1677

Barak Y, Juven T, Haffner R, Oren M (1993) mdm2 expression is induced by wild type p53 activity. Embo J 12:461-468

Barbey R, Baudouin-Cornu P, Lee TA, Rouillon A, Zarzov P, Tyers M, Thomas D (2005) Inducible dissociation of SCF^{Met30} ubiquitin ligase mediates a rapid transcriptional response to cadmium. EMBO J 24:521-532

Bashir T, Dorrello NV, Amador V, Guardavaccaro D, Pagano M (2004) Control of the SCF$^{Skp2-Cks1}$ ubiquitin ligase by the APC/C^{Cdh1} ubiquitin ligase. Nature 428:190-193

Bhattacharya S, Garriga J, Calbo J, Yong T, Haines DS, Grana X (2003) SKP2 associates with p130 and accelerates p130 ubiquitylation and degradation in human cells. Oncogene 22:2443-2451

Bocca SN, Muzzopappa M, Silberstein S, Wappner P (2001) Occurrence of a putative SCF ubiquitin ligase complex in Drosophila. Biochem Biophys Res Commun 286:357-364

Bornstein G, Bloom J, Sitry-Shevah D, Nakayama K, Pagano M, Hershko A (2003) Role of SCF^{Skp2} ubiquitin ligase in the degradation of p21^{Cip1} during S-phase. J Biol Chem 278:25752-25757

Busino L, Donzelli M, Chiesa M, Guardavaccaro D, Ganoth D, Dorrello NV, Hershko A, Pagano M, Draetta GF (2003) Degradation of Cdc25A by beta-TrCP during S phase and in response to DNA damage. Nature 426:87-91

Calhoun ES, Jones JB, Ashfaq R, Adsay V, Baker SJ, Valentine V, Hempen PM, Hilgers W, Yeo CJ, Hruban RH, Kern SE (2003) BRAF and FBXW7 (CDC4, FBW7, AGO, SEL10) mutations in distinct subsets of pancreatic cancer: potential therapeutic targets. Am J Pathol 163:1255-1260

Canman CE, Lim DS, Cimprich KA, Taya Y, Tamai K, Sakaguchi K, Appella E, Kastan MB, Siliciano JD (1998) Activation of the ATM kinase by ionizing radiation and phosphorylation of p53. Science 281:1677-1679

Carrano AC, Eytan E, Hershko A, Pagano M (1999) SKP2 is required for ubiquitin-mediated degradation of the CDK inhibitor p27. Nat Cell Biol 1:193-199

Chan GK, Yen TJ (2003) The mitotic checkpoint: a signaling pathway that allows a single unattached kinetochore to inhibit mitotic exit. Prog Cell Cycle Res 5:431-439

Chan GK, Jablonski SA, Sudakin V, Hittle JC, Yen TJ (1999) Human BUBR1 is a mitotic checkpoint kinase that monitors CENP-E functions at kinetochores and binds the cyclosome/APC. J Cell Biol 146:941-954

Chan GK, Jablonski SA, Starr DA, Goldberg ML, Yen TJ (2000) Human Zw10 and ROD are mitotic checkpoint proteins that bind to kinetochores. Nat Cell Biol 2:944-947

Chen J, Marechal V, Levine AJ (1993) Mapping of the p53 and mdm-2 interaction domains. Mol Cell Biol 13:4107–4114

Chen RH, Waters JC, Salmon ED, Murray AW (1996) Association of spindle assembly checkpoint component XMAD2 with unattached kinetochores. Science 274:242–246

Chen Z, Pickart CM (1990) A 25-kilodalton ubiquitin carrier protein (E2) catalyzes multi-ubiquitin chain synthesis via lysine 48 of ubiquitin. J Biol Chem 265:21835–21842

Ciosk R, Zachariae W, Michaelis C, Shevchenko A, Mann M, Nasmyth K (1998) An ESP1/PDS1 complex regulates loss of sister chromatid cohesion at the metaphase to anaphase transition in yeast. Cell 93:1067–1076

Clarke DJ, Mondesert G, Segal M, Bertolaet BL, Jensen S, Wolff M, Henze M, Reed SI (2001a) Dosage suppressors of pds1 implicate ubiquitin-associated domains in checkpoint control. Mol Cell Biol 21:1997–2007

Clarke DJ, Segal M, Jensen S, Reed SI (2001b) Mec1p regulates Pds1p levels in S phase: complex coordination of DNA replication and mitosis. Nat Cell Biol 3:619–627

Clarke DJ, Segal M, Andrews CA, Rudyak SG, Jensen S, Smith K, Reed SI (2003) S-phase checkpoint controls mitosis via an APC-independent Cdc20p function. Nat Cell Biol 5:928–935

Cohen-Fix O, Koshland D (1997) The anaphase inhibitor of Saccharomyces cerevisiae Pds1p is a target of the DNA damage checkpoint pathway. Proc Natl Acad Sci USA 94:14361–14366

Cohen-Fix O, Peters JM, Kirschner MW, Koshland D (1996) Anaphase initiation in *Saccharomyces cerevisiae* is controlled by the APC-dependent degradation of the anaphase inhibitor Pds1p. Genes Dev 10:3081–3093

Connelly C, Hieter P (1996) Budding yeast SKP1 encodes an evolutionarily conserved kinetochore protein required for cell cycle progression. Cell 86:275–285

Davis ES, Wille L, Chestnut BA, Sadler PL, Shakes DC, Golden A (2002) Multiple subunits of the *Caenorhabditis elegans* anaphase-promoting complex are required for chromosome segregation during meiosis I. Genetics 160:805–813

den Elzen N, Pines J (2001) Cyclin A is destroyed in prometaphase and can delay chromosome alignment and anaphase. J Cell Biol 153:121–136

Donzelli M, Busino L, Chiesa M, Ganoth D, Hershko A, Draetta GF (2004) Hierarchical order of phosphorylation events commits Cdc25A to betaTrCP-dependent degradation. Cell Cycle 3:469–471

Donzelli M, Squatrito M, Ganoth D, Hershko A, Pagano M, Draetta GF (2002) Dual mode of degradation of Cdc25 A phosphatase. EMBO J 21:4875–4884

Draetta G, Luca F, Westendorf J, Brizuela L, Ruderman J, Beach D (1989) Cdc2 protein kinase is complexed with both cyclin A and B: evidence for proteolytic inactivation of MPF. Cell 56:829–838

Dulic V, Kaufmann WK, Wilson SJ, Tlsty TD, Lees E, Harper JW, Elledge SJ, Reed SI (1994) p53-dependent inhibition of cyclin-dependent kinase activities in human fibroblasts during radiation-induced G1 arrest. Cell 76:1013–1023

Ekholm-Reed S, Mendez J, Tedesco D, Zetterberg A, Stillman B, Reed SI (2004) Deregulation of cyclin E in human cells interferes with prereplication complex assembly. J Cell Biol 165:789–800

el-Deiry WS, Harper JW, PM OC, Velculescu VE, Canman CE, Jackman J, Pietenpol JA, Burrell M, Hill DE, Wang Y et al. (1994) WAF1/CIP1 is induced in p53-mediated G1 arrest and apoptosis. Cancer Res 54:1169–1174

Elsasser S, Chandler-Militello D, Muller B, Hanna J, Finley D (2004) Rad23 and Rpn10 serve as alternative ubiquitin receptors for the proteasome. J Biol Chem 279:26817–26822

Evans T, Rosenthal ET, Youngblom J, Distel D, Hunt T (1983) Cyclin: a protein specified by maternal mRNA in sea urchin eggs that is destroyed at each cleavage division. Cell 33:389–396

Fang G, Yu H, Kirschner MW (1998) The checkpoint protein MAD2 and the mitotic regulator CDC20 form a ternary complex with the anaphase-promoting complex to control anaphase initiation. Genes Dev 12:1871–1883

Fang G, Yu H, Kirschner MW (1999) Control of mitotic transitions by the anaphase-promoting complex. Philos Trans R Soc Lond B Biol Sci 354:1583–1590

Fang S, Jensen JP, Ludwig RL, Vousden KH, Weissman AM (2000) Mdm2 is a RING finger-dependent ubiquitin protein ligase for itself and p53. J Biol Chem 275:8945–8951

Feldman RM, Correll CC, Kaplan KB, Deshaies RJ (1997) A complex of Cdc4p, Skp1p, and Cdc53p/cullin catalyzes ubiquitination of the phosphorylated CDK inhibitor Sic1p. Cell 91:221–230

Ferdous A, Gonzalez F, Sun L, Kodadek T, Johnston SA (2001) The 19S regulatory particle of the proteasome is required for efficient transcription elongation by RNA polymerase II. Mol Cell 7:981–991

Ferdous A, Kodadek T, Johnston SA (2002) A nonproteolytic function of the 19S regulatory subunit of the 26S proteasome is required for efficient activated transcription by human RNA polymerase II. Biochemistry 41:12798–12805

Flick K, Ouni I, Wohlschlegel JA, Capati C, McDonald WH, Yates JR, Kaiser P (2004) Proteolysis-independent regulation of the transcription factor Met4 by a single Lys 48-linked ubiquitin chain. Nat Cell Biol 6:634–641

Fuchs SY, Chen A, Xiong Y, Pan ZQ, Ronai Z (1999) HOS, a human homolog of Slimb, forms an SCF complex with Skp1 and Cullin1 and targets the phosphorylation-dependent degradation of IkappaB and beta-catenin. Oncogene 18:2039–2046

Ganoth D, Bornstein G, Ko TK, Larsen B, Tyers M, Pagano M, Hershko A (2001) The cell-cycle regulatory protein Cks1 is required for SCF^{Skp2}-mediated ubiquitinylation of p27. Nat Cell Biol 3:321–324

Geley S, Kramer E, Gieffers C, Gannon J, Peters JM, Hunt T (2001) Anaphase-promoting complex/cyclosome-dependent proteolysis of human cyclin A starts at the beginning of mitosis and is not subject to the spindle assembly checkpoint. J Cell Biol 153:137–148

Ghiara JB, Richardson HE, Sugimoto K, Henze M, Lew DJ, Wittenberg C, Reed SI (1991) A cyclin B homolog in S. cerevisiae: chronic activation of the Cdc28 protein kinase by cyclin prevents exit from mitosis. Cell 65:163–174

Ghislain M, Udvardy A, Mann C (1993) S. cerevisiae 26S protease mutants arrest cell division in G2/metaphase. Nature 366:358–362

Gieffers C, Dube P, Harris JR, Stark H, Peters JM (2001) Three-dimensional structure of the anaphase-promoting complex. Mol Cell 7:907–913

Glotzer M, Murray AW, Kirschner MW (1991) Cyclin is degraded by the ubiquitin pathway. Nature 349:132–138

Goebl MG, Yochem J, Jentsch S, McGrath JP, Varshavsky A, Byers B (1988) The yeast cell cycle gene CDC34 encodes a ubiquitin-conjugating enzyme. Science 241:1331–1335

Goh PY, Surana U (1999) Cdc4, a protein required for the onset of S phase, serves an essential function during G2/M transition in *Saccharomyces cerevisiae*. Mol Cell Biol 19:5512–5522

Goloudina A, Yamaguchi H, Chervyakova DB, Appella E, Fornace AJ, Jr, Bulavin DV (2003) Regulation of human Cdc25A stability by Serine 75 phosphorylation is not sufficient to activate a S phase checkpoint. Cell Cycle 2:473–478

Grossberger R, Gieffers C, Zachariae W, Podtelejnikov AV, Schleiffer A, Nasmyth K, Mann M, Peters JM (1999) Characterization of the DOC1/APC10 subunit of the yeast and the human anaphase-promoting complex. J Biol Chem 274:14500–14507

Guardavaccaro D, Kudo Y, Boulaire J, Barchi M, Busino L, Donzelli M, Margottin-Goguet F, Jackson PK, Yamasaki L, Pagano M (2003) Control of meiotic and mitotic progression by the F box protein beta-Trcp1 in vivo. Dev Cell 4:799–812

Gupta-Rossi N, Le Bail O, Gonen H, Brou C, Logeat F, Six E, Ciechanover A, Israel A (2001) Functional interaction between SEL-10, an F-box protein, and the nuclear form of activated Notch1 receptor. J Biol Chem 276:34371–34378

Hansen DV, Loktev AV, Ban KH, Jackson PK (2004) Plk1 regulates activation of the anaphase promoting complex by phosphorylating and triggering SCFbetaTrCP-dependent destruction of the APC inhibitor Emi1. Mol Biol Cell 15:5623–5634

Haracska L, Torres-Ramos CA, Johnson RE, Prakash S, Prakash L (2004) Opposing effects of ubiquitin conjugation and SUMO modification of PCNA on replicational bypass of DNA lesions in *Saccharomyces cerevisiae*. Mol Cell Biol 24:4267–4274

Hardwick KG, Johnston RC, Smith DL, Murray AW (2000) MAD3 encodes a novel component of the spindle checkpoint which interacts with Bub3p, Cdc20p, and Mad2p. J Cell Biol 148:871–882

Harper JW, Adami GR, Wei N, Keyomarsi K, Elledge SJ (1993) The p21 Cdk-interacting protein Cip1 is a potent inhibitor of G1 cyclin-dependent kinases. Cell 75:805–816

Harrison C, Katayama S, Dhut S, Chen D, Jones N, Bahler J, Toda T (2005) SCF^{Pof1}-ubiquitin and its target Zip1 transcription factor mediate cadmium response in fission yeast. EMBO J 24:599–610

Hartwell LH, Culotti J, Pringle JR, Reid BJ (1974) Genetic control of the cell division cycle in yeast. Science 183:46–51

Hassepass I, Voit R, Hoffmann I (2003) Phosphorylation at serine 75 is required for UV-mediated degradation of human Cdc25A phosphatase at the S-phase checkpoint. J Biol Chem 278:29824–29829

Hatakeyama S, Yada M, Matsumoto M, Ishida N, Nakayama KI (2001) U box proteins as a new family of ubiquitin–protein ligases. J Biol Chem 276:33111–33120

Hattori K, Hatakeyama S, Shirane M, Matsumoto M, Nakayama K (1999) Molecular dissection of the interactions among IkappaBalpha, FWD1, and Skp1 required for ubiquitin-mediated proteolysis of IkappaBalpha. J Biol Chem 274:29641–29647

Hershko A (1983) Ubiquitin: roles in protein modification and breakdown. Cell 34:11–12

Hilioti Z, Chung YS, Mochizuki Y, Hardy CF, Cohen-Fix O (2001) The anaphase inhibitor Pds1 binds to the APC/C-associated protein Cdc20 in a destruction box-dependent manner. Curr Biol 11:1347–1352

Hoege C, Pfander B, Moldovan GL, Pyrowolakis G, Jentsch S (2002) RAD6-dependent DNA repair is linked to modification of PCNA by ubiquitin and SUMO. Nature 419:135–141

Honda R, Yasuda H (2000) Activity of MDM2, a ubiquitin ligase, toward p53 or itself is dependent on the RING finger domain of the ligase. Oncogene 19:1473–1476

Honda R, Tanaka H, Yasuda H (1997) Oncoprotein MDM2 is a ubiquitin ligase E3 for tumor suppressor p53. FEBS Lett 420:25–27

Hornig NC, Knowles PP, McDonald NQ, Uhlmann F (2002) The dual mechanism of separase regulation by securin. Curr Biol 12:973–982

Hsiung YG, Chang HC, Pellequer JL, La Valle R, Lanker S, Wittenberg C (2001) F-box protein Grr1 interacts with phosphorylated targets via the cationic surface of its leucine-rich repeat. Mol Cell Biol 21:2506–2520

Hsu JY, Reimann JD, Sorensen CS, Lukas J, Jackson PK (2002) E2F-dependent accumulation of hEmi1 regulates S phase entry by inhibiting APC^{Cdh1}. Nat Cell Biol 4:358–366

Hunt T, Luca FC, Ruderman JV (1992) The requirements for protein synthesis and degradation, and the control of destruction of cyclins A and B in the meiotic and mitotic cell cycles of the clam embryo. J Cell Biol 116:707–724

Hwang LH, Lau LF, Smith DL, Mistrot CA, Hardwick KG, Hwang ES, Amon A, Murray AW (1998) Budding yeast Cdc20: a target of the spindle checkpoint. Science 279:1041–1044

Jablonski SA, Chan GK, Cooke CA, Earnshaw WC, Yen TJ (1998) The hBUB1 and hBUBR1 kinases sequentially assemble onto kinetochores during prophase with hBUBR1 concentrating at the kinetochore plates in mitosis. Chromosoma 107:386–396

Jaquenoud M, Gulli MP, Peter K, Peter M (1998) The Cdc42p effector Gic2p is targeted for ubiquitin-dependent degradation by the SCF^{Grr1} complex. Embo J 17:5360–5373

Jensen S, Segal M, Clarke DJ, Reed SI (2001) A novel role of the budding yeast separin Esp1 in anaphase spindle elongation: evidence that proper spindle association of Esp1 is regulated by Pds1. J Cell Biol 152:27–40

Jin J, Shirogane T, Xu L, Nalepa G, Qin J, Elledge SJ, Harper JW (2003) SCF$^{\beta-TRCP}$ links Chk1 signaling to degradation of the Cdc25A protein phosphatase. Genes Dev 17:3062–3074

Kaiser P, Sia RA, Bardes EG, Lew DJ, Reed SI (1998) Cdc34 and the F-box protein Met30 are required for degradation of the Cdk-inhibitory kinase Swe1. Genes Dev 12:2587–2597

Kaiser P, Flick K, Wittenberg C, Reed SI (2000) Regulation of transcription by ubiquitination without proteolysis: Cdc34/SCF^{Met30}-mediated inactivation of the transcription factor Met4. Cell 102:303–314

Kallio M, Weinstein J, Daum JR, Burke DJ, Gorbsky GJ (1998) Mammalian p55CDC mediates association of the spindle checkpoint protein Mad2 with the cyclosome/anaphase-promoting complex, and is involved in regulating anaphase onset and late mitotic events. J Cell Biol 141:1393–1406

Kimura T, Gotoh M, Nakamura Y, Arakawa H (2003) hCDC4b, a regulator of cyclin E, as a direct transcriptional target of p53. Cancer Sci 94:431–436

King RW, Peters JM, Tugendreich S, Rolfe M, Hieter P, Kirschner MW (1995) A 20S complex containing CDC27 and CDC16 catalyzes the mitosis-specific conjugation of ubiquitin to cyclin B. Cell 81:279–288

King RW, Glotzer M, Kirschner MW (1996) Mutagenic analysis of the destruction signal of mitotic cyclins and structural characterization of ubiquitinated intermediates. Mol Biol Cell 7:1343–1357

Kishi T, Seno T, Yamao F (1998) Grr1 functions in the ubiquitin pathway in *Saccharomyces cerevisiae* through association with Skp1. Mol Gen Genet 257:143–138

Knuutila S, Aalto Y, Autio K, Bjorkqvist AM, El-Rifai W, Hemmer S, Huhta T, Kettunen E, Kiuru-Kuhlefelt S, Larramendy ML, Lushnikova T, Monni O, Pere H, Tapper J, Tarkkanen M, Varis A, Wasenius VM, Wolf M, Zhu Y (1999) DNA copy number losses in human neoplasms. Am J Pathol 155:683–94

Kobe B, Deisenhofer J (1994) The leucine-rich repeat: a versatile binding motif. Trends Biochem Sci 19:415–421

Kobe B, Deisenhofer J (1995) Proteins with leucine-rich repeats. Curr Opin Struct Biol 5:409–416

Koegl M, Hoppe T, Schlenker S, Ulrich HD, Mayer TU, Jentsch S (1999) A novel ubiquitination factor, E4, is involved in multiubiquitin chain assembly. Cell 96:635–644

Koepp DM, Schaefer LK, Ye X, Keyomarsi K, Chu C, Harper JW, Elledge SJ (2001) Phosphorylation-dependent ubiquitination of cyclin E by the SCF^{Fbw7} ubiquitin ligase. Science 294:173–177

Kominami K, Toda T (1997) Fission yeast WD-repeat protein pop1 regulates genome ploidy through ubiquitin–proteasome-mediated degradation of the CDK inhibitor Rum1 and the S-phase initiator Cdc18. Genes Dev 11:1548–1560

Kominami K, Ochotorena I, Toda T (1998) Two F-box/WD-repeat proteins Pop1 and Pop2 form hetero- and homo-complexes together with cullin-1 in the fission yeast SCF (Skp1-Cullin-1-F-box) ubiquitin ligase. Genes Cells 3:721–735

Kondo T, Kobayashi M, Tanaka J, Yokoyama A, Suzuki S, Kato N, Onozawa M, Chiba K, Hashino S, Imamura M, Minami Y, Minamino N, Asaka M (2004) Rapid degradation of Cdt1 upon UV-induced DNA damage is mediated by SCF^{Skp2} complex. J Biol Chem 279:27315–27319

Kraft C, Herzog F, Gieffers C, Mechtler K, Hagting A, Pines J, Peters JM (2003) Mitotic regulation of the human anaphase-promoting complex by phosphorylation. EMBO J 22:6598–609

Kroll M, Margottin F, Kohl A, Renard P, Durand H, Concordet JP, Bachelerie F, Arenzana-Seisdedos F, Benarous R (1999) Inducible degradation of IkappaBalpha by the proteasome requires interaction with the F-box protein h-betaTrCP. J Biol Chem 274:7941–7945

Kussie PH, Gorina S, Marechal V, Elenbaas B, Moreau J, Levine AJ, Pavletich NP (1996) Structure of the MDM2 oncoprotein bound to the p53 tumor suppressor transactivation domain. Science 274:948–953

Lamb JR, Michaud WA, Sikorski RS, Hieter PA (1994) Cdc16p, Cdc23p and Cdc27p form a complex essential for mitosis. EMBO J 13:4321–4328

Latres E, Chiaur DS, Pagano M (1999) The human F box protein beta-Trcp associates with the Cul1/Skp1 complex and regulates the stability of beta-catenin. Oncogene 18:849–854

Lengronne A, Schwob E (2002) The yeast CDK inhibitor Sic1 prevents genomic instability by promoting replication origin licensing in late G1. Mol Cell 9:1067–1078

Lew DJ, Burke DJ (2003) The spindle assembly and spindle position checkpoints. Annu Rev Genet 37:251–282

Li FN, Johnston M (1997) Grr1 of *Saccharomyces cerevisiae* is connected to the ubiquitin proteolysis machinery through Skp1: coupling glucose sensing to gene expression and the cell cycle. EMBO J 16:5629–5638

Li X, Zhao Q, Liao R, Sun P, Wu X (2003) The SCF^{Skp2} ubiquitin ligase complex interacts with the human replication licensing factor Cdt1 and regulates Cdt1 degradation. J Biol Chem 278:30854–30858

Liu E, Li X, Yan F, Zhao Q, Wu X (2004) Cyclin-dependent kinases phosphorylate human Cdt1 and induce its degradation. J Biol Chem 279:17283–17288

Lisztwan J, Marti A, Sutterluty H, Gstaiger M, Wirbelauer C, Krek W (1998) Association of human CUL-1 and ubiquitin-conjugating enzyme CDC34 with the F-box protein p45^{SKP2}: evidence for evolutionary conservation in the subunit composition of the CDC34-SCF pathway. EMBO J 17:368–383

Littlepage LE, Ruderman JV (2002) Identification of a new APC/C recognition domain, the A box, which is required for the Cdh1-dependent destruction of the kinase Aurora-A during mitotic exit. Genes Dev 16:2274–2285

Lorick KL, Jensen JP, Fang S, Ong AM, Hatakeyama S, Weissman AM (1999) RING fingers mediate ubiquitin-conjugating enzyme (E2)-dependent ubiquitination. Proc Natl Acad Sci USA 96:11364–11369

Lukas C, Sorensen CS, Kramer E, Santoni-Rugiu E, Lindeneg C, Peters JM, Bartek J, Lukas J (1999) Accumulation of cyclin B1 requires E2F and cyclin-A-dependent rearrangement of the anaphase-promoting complex. Nature 401:815–818

Lyapina SA, Correll CC, Kipreos ET, Deshaies RJ (1998) Human CUL1 forms an evolutionarily conserved ubiquitin ligase complex (SCF) with SKP1 and an F-box protein. Proc Natl Acad Sci USA 95:7451–7456

Mai S, Mushinski JF (2003) c-Myc-induced genomic instability. J Environ Pathol Toxicol Oncol 22:179–199

Mailand N, Podtelejnikov AV, Groth A, Mann M, Bartek J, Lukas J (2002) Regulation of G2/M events by Cdc25A through phosphorylation-dependent modulation of its stability. EMBO J 21:5911–5920

Mao JH, Perez-Losada J, Wu D, Delrosario R, Tsunematsu R, Nakayama KI, Brown K, Bryson S, Balmain A (2004) Fbxw7/Cdc4 is a p53-dependent, haploinsufficient tumour suppressor gene. Nature 432:775–779

Margottin-Goguet F, Hsu JY, Loktev A, Hsieh HM, Reimann JD, Jackson PK (2003) Prophase destruction of Emi1 by the SCF$^{\beta\text{-TrCP/Slimb}}$ ubiquitin ligase activates the anaphase promoting complex to allow progression beyond prometaphase. Dev Cell 4:813–826

Maya R, Balass M, Kim ST, Shkedy D, Leal JF, Shifman O, Moas M, Buschmann T, Ronai Z, Shiloh Y, Kastan MB, Katzir E, Oren M (2001) ATM-dependent phosphorylation of Mdm2 on serine 395: role in p53 activation by DNA damage. Genes Dev 15:1067–1077

Meijer L, Arion D, Golsteyn R, Pines J, Brizuela L, Hunt T, Beach D (1989) Cyclin is a component of the sea urchin egg M-phase specific histone H1 kinase. EMBO J 8:2275–2282

Moberg KH, Bell DW, Wahrer DC, Haber DA, Hariharan IK (2001) Archipelago regulates cyclin E levels in Drosophila and is mutated in human cancer cell lines. Nature 413:311–316

Momand J, Zambetti GP, Olson DC, George D, Levine AJ (1992) The mdm-2 oncogene product forms a complex with the p53 protein and inhibits p53-mediated transactivation. Cell 69:1237–1245

Momand J, Wu HH, Dasgupta G (2000) MDM2-master regulator of the p53 tumor suppressor protein. Gene 242:15–29

Morris MC, Kaiser P, Rudyak S, Baskerville C, Watson MH, Reed SI (2003) Cks1-dependent proteasome recruitment and activation of CDC20 transcription in budding yeast. Nature 423:1009–1013

Moshe Y, Boulaire J, Pagano M, Hershko A (2004) Role of Polo-like kinase in the degradation of early mitotic inhibitor 1, a regulator of the anaphase promoting complex/cyclosome. Proc Natl Acad Sci USA 101:7937–7942

Murray AW, Solomon MJ, Kirschner MW (1989) The role of cyclin synthesis and degradation in the control of maturation promoting factor activity. Nature 339:280–286

Nakayama K, Nagahama H, Minamishima YA, Matsumoto M, Nakamichi I, Kitagawa K, Shirane M, Tsunematsu R, Tsukiyama T, Ishida N, Kitagawa M, Nakayama K, Hatakeyama S (2000) Targeted disruption of Skp2 results in accumulation of cyclin E and p27$^{\text{Kip1}}$, polyploidy and centrosome overduplication. EMBO J 19:2069–2081

Nash P, Tang X, Orlicky S, Chen Q, Gertler FB, Mendenhall MD, Sicheri F, Pawson T, Tyers M (2001) Multisite phosphorylation of a CDK inhibitor sets a threshold for the onset of DNA replication. Nature 414:514–521

Nilsson JA, Cleveland JL (2003) Myc pathways provoking cell suicide and cancer. Oncogene 22:9007–9021

Nugroho TT, Mendenhall MD (1994) An inhibitor of yeast cyclin-dependent protein kinase plays an important role in ensuring the genomic integrity of daughter cells. Mol Cell Biol 14:3320–3328

Oberg C, Li J, Pauley A, Wolf E, Gurney M, Lendahl U (2001) The Notch intracellular domain is ubiquitinated and negatively regulated by the mammalian Sel-10 homolog. J Biol Chem 276:35847–35853

Ohta T, Michel JJ, Schottelius AJ, Xiong Y (1999) ROC1, a homolog of APC11, represents a family of cullin partners with an associated ubiquitin ligase activity. Mol Cell 3:535–541

Oliner JD, Pietenpol JA, Thiagalingam S, Gyuris J, Kinzler KW, Vogelstein B (1993) Oncoprotein MDM2 conceals the activation domain of tumour suppressor p53. Nature 362:857–860

Oren M (1999) Regulation of the p53 tumor suppressor protein. J Biol Chem 274:36031–36034

Orian A, Gonen H, Bercovich B, Fajerman I, Eytan E, Israel A, Mercurio F, Iwai K, Schwartz AL, Ciechanover A (2000) SCF$^{\beta\text{-TrCP}}$ ubiquitin ligase-mediated processing of NF-kappaB p105 requires phosphorylation of its C-terminus by IkappaB kinase. EMBO J 19:2580–2591

Orlicky S, Tang X, Willems A, Tyers M, Sicheri F (2003) Structural basis for phosphodependent substrate selection and orientation by the SCF^{Cdc4} ubiquitin ligase. Cell 112:243–256

Patton EE, Peyraud C, Rouillon A, Surdin-Kerjan Y, Tyers M, Thomas D (2000) SCF^{Met30}-mediated control of the transcriptional activator Met4 is required for the G1–S transition. EMBO J 19:1613–1624

Patton EE, Willems AR, Sa D, Kuras L, Thomas D, Craig KL, Tyers M (1998) Cdc53 is a scaffold protein for multiple Cdc34/Skp1/F-box protein complexes that regulate cell division and methionine biosynthesis in yeast. Genes Dev 12:692–705

Peter M, Castro A, Lorca T, Le Peuch C, Magnaghi-Jaulin L, Doree M, Labbe JC (2001) The APC is dispensable for first meiotic anaphase in *Xenopus* oocytes. Nat Cell Biol 3:83–87

Pfleger CM, Kirschner MW (2000) The KEN box: an APC recognition signal distinct from the D box targeted by Cdh1. Genes Dev 14:655–665

Pfleger CM, Lee E, Kirschner MW (2001) Substrate recognition by the Cdc20 and Cdh1 components of the anaphase-promoting complex. Genes Dev 15:2396–2407

Pickart CM, Cohen RE (2004) Proteasomes and their kin: proteases in the machine age. Nat Rev Mol Cell Biol 5:177–187

Pines J, Hunter T (1990) Human cyclin A is adenovirus E1A-associated protein p60 and behaves differently from cyclin B. Nature 346:760–763

Prinz S, Hwang ES, Visintin R, Amon A (1998) The regulation of Cdc20 proteolysis reveals a role for APC components Cdc23 and Cdc27 during S phase and early mitosis. Curr Biol 8:750–760

Prives C, Hall PA (1999) The p53 pathway. J Pathol 187:112–126

Rajagopalan H, Jallepalli PV, Rago C, Velculescu VE, Kinzler KW, Vogelstein B, Lengauer C (2004) Inactivation of hCDC4 can cause chromosomal instability. Nature 428:77–81

Rape M, Kirschner MW (2004) Autonomous regulation of the anaphase-promoting complex couples mitosis to S-phase entry. Nature 432:588–595

Reimann JD, Freed E, Hsu JY, Kramer ER, Peters JM, Jackson PK (2001a) Emi1 is a mitotic regulator that interacts with Cdc20 and inhibits the anaphase promoting complex. Cell 105:645–655

Reimann JD, Gardner BE, Margottin-Goguet F, Jackson PK (2001b) Emi1 regulates the anaphase-promoting complex by a different mechanism than Mad2 proteins. Genes Dev 15:3278–3285

Russell ID, Grancell AS, Sorger PK (1999) The unstable F-box protein p58-Ctf13 forms the structural core of the CBF3 kinetochore complex. J Cell Biol 145:933–950

Salah SM, Nasmyth K (2000) Destruction of the securin Pds1p occurs at the onset of anaphase during both meiotic divisions in yeast. Chromosoma 109:27–34

Sanchez Y, Bachant J, Wang H, Hu F, Liu D, Tetzlaff M, Elledge SJ (1999) Control of the DNA damage checkpoint by chk1 and rad53 protein kinases through distinct mechanisms. Science 286:1166–1171

Schulman BA, Carrano AC, Jeffrey PD, Bowen Z, Kinnucan ER, Finnin MS, Elledge SJ, Harper JW, Pagano M, Pavletich NP (2000) Insights into SCF ubiquitin ligases from the structure of the Skp1–Skp2 complex. Nature 408:381–386

Schwab M, Lutum AS, Seufert W (1997) Yeast Hct1 is a regulator of Clb2 cyclin proteolysis. Cell 90:683–693

Schwab M, Neutzner M, Mocker D, Seufert W (2001) Yeast Hct1 recognizes the mitotic cyclin Clb2 and other substrates of the ubiquitin ligase APC. EMBO J 20:5165–5175

Schwob E, Bohm T, Mendenhall MD, Nasmyth K (1994) The B-type cyclin kinase inhibitor p40^{SIC1} controls the G1 to S transition in *S. cerevisiae*. Cell 79:233–244

Seol JH, Feldman RM, Zachariae W, Shevchenko A, Correll CC, Lyapina S, Chi Y, Galova M, Claypool J, Sandmeyer S, Nasmyth K, Deshaies RJ (1999) Cdc53/cullin and the essential Hrt1 RING-H2 subunit of SCF define a ubiquitin ligase module that activates the E2 enzyme Cdc34. Genes Dev 13:1614–1626

Shirane M, Hatakeyama S, Hattori K, Nakayama K (1999) Common pathway for the ubiquitination of IkappaBalpha, IkappaBbeta, and IkappaBepsilon mediated by the F-box protein FWD1. J Biol Chem 274:28169–28174

Shirayama M, Toth A, Galova M, Nasmyth K (1999) APC^{Cdc20} promotes exit from mitosis by destroying the anaphase inhibitor Pds1 and cyclin Clb5. Nature 402:203–207

Sikorski RS, Boguski MS, Goebl M, Hieter P (1990) A repeating amino acid motif in CDC23 defines a family of proteins and a new relationship among genes required for mitosis and RNA synthesis. Cell 60:307–317

Sitry D, Seeliger MA, Ko TK, Ganoth D, Breward SE, Itzhaki LS, Pagano M, Hershko A (2002) Three different binding sites of Cks1 are required for p27-ubiquitin ligation. J Biol Chem 277:42233–42240

Skowyra D, Craig KL, Tyers M, Elledge SJ, Harper JW (1997) F-box proteins are receptors that recruit phosphorylated substrates to the SCF ubiquitin–ligase complex. Cell 91:209–219

Skowyra D, Koepp DM, Kamura T, Conrad MN, Conaway RC, Conaway JW, Elledge SJ, Harper JW (1999) Reconstitution of G1 cyclin ubiquitination with complexes containing SCF^{Grr1} and Rbx1. Science 284:662–665

Sorensen CS, Lukas C, Kramer ER, Peters JM, Bartek J, Lukas J (2001) A conserved cyclin-binding domain determines functional interplay between anaphase-promoting complex–Cdh1 and cyclin A–Cdk2 during cell cycle progression. Mol Cell Biol 21:3692–3703

Spruck CH, Won KA, Reed SI (1999) Deregulated cyclin E induces chromosome instability. Nature 401:297–300

Spruck C, Strohmaier H, Watson M, Smith AP, Ryan A, Krek TW, Reed SI (2001) A CDK-independent function of mammalian Cks1: targeting of SCF^{Skp2} to the CDK inhibitor p27^{Kip1}. Mol Cell 7:639–650

Spruck CH, Strohmaier H, Sangfelt O, Muller HM, Hubalek M, Muller-Holzner E, Marth C, Widschwendter M, Reed SI (2002) hCDC4 gene mutations in endometrial cancer. Cancer Res 62:4535–4539

Stegmeier F, Visintin R, Amon A (2002) Separase, polo kinase, the kinetochore protein Slk19, and Spo12 function in a network that controls Cdc14 localization during early anaphase. Cell 108:207–220

Stelter P, Ulrich HD (2003) Control of spontaneous and damage-induced mutagenesis by SUMO and ubiquitin conjugation. Nature 425:188–191

Stewart ZA, Pietenpol JA (2001) p53 Signaling and cell cycle checkpoints. Chem Res Toxicol 14:243–263

Strohmaier H, Spruck CH, Kaiser P, Won KA, Sangfelt O, Reed SI (2001) Human F-box protein hCdc4 targets cyclin E for proteolysis and is mutated in a breast cancer cell line. Nature 413:316–322

Sudakin V, Ganoth D, Dahan A, Heller H, Hershko J, Luca FC, Ruderman JV, Hershko A (1995) The cyclosome, a large complex containing cyclin-selective ubiquitin ligase activity, targets cyclins for destruction at the end of mitosis. Mol Biol Cell 6:185–197

Sullivan M, Lehane C, Uhlmann F (2001) Orchestrating anaphase and mitotic exit: separase cleavage and localization of Slk19. Nat Cell Biol 3:771–777

Surana U, Amon A, Dowzer C, McGrew J, Byers B, Nasmyth K (1993) Destruction of the CDC28/CLB mitotic kinase is not required for the metaphase to anaphase transition in budding yeast. Embo J 12:1969–1978

Sutterluty H, Chatelain E, Marti A, Wirbelauer C, Senften M, Muller U, Krek W (1999) p45^{SKP2} promotes p27^{Kip1} degradation and induces S phase in quiescent cells. Nat Cell Biol 1:207–214

Suzuki H, Chiba T, Kobayashi M, Takeuchi M, Suzuki T, Ichiyama A, Ikenoue T, Omata M, Furuichi K, Tanaka K (1999) IkappaBalpha ubiquitination is catalyzed by an SCF-like complex containing Skp1, cullin-1, and two F-box/WD40-repeat proteins, betaTrCP1 and betaTrCP2. Biochem Biophys Res Commun 256:127–132

Suzuki H, Chiba T, Suzuki T, Fujita T, Ikenoue T, Omata M, Furuichi K, Shikama H, Tanaka K (2000) Homodimer of two F-box proteins betaTrCP1 or betaTrCP2 binds to IkappaBalpha for signal-dependent ubiquitination. J Biol Chem 275:2877–2884

Swenson KI, Farrell KM, Ruderman JV (1986) The clam embryo protein cyclin A induces entry into M phase and the resumption of meiosis in *Xenopus* oocytes. Cell 47:861–870

Taieb FE, Gross SD, Lewellyn AL, Maller JL (2001) Activation of the anaphase-promoting complex and degradation of cyclin B is not required for progression from Meiosis I to II in *Xenopus* oocytes. Curr Biol 11:508–513

Tan P, Fuchs SY, Chen A, Wu K, Gomez C, Ronai Z, Pan ZQ (1999) Recruitment of a ROC1–CUL1 ubiquitin ligase by Skp1 and HOS to catalyze the ubiquitination of IkappaB alpha. Mol Cell 3:527–533

Tang Z, Bharadwaj R, Li B, Yu H (2001) Mad2-independent inhibition of APC^{Cdc20} by the mitotic checkpoint protein BubR1. Dev Cell 1:227–237

Tedesco D, Lukas J, Reed SI (2002) The pRb-related protein p130 is regulated by phosphorylation-dependent proteolysis via the protein–ubiquitin ligase SCF^{Skp2}. Genes Dev 16:2946–2957

Terret ME, Wassmann K, Waizenegger I, Maro B, Peters JM, Verlhac MH (2003) The meiosis I-to-meiosis II transition in mouse oocytes requires separase activity. Curr Biol 13:1797–1802

Tetzlaff MT, Yu W, Li M, Zhang P, Finegold M, Mahon K, Harper JW, Schwartz RJ, Elledge SJ (2004) Defective cardiovascular development and elevated cyclin E and Notch proteins in mice lacking the Fbw7 F-box protein. Proc Natl Acad Sci USA 101:3338–3345

Thomas D, Surdin-Kerjan Y (1997) Metabolism of sulfur amino acids in *Saccharomyces cerevisiae*. Microbiol Mol Biol Rev 61:503–532

Thornton BR, Toczyski DP (2003) Securin and B-cyclin/CDK are the only essential targets of the APC. Nat Cell Biol 5:1090–1094

Tsunematsu R, Nakayama K, Oike Y, Nishiyama M, Ishida N, Hatakeyama S, Bessho Y, Kageyama R, Suda T, Nakayama KI (2004) Mouse Fbw7/Sel-10/Cdc4 is required for notch degradation during vascular development. J Biol Chem 279:9417–9423

Tsvetkov LM, Yeh KH, Lee SJ, Sun H, Zhang H (1999) p27^{Kip1} ubiquitination and degradation is regulated by the SCF^{Skp2} complex through phosphorylated Thr187 in p27. Curr Biol 9:661–664

Uhlmann F, Lottspeich F, Nasmyth K (1999) Sister-chromatid separation at anaphase onset is promoted by cleavage of the cohesin subunit Scc1. Nature 400:37–42

Verma R, Oania R, Graumann J, Deshaies RJ (2004) Multiubiquitin chain receptors define a layer of substrate selectivity in the ubiquitin–proteasome system. Cell 118:99–110

Visintin R, Prinz S, Amon A (1997) CDC20 and CDH1: a family of substrate-specific activators of APC-dependent proteolysis. Science 278:460–463

Visintin R, Craig K, Hwang ES, Prinz S, Tyers M, Amon A (1998) The phosphatase Cdc14 triggers mitotic exit by reversal of Cdk-dependent phosphorylation. Mol Cell 2:709–718

Wahl GM, Carr AM (2001) The evolution of diverse biological responses to DNA damage: insights from yeast and p53. Nat Cell Biol 3:E277–286

Waizenegger I, Gimenez-Abian JF, Wernic D, Peters JM (2002) Regulation of human separase by securin binding and autocleavage. Curr Biol 12:1368–1378

Wang H, Liu D, Wang Y, Qin J, Elledge SJ (2001) Pds1 phosphorylation in response to DNA damage is essential for its DNA damage checkpoint function. Genes Dev 15:1361–1372

Wang W, Ungermannova D, Chen L, Liu X (2003) A negatively charged amino acid in Skp2 is required for Skp2–Cks1 interaction and ubiquitination of p27^{Kip1}. J Biol Chem 278:32390–32396

Wang W, Ungermannova D, Chen L, Liu X (2004) Molecular and biochemical characterization of the Skp2-Cks1 binding interface. J Biol Chem 279:51362–51369

Wasch R, Cross FR (2002) APC-dependent proteolysis of the mitotic cyclin Clb2 is essential for mitotic exit. Nature 418:556–562

Watanabe N, Arai H, Nishihara Y, Taniguchi M, Hunter T, Osada H (2004) M-phase kinases induce phospho-dependent ubiquitination of somatic Wee1 by SCFbeta-TrCP. Proc Natl Acad Sci USA 101:4419–4424

Waters JC, Chen RH, Murray AW, Salmon ED (1998) Localization of Mad2 to kinetochores depends on microtubule attachment, not tension. J Cell Biol 141:1181–1191

Waters JC, Chen RH, Murray AW, Gorbsky GJ, Salmon ED, Nicklas RB (1999) Mad2 binding by phosphorylated kinetochores links error detection and checkpoint action in mitosis. Curr Biol 9:649–652

Wei W, Ayad NG, Wan Y, Zhang GJ, Kirschner MW, Kaelin WG, Jr (2004) Degradation of the SCF component Skp2 in cell-cycle phase G1 by the anaphase-promoting complex. Nature 428:194–198

Weinstein J (1997) Cell cycle-regulated expression, phosphorylation, and degradation of p55Cdc. A mammalian homolog of CDC20/Fizzy/slp1. J Biol Chem 272:28501–28511

Welcker M, Singer J, Loeb KR, Grim J, Bloecher A, Gurien-West M, Clurman BE, Roberts JM (2003) Multisite phosphorylation by Cdk2 and GSK3 controls cyclin E degradation. Mol Cell 12:381–392

Welcker M, Orian A, Grim JA, Eisenman RN, Clurman BE (2004a) A nucleolar isoform of the Fbw7 ubiquitin ligase regulates c-Myc and cell size. Curr Biol 14:1852–1857

Welcker M, Orian A, Jin J, Grim JA, Harper JW, Eisenman RN, Clurman BE (2004b) The Fbw7 tumor suppressor regulates glycogen synthase kinase 3 phosphorylation-dependent c-Myc protein degradation. Proc Natl Acad Sci USA 101:9085–9090

Willems AR, Lanker S, Patton EE, Craig KL, Nason TF, Mathias N, Kobayashi R, Wittenberg C, Tyers M (1996) Cdc53 targets phosphorylated G1 cyclins for degradation by the ubiquitin proteolytic pathway. Cell 86:453–463

Winston JT, Strack P, Beer-Romero P, Chu CY, Elledge SJ, Harper JW (1999) The SCFbeta-TRCP-ubiquitin ligase complex associates specifically with phosphorylated destruction motifs in IkappaBalpha and beta-catenin and stimulates IkappaBalpha ubiquitination in vitro. Genes Dev 13:270–283

Wolf DA, McKeon F, Jackson PK (1999) F-box/WD-repeat proteins pop1p and Sud1p/Pop2p form complexes that bind and direct the proteolysis of cdc18p. Curr Biol 9:373–376

Wu G, Xu G, Schulman BA, Jeffrey PD, Harper JW, Pavletich NP (2003) Structure of a beta-TrCP1-Skp1-beta-catenin complex: destruction motif binding and lysine specificity of the SCF$^{\beta\text{-TrCP1}}$ ubiquitin ligase. Mol Cell 11:1445–1456

Wu G, Lyapina S, Das I, Li J, Gurney M, Pauley A, Chui I, Deshaies RJ, Kitajewski J (2001) SEL-10 is an inhibitor of notch signaling that targets notch for ubiquitin-mediated protein degradation. Mol Cell Biol 21:7403–7415

Wu H, Lan Z, Li W, Wu S, Weinstein J, Sakamoto KM, Dai W (2000) p55CDC/hCDC20 is associated with BUBR1 and may be a downstream target of the spindle checkpoint kinase. Oncogene 19:4557–4562

Wu X, Bayle JH, Olson D, Levine AJ (1993) The p53-mdm-2 autoregulatory feedback loop. Genes Dev 7:1126–1132

Xiao Z, Chen Z, Gunasekera AH, Sowin TJ, Rosenberg SH, Fesik S, Zhang H (2003) Chk1 mediates S and G2 arrests through Cdc25A degradation in response to DNA-damaging agents. J Biol Chem 278:21767–21773

Xu K, Belunis C, Chu W, Weber D, Podlaski F, Huang KS, Reed SI, Vassilev LT (2003) Protein–protein interactions involved in the recognition of p27 by E3 ubiquitin ligase. Biochem J 371:957–964

Yada M, Hatakeyama S, Kamura T, Nishiyama M, Tsunematsu R, Imaki H, Ishida N, Okumura F, Nakayama K, Nakayama KI (2004) Phosphorylation-dependent degradation of c-Myc is mediated by the F-box protein Fbw7. EMBO J 23:2116–2125

Yamamoto A, Guacci V, Koshland D (1996a) Pds1p is required for faithful execution of anaphase in the yeast, Saccharomyces cerevisiae. J Cell Biol 133:85–97

Yamamoto A, Guacci V, Koshland D (1996b) Pds1p, an inhibitor of anaphase in budding yeast, plays a critical role in the APC and checkpoint pathway(s). J Cell Biol 133:99–110

Yamano H, Gannon J, Mahbubani H, Hunt T (2004) Cell cycle-regulated recognition of the destruction box of cyclin B by the APC/C in *Xenopus* egg extracts. Mol Cell 13:137–147

Yaron A, Gonen H, Alkalay I, Hatzubai A, Jung S, Beyth S, Mercurio F, Manning AM, Ciechanover A, Ben-Neriah Y (1997) Inhibition of NF-kappa-B cellular function via specific targeting of the I-kappa-B-ubiquitin ligase. EMBO J 16:6486–6494

Yaron A, Hatzubai A, Davis M, Lavon I, Amit S, Manning AM, Andersen JS, Mann M, Mercurio F, Ben-Neriah Y (1998) Identification of the receptor component of the IkappaBalpha-ubiquitin ligase. Nature 396:590–594

Yen JL, Su NY, Kaiser P (2005) The yeast ubiquitin ligase SCF^{Met30} regulates heavy metal response. Mol Biol Cell 16:1872–1882

Yeong FM, Lim HH, Padmashree CG, Surana U (2000) Exit from mitosis in budding yeast: biphasic inactivation of the Cdc28-Clb2 mitotic kinase and the role of Cdc20. Mol Cell 5:501–511

Yu VP, Baskerville C, Grunenfelder B, Reed SI (2005) A kinase-independent function of Cks1 and Cdk1 in regulation of transcription. Mol Cell 17:145–151

Zachariae W, Nasmyth K (1996) TPR proteins required for anaphase progression mediate ubiquitination of mitotic B-type cyclins in yeast. Mol Biol Cell 7:791–801

Zachariae W, Shin TH, Galova M, Obermaier B, Nasmyth K (1996) Identification of subunits of the anaphase-promoting complex of *Saccharomyces cerevisiae*. Science 274:1201–1204

Zachariae W, Schwab M, Nasmyth K, Seufert W (1998a) Control of cyclin ubiquitination by CDK-regulated binding of Hct1 to the anaphase promoting complex. Science 282:1721–1724

Zachariae W, Shevchenko A, Andrews PD, Ciosk R, Galova M, Stark MJ, Mann M, Nasmyth K (1998b) Mass spectrometric analysis of the anaphase-promoting complex from yeast: identification of a subunit related to cullins. Science 279:1216–1219

Zachariae W, Shin TH, Galova M, Obermaier B, Nasmyth K (1996) Identification of subunits of the anaphase-promoting complex of *Saccharomyces cerevisiae*. Science 274:1201–1204

Zhao H, Watkins JL, Piwnica-Worms H (2002) Disruption of the checkpoint kinase 1/cell division cycle 25A pathway abrogates ionizing radiation-induced S and G2 checkpoints. Proc Natl Acad Sci USA 99:14795–14800

Zheng N, Schulman BA, Song L, Miller JJ, Jeffrey PD, Wang P, Chu C, Koepp DM, Elledge SJ, Pagano M, Conaway RC, Conaway JW, Harper JW, Pavletich NP (2002) Structure of the Cul1-Rbx1-Skp1-F boxSkp2 SCF ubiquitin ligase complex. Nature 416:703–709

Zhu G, Spellman PT, Volpe T, Brown PO, Botstein D, Davis TN, Futcher B (2000) Two yeast forkhead genes regulate the cell cycle and pseudohyphal growth. Nature 406:90–94

Zou H, McGarry TJ, Bernal T, Kirschner MW (1999) Identification of a vertebrate sister-chromatid separation inhibitor involved in transformation and tumorigenesis. Science 285:418–422

Zur A, Brandeis M (2001) Securin degradation is mediated by fzy and fzr, and is required for complete chromatid separation but not for cytokinesis. EMBO J 20:792–801

The Retinoblastoma Gene Family in Cell Cycle Regulation and Suppression of Tumorigenesis

Jan-Hermen Dannenberg[1] (✉) · Hein P. J. te Riele[2] (✉)

[1]Department of Medical Oncology, Dana-Farber Cancer Institute
and Harvard Medical School, Boston, Massachusetts, USA
Jan-Hermen_Dannenberg@dfci.harvard.edu

[2]Department of Molecular Biology, Netherlands Cancer Institute, Amsterdam,
The Netherlands
h.t.riele@nki.nl

Abstract Since its discovery in 1986, as the first tumor suppressor gene, the retinoblastoma gene (*Rb*) has been extensively studied. Numerous biochemical and genetic studies have elucidated in great detail the function of the *Rb* gene and placed it at the heart of the molecular machinery controlling the cell cycle. As more insight was gained into the genetic events required for oncogenic transformation, it became clear that the retinoblastoma gene is connected to biochemical pathways that are dysfunctional in virtually all tumor types. Besides regulating the E2F transcription factors, pRb is involved in numerous biological processes such as apoptosis, DNA repair, chromatin modification, and differentiation. Further complexity was added to the system with the discovery of *p107* and *p130*, two close homologs of *Rb*. Although the three family members share similar functions, it is becoming clear that these proteins also have unique functions in differentiation and regulation of transcription. In contrast to *Rb*, *p107* and *p130* are rarely found inactivated in human tumors. Yet, evidence is accumulating that these proteins are part of a "tumor-surveillance" mechanism and can suppress tumorigenesis. Here we provide an overview of the knowledge obtained from studies involving the retinoblastoma gene family with particular focus on its role in suppressing tumorigenesis.

1
Cancer and Genetic Alterations

Cancer can be viewed as a disease of the genome. Sequentially acquired genetic or epigenetic alterations have progressively provided cells with characteristics that allow uncontrolled proliferation and metastasis (Hanahan and Weinberg 2000). Genes modified in cancer are classified as oncogenes and tumor suppressor genes that have been activated by gain-of-function mutations and inactivated by loss-of-function mutations, respectively. The first identified human tumor suppressor gene is the retinoblastoma gene (*Rb*), which was found to be inactivated in hereditary retinoblastoma, a pediatric eye tumor (Friend et al. 1986; Lee et al. 1987). Since the discovery of the *Rb* gene and its product, the pRb protein, numerous studies have shown that most, if not all, human tumors display a deregulated pRb pathway (Sherr 1996). Additionally, many biochemical studies have elucidated the function of pRb in controlling cell cycle progression, providing a platform to understand the relevance of pRb loss in development of cancer (reviewed in Weinberg 1995; Hanahan and Weinberg 2000; Harbour and Dean 2000). The molecular cloning of two other *Rb*-like genes, *p107* and *p130*, defined the retinoblastoma gene family and added to the complexity of cell cycle regulation. This chapter will elaborate on the role of the retinoblastoma gene family in cell cycle regulation and tumor suppression.

2
The pRb Cell Cycle Control Pathway: Components and the Cancer Connection

The retinoblastoma protein, pRb, is a nuclear phosphoprotein that plays a pivotal role in regulation of the cell cycle. pRb can exist in a hyper- or hypophosphorylated state, the latter being able to bind and inhibit E2F transcription factors (Dyson 1998). Mitogenic growth factors induce the sequential activation of cell-cycle-dependent kinase complexes, cyclin D/Cdk4-Cdk6 and cyclin E/Cdk2. This results in the phosphorylation and conformational change of pRb allowing the release of E2Fs. Derepression and activation of E2F target genes then allows progression from G1 into S-phase of the cell cycle (Lundberg and Weinberg 1998; Harbour et al. 1999; Harbour and Dean 2000; Ezhevsky et al. 2001). Conversely, growth-inhibitory signals that promote cell cycle arrest, exert their effect by direct down regulation of cyclin protein levels or by inducing members of the INK4A and/or CIP/KIP family of cyclin dependent kinase inhibitors (CKI), resulting in the down-regulation of cyclin/Cdk activity and inhibition of pRb phosphorylation (Ruas and Peters 1998; Sherr and Roberts 1999; Sherr 2001). Sequestration of active E2Fs subsequently results in repression of E2F target genes and ultimately in a cell cycle arrest or exit from the cell cycle (see Fig. 1). Thus, pRb can be viewed as a molecular cell cycle switch that is either turned on by growth-inhibiting signals or turned off by growth promoting signals, resulting in cell cycle exit/arrest and cell cycle entry/progression, respectively.

Inactivation of this proliferation controlling pathway seems to be an essential step in the transition of a normal cell into a cancer cell. Inactivation of pRb has been found in many tumor types in humans, including hereditary retinoblastoma and sporadic breast, bladder, prostate and small cell lung carcinomas (Friend et al. 1986; Harbour et al. 1988; Lee et al. 1987; T'Ang et al. 1988; Bookstein et al. 1990; Horowitz et al. 1990). Since pRb/E2F function is controlled at different levels, its deregulation can also occur at different levels. Besides loss of pRb function by inactivating mutations or sequestration by viral oncoproteins like adenovirus E1A, simian virus 40 (SV40) large T antigen or human papillomavirus 16 (HPV-16) E7 (DeCaprio et al. 1988; Whyte et al. 1988; Dyson et al. 1989; Ludlow et al. 1989), the pRb pathway can be compromised by over-expression of D-type cyclins, mutations rendering Cdk4

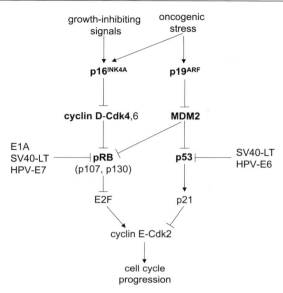

Fig. 1 The p16^{INK4A}-pRb and the p19^{ARF}-p53 pathway involved in cell cycle progression and tumorigenesis. Components of these pathways frequently found inactivated (p16^{INK4A}, p19^{ARF}, pRb, p53) or overexpressed (cyclin D, Cdk4) in human cancer are indicated in *bold*. pRb inactivation can also be achieved by viral proteins like SV40-LargeT, adenovirus-E1A or HPV-E7. p53 is inactivated by SV40-LargeT and HPV-E6. We envisage that growth-stimulating or inhibiting signals generally impinge on the activity of cyclin E/Cdk2. We speculate that the pRb pathway regulates the level of cyclin E/Cdk2 while the p53-pathway regulates the cyclin E/Cdk2 activity by controlling the levels of p21^{CIP1}. In the absence of pocket proteins, cyclin E is induced to a level that is refractory to p21^{CIP1}-mediated inhibition. In the absence of p19^{ARF} or p53, p21^{CIP1} levels are too low to effectively inhibit cyclin E/Cdk2 activity. Hence both pathways are required for replicative or oncogene-induced senescence

resistant to CKIs, deletion of CKIs or over-expression of E2F transcription factors. In accordance with this many human tumors show genetic aberrations affecting the p16^{INK4A}-cyclin D-pRb/E2F pathway: *p16^{INK4A}* loss of function in melanoma, T-cell leukemias, pancreatic and bladder carcinomas, amplification of *cyclin D* in breast, oesophagus and head and neck cancer, *Cdk4* amplification or mutational activation in melanoma (reviewed in: Sherr 1996; Malumbres and Barbacid 2001; see Fig. 1).

3
Regulation of E2F Responsive Genes by pRb

E2F transcription factors, named for their activity to mediate transcriptional activation of the adenovirus E2 promoter, recognize and bind together with their dimerization partners DP-1 or DP-2 to recognition sequences present in many E2F-responsive genes (Trimarchi and Lees 2001). An intriguing finding was that these target genes are involved in a variety of biological processes such as cell cycle regulation (*Rb, p107, E2F1, cyclin A2, cyclin E1, Cdc2*), DNA replication (*DHFR, MCM, Cdc6, PCNA, DNA polymerase α*), DNA repair (*RAD54, BARD1*), G2/M-checkpoints (*CHK1, MAD2, BUB3, SECURIN*) and differentiation (*EED, EZH2*) (Dyson 1998; Harbour and Dean 2000; Ishida et al. 2001; Kalma et al. 2001; Müller et al. 2001; Ren et al. 2002), suggesting that pRb/E2F function is not only restricted to regulation of the G1/S transition of cell cycle.

Whether an E2F target gene is transcriptionally activated or repressed depends on binding of pRb to E2F. pRb inhibits the transcriptional activity of E2F by binding to its carboxy-terminal transactivation domain, thereby preventing the interaction of E2F with the basal transcription machinery (Helin et al. 1992, 1993; Flemington et al. 1993). However, expression of an E2F variant containing the DNA binding motif but not the pRb-binding or transactivation domain or introduction of a competitor plasmid containing multiple E2F binding sites, preventing the binding of E2F and pRb/E2F complexes to cellular promoters, alleviated growth suppression by pRb (Zhang et al. 1999; He et al. 2000). Active repression of gene transcription thus seems an important mechanism by which pRb arrests the cell cycle. pRb bound to E2F recruits chromatin-remodeling proteins that influence the accessibility of a locus for the transcriptional machinery. Among these remodeling proteins are histone deacetylases (HDAC1-3), SWI/SNF family proteins (BRG1, Brm), polycomb group proteins (HPC2, Ring1) and histone methyltransferases (SUV39H1, RIZ-1) (Buyse et al. 1995; Brehm et al. 1998; Luo et al. 1998; Magnaghi et al. 1998; Lai et al. 1999; Dahiya et al. 2001; Nielsen et al. 2001). Since E2F-1 has been shown to interact with co-activators that have histone acetyltransferase (HAT) activity, which promotes an open chromatin structure and transcriptionally active genomic loci, it seems likely that inhibition of E2F requires HDAC activity, provided by histone deacetylases HDAC1-3. This active repression could result in silencing of a whole locus by recruitment of SUV39H1 and RIZ-1 methyltransferases, allowing tight repression of E2F target genes upon a variety of growth-inhibitory signals. Finally, it was shown in a reconstitution transcription assay that chromatin is an essential component for pRb to actively repress transcription, although HDACs did not seem to play a role in this setting (Ross et al. 2001). In summary, pRb is able to repress gene transcription by means of direct inhibition of the transcription machinery, direct binding and inhibition of E2F transactivation capacity or by recruiting histone modification proteins. It is very likely that the genetic locus, signaling and other (unknown) cellular conditions determine which particular pRb-dependent inhibitory program will be used.

4
The Retinoblastoma Gene Family

4.1
Rb Gene Family Members

The retinoblastoma gene family comprises, besides *Rb*, the structurally and functionally related *Rb*-like genes *p107* (*RBL1*) and *p130* (*RBL2*). Whereas the *Rb* gene was identified as the tumor suppressor gene on the deleted chromosomal region 13q14 in hereditary retinoblastoma, *p107* and *p130* were cloned by their ability to bind viral oncoproteins, cyclin A and E and Cdk2. *p107* is located on human chromosome 20q11, *p130* on chromosome 16q12 (Ewen et al. 1991; Hannon et al. 1993; Li et al. 1993; Yeung et al. 1993)

4.2
pRb Family Protein Structure

The *Rb* proteins share a high degree of homology within two sub-domains (A and B), which make up the so-called "pocket" domain (Chow and Dean 1996; Lipinski and Jacks 1999; Harbour and Dean 2000; see Fig. 2). This region defines the minimal region essential for binding to proteins containing a LXCXE motif, such as the viral oncoproteins adenovirus E1A, SV40 large T antigen and HPV-16 E7, as well as many cellular proteins. Although the binding site for LXCXE motif containing proteins is present in the B sub-domain, the crystal structure of the pRb A/B pocket bound to the LXCXE-containing part of HPV-16 E7 revealed that sub-domain A is required for an active conformation of sub-domain B (Lee et al. 1998). The functional importance of this region is emphasized by the fact that it is highly conserved between species ranging from *C. elegans* to mammals (Lu and Horvitz 1998). Furthermore, the A/B pocket is sufficient for stable interaction with E2F1 and several transcriptional repressor complexes (Qin et al. 1992; Trouche et al. 1997; Brehm et al. 1998; Magnaghi et al. 1998). Studies have shown that the interaction between the A/B pocket region of the pocket protein family and histone modifying enzymes such as histone deacetylase is not direct but is mediated by RBP1 (Lai et al. 2001). Outside the pocket domain p107 and p130 are more similar to each other than to pRb. C-terminal of the pocket domain in pRb, a region known as the C-domain can bind the proto-oncogene products C-ABL and MDM2, thereby inhibiting C-ABL tyrosine kinase activity and pRb growth suppression functions (Welch et al. 1993; Xiao et al. 1995). Underscoring the complexity of the interaction between pocket proteins and E2Fs, it was shown that the C-terminal region of pRb contains a E2F1 specific binding site that is sufficient to inhibit E2F1 mediated apoptosis, independent of its transcriptional function (Dick and Dyson 2003). An amino acid sequence identified in sub-domain B of p130 and named the Loop, was shown to be specifically phosphorylated when cells are in quiescence (Canhoto et al. 2000; Hansen et al. 2001). This indicates that p107 and p130 harbor regions that are not homologous to each other or to pRb, suggesting that besides similar, each protein also has specific functions.

4.3
Similar and Distinct Functions of the pRb Protein Family

A similar function of all three pocket proteins is their ability to inhibit E2F-responsive promoters, recruit HDACs and repress transcription (Zamanian and La 1993; Bremner et al. 1995; Starostik et al. 1996; Ferreira et al. 1998). pRb, p107 and p130 undergo cell-cycle-dependent phosphorylation (Graña et al. 1998; Lundberg and Weinberg 1998; Canhoto et al. 2000; Hansen et al. 2001). Over-expression of each of the pocket proteins results in growth suppression, although not every (tumor) cell-type is equally sensitive to each pRb family member (Zhu et al. 1993; Claudio et al. 1994; Beijersbergen et al. 1995; Ashizawa et al. 2001).

Besides these similarities, the pRb family members also have unique properties. The spacer region that links the A and B domains shows significantly more homology between p107 and p130 than between p107/p130 and pRb. This spacer region was shown to contain a p21-like sequence that can recruit and inhibit cyclin A/Cdk2 and cyclin E/Cdk2 kinase complexes. Although all pocket proteins are (de)phosphorylated in a cell cycle-dependent manner, pRb and p107 predominantly are phosphorylated during mid-G1 and G1-S phase transition by cyclin D/Cdk4 complexes and subsequently hyperphosphorylated by cyclin E/Cdk2 and cyclin A/Cdk2 (Graña et al. 1998; Lundberg and Weinberg 1998). In contrast, p130 is specifically phosphorylated in quiescencent cells in the Loop by Cdk2 and glycogen synthase kinase 3 (Canhoto et al. 2000; Hansen et al. 2001; Litovchick et al. 2004; see Fig. 2). Since the phosphorylation sites in the Loop region are largely dispensable for regulation of E2F4 activity it is likely that phosphorylation of these sites are involved in the regulation of p130 specific functions and interactions. The difference in phosphorylation sites and kinases involved in the phosphorylation of these sites between p107 and p130 further support specific functions for p107 and p130. (Farkas et al. 2002; Litovchick et al. 2004). Furthermore, the different retinoblastoma protein family members bind to distinct E2F family members. The E2F family of transcription factors consists of six members, E2F1-6. They can be divided into two subgroups on the basis of their activity in regulating transcription. E2F1, E2F2 and E2F3 are viewed as "activating" E2Fs, since they are potent transcriptional activators. Inactivation of *E2f3* impairs the proliferation of mouse embryonic fibroblasts (MEFs) while combined inactivation of *E2f1*, *E2f2* and *E2f3* completely blocks proliferation of these cells (Humbert et al. 2000; Wu et al. 2001), indicating that the members of this

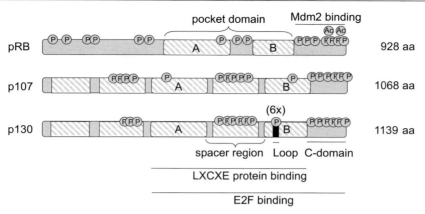

Fig. 2 Protein structure and modifications of *pRb*, *p107* and *p130*. Within the Rb protein family p107 and p130 share the highest degree of homology (indicated by *shaded areas*). Within the pocket domain (pocket subdomains A and B and the spacer region) the highest homology between the pRb protein family is found in the A and B subdomains. The pocket-domain is responsible for binding to proteins containing LXCXE motifs while the pocket-domain and the C-domain are involved in binding E2F proteins. Mdm2 (as well as c-Abl) binds to the C-domain. All pocket proteins are subject to phosphorylation (indicated with "P") although the phosphorylation sites are not all conserved (for detailed information see Canhoto et al. 2000; Hansen et al. 2001; Farkas et al. 2002; Litovchick et al. 2004). In p130 the Loop region, a part of the B-pocket subdomain, which is not shared with pRb nor p107, is in particular subject to phosphorylation by GSK3β. The Loop region contains 6 phosphorylation sites. Besides phosphorylation, pRb is also subject to acetylation (indicated with "Ac") in its C-domain, a modification that is thought to be involved in the interaction with Mdm2. The size of the pocket proteins is indicated on the *right*

class of E2Fs have overlapping functions and play an essential role in cell cycle progression. E2F4, E2F5 and E2F6 form the class of "active repressor" E2Fs. Whereas E2F4 and E2F5 execute their function by binding to pocket proteins, E2F6 confers active repression in a pocket protein-independent manner (reviewed in Dyson 1998; Trimarchi and Lees 2001; Cobrinick 2005). Recently, two additional E2F proteins have been identified, E2F7 and E2F8. Similar to E2F6 these proteins seem to repress transcription independently of the pRb protein family (de Bruin et al. 2003b; DiStefano et al. 2003; Logan et al. 2004; Maiti et al. 2005). Whereas pRb predominantly binds E2F1, E2F2 and E2F3, p107 and p130 bind specifically E2F4 and E2F5 (Dyson 1998, see Fig. 3). The different functionality of the pocket protein/E2F complexes is emphasized by the fact that p107/E2F and p130/E2F complexes act as transcriptional repressors of a set of genes different from that regulated by pRb/E2F complexes (Hurford et al. 1997). Upon re-entering the cell cycle and progression through G1 into S phase the levels of p130 protein decrease while p107 protein expression increases, indicating that p107/E2F4 and p130/E2F4 complex formation is temporally regulated (Graña 1998). Indeed, each of the pocket proteins appears in complex with E2Fs at different stages of the cell cycle: p130/E2F4 complexes are predominantly found in G0, pRb bound to E2F in G0 and G1, while p107 complexes with E2F in the S-phase of the cell cycle (Dyson 1998). This might reflect the not yet fully understood specific functions of these proteins at these specific stages of the cell cycle.

4.4
pRb Family Mediated Regulation of E2F by Cellular Localization

Another level of control of the E2F transcriptional activity is added by the cellular compartmentalization of E2F transcription factors. E2F1, E2F2 and E2F3 are constitutively nuclear, whereas E2F4 and E2F5 are predominantly cytoplasmic. Upon progression from G0 to S-phase, E2F4 and E2F5 are translocated from the nucleus to the cytoplasm (Verona et al. 1997). Since E2F4 and E2F5, in contrast to the activating E2Fs, do not contain a nuclear localization signal (NLS), other proteins must be involved in their translocation. Interaction of these E2Fs with p107 and p130 has been proposed to be required for their nuclear localization (Lindeman et al. 1997; Verona et al. 1997). As a consequence, p107 and p130 should be able to translocate from the nucleus to the cytoplasm. Indeed, besides the presence of nuclear localization signals in the carboxy-terminal region and pocket domain of pRb, p107 and p130 and an additional NLS in the Loop region of p130, a nuclear export signal (NES) is present in the N-terminal region of p130, which is conserved in p107 and pRb (Zacksenhaus et al. 1999; Cinti et al. 2000; Chestukhin et al. 2002). Nucleocytoplasmic shuttling of p130 and p107 might regulate the transcriptional repression activity of E2F4 and/or E2F5 different from phosphorylation mediated disruption of pocket/E2F repression complexes. Besides the reliance on these nuclear import and export signals present in the Rb protein family, translocation of p107/E2F repressor complexes to the nucleus has also been observed by usage of other signaling molecules. Upon TGF-β signaling cytoplasmic complexes consisting of Smad3 and specifically p107 and E2F4/5 can translocate to the nucleus. These complexes subsequently bind to Smad4 and repress *Myc* transcription, thereby blocking cell cycle progression (Chen et al. 2002).

4.5
Regulation of E2F Mediated Gene Expression

All three pocket proteins have the ability to repress transcription of E2F responsive genes. However, which of the pocket proteins is actually assembled on the promoter of a particular gene seems both gene-specific and condition-specific. Detection of protein complexes associated with promoters of E2F-responsive genes *in vivo* by chromatin immuno-precipitation (ChIP) assays,

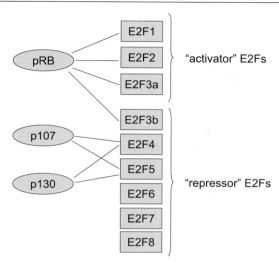

Fig. 3 Interaction of pRb family members with E2F transcription factors. Whereas pRb primarily binds to "activator" E2Fs (E2F1, 2, 3a), p107 and p130 interact with the "repressor" E2Fs (E2F4 and E2F5). E2F6, E2F7, E2F8 are involved in pRb family-independent repression of gene transcription

revealed that in serum-starved G0 cells these promoters were predominantly occupied by E2F4 and p130. Upon re-entry into the cell cycle these repressive complexes were replaced by activating E2F1, E2F2 and E2F3. In these assays, pRb could not be detected on promoters of a selected group of E2F-target genes in cycling cells (Takahashi et al. 2000; Wells et al. 2000; Dahiya et al. 2001). However, the observation that *cyclin E* is de-repressed in $Rb^{-/-}$ MEFs and not in $p107^{-/-}p130^{-/-}$ MEFs, suggests that pRb and not p107 and p130, is primarily involved in suppression of *cyclin E* transcription (Herrera et al. 1996; Hurford et al. 1997). Indeed, pRb could be detected on the promoters of *cyclin E* as well as *cyclin A* upon ectopic expression of p16^{INK4A} or serum withdrawal indicating that pRb/E2F mediated repression of E2F-responsive genes may play a role in establishing cell cycle arrest (Dahiya et al. 2001; Morrison et al. 2002). This view was further supported by the observation that in senescent cells pRb, together with heterochromatin proteins, could be found in senescence associated heterochromatin foci (SAHF) that included E2F-responsive promoters (Narita et al. 2003). However, it should be noted that under growth inhibiting conditions such as cell-cell contact, serum deprivation and p16^{INK4A} over-expression, p130 and E2F4 can be found on the promoters of a common set of genes. Surprisingly, most of these genes are not involved in cell cycle regulation but in mitochondrial biogenesis and metabolism (Cam et al. 2004). Furthermore, many recently identified E2F-responsive genes were de-repressed in $p107^{-/-}p130^{-/-}$ MEFs, suggesting that p107 and p130 bound to E2F4/E2F5 are important repressors, and that pRb/E2F complexes cannot compensate in repressing the transcription of these genes (Ren et al. 2002). Strikingly, MEFs deficient for E2F4 and E2F5 did not show de-repression of E2F-responsive genes, suggesting that p107 and p130 can repress transcription in an E2F4/5 independent fashion. A specific function was found for p130 in the regulation of neuronal survival and death by repressing pro-apoptotic genes through recruitment of histone modifiers such as HDAC1 and Suv39H1 (Liu et al. 2005). The observation that only p107 together with E2F4/5 and Smad proteins was found on the promoter of c-Myc upon TGFβ-signaling underscores the specific functions of the different pRb gene family members in repression of specific genes upon activation of specific signaling pathways (Chen et al. 2002)

4.6
The pRb Family and the Cellular Response Towards Growth-Inhibitory Signals

Many growth-inhibitory conditions such as lack of growth factors, cell-cell contact, DNA damage, lack of anchorage and differentiation are accompanied by the induction of cyclin dependent kinase inhibitors and result in the accumulation of hypophosphorylated pocket proteins and (temporal or definitive) cell cycle arrest. This led to the model that pocket proteins are mediators of growth-inhibitory signals (Weinberg 1995). Indeed, analysis of mouse embryonic fibroblasts deficient for combinations of pocket proteins revealed that the *Rb* gene family members have overlapping roles in controlling cell cycle exit upon growth-inhibiting signals. Only ablation of all pocket proteins fully alleviated a cell cycle arrest upon serum withdrawal, cell-cell contact inhibition, DNA damage, differentiation and prolonged culturing (Dannenberg et al. 2000; Sage et al. 2000). The functional redundancy of the pocket proteins is also manifested by the upregulation of p107 and to a lesser extent of p130 in pRb-deficient cells (Hurford et al. 1997; Dannenberg et al. 2000, 2004; MacPherson et al. 2004). Indeed, MEFs lacking either pRb and p107 or pRb and p130 are more resistant to growth inhibitory stimuli than MEFs lacking only pRb (Dannenberg et al. 2000, 2004, Sage et al. 2000; Peeper et al. 2001). Interestingly, whereas MEFs deficient for either pRb or p107 require serum to enter S-phase, MEFs lacking both pRb and p107 lack this serum requirement. In contrast, *Rb/p107* deficient MEFs still require cell anchorage in order to progress into S-phase, suggesting that pRb and p107 constitute the serum restriction point whereas the cell-anchorage restriction point extends beyond these retinoblastoma gene family members (Gad et al. 2004).

$p16^{INK4A}$ requires functional pRb to impose a G1 arrest (Lukas et al. 1995; Medema 1995). Unexpectedly, MEFs lacking either p107 and p130 or E2F4 and E2F5, were also refractory to $p16^{INK4A}$-induced G1 arrest (Bruce et al. 2000; Gaubatz et al. 2000), suggesting that $p16^{INK4A}$-mediated growth arrest requires repression of specific genes by p107 and p130. Alternatively, the pocket protein/E2F complexes may target the same set of genes, but the total level of their activity needs to accumulate above a certain threshold that cannot be reached by ectopic expression of p16^{INK4A} in $Rb^{-/-}$ or $p107^{-/-}p130^{-/-}$ cells. Cell cycle studies performed with isogenic sets of MEFs deficient for combinations of pocket proteins indicate that although p107 and p130 can to some extent functionally compensate, pRb is the critical regulator of most cell cycle responses. While each single pocket protein can mediate cell cycle arrest upon cell-cell contact, pRb is the critical mediator of cell cycle arrest upon growth factor deprivation and irradiation. p107 can partially compensate for the absence of pRb under both conditions. p130 can mediate a modest response upon serum withdrawal which is additive to that of p107, but does not play a role in the response of cells to ionizing radiation (Dannenberg and Te Riele; unpublished observations). The latter observation is consistent with a proposed role for p107 in establishing a cell cycle arrest upon ionizing irradiation (Voorhoeve et al. 1998; Kondo et al. 2001). In view of the previously mentioned transcriptional derepression of many E2F-responsive genes in $p107^{-/-}p130^{-/-}$ MEFs, it is striking that these MEFs respond like wild-type MEFs to various growth-inhibiting conditions, except ectopic expression of p16^{INK4A} (Bruce et al. 2000). In concordance with our data, this may suggest that regulation of cyclin E expression and therefore cyclin E/Cdk2 kinase activity by pRb/E2F protein complexes is critical in implementing a cell cycle arrest upon growth-inhibitory signals.

5
The pRb and p53 Pathway in Senescence and Tumor Surveillance

5.1
Replicative Senescence

In both human and mouse primary fibroblasts, prolonged culturing generates a growth-inhibiting signal that ultimately leads to a state of replicative senescence reflected by an enlarged, flattened morphology and the absence of DNA synthesis (Hayflick and Moorhead 1961; Todaro and Green 1963; Campisi 1997). The growth-inhibiting signal is very likely generated by the non-physiological tissue culture conditions such as incorrect media and growth factors, very high oxygen tension and artificial adherence substrates, since culturing primary cells under optimized conditions prevents replicative senescence (Sherr and DePinho 2000; Mathon et al. 2001; Ramirez et al. 2001; Tang et al. 2001). On the other hand, evidence is accumulating that senescence may also play a role *in vivo* as a "fail-safe" mechanism to prevent tumorigenesis (Schmitt et al. 2002). This type of growth arrest is accompanied by gradually increasing levels of the Cdk2/Cdk4 inhibitors, p21^{CIP1} and p16^{INK4A}, the cell cycle inhibitor p19^{ARF}, and p53 (Lloyd et al. 1997; Palmero et al. 1997; Zindy et al. 1997, 1998; see Fig. 1). p16^{INK4A} and p19^{ARF} are encoded by one genetic locus, *Ink4a/Arf*, whereby p19^{ARF} is expressed from an alternative reading frame (Quelle et al. 1995). While p16^{INK4A} was shown to act upstream of pRb to promote cell cycle arrest (Serrano et al. 1993; Lukas et al. 1995; Medema et al. 1995), p19^{ARF} can physically interact with p53 and/or MDM2 thereby antagonizing the function of MDM2 and ultimately stabilizing p53 (Kamijo et al. 1998; Pomerantz et al. 1998; Zhang et al. 1998). Spontaneous immortalization of MEFs is usually accompanied by either deletion of the *Ink4a/Arf* locus (Kamb et al. 1994; Nobori et al. 1994; Kamijo et al. 1997; Zindy et al. 1998) or loss of p53 function (Harvey and Levine 1991; Rittling and Denhardt 1992).

Although the *Ink4a/Arf* locus encodes both p16^{INK4A} and p19^{ARF}, only ablation of p19^{ARF} alleviates a replicative senescence response (Kamijo et al. 1997; Krimpenfort et al. 2001, Sharpless et al. 2001). Thus, in MEFs p19^{ARF} and not p16^{INK4A} seems to be the critical component required to impose a replicative arrest. However, analysis of murine bone-marrow derived cell types revealed that replicative senescence in pre-B cells depends on *p19^{Arf}* inactivation, whereas macrophages can become immortal by silencing *p16^{INK4A}* and retaining p19^{ARF} expression, suggesting a differential requirement for inactivation of *p19^{Arf}* or *p16^{INK4A}* dependent on the cell type (Randle et al. 2001). p21^{CIP1} and p27^{KIP1}, members of the CIP/KIP cyclin dependent kinase inhibitor family seem to be irrelevant for establishing replicative senescence, since their inactivation still renders MEFs sensitive to replicative arrest upon subsequent passaging (Pantoja and Serrano 1999; Groth et al. 2000; Modestou et al. 2001).

The retinoblastoma gene family seems to be a critical down-stream mediator of p19^{ARF}-p53 induced replicative senescence. Inactivation of *Rb*, *p107* and *p130* in MEFs fully alleviated a senescence response upon prolonged passaging and allowed retaining an intact p19^{ARF}-p53 pathway. Furthermore, p19^{ARF} over-expression in *TKO* MEFs did not result in a G1-arrest. Although p19^{ARF} was still able to restrain proliferation in MEFs lacking pRb and p107 or pRb and p130, upon prolonged passaging these cells did not senesce (Dannenberg et al. 2000, 2004; Peeper et al. 2001). These data indicate that p107 *and* p130 together can compensate for the loss of pRb in cellular senescence. This is further supported by the observation that in pRb-deficient MEFs, p107 and p130 are upregulated. Recently, pRb was shown to be part of a heterochromatic structure that is specifically observed in senescent cells and therefore designated senescence-associated heterochromatic foci (SAHF) (Narita et al. 2003). pRb is thought to be responsible for the enucleation of heterochromatin on E2F responsive promoters by recruiting pRb binding proteins such as heterochromatin protein 1 (HP1), macroH2A and histone methyltransferase Suv39h, resulting in lysine 9 methylation of histone H3 (Narita et al. 2003; Ait-Si-Ali et al. 2004; Zhang et al. 2005). Ablation of the pRb protein family by expression of E1A totally abolished SAHF formation upon induction of senescence, further establishing a role for the *Rb* gene family in replicative senescence.

5.2
Tumor Surveillance

Oncogenes such as RAS^{V12}, c-Myc, v-Abl and E2F-1 are known to activate p19^{ARF} expression, leading to a p53-dependent cell cycle arrest or apoptosis, thereby withdrawing cells carrying oncogenic potential from the cell cycle. Inactivation of either p19^{ARF} or p53 eliminates this "fail-safe" mechanism, leading to infinitive and oncogene driven proliferation (Sherr 2001; see Fig. 1).

In vivo, this "fail safe" mechanism appears to play an important role in tumor suppression as evidenced by the cancer prone phenotypes of p19^{Arf} and p53 deficient mice. Transgenic expression of the oncoprotein C-MYC under control of the immuno-globulin heavy chain enhancer $E\mu$ in a wild-type background results in B-cell lymphomas, which invariably show loss of the Ink4a/Arf locus, Mdm2 induction or p53 mutation. In a p19^{Arf} null, but not in a p16^{Ink4a} mutant background, lymphomagenesis is strongly accelerated, indicating that an intact p19^{ARF}-Mdm2-p53 pathway functions as a tumor surveillance pathway in vivo (Eischen et al. 1999; Krimpenfort et al. 2001; Sharpless et al. 2001). Inducible transgenic expression of K-Ras4b^{G12D} in alveolar type-II pneumocytes in the presence of an intact p19^{ARF}-Mdm2-p53 pathway rapidly induced proliferation and the development of adenocarcinomas instead of a cell cycle arrest. Although this might suggest that the observed in vitro "fail-safe" mechanism upon oncogene expression is not activated in this cell type, inactivation of the Ink4a/Arf locus or p53 accelerated K-Ras-driven tumorigenesis and resulted in more aggressive adenocarcinomas (Fisher et al. 2001).

Evidence is accumulating that the Rb gene family is also part of such a tumor surveillance pathway. MEFs deficient for either pRb and p107, pRb and p130 or all pocket proteins sustained high levels of ectopically expressed oncogenic RAS and continued proliferation in the presence of a functional p19^{ARF}-p53 pathway, suggesting a defective "fail-safe" mechanism (Sage et al. 2000; Peeper et al. 2001; Dannenberg et al. 2004). Surprisingly, in contrast to inactivation of the p19^{ARF}-p53 pathway, inactivation of the Rb gene family members did not result in oncogenic transformation upon RAS^{V12} expression, as judged by the incapacity to grow anchorage-independently. This suggests that immortalization and oncogenic transformation are two independent processes. Apparently, disruption of the p19^{ARF}-Mdm2-p53 pathway simultaneously deregulates both anti-immortalizing and anti-tumorigenic mechanisms, whereas loss of the Rb gene family only causes immortalization (Peeper et al. 2001; Dannenberg et al. 2004). MEFs lacking all pocket proteins and expressing RAS^{V12} were able to grow in nude mice, suggesting that in this assay loss of the Rb gene family is sufficient to allow oncogenic transformation. On the other hand, in view of the lack of anchorage-independent growth of these cells, it remains possible that additional oncogenic mutations were quickly obtained and selected and allowed tumor development (Sage et al. 2000; Dannenberg and Te Riele, unpublished observations).

In concordance with a role for the pocket proteins in tumor surveillance, $Rb^{-/-}p107^{-/-}$ and $Rb^{-/-}p130^{-/-}$ chimeric mice were tumor prone at early age (see below; Robanus-Maandag et al. 1998; Chen et al. 2004; Dannenberg et al. 2004; McPherson et al. 2004). Interestingly, mice carrying the Cdk4^{C24R} mutation resulting in hyperphosphorylated pocket proteins, showed a broad spectrum of tumors, while MEFs isolated of these mice retained functional p19^{ARF} and p53 upon prolonged passaging and were susceptible to RAS^{V12}-induced transformation (Sotillo et al. 2001; Rane et al. 2002). In contrast, deletion of Cdk4 rendered MEFs resistant to oncogenic transformation by Ras^{V12}, even in the absence of p16^{Ink4a} and p19^{Arf}. Inactivation of either p21^{CIP1} or inactivation of the pocket protein family by expression of HPV-E7 restored the immortalization and Ras^{V12}-mediated transformation of $Cdk4^{-/-}Ink4a^{-/-}p19^{Arf-/-}$ MEFs, suggesting again a role for the Rb gene family downstream of p19^{Arf}/p53 in preventing oncogenic transformation (Zou et al. 2002). These data suggest that the repressor function of pocket protein/E2F complexes is essential for imposing a replicative cell cycle arrest and tumor suppression.

Loss of p16^{Ink4a} does not lead to evident immortalization, collaboration in RAS^{V12}-induced transformation and tumor predisposition in mice. Tumor incidence was strongly increased in $p16^{Ink4a}$ null mice upon carcinogen treatment or $p19^{Arf}$ heterozygosity, indicating that additional mutations or reduced p19^{ARF} dosage levels can strongly collaborate in tumorigenesis (Krimpenfort et al. 2001; Sharpless et al. 2001). Furthermore it suggests that in mice, the "p19^{ARF}-Mdm2-p53-pRb protein family" pathway, rather than the "p16^{INK4A}-Cyclin D/Cdk-pRb protein family" pathway plays a predominant role in preventing uncontrolled oncogene-driven proliferation and suppression of tumorigenesis.

The picture that now emerges is that p19^{ARF} acts as a sensor of abnormal or conflicting mitogenic signaling, and activates a p53-dependent response that can either cause cell cycle arrest or sensitize cells to apoptosis. The behavior of triple knockout cells indicates that this decision depends on pocket protein functions. In their presence, cells arrest; in their absence, e.g. by genetic ablation, sequestration by E1A or inhibition following over-expression of Myc (Berns et al. 2000; Lasorella et al. 2000), cells become immortal but also highly sensitive to apoptosis.

6
Interconnectivity between the pRb and p53 Pathway

Although the pRb and p53 pathways in cell cycle control and checkpoint control are mostly depicted as two separate pathways, there are multiple interactions that connect the two pathways, resulting in a highly intertwined network that is regulated via complex feedback loop mechanisms (Fig. 1). The observation that p19^{ARF}-induced senescence and cell cycle arrest require pRb family members, suggests an interaction between the pRb protein family and p19^{ARF}. How p19^{ARF} signals to the pRb family is still unclear. The immortal phenotype of *p53*$^{-/-}$ MEFs suggests this pathway to be p53-dependent. However, p19^{ARF}-p53 induced senescence is not implemented via p21^{CIP}-induced inhibition of cyclin E/Cdk2 as *p21*$^{-/-}$ and *p21*$^{-/-}$*p27*$^{-/-}$ MEFs still undergo senescence and/or are responsive to over-expression of p19^{ARF} (Pantoja and Serrano 1999; Groth et al. 2000; Modestou et al. 2001). Furthermore, TKO MEFs are not blocked in G1 by p19^{ARF} over-expression, although they can still be blocked by inhibition of Cdk2 activity upon expression of a dominant-negatively acting Cdk2 mutant. A picture thus emerges wherein p19^{ARF} induces senescence via the pRb-family, but independently of p21^{CIP1}-mediated inhibition of cyclin E/Cdk2 activity. Whether the link between p19^{ARF} and the *Rb* gene family is p53-dependent, remains obscure. In contrast to one report (Kamijo et al. 1997) another showed that p19^{ARF} could induce a cell cycle arrest in *p53*$^{-/-}$ MEFs, which can be relieved by over-expression of E2F-1 or by blocking p16^{INK4A} function (Carnero et al. 2000). These data suggest that p19^{ARF} can target the pRb pathway independently of p53.

p19^{Arf} is induced by E2F1, leading to the idea that p19^{ARF} connects pRb and p53 (DeGregory et al. 1997; Bates et al. 1998). Over-expression of the oncoproteins E1A or Myc in MEFs induces apoptosis in a p53-dependent fashion (Evan et al. 1992). It seems that both proteins act so by releasing or inducing E2F1, consistent with the observation that E2F1 by it self can induce apoptosis (Lowe and Ruley 1993; Wagner et al. 1994; Leone et al. 2001; Trimarchi and Lees 2001). MEFs lacking p53 or p19^{ARF} resist E1A- or Myc-induced apoptosis, supporting the hypothesis that p19^{ARF} is an important mediator of oncogene-mediated apoptosis (De Stanchina et al. 1998; Zindy et al. 1998). Furthermore, *Eμ-Myc* transgenic animals develop B-cell lymphomas in which Myc-induced apoptosis was suppressed by deletion of the *Ink4a/Arf* locus, Mdm2 induction or p53 mutation. In a *p19*^{Arf} null, but not in a *p16*^{Ink4a} mutant background, lymphomagenesis is strongly accelerated, indicating that an intact p19^{ARF}-Mdm2-p53 pathway functions as a tumor surveillance pathway *in vivo* by counteracting oncogene-induced apoptosis (Eischen et al. 1999). On the other hand, others have shown that p19^{ARF} is dispensable for E2F-1 mediated apoptosis. The absence of p19^{ARF} even enhanced the ability of E2F-1 to induce apoptosis, suggesting that p19^{ARF} is a negative regulator of E2F-1 (Russel et al. 2002). In a tumor model engineered by tissue-specific expression of the TgT$_{121}$ variant of SV40 large T antigen, which binds all members of the pRb family but not p53, the formation of choroid plexus tumors (Saenz-Robles et al. 1994) is accompanied by a high cell turn-over as a result of p53-dependent apoptosis. Expression of the T$_{121}$ variant of T-antigen in a *Bax-*, *p53-* or *E2F-1* deficient background results in acceleration of tumorigenesis due to inhibition of apoptosis (Symonds et al. 1994; Yin et al. 1997; Pan et al. 1998). This indicates that Bax, E2F-1 and p53 function in a tumor surveillance pathway that mediates SV40 large T-antigen induced apoptosis. Interestingly, despite the induction of p19^{ARF} by T$_{121}$ expression in a wild-type background, inactivation of *p19^{Arf}* in this system does not have any effect on cell proliferation, the level of apoptosis or tumor formation in this tumor model system, suggesting that E2F-1 induces apoptosis in a p19^{ARF}-independent manner (Tolbert et al. 2002).

Although it remains possible that p19^{ARF} is a critical target for E2F-1 mediated apoptosis in some, yet unknown, settings, E2F1-induced apoptosis is more likely to be the result of direct activation of other apoptosis-inducing genes like *p53*, its homologue *p73* and the apoptosis protease-activating factor 1 (*Apaf-1*), which are all shown to be direct E2F targets and are required for E2F-1 induced apoptosis *in vitro* and *in vivo* (Irwin et al. 2000; Lissy et al. 2000; Stiewe and Pützer 2000; Moroni et al. 2001; Ren et al. 2002; Russel et al. 2002).

In addition, the significance of the pRb-MDM2 interaction is not very clear. MDM2 is a potent negative regulator of p53. As a transcriptional target of p53, MDM2 participates in an auto-regulatory feedback loop to antagonize p53 function. MDM2 binds to p53 and blocks its transcriptional activity, acts as an E3 ubiquitin ligase to target p53 for degradation in cytoplasmic proteasomes and accelerates p53 nuclear export. p19^{ARF} is able to block MDM2 function by binding to MDM2 and antagonizing MDM2-mediated ubiquitination and nuclear export of p53. A less well characterized function of MDM2 is the regulation of pRb protein family function. First, MDM2 is able to bind pRb in its C-terminus, an interaction that is enhanced by the p300/CBP and pCAF mediated acetylation of pRb (Xiao et al. 1995; Chan et al. 2001). Acetylation of pRb occurs primarily upon differentiation in the C-terminal region on amino acids that are not conserved in p107 and p130 (see Fig. 2). Studies with acetylation-impaired pRb mutants showed that acetylation of pRb is required for pRb-mediated cell cycle exit and induction of late myogenic gene expression, possibly by degradation of EID-1, an inhibitor of differentiation (Nguyen et al. 2004). Other reports suggested that MDM2 directly inhibits pRb function by ubiquitination-mediated degradation of pRb through the E3-ligase function of MDM2 (Uchida et al. 2005). MDM2 is also able to modulate E2F-1 transcriptional activity by binding the C-terminus of E2F-1 and to reduce E2F-1 levels (Martin et al. 1995; Xiao et al. 1995; Loughran et al. 2000). Adding to the complexity, E2F-1 on its turn can reduce MDM2 protein levels by proteolytic degradation, suggesting a regulatory feedback loop between E2F-1 and MDM2 (Strachan et al. 2001). pRb can form a trimeric complex with MDM2 and p53, thereby blocking MDM2-mediated degradation of p53 (Xiao et al. 1995; Hsieh et al. 1999). The identification of MDM2 as an inhibitor of transforming growth factor-β (TGF-β)-mediated cell cycle arrest may provide a clue for the MDM2/pRb connection (Sun et al. 1998).

Ectopic expression of MDM2 rescued TGF-β-induced growth arrest in a p53-independent manner by interference with pRb, indicating that MDM2, by binding to pRb can alleviate its growth suppressing function independently of p53. p19^{ARF} may act as an antagonist of the MDM2-mediated inactivation of pRb, since it can bind and inactivate MDM2. Therefore, it would be interesting to see whether p19^{ARF} can disrupt the interaction between pRb and MDM2. The ability of MDM2 to alleviate a p107-mediated G1-arrest suggests that MDM2 may modulate the function of all *Rb* gene family members, rather than pRb alone (Dubs-Poterszman et al. 1995). Although a clear mechanism of pRb/E2F regulation by MDM2 is lacking at the moment, it seems that MDM2 is able to facilitate cell cycle progression by inactivation of the repressor function of the pocket protein/E2F complexes (Fig. 2).

Finally, the pocket proteins might be direct targets of p19^{ARF}-p53 signaling, since over-expression of p19^{ARF} in MEFs deficient for all pocket proteins does not inhibit proliferation (Dannenberg et al. 2000). Moreover, transformation of MEFs by SV40 large T antigen was shown to be dependent on inactivation of the *Rb* gene family and p53. Transformation of MEFs lacking p19^{ARF} did no longer depend on inactivation of the *Rb* gene family or p53, indicating that the pocket proteins are functionally inactivated by loss of p19^{ARF} (Chao et al. 2000). Whether p19^{ARF} requires p53 or can directly induce the formation of pocket protein/E2F repressor complexes in this context remains elusive. Functional inactivation of the pRb protein family by expression of HPV-16 E7 or genetic inactivation, can bypass a p53-mediated G1-arrest, upon DNA damage or ectopic expression of p53, showing that pocket proteins are downstream targets of p53 (Slebos et al. 1994; Demer et al. 1996; Dannenberg et al. 2000; Sage et al. 2000). Since the pocket proteins are downstream of p19^{ARF} and p16^{INK4A}, regulation of pRb/p107/p130 function by both proteins might be essential to create sufficient levels of repressor complexes in order to induce a sustained cell cycle arrest upon oncogenic signaling. Inactivation of either p19^{ARF} or p16^{INK4A} might result in insufficient repressor capacity under such circumstances and might predispose cells to acquire additional mutations. The accelerated tumorigenesis in mice lacking p16^{INK4A} and expressing reduced levels of p19^{ARF} due to heterozygosity for *p19^{Arf}*, suggest that *Ink4a* and *Arf* pathways might be connected in regulating the retinoblastoma protein family function (Carnero et al. 2000; Krimpenfort et al. 2001).

7
The *Rb* Gene Family in Tumor Suppression in Mice

Genetic inactivation of *Rb* through the germ line results in embryonic lethality predominantly due to widespread apoptosis in the liver and central nervous system (Clarke et al. 1992; Jacks et al. 1992; Lee et al. 1992; see Table 1). These phenotypes could be rescued by providing the embryo with a pRb-proficient placenta indicating non-cell autonomy. Furthermore, conditional deletion of pRb in the CNS did not result in apoptosis (Ferguson et al. 2002; de Bruin et al. 2003a; MacPherson et al. 2003; Wu et al. 2003). Ultimately pRb deficient embryos provided with a wild-type placenta succumb around birth, probably due to the dysfunctional musculature resulting in failure to breath (Wu et al. 2003). Whereas *Rb* heterozygosity leads to pituitary gland and thyroid gland tumors, $p107^{-/-}$ or $p130^{-/-}$ mice have no overt phenotype and show no higher incidence of tumors compared with wild-type mice (Cobrinik et al. 1996; Lee et al. 1996). However, *p107* or *p130* inactivation in a Balb/cJ genetic background results in embryonic lethality and in runted mice that are susceptible to myeloid hyperplasia, respectively (LeCouter et al. 1998a,b). Since Balb/cJ mice were shown to carry a point mutation in the $p16^{INK4A}$ allele resulting in a less efficient block of pRb (and possibly p107 or p130) phosphorylation, this suggests that only in the presence of sufficient pocket protein/E2F complexes can the retinoblastoma gene family members functionally compensate for each other (Zhang et al. 1998a). Combined inactivation of *p107* and *p130* in a C57BL/6 background results in a neonatal lethal phenotype, whereas inactivation of *Rb* together with *p107* in the same genetic background results in a more severe embryonic lethal phenotype compared with *Rb* deficiency alone. These phenotypes precluded the analysis of the role of pocket proteins in tumorigenesis. $p107^{+/-}p130^{-/-}$ or $p107^{-/-}p130^{+/-}$ mice did not show a higher incidence of tumors compared with wild-type mice, suggesting that p107 or p130 are not involved in murine tumorigenesis (Cobrinik et al. 1996; Lee et al. 1996). However, concomitant ablation of pRb and p107 or pRb and p130 in chimeric mice or conditional knockout mice resulted in development of retinoblastoma, resembling characteristics of the inner nuclear layer of the retina (Robanus-Maandag et al. 1998; Chen et al. 2004; Dannenberg et al. 2004; McPherson et al. 2004). Furthermore, $Rb^{-/-}p130^{-/-}$ chimeras developed adrenal medullary tumors and bronchial epithelial neuro-endocrine dysplasia (Dannenberg et al. 2004). Interestingly, $Rb^{+/-}p107^{-/-}$ chimeric mice developed a variety of tumors, most predominantly pituitary gland tumors, adenocarcinomas of the caecum, osteosarcomas and lymphomas, which show in 70% of cases loss of the remaining wild-type *Rb* allele (Dannenberg et al. 2004). These data strongly indicate that p107 and p130 act as potent suppressors of oncogenic transformation of pRb-deficient cells. Since the tumor spectrum in *Rb/p107*- and *Rb/p130*-deficient chimeras did not entirely overlap, these data suggest that the requirement for pocket proteins in tumor suppression is cell-type dependent. In the retina, similar to MEF cultures, p107 and p130 are *both* required to suppress proliferation of pRb-deficient cells, while in the adrenal gland, p130 *alone* can compensate for loss of pRb. This may indicate that the suppression of tumorigenesis by p107 and p130 involves other, tissue-specific functions besides regulation of E2Fs. For example, the occurrence of osteosarcomas in $Rb^{+/-}p107^{-/-}$ chimeras but not in $Rb^{+/-}p130^{-/-}$ chimeras may be

Table 1 Overview of phenotypes in mice containing inactivated alleles of the *Rb* gene family members. *Rb*-gene family mutant mice are divided in three groups; (1) mice containing conventional germ-line knockout alleles, (2) chimeric mice (mice generated by injection of mutant ES cells into wild-type blastocysts) and (3) mice containing conditional knockout alleles. Tissues in which pRb is ablated by expression of Cre-recombinase are indicated between brackets. CNS (central nervous system), PNS (peripheral nervous system), ONL (outer nuclear layer of the retina), PNEC (pulmonary neuro-endocrine cells)

Genotype	Phenotype	Refs.
Conventional knock-out		
$Rb^{+/-}$	Pituitary and thyroid tumors	Robanus-Maandag et al. 1994; Hu et al. 1994; Williams et al. 1994
$Rb^{-/-}$	E13.5–E15.5 embryonic lethal, apoptosis in CNS and liver, nucleated erythrocytes, placental defect	Clarke et al. 1992; Jacks et al. 1992; Lee et al. 1992; Wu et al. 2003
$p107^{-/-}$	No phenotype	Cobrinik et al. 1996
$p107^{-/-}$ (*Balb/c*)	Impaired growth, myeloid hyperplasia	LeCouter et al. 1998
$Rb^{+/-}p107^{-/-}$	Growth retardation, increased mortality, retinal dysplasia in ONL, pituitary gland tumors	Lee et al. 1996
$Rb^{-/-}p107^{-/-}$	E11.5–E12.5 embryonic lethal, aggravated phenotype compared with $Rb^{-/-}$	Lee et al. 1996
$p130^{-/-}$	No phenotype	Cobrinik et al. 1996
$p130^{-/-}$ (*Balb/c*)	Embryonic lethal E11–E13, neuronal and dermamyotomal structural defects	LeCouter et al. 1998
$p107^{-/-}p130^{-/-}$	Perinatal lethal, defective endochondral bone development, shortened limbs	Cobrinik et al. 1996
$Rb^{+/-}p53^{+/-}$	Pinealoblastoma, thyroid and islet cell tumors, bronchial epithelial hyperplasia, retinal dysplasia	Williams et al. 1994; Harvey et al. 1995

Table 1 (continued)

Genotype	Phenotype	Reference
Chimeras		
$Rb^{+/-}$	Pituitary gland tumors	Robanus-Maandag et al. 1994; Williams et al. 1994
$Rb^{-/-}$	Pituitary gland tumors, lens cataracts	Robanus Maandag et al. 1994; Williams et al. 1994
$Rb^{+/-}p107^{-/-}$	Tumors originating form pituitary gland, bone, caecum, adrenal gland, lung	Dannenberg et al. 2004
$Rb^{-/-}p107^{-/-}$	Retinoblastoma, ectopic proliferation in retina, retinal dysplasia, pituitary gland tumors	Robanus-Maandag et al. 1998
$Rb^{+/-}p130^{-/-}$	Adrenal gland tumor	Dannenberg et al. 2004
$Rb^{-/-}p130^{-/-}$	Ectopic proliferation in INL; retinoblastoma, bronchial epithelial hyperplasia, adrenal gland tumors	Dannenberg et al. 2004
Conditional knock-out		
$GFAP\text{-}Cre\text{-}Rb^{lox/lox}$	(Brain; astrocytes/neurons; intestine, testis) no phenotype	Marino et al. 2000
$GFAP\text{-}Cre\text{-}Rb^{lox/lox}p53^{-/-}$	(Brain; astrocytes/neurons; intestine, testis) medulloblastoma	Marino et al. 2000
$En2\text{-}Cre\text{-}Rb^{lox/lox}$	(Dorsal mid-hindbrain junction) subset of granule cell precursors show ectopic proliferation and apoptosis	Marino et al. 2003
$En2\text{-}Cre\text{-}Rb^{lox/lox}p107^{-/-}$	(Dorsal mid-hindbrain junction) ataxia, impaired terminal differentiation and migration of granule cell precursors, apoptosis of granule cell	Marino et al. 2003
$Mox2\text{-}Cre\text{-}Rb^{lox/lox}$	(All embryonic tissues), PNS and lens apoptosis/ectopic proliferation and skeletal muscle defects	Wu et al. 2003; deBruin et al. 2003
$Foxg1\text{-}Cre\text{-}Rb^{lox/lox}$	(Telencephalon) ectopic proliferation of progenitors, increased brain cellularity	Ferguson et al. 2002

Table 1 (continued)

Genotype	Phenotype	Reference
Conditional knock-out (continued)		
Nes-Cre1-Rb$^{lox/lox}$	(Midgestation brain, germ-line, muscle) apoptosis in PNS and lens, ectopic proliferation in CNS, PNS and lens, skeletal muscle defect, ectopic proliferation in retina, apoptosis of rods, ganglion and bipolar cells	MacPherson et al. 2003, 2004
Nes-Cre1Rb$^{lox/lox}$p107$^{-/-}$	(Midgestation brain, germ-line, muscle) embryonic lethal; retinal dysplasia	MacPherson et al. 2004
Nes-Cre1-Rb$^{lox/lox}$p130$^{-/-}$	(Midgestation brain, germ-line, muscle) ectopic proliferation in retina retinoblastoma	MacPherson et al. 2004
α-Cre-Rb$^{lox/lox}$	(Peripheral retinal precursors), apoptosis of rods, ganglion and bipolar cells/ectopic proliferation of all retinal precursors	Chen et al. 2004
α-Cre-Rb$^{lox/lox}$p107$^{-/-}$	(Peripheral retinal precursors) apoptosis of rods/cones, ganglion and bipolar cells; ectopic proliferation of all retinal precursor cells; retinoblastoma	Chen et al. 2004
Chx10-Cre-Rb$^{lox/lox}$	(Retinal progenitor cells) loss of rods	Zhang et al. 2004; Donovan and Dyer 2004
Chx10-Cre-Rb$^{lox/lox}$p107$^{-/-}$	(Retinal progenitor cells) retinoblastoma	Zhang et al. 2004
IRBP-Cre-Rb$^{lox/lox}$	(Rod/cone photoreceptor cells in ONL) pituitary gland tumors	Vooijs et al. 2002
IRBP-Cre-Rb$^{lox/lox}$p107$^{-/-}$	(Rod/cone photoreceptor cells in ONL) pituitary gland tumors	Vooijs et al. 2002
IRBP-Cre-Rb$^{lox/lox}$p53$^{-/-}$	(Rod/cone photoreceptor cells in ONL) pituitary gland tumors, pinealoblastoma	Vooijs et al. 2002
K14-Cre-Rb$^{lox/lox}$	(Striated epithelium) hyperplasia and hyperkeratosis	Balsitis et al. 2003; Ruiz et al. 2004
K14-Cre-Rb$^{lox/lox}$p107$^{-/-}$	(Striated epithelium) severe hyperplasia and hyperkeratosis	Ruiz et al. 2004
AdenoCre-RbRb$^{lox/lox}$	(Ovarian surface epithelium) no phenotype	Flesken-Nikitin et al. 2003

Table 1 (continued)

Genotype	Phenotype	Reference
Conditional knock-out (continued)		
AdenoCre-$Rb^{lox/lox}p53^{lox/lox}$	(Ovarian surface epithelium) ovarian epithelial carcinogenesis	Flesken-Nikitin et al. 2003
AdenoCre-$Rb^{lox/lox}$	(Bronchus/lung) no phenotype	Meuwissen et al. 2004
AdenoCre-$Rb^{lox/lox}p53^{lox/lox}$	(Bronchus/lung) neuro-endocrine lung tumors	Meuwissen et al. 2004
AdenoCre-$Rb^{lox/lox}$	(Liver) hepatocytes proliferate ectopically and show polyploidy	Mayhew et al. 2005
Alb-Cre-$Rb^{lox/lox}$	(Hepatocytes) hepatocytes proliferate ectopically and show polyploidy	Mayhew et al. 2005
CC10-rTA-Cre-$Rb^{lox/lox}$	(Lung epithelial progenitor cells) increased proliferation and apoptosis in PNEC, hypercellular neuroendocrine lesions	Wikenheiser-Brokamp et al. 2004
αMHC-Cre-$Rb^{lox/lox}$	(Cardiac myoblasts) no phenotype	MacLellan et al. 2005
αMHC-Cre-$Rb^{lox/lox}p130^{-/-}$	(Cardiac myoblasts) increased heart weight, ectopic myocyte cycling	MacLellan et al. 2005
PB-Cre-$Rb^{lox/lox}$	(Prostate epithelium) focal hyperplasia, increased proliferation, in prostate epithelium	Maddison et al. 2004
Col1A1-Cre-$Rb^{lox/lox}$	(Osteoblast cell lineage) ectopic proliferation of inner ear hair cells, perinatal lethality	Sage et al. 2005
TEC1-Cre-$Rb^{lox/lox}$	(Melanocyte lineage); no phenotype	Tonks et al. 2005

related to a specific role of p107 in bone development (Thomas et al. 2001; Laplantine et al. 2002).

7.1
Mechanistic Insights in the Tumor Suppressive Role of the Rb Gene Family

Several mechanisms for the contribution of pRb deficiency to tumorigenesis have been postulated. In general, a picture is emerging that loss of pRb either leads to increased apoptosis or a cell cycle arrest in a cell-type specific manner. To overcome these impediments to the progression of pre-neoplastic cells into neoplasia, additional mutations are required that will block apoptosis or overcome a cell cycle arrest. One possible mechanism to acquire these additional mutations might be the increased cell turn-over that accompanies pRb inactivation *in vivo*. For example, conditional inactivation of *Rb* in the pituitary gland revealed that tumorigenesis is preceded by induction of apoptosis and proliferation (Vooijs et al. 1998). Similarly, selective ablation of *Rb* from neuro-endocrine cells in the respiratory epithelium resulted in increased proliferation and apoptosis (Wikenheiser-Brokamp et al. 2004). Also, inactivation of all pocket proteins in MEFs exposed to growth-inhibitory signals such as growth factor deprivation and cell-cell contact results in increased cell turnover (Dannenberg et al. 2000; Sage et al. 2000). This increased cell turnover may accelerate the acquisition of additional mutations required for oncogenic transformation. Indeed, pRb-deficient pituitary gland tumors appeared to arise from cells in which apoptosis was blocked, presumably through an additional mutation (Vooijs et al. 1998). In addition, inactivation of the pRb family members in the mouse brain epithelium by the SV40 large T variant T_{121} induced aberrant proliferation and p53-dependent apoptosis. Inactivation of p53 or E2F-1 in this setting resulted in a dramatic reduction in apoptosis (Saenz-Robles et al. 1994; Yin et al. 1997; Pan et al. 1998). Although p53 deficiency was able to rescue pRb-deficient cells in the lens and CNS from apoptosis, and accelerated tumorigenesis of the pituitary gland in $Rb^{+/-}p53^{+/-}$ mice, no loss of p53 function was observed (Morgenbesser et al. 1994; Williams et al. 1994a; Macleod et al. 1996). Furthermore, deficiency of E2F-1 also rescued the p53-dependent apoptosis of pRb deficient cells and lowered, but not abolished, the penetrance of pituitary gland tumor formation in $Rb^{+/-}$ mice (Tsai et al. 1998; Yamasaki et al. 1998). E2F-1 deficiency did, however, completely inhibit tumor formation in the thyroid gland. This indicates that tumorigenesis of pRb deficient cells is modulated only in part, in a tissue-specific manner, by E2F1 and p53. Indeed, tumors arising in Rb/p107 or Rb/p130 deficient chimeras did not acquire p53 mutations. Furthermore, combined loss of pRb and p53 also resulted in bronchial epithelial hyperplasia and SCLC (Williams et al. 1994b; Meuwissen et al. 2003). Although this could suggest that *p53* and the *Rb* gene family function in similar tumor suppressor pathways, it is more likely that other p53- and E2F-1 independent pathways are inactivated to allow escape from apoptosis.

While the increased cell turnover hypothesis seems to be a valid mechanism in some settings, recent discoveries showed that cell turnover due to deficiency for pRb and its family members is tissue and cell-type dependent. In the lens, enhanced proliferation concomitant with enhanced apoptosis was observed (de Bruin et al. 2003; Wu et al. 2003). Furthermore, conditional inactivation of *Rb* in the brain resulted in enhanced proliferation in the absence of apoptosis (Ferguson et al. 2002; MacPherson et al. 2003). Conditional inactivation of *Rb* in the skin using K14-*Cre* transgenic mice resulted in increased proliferation in interfollicular epidermis, while increased apoptosis was only observed in follicular cells lacking both pRb and p107 (Ruiz et al. 2004). Interesting observations were made in cells lacking pRb in the retina. Somatic inactivation of pRb, even in the absence of p107 and p53, in the outer nuclear layer had no obvious consequences, indicating that in some settings acute loss of pRb in post-mitotic cells does not lead to re-entering of the cell cycle (Vooijs et al. 2002). Inactivation of pRb in the retinal progenitors resulted in the selective apoptosis of many retinal cells (rods, cones, ganglion and bipolar cells) but extended the proliferative capacity of amacrine precursors, which was exacerbated in the absence of p107. Retinal amacrine cells were shown to be the major component of *Rb/p107* and *Rb/p130* deficient retinoblastomas (Robanus-Maandag et al. 1998; Chen et al. 2004; Dannenberg et al. 2004; MacPherson et al. 2004) suggesting that the extended proliferative capacity of amacrine precursors allows accumulation of additional mutations required for retinoblastomagenesis. As amacrine precursors lacking pRb or pRb and p107, despite their extended proliferative capacity, ultimately terminally differentiate it was suggested that retinoblastoma in mice requires additional oncogenic mutations aimed at circumventing differentiation rather than apoptosis (Chen et al. 2004). Based on *in vitro* data, the delayed cell cycle exit of double knockout cells may indicate that oncogenic signals can induce a bypass of differentiation more easily in cells that lack both *Rb* and *p107* or *Rb* and *p130*. In cells lacking only *Rb* this would lead to p107- and p130-mediated cell cycle arrest (Dannenberg et al. 2000, 2004; Sage et al. 2000, 2003; Peeper et al. 2001). How p107 and p130 compensate for the loss of pRb, was shown in a series of *in vitro* and *in vivo* experiments showing that in the absence of pRb, p107 and p130 confer tumor suppressor function through binding of "activator" E2Fs (E2F1-3) and thereby repress pRb regulated genes (Lee et al. 2002). Finally, other studies have shown that pRb is involved in preventing chromosome instability by transcriptional regulation of genes involved in G2/M checkpoints such as *Mad2*, adding another condition that fuels the production of additional oncogenic events (Ren et al. 2002; Hernando et al. 2004). Others have shown that the pRb protein family is involved in the stabilization of histone methylation at constitutive heterochromatin. Ablation of all pocket proteins leads to aberrant centromeres and telomere abnormalities and subsequent chromosome instability (Gonzalo et al. 2005).

In conclusion, we speculate that the consequences of loss of pRb are cell-type specific but almost invariably leads to functional compensation by p107 and/or p130. On top of that the redundancy of pocket proteins in itself is cell-type specific. For example, in the mouse pituitary gland p107 and p130 may not be expressed or functional in compensating for loss of pRb, whereas in the mouse retina both p107 and p130 are active and can together compensate for pRb. In this case loss of either p107 or p130 is required to sustain prolonged cell proliferation ultimately leading to retinoblastoma development. Alternatively, specific functions of each of the *Rb* gene family in the differentiation of tissues might add to the tissue-specific tumor development in the different *Rb* gene family deficient mice (see below).

8
Role of *p107* and *p130* in Human Cancer

In human tumors *p107* or *p130* inactivation is rarely found, in contrast to the frequent loss of pRb function. So far, only one report shows that p107 is inactivated by an intragenic deletion in a B-cell lymphoma cell line (Ichimura et al. 2000). Interestingly, $Rb^{+/-}p107^{-/-}$ chimeric mice develop osteosarcoma and leiomyosarcoma, which are well-documented secondary malignancies in retinoblastoma patients (Moll et al. 1997; Ryan et al. 2003). More evidence is available for the involvement of p130 in human tumorigenesis. First, *p130* is located at a chromosomal region, 16q12.2-13 that is frequently lost in breast, ovarian, hepatic and prostate cancers (Yeung et al. 1993). Small cell lung carcinomas, which show in more than 90% of the cases deletion of chromosome region 13q14, harboring *Rb*, show in more than 70% of the cases also loss of the p130 containing chromosome arm 16q. This suggests that a large fraction of this type of tumors has lost both pRb and p130 expression. A smaller fraction of these tumors (10%) had lost chromosome arm 20q harboring p107 (Stanton et al. 2000). Analysis of additional non-*Rb* mutations in human retinoblastoma revealed that nearly all human retinoblastomas display either gain of chromosomal regions 6p22, 1q31, 2p24-25 and 13q32-q34 or loss of 16q (Potluri et al. 1986; Mairal et al. 2000; Chen et al. 2001). Interestingly, one study showed loss of the chromosomal region 16q, on which *p130* and *E2F4* are located in almost 50% of the cases (Marial et al. 2000). Others found a strong correlation between lack of p130 expression in retinoblastomas and poor differentiation and low apoptotic index (Bellan et al. 2002). Second, mutational analysis of p130 in primary nasopharyngeal carcinoma (Claudio et al. 2000b), Burkitt's lymphoma (Cinti et al. 2000), lung tumors (Claudio et al. 2000a) and the small cell lung cancer cell line GLC2 (Helin et al. 1997), suggested loss of p130 function either by mutations in its bipartite NLS or a mutation leading to complete absence of the protein. Contradicting data were obtained by others showing that p130 is not affected at the nucleotide and protein level in primary nasopharyngeal carcinomas and primary lung cancers (Modi et al. 2000; Gray and Guo 2001). Loss of pRb and p107 or p130 in mice results in tumorigenesis, indicating that loss of p107 or p130 collaborates with pRb deficiency in tumorigenesis (Robanus-Maandag et al. 1998; Dannenberg et al. 2004; MacPherson et al. 2004). *In vitro* data have indicated that p107 *plus* p130 can only partially accomplish the function of pRb as the critical regulator of most cell cycle responses (Dannenberg and Te Riele, unpublished observations). Also *in vivo*, the consequences of loss of pRb may only partially be suppressed by combined action of p107 and p130. Therefore, uncontrolled proliferation of pRb-defective cells may not necessarily require complete loss of p107 and p130, but already occur upon partial loss of p107/p130 function. For example, loss of p107/p130 function as a result of enhanced phosphorylation or reduced protein levels may be sufficient. In addition, evidence is accumulating that either deregulated cyclin D/Cdk4 or cyclin E/Cdk2 activity or over-expression of Myc indirectly inactivates p107 and p130 in human tumors (Cheng et al. 2000; Lasorella et al. 2000; Ashizawa et al. 2001). The identification of genetic alterations that indirectly or directly cause functional inactivation of p107 or p130 will lend support to this hypothesis.

9
The Retinoblastoma Gene Family in Differentiation and Tumorigenesis

Regulation of the E2F transcription factors is a key role of the retinoblastoma gene family and disruption of that function is critical for tumorigenesis. Nevertheless, as mentioned before, there seems to be differential sensitivity of cell types for oncogenic transformation upon loss of the *Rb* gene. Predominantly retinoblasts and osteoblasts will undergo oncogenic transformation in individuals born with one inactivated *Rb* allele, leaving other cell types unaffected. Several factors may play a role in determining whether a certain cell type that loses pRb will become a tumor cell. First, in certain cell-types expression of p107 and/or p130 may compensate for the loss of pRb. Second, besides its function as a regulator of E2F activity, the retinoblastoma protein family is clearly involved in (terminal) differentiation of certain cell types. As a consequence, inactivation of pRb may impair the (terminal) differentiation of these cells, leaving others unaffected. Probably a combination of these factors will decide whether a cell will ultimately become a tumor cell upon loss of pRb.

9.1
A Link between Pax, bHLH and Pocket Proteins in Differentiation and Tumorigenesis

A substantial portion of the tumors in *Rb-*, *Rb/p107-* or *Rb/p130*-deficient mice are of neuro-ectoderm and neuro-endocrine origin. These tissues share a common developmental program involving the Pax and the basic helix-loop-helix (bHLH) family of transcription factors (Pearse and Polak 1978). Interestingly, the Pax family protein Pax-3 was shown to interact with Rb, p107 and p130. Another Pax family member, Pax-5, is able to interact with pRb but not with p107 (Wiggan et al. 1998; Ebenhard and Busslinger 1999). The interaction between these proteins is dependent on the large pocket (sub-domains A and B and the C-terminus) of the pRb protein family and a consensus sequence present in the paired-like homeodomain of the Pax proteins. Since most Pax family members contain this consensus sequence, except Pax-1 and Pax-9, it is likely that besides Pax-3 and -5 also Pax-2, Pax-6 and Pax-8 can bind to pRb protein family members. A striking similarity was observed between the pRb binding site in the transactivation domain of E2F1-5 and the paired-like homeodomain consensus sequence, indicating that E2F might compete with the Pax proteins for binding to the pRb protein family. Transient co-transfection studies showed that the pRb protein family represses transcription activation by Pax-3 from promoters containing paired-like homeodomain protein binding sites. The observation that pRb can bind simultaneous Pax-5 and the TATA-binding protein TBP, suggested a mechanism for inhibition of Pax family protein-mediated transcription of target genes by the pRb protein family (Ebenhard and Busslinger 1999).

9.2
Pax and bHLH Proteins in Retina and Pulmonary Epithelium Development

Interestingly, genes encoding basis helix-loop-helix transcription factors have been identified as paired-like homeodomain protein targets, which are directly involved in controlling the developmental fate of neuro-ectoderm tissues. For example, conditional inactivation of Pax-6 in the retinal progenitor cells (RPC), which are able to generate almost all principal cell types of the vertebrate retina, results in the absence of a subset of retinogenic bHLH transcription factors such as Mash1, Ngn2 and Math5. In contrast, expression of another bHLH transcription factor involved in terminal differentiation, NeuroD, was induced in the absence of Pax-6. Consistent with the observation that over-expression of NeuroD increased the amacrine cell population in the retina, loss of Pax-6 activity in the RPC lead to the exclusive generation of amacrine neuronal cells (Morrow et al. 1999; Marquardt et al. 2001). Since loss of pRb and p130 or pRb and p107 in mice results in retinoblastoma originating from a retinal cell compartment committed to amacrine cell differentiation, one can envision a scenario in which loss of Pax-6/pocket protein complex mediated transcriptional suppression of retinogenic bHLH transcription factors, such as NeuroD, will lead to an expansion of the amacrine cell compartment. Interference with the amacrine differentiation program (e.g. by inactivation of *NeuroD*), in pRb/p107- or pRb/p130-deficient mice should suppress tumorigenesis. This hypothesis also provides an explanation for the absence of retinoblastoma in $p16^{Ink4a-/-}$, $p19^{Arf-/-}$, $p53^{-/-}$ mice or mice expressing the constitutive active Cdk4^{R24C} mutant, since these mutations will not lead to the inactivation of the differentiation function of the pRb protein family. Retinoblastoma development upon conditional inactivation of Pax-6 in these mice would support this hypothesis.

NeuroD and other bHLH transcription factors are also involved in the development of pulmonary neuro-endocrine cells (PNEC), the cell type that gave rise to hyperplasia in $Rb^{-/-}p130^{-/-}$ chimeric mice (Ito et al. 2000; Dannenberg et al. 2004). Ablation of the bHLH transcription factor Hes-1 results in elevated levels of NeuroD and an excess of PNECs, which develop into nodular lesions. Inactivation of the bHLH transcription factor achaete-scute homolog-1 (ASH1 or Mash1), results in the complete absence of PNECs (Ito et al. 2000). Over-expression of Mash1, a cardinal feature of SCLC, results in airway epithelial proliferation and collaborates with SV40 large T in lung tumorigenesis (Linnoila et al. 2000; Bhattacharjee et al. 2001; Garber et al. 2001). Another class of bHLH transcription factors, Inhibitor of differentiation (Id) proteins, of which Id2 is known to interact with pRb, p107 and p130, might be involved in the differentiation of PNECS, since Id2 and Id4 expression have been observed in bronchial epithelium (Jen et al. 1996; Lasorella et al. 2000).

In summary, these data may indicate that bHLH transcription factors encoding genes, transcriptionally regulated by Pax/pocket protein complexes, are critical for the terminal differentiation of various neuro-ectoderm tissues. As a consequence, inactivation of the *Rb* gene family will abolish the induction or repression of bHLH transcription factors required for cell-type specific differentiation, which subsequently impairs terminal differentiation. On top of that, loss of pocket protein-E2F repressor complexes will allow these cells to proliferate and ultimately, probably by acquiring additional mutations, lead to oncogenic transformation of these tissues. The dual function of the *Rb* gene family in differentiation and cell cycle control explains the frequent involvement of loss of pocket protein function in a subset of human cancers.

10
Conclusion

A picture emerges in which pRb and its family members p107 and p130 are crucial in the regulation of the cell cycle in response to growth-inhibiting conditions such as cell culture stress, serum starvation, contact inhibition, oncogene activation and differentiation. These observations provide an understanding for the almost universal inactivation of the p16-cyclin D/Cdk-pRb-E2F pathway across tumor types. Loss of function of pRb will not only lead to loss of cell cycle regulation under growth restricting conditions (e.g. differentiation), but also create an environment that favors the acquisition of additional transforming mutations through increased cell turnover or increased chromosomal instability. Tissue specific tumor development due to inactivation of the *Rb* gene family may reflect its role beyond regulation of E2F such as activation of tissue specific differentiation inducers in differentiating cells. Furthermore, direct or indirect inactivation of p107 or p130 function may be required on top of pRb deficiency as indicated by the tumor prone phenotype of chimeric mice lacking either pRb and p107 or Rb and p130. Thus, the oncogenic potential of loss of *Rb* is likely to be decided by the expression of each of the *Rb* family members and the role these family members play in the differentiation of the cell type involved. Future research will unquestionably reveal more specific functions for each of the pRb protein family members in biological processes relevant to tumor suppression. Ultimately, this knowledge will help to define targets for therapeutic intervention.

Acknowledgements We thank our colleagues Floris Foijer, Jacob Hansen, René Medema, Daniel Peeper, Rob Wolthuis and many others for helpful discussions related to this chapter. Work in the Te Riele lab is supported by grants from the Dutch Cancer Society, the European Commission and the Netherlands Genomics Initiative.

References

Ashizawa S, Nishizawa H, Yamada M, Higashi H, Kondo T, Ozawa H, Kakita A, Hatakeyama M (2001) Collective inhibition of pRB family proteins by phosphorylation in cells with p16^{INK4A} loss or cyclin E overexpression. J Biol Chem 276:11362–11370

Balsitis SJ, Sage J, Duensing S, Munger K, Jacks T, Lambert PF (2003) Recapitulation of the effects of the human papillomavirus type 16 E7 oncogene on mouse epithelium by somatic Rb deletion and detection of pRb-independent effects of E7 in vivo. Mol Cell Biol 23:9094–9103

Bates S, Phillips AC, Clark PA, Stott F, Peters G, Ludwig RL, Vousden KH (1998) p14^{ARF} links the tumour suppressors RB and p53. Nature 395:124–125

Beijersbergen RL, Carlée L, Kerkhoven RM, Bernards R (1995) Regulation of the retinoblastoma protein-related p107 by G1 cyclin complexes. Genes Dev 9:1340–1353

Bellan C, De Falco G, Tosi GM, Lazzi S, Ferrari F, Mobini G, Bartolomei S, Toti P, Mangiavacchi P, Cevenini G (2002) Missing expression of pRb2/p130 in human retinoblastomas is associated with reduced apoptosis and lesser differentiation. Invest Ophthalmol Vis Sci 43:3602–3608

Berns K, Martins C, Dannenberg JH, Berns A, Te Riele H, Bernards R (2000) p27^{kip}-independent cell cycle regulation by MYC. Oncogene 19:4822–4827

Bhattacharjee A, Richards WG, Staunton J, Li C, Monti S, Vasa P, Ladd C, Beheshti J, Bueno R, Gillette M, Loda M, Weber G, Mark EJ, Lander ES, Wong, W, Johnson, BE, Golub TR, Sugarbaket DJ, Meyerson M (2001) Classification of human lung carcinomas by mRNA expression profiling reveals distinct adenocarcinomas subclasses. Proc Natl Acad Sci USA 98:13790–13795

Bookstein R, Rio P, Madreperla SA, Hong F, Allred C, Grizzle WE, Lee WH (1990) Promoter deletion and loss of retinoblastoma gene expression in human prostate carcinoma. Proc Natl Acad Sci USA 87:7762–7766

Brehm A, Miska EA, McCance DJ, Reid JL, Bannister AJ, Kouzarides T (1998) Retinoblastoma protein recruits histone deacytelase to repress transcription. Nature 391:597–601

Bremner R, Cohen BL, Sopta M, Hamel PA, Ingles CJ, Gallie BL, Phillips RA (1995) Direct transcriptional repression by pRB and its reversal by specific cyclins. Mol Cell Biol 15:3256–3265

Bremner R, Du DC, Connolly-Wilson MJ, Bridge P, Ahmad KF, Mostachfi H, Rushlow D, Dunn JM, Gallie BL (1997) Deletion of RB exons 24 and 25 causes low-penetrance retinoblastoma. Am J Hum Genet 61:556–570

Bruce JL, Hurford RK, Classon M, Koh J, Dyson N (2000) Requirements for cell cycle arrest by p16^{INK4A}. Mol Cell 6:737–742

de Bruin A, Wu L, Saavedra HI, Wilson P, Yang Y, Rosol TJ, Weinstein M, Robinson ML, Leone G (2003a) Rb function in extraembryonic lineages suppresses apoptosis in the CNS of Rb-deficient mice. Proc Natl Acad Sci USA 100:6546–6551

de Bruin A, Maiti B, Jakoi L, Timmers C, Buerki R, Leone G (2003b) Identification and characterization of E2F7, a novel mammalian E2F family member capable of blocking cellular proliferation. J Biol Chem 278:42041–42049

Buyse IM, Shao G, Huang S (1995) The retinoblastoma protein binds to RIZ, a zinc finger protein that shares an epitope with the adenovirus E1A protein. Proc Natl Acad Sci USA 92:4467–4471

Cam H, Balciunaite E, Blais A, Spektor A, Scarpulla RC, Young R, Kluger Y, Dynlacht BD (2004) A common set of gene regulatory networks links metabolism and growth inhibition. Mol Cell 16:399–411

Campisi J (1997) The biology of replicative senescence. Eur J Cancer 33:703–709

Canhoto AJ, Chestukhin A, Litovchick L, DeCaprio JA (2000) Phosphorylation of the retinoblastoma-related protein p130 in growth-arrested cells. Oncogene 19:5116–5122

Carnero A, Hudson JD, Price CM, Beach DH (2000) p16^{INK4A} and p19^{ARF} act in overlapping pathways in cellular immortalization. Nat Cell Biol 2:148–155

Chan HM, Krstic-Demonacos M, Smith L, Demonacos C, La Thangue NB (2001) Acetylation control of the retinoblastoma tumour-suppressor protein. Nat Cell Biol 3:667–674

Chao HHA, Buchmann AM, DeCaprio JA (2000) Loss of p19^{ARF} eliminates the requirement for the pRB-binding motif in simian virus 40 large T antigen-mediated transformation. Mol Cell Biol 20:7624–7633

Chen CR, Kang Y, Siegel PM, Massague J (2002) E2F4/5 and p107 as Smad cofactors linking the TGFβ receptor to c-Myc repression. Cell 110:19–32

Chen D, Gallie BL, Squire JA (2001) Minimal regions of chromosomal imbalance in retinoblastoma detected by comparative genomic hybridization. Cancer Genetics and Cytogenetics 129:57–63

Chen D, Livne-bar I, Vanderluit JL, Slack RS, Agochiya M, Bremner R (2004) Cell-specific effects of RB or RB/p107 loss on retinal development implicate an intrinsically death-resistant cell-of-origin in retinoblastoma. Cancer Cell 5:539–551

Cheng L, Rossi F, Fang W, Mori T, Cobrinik D (2000) Cdk2-dependent phosphorylation and functional inactivation of the pRB-related protein in pRB(-), p16^{INK4A}(+) tumor cells. J Biol Chem 275:30317–30325

Chestukhin A, Litovchick L, Rudich K, DeCaprio JA (2002) Nucleocytoplasmic shuttling of p130/RBL2: novel regulatory mechanism. Mol Cell Biol 22:453–468

Chow KN, Dean DC (1996) Domains A and B in the Rb pocket interact to form a transcriptional repressor motif. Mol Cell Biol 16:4862–4868

Cinti C, Claudio PP, Howard CM, Neri LM, Fu Y, Leoncini L, Tosi GM, Maraldi NM, Giordano A (2000) Genetic alterations disrupting the nuclear localization of the retinoblastoma-related gene RB2/p130 in human tumor cell lines and primary tumors. Cancer Res 60:383–389

Clarke AR, Robanus Maandag E, Van Roon M, Van der Lugt NMT, Van der Valk M, Hooper ML, Berns A, Te Riele H (1992) Requirement for a functional Rb-1 gene in murine development. Nature 359:328–330

Claudio PP, Howard CM, Baldi A, De Luca A, Fu Y, Condorelli G, Sun Y, Colburn N, Calabretta B, Giordano A (1994) p130/RB2 has growth suppressive properties similar to yet distinctive from those of retinoblastoma family members pRB and p107. Cancer Res 54:5556–5560

Claudio PP, Howard CM, Pacilio C, Cinti C, Romano G, Minimo C, Maraldi NM, Minna JD, Gelbert L, Leoncini L, Tosi GM, Hicheli P, Caputi Giordano GG, Giordano A (2000a) Mutations in the retinoblastoma-related gene RB2/p130 in lung tumors and suppression of tumor growth in vivo by retrovirus-mediated gene transfer. Cancer Res 60:372–382

Claudio PP, Howard CM, Fu Y, Cinti C, Califano L, Micheli P, Mercer EW, Caputi M, Giordano A (2000b) Mutations in the retinoblastoma-related gene RB2/p130 in primary nasopharyngeal carcinoma. Cancer Res 60:8–12

Cobrinik D (2005) Pocket proteins and cell cycle control. Oncogene 24:2796–2809

Cobrinik D, Lee MH, Hannon G, Mulligan G, Bronson RT, Dyson N, Harlow E, Beach D, Weinberg RA, Jacks T (1996) Shared role of the pRB-related p130 and p107 proteins in limb development. Genes Dev 10:1633–1644

Dannenberg JH, Rossum A van, Schuijff L, Te Riele H (2000) Ablation of the retinoblastoma gene family deregulates G_1 control causing immortalization and increased cell turnover under growth restricting conditions. Genes Dev 14:3051–3064

Dannenberg JH, Schuijff L, Dekker M, van der Valk M, Te Riele H (2004) Tissue-specific tumor suppressor activity of retinoblastoma gene homologs p107 and p130 Genes Dev 18:2952–2962

Dahiya A, Wong S, Gonzalo S, Gavin M, Dean DC (2001) Linking the Rb and Polycomb pathways. Mol Cell 8:557–568

DeCaprio JA, Ludlow JW, Figge J, Shew JY, Huang CM, Lee W-H, Marsilio E, Paucha E, Livingston DM (1988) SV40 large tumor antigen forms a specific complex with the product of the retinoblastoma susceptibility gene. Cell 54:275–283

DeGregory J, Leone G, Miron A, Jakoi L, Nevins JR (1997) Distinct roles for E2F proteins in cell growth control apoptosis. Proc Natl Acad Sci USA 94:7245–7250

Demers GW, Foster SA, Halbert CL, Galloway DA (1994) Growth arrest by induction of p53 in DNA damaged keratinocytes is bypassed by human papillomavirus 16 E7. Proc Natl Acad Sci USA 91:4382–4386

De Stanchina E, McCurrach ME, Zindy F, Shieh SY, Ferbeyre G, Samuelson AV, Prives C, Roussel MF, Sherr CJ, Lowe SW (1998) E1A signaling to p53 involves the p19^{ARF} tumor suppressor. Genes Dev 12:2434–2442

Dick FA, Dyson N (2003) pRB contains an E2F1 specific binding domain that allows E2F-1 induced apoptosis to be regulated separately from other E2F activities. Mol Cell 12:639–649

Di Cristofano A, De Acetis M, Koff A, Cordon-Cardo C, Pandolfi PP (2001) Pten and p27^{KIP1} cooperate in prostate cancer tumor suppression in the mouse. Nat Genet 27:222–224

Di Stefano L, Jensen MR, Helin K (2003) E2F7, a novel E2F featuring DP-independent repression of a subset of E2F-regulated genes. EMBO J 22:6289–6298

Donehower LA, Harvey M, Slagte BL, McArthur MJ, Montgomery CA Jr, Butel JS, Bradley A (1992) Mice deficient for p53 are developmentally normal but susceptible to spontaneous tumours. Nature 356:215–221

Donovan SL, Dyer MA (2004) Developmental defects in Rb-deficient retinae. Vision Res 44:3323–3333

Dubs-Poterszman MC Tocque B, Wasylyk B (1995) MDM2 transformation in the absence of p53 and abrogation of the p107 G_1 cell-cycle arrest. Oncogene 11:2445–2449

Dyson N (1998) The regulation of E2F by pRB-family proteins. Genes Dev 12:2245–2262

Dyson N, Howley PM, Munger K, Harlow E (1989) The human papillomavirus-16 E7 oncoprotein is able to bind to the retinoblastoma gene product. Science 243:934–936

Ebenhard D, Busslinger M (1999) The partial homeodomain of the transcription factor Pax-5 (BSAP) is an interaction motif for the retinoblastoma and TATA-binding proteins. Cancer Res 59:1716s–1725s

Eischen CM, Weber JD, Roussel MF, Sherr CJ, Cleveland JL (1999) Disruption of the ARF-Mdm2-p53 tumor suppressor pathway in Myc-induced lymphomagenesis. Genes Dev 13:2658–2669

Evan GI, Wyllie AH, Gilbert CS, Littlewood TD, Land H, Brooks M, Waters C, Penn LZ, Hancock DC (1992) Induction of apoptosis in fibroblasts by c-myc protein. Cell 69:119–128

Ewen ME, Xing YG, Lawrence JB, Livingston DM (1991) Molecular cloning, chromosomal mapping, and expression of the cDNA for p107, a retinoblastoma gene product-related protein. Cell 66:1155–1164

Ezhevsky SA, Ho A, Becker-Hapak M, Davis PK, Dowdy SF (2001) Differential regulation of retinoblastoma tumor suppressor protein by G_1 cyclin-dependent kinase complexes in vivo. Mol Cell Biol 21:4773–4784

Farkas T, Hansen K, Holm K, Lukas J, Bartek J (2002) Distinct phosphorylation events regulate p130- and p107-mediated repression of E2F-4. J Biol Chem 277:26741–26752

Ferguson KL, Vanderluit JL, Hebert JM, McIntosh WC, Tibbo E, MacLaurin JG, Park DS, Wallace VA, Vooijs M, McConnell SK, Slack RS (2002) Telencephalon-specific Rb knockouts reveal enhanced neurogenesis, survival and abnormal cortical development. EMBO J 21:3337–3346

Ferreira R, Magnaghi-Jaulin L, Robin P, Harel-Bellan A, Trouche D (1998) The three members of the pocket proteins family share the ability to repress E2F activity through recruitment of a histone deacetylases. Proc Natl Acad Sci USA 95:10493–10481

Fisher GH, Wellen SL, Klimstra D, Lenczowski JM, Tichelaar JW, Lizak MJ, Whitsett JA, Koresky A, Varmus HE (2001) Induction and apoptotic regression of lung adenocarcinomas by regulation of a K-Ras transgene in the presence and absence of tumor suppressor genes. Genes Dev 15:3249–3262

Flesken-Nikitin A, Choi KC, Eng JP, Shmidt EN, Nikitin AY (2003) Induction of carcinogenesis by concurrent inactivation of p53 and Rb1 in the mouse ovarian surface epithelium. Cancer Res 63:3459–3463

Flemington EK, Speck SH, Kaelin WG Jr (1993) E2F-1-meidated transactivation is inhibited by complex formation with the retinoblastoma susceptibility gene product. Proc Natl Acad Sci USA 90:6914–6918

Friend SH, Bernards R, Rogelj S, Weinberg RA, Rapaport JM, Albert DM, Drya TP (1986) A human DNA segment with properties of the gene that predisposes to retinoblastoma and osteosarcoma. Nature 323:643–646

Friend SH, Horowitz JM, Gerber MR, Wang XF, Bogenmann E, Lif P, Weinberg RA (1987) Deletion of a DNA sequence in retinoblastomas and mesenchymal tumours: organization of the sequence and its encoded protein. Proc Natl Acad Sci USA 84:9059–9063

Gad A, Thullberg M, Dannenberg JH, Te Riele H, Stromblad S (2004) Retinoblastoma susceptibility gene product (pRb) and p107 functionally separate the requirements for serum and anchorage in the cell cycle G1-phase. J Biol Chem 279:13640–13644

Gallie BL, Campbell C, Devlin H, Duckett A, Squire JA (1999) Developmental basis of retinal-specific induction of cancer by RB mutation. Cancer Res 59:1731s–1735s

Garber ME, Troyanskaya OG, Schluens K, Petersen S, Thaesler Z, Pacyna-Gengelbach van de Rijn M, Rosen GD, Perou CM, Whyte RI, Altman RB, Brown PO, Botstein D, Petersen I (2001) Diversity of gene expression in adenocarcinomas of the lung. Proc Natl Acad Sci USA 98:13784–13789

Gaubatz S, Lindeman GJ, Ishida S, Jakoi L, Nevins JR, Livingston DM, Rempel RE (2000) E4F4 and E2F5 play and essential role in pocket protein-mediated G_1 control Mol Cell 6:729–735

Gonzalo S, Garcia-Cao M, Fraga MF, Schotta G, Peters AHFM, Cotter SE, Eguia R, Dean DC, Esteller M, Jenuwein T, Blasco M (2005) Role of the RB1 family in stabilizing histone methylation at constitutive heterochromatin. Nat Cell Biol 7:420–428

Graña X, Garriga J, Mayol X (1998) Role of the retinoblastoma protein family, pRB, p107 and p130 in the negative control of cell growth. Oncogene 17:3365–3383

Gray SG, Guo X (2001) Correspondence re: Claudio PP et al. Mutation in the retinoblastoma-related gene RB2/p130 in primary nasopharyngeal carcinoma. Cancer Res 60:8–12

Groth A, Weber JD, Willumsen BM, Sherr CJ, Roussel MF (2000) Oncogenic Ras induces p19^{ARF} and growth arrest in mouse embryo fibroblasts lacking p21^{CIP1} and p27^{KIP1} without activating cyclin D-dependent kinases. J Biol Chem 275:27473–27480

Guo Z, Yikang S, Yoshida H, Mak TW, Zacksenhaus E (2001) Inactivation of the retinoblastoma tumor suppressor induces apoptosis protease-activating factor-1 dependent and independent apoptotic pathways during embryogenesis. Cancer Res 61:8395–8400

Hanahan D, Weinberg RA (2000) The hallmarks of cancer. Cell 100:57–70

Hannon GJ, Demetrick D, Beach D (1993) Isolation of the Rb-related p130 through its interaction with Cdk2 and cyclins. Genes Dev 7:2378–2391

Hansen K, Farkas T, Lukas J, Holm K, Rönnstrand L, Bartek J (2001) Phosphorylation-dependent and -independent functions of p130 cooperate to evoke a sustained G_1 block. EMBO J 20:422–432

Harbour JW, Dean DC (2000) The Rb/E2F pathway: expanding roles and emerging paradigms. Genes Dev 14:2393–2409

Harbour JW, Lai SL, Whang-Peng J, Gazdar AF, Minna JD, Kaye FJ (1988) Abnormalities in structure and expression of the human retinoblastoma gene in SCLC. Science 241:353–357

Harbour JW, Luo RX, Dei Santi A, Postigo AA, Dean DC (1999) Cdk phosphorylation triggers sequential intramolecular interactions that progressively block Rb functions as cells move through G_1. Cell 98:859–869

Harvey DM, Levine AJ (1991) p53 alteration is a common event in the spontaneous immortalization of primary BALB/c murine fibroblasts. Genes Dev 5:2375–2385

Harvey M, Sands AT, Weiss RS, Hegi ME, Wiseman RW, Pantazis P, Giovanella BC, Tainsky MA, Bradley A, Donehower LA (1993) In vitro growth characteristics of embryo fibroblasts isolated of p53-deficient mice. Oncogene 8:2457–2467

Harvey M, Vogel H, Lee EY, Bradley A, Donehower LA (1995) Mice deficient in both p53 and Rb develop tumors primarily of endocrine origin. Cancer Res 55:1146–1151

Hayflick L, Moorhead PS (1961) The serial cultivation of human diploid cell strains. Exp Cell Res 25:585–621

He S, Cook BL, Deverman BE, Wehe U, Zhang F, Prachand V, Zheng J, Weintraub SJ (2000) E2F is required to prevent inappropriate S-phase entry of mammalian cells. Mol Cell Biol 20:363–371

Helin K, Lees JA, Vidal M, Dyson N, Harlow E, Fattaey A (1992) A cDNA encoding a pRB-binding protein with properties of the transcription factor E2F. Cell 70:337–350

Helin K, Harlow E, Fattaey A (1993) Inhibition of E2F-1 transactivation by direct binding of the retinoblastoma protein. Mol Cell Biol 13:6501–6508

Helin K, Holm K, Niebuhr A, Eiberg H, Tommerup N, Hougaard S, Poulse HS, Spang-Thomsen M, Norgaard P (1997) Loss of the retinoblastoma protein-related p130 protein in small cell lung carcinoma. Proc Natl Acad Sci USA 94:6933–6938

Hernando E, Nahle Z, Juan G, Diaz-Rodrigriguez E, Alaminos Hermann M, Michel L, Mittal V, Gerald W, Benezra R, Lowe SW, Cordon-Cardo C (2004) Rb inactivation promotes genomic instability by uncoupling cell cycle progression from mitotic control. Nature 430:797–802

Herrera RE, Sah VP, Williams BO, Weinberg RA, Jacks T (1996) Altered cell cycle kinetics, gene expression and G_1 restriction point regulation in Rb-deficient fibroblasts. Mol Cell Biol 16:2402–2407

Horowitz JM, Park SH, Bogenmann E, Cheng JC, Yandell DW, Kaye FJ, Minna JD, Drya TP, Weinberg RA (1990) Frequent inactivation of the retinoblastoma anti-oncogene is restricted to a subset of human tumor cells. Proc Natl Acad Sci USA 87:2775–2779

Hsieh JK, Chan FS, O'Connor DJ, Mittnacht S, Zhong S, Lu X (1999) RB regulates the stability and the apoptotic function of p53 and MDM2. Mol Cell 3:181–193

Hu N, Gutsmann A, Herbert DC, Bradley A, Lee WH, Lee EY (1994) Heterozygous Rb-1 delta 20/+ mice are predisposed to tumors of the pituitary gland with a nearly complete penetrance. Oncogene 9:1021–1027

Humbert PO, Verona R, Trimarchi JM, Rogers C, Dandapani S, Lees JA (2000) E2F3 is critical for normal cellular proliferation. Genes Dev 14:690–703

Hurford RK, Cobrinik D, Lee MH, Dyson N (1997) pRB and p107/p130 are required for the regulated expression of different sets of E2F responsive genes. Genes Dev 11:1447–1463

Ichimura K, Hanafusa H, Takimoto H, Ohgma Y, Akagi T, Shimizu K (2000) Structure of the human retinoblastoma-related p107 gene and its intragenic deletion in a B-cell lymphoma cell line. Gene 251:37–43

Irwin M, Marin MC, Phillips AC, Seelan RS, Smith DI, Liu W, Flores ER, Tsai KY, Jacks T, Vousden KH, Kaelin WG Jr (2000) Role for the p53 homologue p73 in E2F-1 induced apoptosis. Nature 407:642–645

Ishida S, Huang E, Zuzan H, Spang R, Leone G, West M, Nevins JR (2001) Role for E2F in control of both DNA replication and mitotic functions as revealed from DNA microarray analysis. Mol Cell Biol 21:4684–4699

Ito T, Udak N, Yazawa T, Okudela K, Hayashi H, Sudo T, Guillemot F, Kageyama R, Kitamura H (2000) Basic helix-loop-helix transcription factors regulate the neuroendocrine differentiation of fetal mouse pulmonary epithelium. Development 127:3913–3921

Jacks T, Fazeli A, Schmitt EM, Bronson RT, Goodell MA, Weinberg RA (1992) Effects of an Rb mutation in the mouse. Nature 359:295–300

Jacks T, Remington L, Williams BO, Schmitt EM, Halachmi S, Bronson RT, Weinberg RA (1994) Tumor spectrum analysis in p53 mutant mice. Curr Biol 4:1–7

Jen Y, Manova K, Benezra R (1996) Expression patterns of Id1, Id2 and Id3 are highly related but distinct from that of Id4 during mouse embryogenesis. Dev Dyn 207:235–252

Kalma Y, Marash L, Lamed Y, Ginsberg D (2001) Expression analysis using DNA microarrays demonstrates that E2F-1 up-regulates expression of DNA replication genes including replication protein A2. Oncogene 20:1379–1387

Kamb A, Gruis NA, Weaver-Feldhaus J, Liu Q, Harshman K, Tavtigian SV, Stockert E, Day RS III, Johnson BE, Skolnick MH (1994) A cell cycle regulator involved in genesis of many tumor types. Science 264:436–440

Kamijo T, Zindy F, Roussel MF, Quelle DE, Downing JR, Ashmun RA, Grosveld G, Sherr CJ (1997) Tumor suppression at the mouse INK4A locus mediated by the alternative reading frame product p19^{ARF}. Cell 91:649–659

Kamijo T, Weber JD, Zambetti G, Zindy F, Roussel MF, Sherr CJ (1998) Functional and physical interactions of the ARF tumor suppressor with p53 and Mdm2. Proc Natl Acad Sci USA 95:8292–8287

Kondo T, Higashi H, Nishizawa H, Ishikawa S, Ashizawa S, Yamada M, Makita Z, Koike T, Hatakeyama M (2001) Involvement of pRB-related p107 protein in the inhibition of S-phase progression in response to genotoxic stress. J Biol Chem 276:17559–17567

Krimpenfort P, Quon KC, Mooi WJ, Loonstra A, Berns A (2001) Loss of p16^{Ink4a} confers susceptibility to metastatic melanoma in mice. Nature 413:83–86

Lai A, Lee JM, Yang WM, DeCaprio JA, Kaelin Jr WG, Seto E, Branton PE (1999) RBP1 recruits both histone deacetylases-dependent and -independent repression activities to retinoblastoma family proteins. Mol Cell Biol 19:6632–6641

Lai A, Kennedy BK, Barbie DA, Bertos NR, Yang JX, Theberge MC, Tsai SC, Seto E, Zhang Y, Kuzmichev A, Lane WS, Reinberg D, Harlow E, Branton PE (2001) RBP1 Recruits the mSin3-histone deacetylase complex to the pocket of retinoblastoma tumor suppressor family proteins found in limited discrete regions of the nucleus at growth arrest. Mol Cell Biol 21:2918–2932

Lasorella A, Noseda M, Beyna M, Iavarone A (2000) Id2 is a retinoblastoma protein target and mediates signaling by Myc oncoproteins. Nature 407:592–598

LeCouter J, Kablar B, Hardy WR, Ying C, Megeney LA, May LL, Rudnicki MA (1998a) Strain-dependent myeloid hyperplasia, growth deficiency and accelerated cell cycle in mice lacking the Rb-related p107 gene. Mol Cell Biol 18:7455–7465

LeCouter J, Kablar B, Whyte PFM Ying C, Rudnicki MA (1998b) Strain-dependent embryonic lethality in mice lacking the retinoblastoma-related p130 gene. Development 125:4669–4679

Lee EY, Cam H, Ziebold U, Rayman JB, Lees JA, Dynlacht BD (2002) E2F4 loss suppresses tumorigenesis in Rb mutant mice. Cancer Cell 2:463–472

Lee EYHP, To H, Shew JY, Bookstein R, Scully P, Lee WH (1987) Inactivation of the retinoblastoma susceptibility gene in human breast cancers. Science 241:218–221

Lee EYHP, Chang CY, Hu N, Wang YC, Lai CC, Herrup K, Lee WH, Bradley A (1992) Mice deficient for Rb are nonviable and show defects in neurogenesis and haematopoiesis. Nature 359:288–294

Lee JO, Russo AA, Pavlitch NP (1998) Structure of the retinoblastoma tumour-suppressor pocket domain bound to a peptide from HPV E7. Nature 391:859–865

Lee MH, Williams BO, Mulligan G, Mukai S, Bronson RT, Dyson N, Harlow E, Jacks T (1996) Targeted disruption of p107: functional overlap between p107 and Rb. Genes Dev 10:1621–1632

Li Y, Graham C, Lacy S, Duncan AMV, Whyte P (1993) The adenovirus E1A-associated 130-kD protein is encoded by a member of the retinoblastoma gene family and physically interacts with cyclins A and E. Genes Dev 7:2366–2377

Lindeman GJ, Gaubatz S, Livingston DM, Ginsberg D (1997) The subcellular localization of E2F-4 is cell-cycle dependent. Proc Natl Acad Sci USA 94:5095–5100

Lipinski MM, Jacks T (1999) The retinoblastoma gene family in differentiation and development. Oncogene 18:7873–7882

Linnoila IR, Zhao B, DeMayo JL, Nelkin BD, Baylin SB, DeMayo FJ, Ball DW (2000) Constitutive achaete-scute homologue-1 promotes airway dysplasia and lung neuroendocrine tumors in transgenic mice. Cancer Res 60:4005–4009

Lissy NA, Davis PK, Irwin M, Kaelin WG, Dowdy SF (2000) A common E2F-1 and p73 pathway mediates cell death induced by TCR activation. Nature 407:642–645

Litovchick L, Chestukhin A, DeCaprio J (2004) Glycogen synthase kinase 3 phosphorylates RBL2/p130 during quiescence. Mol Cell Biol 24:8970–8990

Liu DX, Nath N, Chellappan SP, Greene LA (2005) Regulation of neuron survival and death by p130 and associated chromatin modifiers. Genes Dev 19:719–732

Lloyd AC, Obermuller F, Staddon S, Barth CF, McMahon M, Land H (1997) Cooperating oncogenes converge to regulate cyclin/Cdk complexes. Genes Dev 11:663–677

Logan N, Delavaine L, Graham A, Reilly C, Wilson J, Brummelkamp TR, Hijmans EM, Bernards R, La Thangue NB (2004) E2F-7: a distinctive E2F family member with an unusual organization of DNA-binding domains. Oncogene 23:5138–5150

Lowe SW, Ruley HE (1993) Stabilization of the p53 tumor suppressor is induced by adenovirus 5 E1A and accompanies apoptosis. Genes Dev 7:535–545

Loughran O, La Thangue NB (2000) Apoptotic and growth-promoting activity of E2F modulated by MDM2. Mol Cell Biol 20:2186–2197

Lu X, Horvitz HR (1998) lin-35 and lin-53, two genes that antagonize a C elegans Ras pathway, encode proteins similar to Rb and its binding protein RbAp48. Cell 95:981–991

Ludlow JW, DeCaprio JA, Huang CM, Lee WH, Paucha E, Livingston DM (1989) SV40 large T antigen binds preferentially to an underphosphorylated member of the retinoblastoma susceptibility gene product family. Cell 56:57–65

Lukas J, Parry D, Aagaard L, Mann DJ, Bartkova J, Strauss M, Peters G, Bartek J (1995) Retinoblastoma protein-dependent cell cycle inhibition by the tumor suppressor p16. Nature 375:503–506

Lundberg AS, Weinberg RA (1998) Functional inactivation of the retinoblastoma protein requires sequential modification by at least two distinct cylin-Cdk complexes. Mol Cell Biol 18:753–761

Luo RX, Postigo AA, Dean DC (1998) Rb interacts with histone deacytelase to repress transcription. Cell 92:463–473

MacLellan WR, Garcia A, Oh H, Frenkel P, Jordan MC, Roos KP, Schneider MD (2005) Overlapping roles of pocket proteins in the myocardium are unmasked by germ line deletion of p130 plus heart-specific deletion of Rb. Mol Cell Biol 25:2486–2497

Macleod KF, Hu Y, Jacks T (1996) Loss of Rb activates both p53-dependent and independent cell death pathways in the developing mouse nervous system. EMBO J 15:6178–6188

MacPherson D, Sage J, Crowley D, Trumpp A, Bronson RT, Jacks T (2003) Conditional mutation of Rb causes cell cycle defects without apoptosis in the central nervous system. Mol Cell Biol 23:1044–1053

MacPherson D, Sage J, Kim T, Ho D, McLaughlin ME, Jacks T (2004) Cell type-specific effects of Rb deletion in the murine retina. Genes Dev 18:1681–1694

Maddison LA, Sutherland BW, Barrios RJ, Greenberg NM (2004) Conditional deletion of Rb causes early stage prostate cancer. Cancer Res 64:6018–6025

Magnaghi JL, Groisman R, Naguibneva I, Robin P, Lorain S, Le VJ, Troalen F, Trouche D, Harel BA (1998) Retinoblastoma protein represses transcription by recruiting a histone deacytelase. Nature 391:601–605

Mairal A, Pinglier E, Gilbert E, Peter M, Validire P, Desjardins L, Doz F, Aurias A, Couturier J (2000) Detection of chromosome imbalances in retinoblastoma by parallel karyotype and CGH analyses. Genes Chrom Cancer 28:370–379

Maiti B, Li J, de Bruin A, Gordon F, Timmers C, Opavsky R, Patil K, Tuttle J, Cleghorn W, Leone G (2005) Cloning and characterization of mouse E2F8, a novel mammalian E2F family member capable of blocking cellular proliferation. J Biol Chem 280:18211–18220

Malumbres M, Barbacid M (2001) To cycle or not to cycle: a critical decision in cancer. Nature Reviews Cancer 1:222–231

Martelli F, Hamilto T, Silver DP, Sharpless NE, Bardeesy N, Rokas M, DePinho RA, Livingston DM, Grossman SR (2001) p19^{ARF} targets certain E2F species for degradation. Proc Natl Acad Sci USA 98:4455–4460

Martin K, Trouche D, Hagemeier D, Sorensen TS, La Thangue NB, Kouzarides T (1995) Stimulation of E2F1/DP1 transcriptional activity by MDM2 oncoprotein. Nature 375:691–694

Marguardt T, Ashery-Padan R, Andrejewski N, Scardigli F, Guillemot F, Gruss P (2001) Pax6 is required for the multipotent state of retinal progenitor cells. Cell 105:43–55

Marino S, Vooijs M, Van Der Gulden H, Jonkers J, Berns A (2000) Induction of medulloblastomas in p53-null mutant mice by somatic inactivation of Rb in the external granular layer cells of the cerebellum. Genes Dev 14:994–1004

Marino S, Hoogervoorst D, Brandner S, Berns A (2003) Rb and p107 are required for normal cerebellar development and granule cell survival but not for Purkinje cell persistence. Development 130:3359–3368

Mathon NF, Malcolm DS, Harrisingh MC, Cheng L, Lloyd C (2001) Lack of replicative senescence in normal rodent glia. Science 291:872–875

Mayhew CN, Bosco EE, Fox SR, Okaya T, Tarapore P, Schwemberger SJ, Babcock GF, Lentsch AB, Fukasawa K, Knudsen ES (2005) Liver-specific pRB loss results in ectopic cell cycle entry and aberrant ploidy. Cancer Res 65:4568–4577

Medema RH, Herrera RE, Lam F, Weinberg RA (1995) Growth suppression by the p16^{Ink4a} requires functional retinoblastoma protein. Proc Natl Acad Sci USA 92:6289–6293

Meuwissen R, Linn SC, Linnoila RI, Zevenhoven J, Mooi WJ, Berns A (2003) Induction of small cell lung cancer by somatic inactivation of both Trp53 and Rb1 in a conditional mouse model. Cancer Cell 3:181–189

Modestou M, Antich VP, Korgaonkar C, Eapen A, Quelle DE (2001) The alternative reading frame tumor suppressor inhibits growth through p21-dependent and p21-independent pathways. Cancer Res 61:3145–3150

Modi S, Kubo A, Oie H, Coxon AB, Rehmatulla A, Kaye FJ (2000) Protein expression of the RB-related gene family and SV40 large T antigen in mesothelioma and lung cancer. Oncogene 19:4632–4639

Moll AC, Imhof SM, Bouter LM, Tan KE (1997) Second primary tumors in patients with retinoblastoma. A review of the literature. Opthalmic Genet 1:27–34

Morgenbesser SD, Williams BO, Jacks T, DePinho RA (1994) p53-dependent apoptosis produced by Rb-deficiency in the developing mouse lens. Nature 371:72–74

Moroni MC, Hickman ES, Denchi EL, Carprara G, Colli E, Cecconi F, Muller H, Helin K (2001) Apaf-1 is a transcriptional target for E2F and p53. Nat Cell Biol 3:552–558

Morrison AJ, Sardet C, Herrera RE (2002) Retinoblastoma protein transcriptional repression through histone deacetylation of a single nucleosome. Mol Cell Biol 22:856–865

Morrow E, Furukawa T, Lee JE, Cepko CL (1999) NeuroD regulates multiple functions in the developing neural retina in rodent. Development 126:23–36

Müller H, Bracken AP, Vernell R, Moroni MC, Christians F, Grassili E, Prosperini E, Vigo E, Oliner JD, Helin K (2001) E2Fs regulate the expression of genes involved in differentiation, development, proliferation and apoptosis. Genes Dev 15:267–285

Narita M, Nunez S, Heard E, Narita M, Lin AW, Hearn SA, Spector DL, Hannon GJ, Lowe SW (2003) Rb-mediated heterochromatin formation and silencing of E2F target genes during cellular senescence. Cell 113:703–716

Nguyen DX, Baglia LA, Huang SM, Baker CM, McCance JD (2004) Acetylation regulates the differentiation-specific functions of the retinoblastoma protein. EMBO J 23:1609–1618

Nielsen SJ, Schneider R, Bauer U-M, Bannister AJ, Morrison A, O'Caroll D, Firestein R, Cleary M, Jenuwein T, Herrera RE, Kouzarides T (2001) Rb targets histone H3 methylation and HP1 to promoters. Nature 412:561–565

Nobori T, Miura K, Wu DJ, Lois A, Tkabayashi K, Carson DA (1994) Deletions of the cyclin-dependent kinase-4 inhibitor gene in multiple human cancers. Nature 368:753–756

Nork TM, Schwartz TL, Doshi HM, Millecchia LL (1995) Retinoblastoma: Cell of origin. Arch Ophthalmol 113:791–802

Nork TM, Poulsen GL, Millechia LL, Jantz RG, Nickells RW (1997) p53 regulates apoptosis in human retinoblastoma. Arch Opthalmol 115:213–219

Novitch BG, Mulligan GJ, Jacks T, Lassar AB (1996) Skeletal muscle cells lacking the retinoblastoma protein display defects in muscle gene expression and accumulate in S and G2 phases of the cell cycle. J Cell Biol 135:441–456

Novitch BG, Spicer DB, Kim PS, Cheung WL, Lassar AB (1999) pRb is required for MEF2-dependent gene expression as well as cell-cycle arrest during skeletal muscle differentiation. Curr Biol 9:449–459

Pan H, Yin C, Dyson N, Harlow E, Yamasaki L, Van Dyke (1998) A key role for E2F1 in p53-dependent apoptosis and cell division within developing tumors. Mol Cell 2:283–292

Palmero I, Pantoja C, Serrano M (1998) p19^{ARF} links the tumour suppressor p53 to Ras. Nature 395:125–126

Pantoja C, Serrano M (1999) Murine fibroblasts lacking p21 undergo senescence and are resistant to transformation by oncogenic Ras. Oncogene 18:4974–4982

Paramio JM, Navarro M, Segrelles C, Gomez-Casero E, Jorcano JL (1999) PTEN tumour suppressor is linked to the cell cycle control through the retinoblastoma protein. Oncogene 18:7462–7468

Pearse AGE, Polak JM (1978) The diffuse neuroendocrine system and the APUD concept In: Bloom SR (ed) Gut hormones. Churchill Livingstone, London New York, pp 33–39

Peeper DS, Dannenberg JH, Douma S, Te Riele H, Bernards R (2001) Escape from premature senescence is not sufficient for oncogenic transformation. Nat Cell Biol 3:198–203

Pomerantz J, Schreiber-Agus N, Liegeois NJ, Silverman A, Alland L, Chin L, Potes J, Chen K, Orlow I, Lee HW, Cordon-Cardo C, DePinho RA (1998) The Ink4a tumor suppressor gene product, p19^{ARF}, interacts with MDM2 and neutralizes MDM2's inhibition of p53. Cell 92:713–723

Potluri VR, Helson L, Ellsworth RM, Reid T, Gilbert F (1986) Chromosomal abnormalities in human retinoblastoma: a review. Cancer 58:663–671

Qin XQ, Chittenden T, Livingston DM, Kaelin WG Jr (1992) Identification of a growth suppression domain within the retinoblastoma gene product. Genes Dev 6:953–964

Quelle DE, Zindy F, Ashmun RA, Sherr CJ (1995) Alternative reading frames of the INK4A tumor suppressor gene encode two unrelated proteins capable of inducing cell cycle arrest. Cell 83:993–1000

Ramirez RD, Morales CP, Herbert BS, Rohde JM, Passons C, Shay JW, Wright WE (2001) Putative telomere-independent mechanisms of replicative aging reflect inadequate growth conditions. Genes Dev 15:398–403

Radfar A, Unnikrishnan I, Lee HW, DePinho RA, Rosenberg N (1998) p19^{Arf} induces p53-dependent apoptosis during abelson virus-mediated pre-B cell transformation. Proc Natl Acad Sci USA 95:13194–13197

Randle DH, Zindy F, Sherr CJ, Roussel MF (2001) Differential effects of p19^{Arf} and p16^{Ink4a} loss on senescence of murine bone marrow-derived preB cells and macrophages. Proc Natl Acad Sci USA 98:9654–9659

Rane GS, Cosenza SC, Mettus RV, Reddy EP (2002) Germ line transmission of the Cdk4^{R24C} mutation facilitates tumorigenesis and escape from cellular senescence. Mol Cell Biol 22:644–656

Ren B, Cam H, Takahashi Y, Volkert T, Terragni J, Young RA, Dynlacht D (2002) E2F integrates cell cycle progression with DNA repair, replication, and G$_2$/M-checkpoints. Genes Dev 16:245–256

Rittling SR, Denhardt DT (1992) p53 mutations in spontaneously immortalized 3T12 but not 3T3 mouse embryo cells. Oncogene 7:935–942

Robanus Maandag EC, Van der Valk M, Vlaar M, Feltkamp C, O'Brien J, Van Roon M, Van der Lugt N, Berns A, Te Riele H (1994) Developmental rescue of an embryonic-lethal mutation in the retinoblastoma gene in chimeric mice. EMBO J 13:4260–4268

Robanus-Maandag E, Dekker M, Van der Valk M, Carrozza ML, Jeanny JC, Dannenberg JH, Berns A, Te Riele H (1998) p107 is a suppressor of retinoblastoma development in pRb-deficient mice. Genes Dev 12:1599–1609

Ross JF, Näär A, Cam H, Gregory R, Dynlacht D (2001) Active repression and E2F inhibition by pRB are biochemically distinguishable. Genes Dev 15:392–397

Ruas M, Peters G (1998) The p16^{INK4A}/CDKN2A tumor suppressor and its relatives. Biochem Biophys Acta Rev Cancer 1378:F115–F177

Russell JL, Powers JT, Rounbehler RJ, Rogers PM, Conti CJ, Johnson DG (2002) ARF differentially modulates apoptosis induced by E2F1 and Myc. Mol Cell Biol 22:1360–1368

Ruiz S, Santos M, Segrelles C, Hugo L, Jorcano JL, Berns A, Paramio JM, Vooijs M (2004) Unique and overlapping functions of pRb and p107 in the control of proliferation and differentiation in epidermis. Development 131:2737–2748

Ryan RS, Gee R, O'Connell JX, Harris AC, Munk PL (2003) Leiomyosarcoma of the distal femur in a patient with a history of bilateral retinoblastoma: a case report and review of the literature. Skeletal Radiol 32:476–480

Saenz-Rjobles MT, Symonds H, Chen J, Van Dyke T (1994) Induction versus progression of brain tumor development: differential functions for the pRB- and p53-targeting domains of simian virus 40 T antigen. Mol Cell Biol 14:2686–2698

Sage C, Huang M, Karimi K, Gutierrez G, Vollrath MA, Zhang DS, Garcia-Anoveros J, Hinds PW, Corwin JT, Corey DP, Chen ZY (2005) Proliferation of functional hair cells in vivo in the absence of the retinoblastoma protein. Science 307:1114–1118

Sage J, Mulligan GJ, Attardi LD, Miller A, Chen S, Williams B, Theorou E, Jacks T (2000) Targeted disruption of the three Rb-related genes leads to loss of G_1 control and immortalization. Genes Dev 14:3037–3050

Sage J, Miller AL, Perez-Mancera P, Wysocki JM, Jacks T (2003) Acute mutation of retinoblastoma gene function is sufficient for cell cycle entry. Nature 424:223–228

Schmitt CA, Fridman JS, Yang M, Lee S, Baranov E, Hoffman RM, Lowe SW (2002) A senescence program controlled by p53 and p16^{INK4A} contributes to the outcome of cancer therapy. Cell 109:335–346

Sellers WR, Novitch BG, Miyake S, Heith A, Otterson GA, Kaye FJ, Lassar AB, Kaelin WG (1998) Stable binding to E2F is not required for the retinoblastoma protein to activate transcription, promote differentiation, and suppress tumor cell growth. Genes Dev 12:95–106

Serrano M, Hannon GJ, Beach D (1993) A new regulatory motif in cell-cycle control causing specific inhibition of cyclin D/CDK4. Nature 366:704–707

Serrano M, Lee HW, Chin L, Cordon-Cardo C, Beach D, DePinho RA (1996) Role of the INK4A locus in tumor suppression and cell mortality. Cell 85:27–37

Serrano M, Lin AW, Mila EM, Beach D, Lowe SW (1997) Oncogenic ras provokes premature cell senescence associated with accumulation of p53 and p16^{Ink4a}. Cell 88:593–602

Schneider JW, Gu W, Zhu L, Mahdavi V, Nadal-Ginard B (1994) Reversal of terminal differentiation mediated by p107 in Rb$^{-/-}$ muscle cells. Science 264:1467–1471

Sharpless NE, Bardeesy N, Lee KH, Carrasco D, Castrillon DH, Aguirre AJ, Wu EA, Horner JW, DePinho RA (2001) Loss of p16^{Ink4a} with retention of p19^{Arf} predisposes mice to tumorigenesis. Nature 413:86–91

Sherr CJ (1996) Cancer cell cycles. Science 274:1672–1677

Sherr CJ (1998) Tumor surveillance by the ARF-p53 pathway. Genes Dev 12:2984–2991

Sherr CJ (2001a) Parsing Ink4a/Arf: pure p16-null mice. Cell 106:531–534

Sherr CJ (2001b) The Ink4a/ARF network in tumour suppression. Nature Reviews Molecular Biology 2:731–737

Sherr CJ, DePinho RA (2000) Culture clock or culture shock? Cell 102:407–410

Sherr CJ, Roberts JM (1999) CDK inhibitors: positive and negative regulators of G_1-phase progression. Genes Dev 13:1501–1512

Slebos RJC, Lee MH, Plunkett BS, Kessis TD, Williams BO, Jacks T, Hedrick L, Kastan MB, Cho KR (1994) p53-dependent G1-arrest involves pRB-related proteins and is disrupted by the human papillomavirus 16 E7 oncoprotein. Proc Natl Acad Sci USA 91:5320–5324

Sotillo R, Dubus P, Martin J, de la Cueva E, Ortega S, Malumbres M, Barbacid M (2001) Wide spectrum of tumors in knock-in mice carrying a Cdk4 protein insensitive to INK4 inhibitors EMBO J 20:6637–6647

Stanton SE, Shin SW, Johnson BE, Meyerson M (2000) Recurrent allelic deletions of chromosome arms 15q and 16q in human small cell lung carcinomas. Genes, Chrom Cancer 27:323–331

Starostik P, Chow KN, Dean DC (1996) Transcriptional repression and growth suppression by the p107 pocket protein. Mol Cell Biol 16:3606–3614

Steele-Perkins G, Fang W, Yang XH, Van Gele M, Carling T, Gu J, Buyse IM, Fletcher JA, Liu J, Bronson R, Chadwick RB, de la Chapelle A, Zhan X, Speleman F, Huang S (2001) Tumor formation and inactivation of RIZ1, an Rb-binding member of a nuclear protein-methyltransferase superfamily. Genes Dev 15:2250–2262

Stiewe T, Pützer BM (2000) Role of the p53-homologue p73 in E2F1-induced apoptosis. Nat Genet 26:464–469

Strachan GD, Rallapalli R, Pucci B, Toulouse PL, Hall DJ (2001) A transcriptionally inactive E2F-1 targets the MDM family of proteins for proteolytic degradation. J Biol Chem 276:45677–45685

Sun P, Dong P, Dai K, Hannon GJ, Beach D (1998) p53-independent role of MDM2 in TGF-beta1 resistance. Science 282:2270–2272

Symonds H, Krall L, Remington L, Saenz-Robles M, Lowe S, Jacks T, Van Dyke T (1994) p53-dependent apoptosis suppresses tumor growth and progression in vivo. Cell 78:703–711

Tajima Y, Munakata S, Ishida Y, Nakajima T, Sugano I, Nagao K, Minoa K, Kondo Y (1994) Photoreceptor differentiation of retinoblastoma: an electron microscopic study of 29 retinoblastomas. Pathol Int 44:837–843

Takahashi Y, Rayman JB, Dynlacht BD (2000) Analysis of promoter binding by the E2F and pRB families in vivo: distinct E2F proteins mediate activation and repression. Genes Dev 14:804–816

T'Ang A, Varley JM, Chakraborty S, Murphree AL, Fung YK (1988) Structural rearrangement of the retinoblastoma gene in human breast carcinoma. Science 242:263–266

Tang DG, Tokumoto YM, Apperly JA, Lloyd AC, Raff MC (2001) Lack of replicative senescence in cultured rat oligodendrocyte precursor cells. Science 291:868–871

Te Riele H, Robanus-Maandag E, Berns A (1992) Highly efficient gene targeting in embryonic stem cells through homologous recombination with isogenic DNA constructs. Proc Natl Acad Sci USA 89:5128–5132

Thomas DM, Carty SA, Piscopo DM, Lee JS, Wang WF, Forrester WC, Hinds PW (2001) The retinoblastoma protein acts as a transcriptional co-activator required for osteogenic differentiation. Mol Cell 8:303–316

Todaro GJ, Green H (1963) Quantitative studies of the growth of mouse embryo cells in culture and their development into established lines. J Cell Biol 17:299–313

Tolbert D, Lu X, Yin C, Tantama M, Van Dyke T (2002) p19^{ARF} is dispensable for oncogenic stress-induced p53-mediated apoptosis and tumor suppression in vivo. Mol Cell Biol 22:370–377

Tonks ID, Hacker E, Irwin N, Muller HK, Keith P, Mould A, Zournazi A, Pavey S, Hayward NK, Walker G, Kay GF (2005) Melanocytes in conditional Rb-/- mice are normal in vivo but exhibit proliferation and pigmentation defects in vitro. Pigment Cell Res 18:252–264

Trimarchi JM, Lees JA (2001) Sibling rivalry in the E2F family. Nat Rev Mol Cell Biol 3:11–20

Trouche D, Le Chalony C, Muchardt C, Yaniv M, Kouzarides T (1997) RB and hBrm cooperate to repress the activation functions of E2F1. Proc Natl Acad Sci USA 94:11268–11273

Tsai KY, Hu Y, Macleod KF, Crowley D, Yamasaki L, Jacks T (1998) Mutation of E2F-1 suppresses apoptosis and inappropriate S phase entry and extends survival of Rb-deficient mouse embryos. Mol Cell 2:293–304

Tsai KY, MacPherson D, Rubinson DA, Crowley D, Jacks T (2002) ARF is not required for apoptosis in Rb mutant mouse embryos. Curr Biol 12:159–163

Uchida C, Miwa S, Kitagawa K, Hattori T, Isobe T, Otani S, Oda T, Sugimura H, Kamijo T, Ookawa K, Yasuda H, Kitagawa M (2005) Enhanced Mdm2 activity inhibits pRB function via ubiquitin-dependent degradation. EMBO J 24:160–169

Verona R, Moberg K, Estes S, Starz M, Vernon JP, Lees JA (1997) E2F activity is regulated by cell cycle-dependent changes in subcellular localization. Mol Cell Biol 17:7268–7282

Vooijs M, Berns A (1999) Developmental defects and tumor predisposition in Rb mutant mice. Oncogene 18:5293–5303

Vooijs M, van der Valk M, Te Riele H, Berns A (1998) Flp-mediated tissue-specific inactivation of the retinoblastoma tumor suppressor gene in the mouse. Oncogene 17:1–12

Vooijs M, Te Riele H, Van Der Valk M, Berns A (2002) Tumor formation in mice with somatic inactivation of the retinoblastoma gene in interphotoreceptor retinal binding protein-expressing cells. Oncogene 21:4635–4645

Voorhoeve PM, Watson RJ, Farlie PG, Bernards R, Lam EWF (1998) Rapid dephosphorylation of p107 upon UV irradiation. Oncogene 18:679–688

Weber JD, Jeffers JR, Rehg JE, Randle DH, Lozano G, Roussel MF, Sherr CJ, Zambetti GP (2000) p53-independent functions of the p19^{ARF} tumor suppressor. Genes Dev 14:2358–2365

Weinberg RA (1995) The retinoblastoma protein and cell cycle control. Cell 81:323–330

Welch PJ, Wang JY (1993) A C-terminal protein-binding domain in the retinoblastoma protein regulates nuclear c-Abl tyrosine kinase in the cell cycle. Cell 75:779–790

Wells J, Boyd KE, Fry CJ, Bartley SM, Farnham PJ (2000) Target gene specificity of E2F and pocket protein family members in living cells. Mol Cell Biol 20:5797–5807

White E (1996) Life, death, and the pursuit of apoptosis. Genes Dev 10:1–15

Whyte P, Buchkovich KJ, Horowitz JM, Friend SH, Raybuck M, Weinberg RA, Harlow E (1988) Association between an oncogene and an anti-oncogene: The adenovirus E1A proteins bind to the retinoblastoma gene product. Nature 334:124–129

Wiggan O, Taniguchi-Sidle A, Hamel PA (1998) Interaction of the pRB-family proteins with factors containing paired-like homeodomains. Oncogene 16:227–236

Wikenheiser-Brokamp KA (2004) Rb family proteins differentially regulate distinct cell lineages during epithelial development. Development 131:4299–4310

Williams BO, Remington L, Albert DM, Mukai S, Bronson RT, Jacks T (1994a) Cooperative tumorigenic effects of germline mutations in Rb and p53. Nat Genet 7:480–484

Williams BO, Schmitt EM, Remington L, Bronson RT, Jacks T (1994b) Extensive contribution of Rb-deficient cell to adult chimeric mice with limited histopathological consequences. EMBO J 13:4251–4259

Wu L, Timmers C, Maiti B, Saavedra HI, Sang L, Chong GT, Nuckolis F, Giangrande P, Wright FA, Field SJ, Greenberg M, Orkin S, Nevins JR, Robinson ML, Leone G (2001) The E2F1-3 transcription factors are essential for cellular proliferation. Nature 414:457–461

Wu L, de Bruin A, Saavedra HI, Starovic M, Trimboli A, Yang Y, Opavska J, Wilson P, Thompson JC, Ostrowski MC, Rosol TJ, Woollett LA, Weinstein M, Cross JC, Robinson ML, Leone G (2003) Extra-embryonic function of Rb is essential for embryonic development and viability. Nature 421:942–947

Xiao Z, Chen J, Levine AJ, Modjtahedi N, Xing J, Sellers WR, Livingston DM (1995) Interaction between the retinoblastoma protein and the oncoprotein MDM2. Nature 375:694–698

Yamasaki L, Bronson R, Williams BO, Dyson NJ, Harlow E, Jacks T (1998) Loss of E2F-1 reduces tumorigenesis and extends the lifespan of Rb1(+/–) mice. Nat Genet 18:360–364

Yeung RS, Bell DW, Testa JR, Mayol X, Baldi A, Graña X, Klinga-Levan K, Knudson AG, Giordano A (1993) The retinoblastoma-related gene, Rb2, maps to a human chromosome 16q12 and rat chromosome 19. Oncogene 8:3465–3468

Yin C, Knudson CM, Korsmeyers JS, Van Dyke T (1997) Bax suppresses tumorigenesis and stimulates apoptosis in vivo. Nature 385:637–640

Zamanian M, La TN (1993) Transcriptional repression by the Rb-related protein p107. Mol Biol Cell 4:389–396

Zacksenhaus E, Jiang Z, Chung D, Marth JD, Phillips RA, Gallie BL (1996) pRb controls proliferation, differentiation and death of skeletal muscle cells and other lineages during embryogenesis. Genes Dev 10:3051–3064

Zacksenhaus E, Jiang Z, Hei YJ, Philips RA, Gallie B (1999) Nuclear localization conferred by the pocket domain of the retinoblastoma gene product. Biochim Biophys Acta 1451:288–296

Zhang HS, Postigo AA, Dean DC (1999) Active transcriptional repression by the Rb-E2F complex mediates G_1 arrest triggered by p16^{INK4A}, TGFβ, and contact inhibition. Cell 97:53–61

Zhang J, Schweers B, Dyer MA (2004) The first knockout mouse model of retinoblastoma. Cell Cycle 3:952–959

Zhang S, Ramsay ES, Mock B (1998a) Cdkn2a, the cyclin-dependent kinase inhibitor encoding p16^{INK4A} and p19^{ARF}, is a candidate for the plasmacytoma susceptibility locus, Pctr1. Proc Natl Acad Sci USA 95:2429–2434

Zhang Y, Xiong Y, Yarbrough WG (1998b) ARF promotes MDM2 degradation and stabilizes p53: ARF-INK4A locus deletion impairs both the Rb and p53 tumor suppression pathways. Cell 92:725–734

Zhu L, Van den Heuvel S, Helin K, Fattaey A, Ewen M, Livingston DM, Dyson N, Harlow E (1993) Inhibition of cell proliferation by p107, a relative of the retinoblastoma protein. Genes Dev 7:1111–1125

Ziebold U, Reza T, Caron A, Lees JA (2001) E2F3 contributes both to the inappropriate proliferation and to the apoptosis arising in Rb mutant embryos. Genes Dev 15:386–391

Zindy F, Quelle DE, Roussel MF, Sherr CJ (1997) Expression of the p16^{INK4A} tumor suppressor versus other INK4 family members during mouse development and aging. Oncogene 15:203–211

Zindy F, Eischen CM, Randle D, Kamijo T, Cleveland JL, Sherr CJ, Roussel MF (1998) Myc signaling via the ARF tumor suppressor regulates p53-dependent apoptosis and immortalization. Genes Dev 12:2424–2433

Zou X, Ray D, Aziyu A, Christov K, Boiko AD, Gudkov AV, Kiyokawa H (2002) Cdk4 disruption renders primary mouse cells resistant to oncogenic transformation, leading to Arf/p53-independent senescence. Genes Dev 16:2923–2934

# Modeling Cell Cycle Control and Cancer with pRB Tumor Suppressor

Lili Yamasaki

Department of Biological Sciences, Columbia University, New York, NY 10027, USA

Abstract Cancer is a complex syndrome of diseases characterized by the increased abundance of cells that disrupts the normal tissue architecture within an organism. Defining one universal mechanism underlying cancer with the hope of designing a magic bullet against cancer is impossible, largely because there is so much variation between various types of cancer and different individuals. However, we have learned much in past decades about different journeys that a normal cell takes to become cancerous, and that the delicate balance between oncogenes and tumor suppressor is upset, favoring growth and survival of the tumor cell. One of the most important cellular barriers to cancer development is the retinoblastoma tumor suppressor (pRB) pathway, which is inactivated in a wide range of human tumors and controls cell cycle progression via repression of the E2F/DP transcription factor family. Much of the clarity with which we view tumor suppression via pRB is due to our belief in the universality of the cell cycle and our attempts to model tumor pathways in vivo, nowhere so evident as in the multitude of data emerging from mutant mouse models that have been engineered to understand how cell cycle regulators limit growth in vivo and how deregulation of these regulators facilitates cancer development. In spite of this clarity, we have witnessed with incredulity several stunning results in the last 2 years that have challenged the very foundations of the cell cycle paradigm and made us question seriously how important these cell cycle regulators actually are.

1
Introduction and Background

1.1
Epidemiology

Cancer is the most frequent cause of death of individuals under 85 years of age in the United States, now even surpassing heart disease (American Cancer Society, Cancer Facts and Figures 2005, http://www.cancer.org). Mutations in *RB* or in genes encoding upstream regulators of pRB (i.e. *INK4A*, *CCND1*, *CDK4*) are found frequently in a mutually exclusive pattern in almost all human tumors (see Sect. 3 below and Palmero and Peters 1996, Sherr 1996). Two examples of human tumors that illustrate the impact of inactivating the *p*RB tumor suppressor pathway include lung cancer and cervical cancer, and are highlighted below.

The most frequent types of cancer diagnosed in the United States in decreasing incidence are breast cancer, prostate cancer, lung cancer and colon cancer; yet lung cancer is the deadliest form of cancer in the United States (163 500 deaths annually and 172 500 new cases annually). Twenty percent of lung cancer is classified as SCLC (small-cell lung cancer; 32 700 deaths annually); the vast majority ($\sim$ 90%) of cases carry mutations that directly inactivate the *RB* locus (encoding *p*RB)(Kaye 2002, Minna et al. 2004). It is indeed sobering to consider the number of deaths due to lung cancer that are largely preventable with abstinence from or cessation of smoking, and the fact that while smoking is on the decline in the United States, smoking and its associated lung cancer has been exported heavily to the developing world in the past 30 years. Moreover, the list of cancers for which smoking is a causative factor has grown to include cancers of the respiratory, gastrointestinal and genitourinary tracts (US Department of Health and Human Services, Surgeon General's Report on Smoking, http://www.surgeongeneral.gov).

Human cancer caused by viral infection remains a significant cause of suffering and death worldwide. Human papillomavirus (HPV) infection is a prominent sexually transmitted disease (20 million infected in the United States and 630 million infected worldwide), and its striking association ($\sim$ 99%) with cervical cancer (15 000 new cases annually in the US and 470 000 new cases annually worldwide) is most prevalent in the developing world where screening (i.e. Pap smear) is not routinely performed (zur Hausen 2002 and World Health Organization, http://www.who.it). The association of high-risk HPV (types 16 and 18) with cervical cancer is due to its ability to inactivate pRB growth suppression by direct binding of the HPV-E7 oncoprotein to pRB. Although a substantial time after primary viral infection is observed before cancer develops, the evidence that HPV causes cervical cancer is strong enough to classify it as a carcinogen by IARC (International Agency for Research on Cancer) and by the Federal Department of Health and Human Services.

1.2
Modeling Human Cancer in the Mouse

The ability of researchers to model cancer has grown substantially in the past decade, particularly in the mouse, in which numerous models of tumor development are now available for study. The eventual goals of such modeling are to faithfully reproduce the complexity of human tumorigenesis in the model organism, and then exploit the model system to uncover the regulatory points or Achilles heel of cancer, such that new therapies can be designed to attack the clinical disease. The many engineered strains of mutant mice have provided an excellent genetic model system in which to pursue these goals, and in fact, our efforts to model human cancer in the mouse, are far from exhaustive.

Many transgenic mice overexpressing an oncogene of interest have been made using tissue-specific promoters (e.g. MMTV LTR for breast, keratin 14 promoter for skin, and CD4 promoter for T-cells) to model a particular form of cancer. Knockout mice lacking specific tumor suppressors have been engineered to define the requirement of these genes during development and for inhibition of tumor development throughout life. Multistep tumorigenesis can be studied by treating these mutant mice with carcinogens, by combining such transgenic mice with knockout mice or by using proviral tagging to enhance the frequency or severity of specific cancers. Of course, the ability to generate these mice from genetically pure or inbred backgrounds, and to examine time points during embryogenesis or life after birth gives us greater insight into the different mechanisms of tumor development than that possible from studying human tumors alone. Additionally, mutation of tumor suppressors in mice leads to a restricted set of tumor types analogous to the narrow spectrum of tumors seen in inherited human tumor syndromes, the mechanistic basis for which is still elusive.

Nowhere has hope for modeling cancer in the mouse been as great as in the large number of mutants engineered in genes encoding components of the cell cycle machinery and the pRB and p53 tumor suppressor pathways. The universality of the cell cycle in all eukaryotes strongly suggested that anti-cancer therapy aimed at regulators of the cell cycle would be of great clinical benefit. This review will discuss the phenotypes of such mutant mice and the surprising recent results that suggest the cell cycle paradigm may greatly underestimate the complexity of the wiring of the cell cycle at least during embryonic development (see Sect. 5 below).

While the existing mutant mouse models of cancer are both impressive and powerful, they do not necessarily reflect the spectrum or complexity of human cancer. Only a small amount of human cancer can be attributed to the inheritance of dominantly acting mutant alleles of known tumor suppressors (Balmain et al. 2003). It has been estimated that only 12% of human breast cancer patients carry mutations in either the *BRCA1* or the *BRCA2* tumor suppressor gene, and that the majority of human, cancer susceptibility is due to the combined action of common, low penetrance cancer predisposing alleles or genetic modifiers of tumor susceptibility that have not yet been identified (Pharoah et al. 2002). Recently, the *Stk6* locus encoding the Aurora-2 centrosome-associated kinase was identified as a weak modifier of skin tumorigenesis in mice, and a polymorphism in the human Aurora homologue, *STK15* (Phe311), is found frequently amplified in human colon cancer (Ewart-Toland et al. 2003). Eventually, the hope is that as mutant mouse models of cancer improve, we can pursue more of these weak tumor predisposing alleles, for instance, by using our dominantly acting mutant mouse models of cancer to screen for enhancers or suppressors of these phenotypes. Perhaps only then can more clinically relevant and beneficial information be forthcoming from mutant mouse models of cancer.

2
The Universality of the Cell Cycle

The discovery of the cell cycle emerged from distinct studies in yeast, flies, clams, and frogs, and eventually the conclusions drawn were tested and found to be operative also in mice and humans. The power of the cell cycle theory is that it unified eukaryotic biology, for which Hartwell, Nurse and Hunt were recognized with the Nobel Prize in Medicine in 2001 (http://nobelprize.org/medicine/laureates/2001). The model of the cell cycle has given researchers the chance to understand complex mammalian systems and human disease, using genetic studies in evolutionarily lower organisms.

From studying temperature sensitive *cdc* mutants of budding yeast (*Saccharomyces cerevisiae*) with abnormal budded morphology at the non-permissive temperature, Hartwell and colleagues proposed that a simple order of dependent events (e.g. budding, DNA synthesis, cytokinesis), completion of which were necessary for cell division. In this way, the many of the primary components of the cell division cycle were identified and placed in a genetic pathway, in particular, *Cdc28*, which held precedence over all the *cdc* mutants as a master regulator in G1. Importantly, this work initiated the concept of checkpoints, non-essential genes that ensure these fundamental processes were completed. Nurse and co-workers built on these seminal concepts using fission yeast (*S. pombe*), first identifying temperature sensitive *wee1* mutants and then *Cdc2* as a master regulator of the cell cycle in G2/M. Nurse showed that *Cdc2* was required also in G1, and complementation experiments then showed that *Cdc2* was the *S. pombe* homologue of the *S. cerevisiae Cdc28* regulator. The identification of Cdc28 and then Cdc2 as kinases, helped define the function of other *cdc* genes (e.g. Cdc25 and Wee1) that modify the kinase activity of these master switches. Importantly, Masui's early studies of MPF (maturation promoting factor) activation in frog oocytes were crucial for understanding the commonality of these mechanisms even in vertebrates. Nurse and colleagues then cloned human *Cdc2* through complementation of the yeast *cdc2* mutant.

By studying changes in protein expression ongoing in the fertilization of clam and sea urchin eggs, Hunt and colleagues identified the first cyclin proteins, the abundance of which fluctuates with the division of the eggs. The identification of classes of yeast cyclins (Clns in G1 and Clbs in G2) that control Cdc28 and classes of vertebrate cyclins (D- and E-type cyclins in G1 and A-and B-type cyclins in G2/M) that control a family of Cdks (cyclin-dependent kinases) greatly enhanced our understanding of the complexity underlying control of the cell cycle. Control of Cdk activity through cyclin binding and degradation, inhibitory and activating Cdk phosphorylations and association of Cdk inhibitors outlined a range of regulatory mechanisms for controlling cell cycle progression (Morgan 1997; Zachariae and Nasmyth 1999). These studies and others demonstrated the universality of the cell cycle, and strongly suggested that deregulation of the mechanisms controlling cell cycle progression could result in cancer.

3
The pRB Tumor Suppressor Pathway

The concept that chromosomes could suppress malignancy is attributed to Boveri's writings from 1914 (Balmain 2001; Knudson 2001). In 1969, somatic cell hybridization experiments between normal and transformed cells resulted in phenotypically normal hybrids that often reverted to being transformed, indicating the existence of cellular genes that normally suppressed transformation (Ephrussi et al. 1969). Below is outlined the compelling evidence that the pRB tumor suppressor pathway is a crucial target that must be inactivated during the progression of normal cells into tumors.

3.1
The Discovery of pRB

The path of discovery for the prototypic tumor suppressor, *RB*, has been repeated many times with the identification of human tumor suppressors, mutation of which leads to cancer development. In 1971, Knudson considered the clear differences between the clinical presentation of inherited and sporadic retinoblastoma cases (i.e. frequency, age of onset, unilateral vs. bilateral, unifocal vs. multifocal lesions), and proposed the "two-hit" hypothesis for pediatric retinoblastoma development that put forward the following explanation for these clinical differences (Knudson 1971). Inherited retinoblastoma patients must carry a germ-line loss-of-function mutation in a putative retinoblastoma tumor suppressor gene (*RB*), and sometime during development or shortly after birth, a somatically acquired mutation specifically in a few retinal cells would inactivate the remaining normal *RB* allele, giving rise to multiple retinoblastomas per patient. In contrast, sporadic retinoblastoma patients must acquire two somatic *RB* mutations within the same retinal cell, an extremely rare event, giving rise to a single retinoblastoma per patient. The later identification of cytogenetic abnormalities involving deletions of Chr13q14 in normal blood cells from inherited retinoblastoma patients and in sporadic retinoblastoma tumor samples, strongly suggested the location of the *RB* gene, facilitating its subsequent positional cloning of the *RB* gene in 1986. Shortly thereafter, *RB* mutations were found frequently in osteosarcomas, small cell lung carcinomas and carcinomas of the prostate, bladder and breast. It has been estimated that 40–50% of human tumors contain direct inactivation of the *RB* gene (Palmero and Peters 1996; Sherr 1996).

Importantly, re-expression of pRB in tumor cells can revert the transformed phenotype (Huang et al. 1988), giving support for cancer therapies that restore pRB function. In 1988, interactions of pRB with viral oncoproteins from three distinct DNA tumor virus families (i.e. adenovirus E1A, SV40-T and HPV-E7) were demonstrated to be necessary for cellular transformation by these viruses (DeCaprio et al. 1988; Dyson et al. 1989). Binding of viral oncoproteins requires a large central region of pRB, known as the "pocket" domain, and tumor-derived *RB* mutants often had sustained deletions of exons encoding pieces of this "pocket" region of pRB. Two pRB homologues (p107 and p130) have been identified that show extensive homology through the central "pocket" domain of pRB and also to each other (reviewed in Classon and Harlow 2002). Although both pRB homologues can inhibit growth when overexpressed, mutations in genes encoding p107 or p130 are only rarely found in human tumors, and thus, these pRB homologues are not generally considered to be human tumor suppressors. Importantly, pRB suppresses growth and promotes lineage-specific differentiation; yet p107 and p130 appear to act together to suppress growth. It may be important to reassess the status of p107 and p130 mutations in tumors bearing RB mutations, given the recent evidence that these pRB family members act as tumor suppressors in conjunction with pRB (see Sect. 7).

3.2
Upstream Regulators of pRB

Cell cycle-dependent phosphorylation of pRB occurs in G1 by cyclin-dependent kinases that sequentially inactivate the tumor suppressive properties of pRB. Hyper-phosphorylation of pRB prevents binding of viral oncoproteins and cellular proteins to the central "pocket" region of pRB. Non-phosphorylatable pRB mutants suppress growth more efficiently than wild-type pRB. Normally, D-type cyclin/Cdk4 or cyclin/Cdk6 complexes phosphorylate pRB in early G1, while cyclin E1–2/Cdk2 complexes phosphorylate pRB at the G1/S transition, stimulating S-phase entry and cell cycle progression. Overexpression of D- and E-type cyclins and Cdk4 is observed frequently in human tumors (see Sect. 5), supporting the notion that these cell cycle regulators are critical for cell cycle progression. Inactivation of the complex locus at Chr9p21 containing the *INK4A* gene encoding p16, the cyclin-dependent kinase inhibitor specific for Cdk4 or Cdk6, is commonly seen in about half of human tumors (for discussion of the ARF tumor suppressor also residing at Chr9p21, see Sect. 6). Mutation of genes encoding these upstream regulators of pRB is observed in approximately 50% of all human tumors, in a mutually exclusive pattern to those tumors carrying *RB* mutations (Palmero and Peters 1996; Sherr 1996). Thus, inactivation of the pRB tumor suppressor pathway directly (*RB* mutations) or indirectly (*CCND1*, *INK4A* or *CDK4* mutations) occurs in almost all human tumors, emphasizing the importance of overcoming pRB-mediated growth control for tumor progression.

Two classes of human tumors that do not contain mutations of RB or mutations in genes encoding upstream regulators of pRB have derived other ways to circumvent the pRB tumor suppressor pathway. Colon carcinomas increase transcription of CCND1 via increases in β-catenin/TCF signaling resulting from loss of the *APC* tumor suppressor (Tetsu and McCormick 1999). Neuroblastomas increase transcription of Id2 via amplification of N-MYC that antagonizes pRB-mediated tumor suppression (Lasorella et al. 2000, and see Sect. 3.4).

3.3
Phenotype of Mice Lacking pRB Family Members

Mice lacking pRB die in mid-gestation with extensive defects in the central and peripheral nervous systems, fetal liver and lens (Clarke et al. 1992; Jacks et al. 1992; Lee et al. 1992) (Table 1). Chimaeras made with *Rb*-deficient ES cells develop surprisingly well, demonstrating that many tissues do not require pRB for normal function (Maandag et al. 1994; Williams et al. 1994b). Bypass of the mid-gestational lethality in *Rb*-deficient embryos allowed defects in muscle differentiation to be observed later in development (Zacksenhaus et al. 1996). The extensive apoptosis evident in the *Rb*-deficient embryos has now been shown to be due to placental insufficiency, specifically due to the hyperproliferation of the spongiotrophoblast layer of the placenta at E11.5 (Wu et al. 2003). Hyperproliferative and/or differentiation defects in the CNS, lens, and muscle are still present in *Rb*-deficient embryos once the placental requirement for *Rb* is circumvented through the use of chimaeras or through conditional deletion of *Rb* (Lipinski et al. 2001; Ferguson et al. 2002; de Bruin et al. 2003; MacPherson et al. 2003). Similarly, erythropoietic defects are apparent in the fetal liver of *Rb*-deficient embryos, and *Rb*-deficient erythroblasts fail to fully mature in vitro or reconstitute irradiated wild-type donors (Iavarone et al. 2004; Spike et al. 2004).

Rb+/- mice develop neuroendocrine tumors of the pituitary, thyroid, and adrenals (Jacks et al. 1992; Hu et al. 1994; Harrison et al. 1995) (Table 2). Neuroendocrine tumorigenesis in *Rb*+/- mice is dependent on LOH of the wild-type *Rb* allele similar to the retinoblastomas and osteosarcomas developing in germ-line retinoblastoma patients. This system has been used extensively to test the functional significance of numerous cell cycle regulators and interactors of pRB, including *E2F* family members and CKIs (Table 1 and Sects. 3.4 and 4). Interestingly, the spectrum of neuroendocrine tumors observed in *Rb*+/- mice is dependent on strain-specific modifiers, and specifically, inherent abnormalities of the 129Sv strain enhance the development of tumors in the intermediate lobe of the pituitary (Leung et al. 2004).

In contrast to phenotypes of the *Rb* mutant mice, *p*107-deficient or *p*130-deficient mice live to be viable adults without tumor predisposition on a mixed genetic background with the C57LB/6 strain (Cobrinik et al. 1996;

Table 1 Phenotypes of mice lacking *Rb* family members and rescue of *Rb* deficiency

Genotype	Phenotype	References
Rb−/−	Mid-gestational lethality at E13.5-E15.5 with widespread apoptosis	(Clarke et al. 1992; Jacks et al. 1992; Lee et al. 1992)
	Placental bypass required for survival to late gestation	(Maandag et al. 1994; Williams et al. 1994b; Wu et al. 2003; Zacksenhaus et al. 1996)
	Defects then found in CNS/PNS, fetal liver, muscle and lens	(de Bruin et al. 2003; Ferguson et al. 2002; Lipinski et al. 2001; MacPherson et al. 2003; Spike et al. 2004)
p107−/−	Myeloid hyperplasia and growth deficiency on Balb/c background	(LeCouter et al. 1998a; Lee et al. 1996)
	No obvious phenotype on mixed genetic background	
p130−/−	Embryonic lethality E11-E13 on a Balb/c background	(Cobrinik et al. 1996; LeCouter et al. 1998b)
	No obvious phenotype on mixed genetic background	
p107−/−; *p130*−/−	Perinatal lethality with endochondral bone defects on mixed background	(Cobrinik et al. 1996)
Rb−/−; *p107*−/−	Embryonic death prior to E12.5	(Lee et al. 1996)
Rb−/−; *E2f1*−/−	Rescue of mid-gestational lethality until late gestation	(Tsai et al. 1998)
Rb−/−; *E2f3*−/−	Rescue of mid-gestational lethality until late gestation	(Ziebold et al. 2001)
Rb−/−; *Id2*−/−	Rescue of mid-gestational lethality until late gestation and RBC enucleation in fetal liver	(Iavarone et al. 2004; Lasorella et al. 2000)
Rb−/−; *p53*−/−	Reduction of cell death in CNS and lens, but not PNS	(Macleod et al. 1996; Morgenbesser et al. 1994)
Rb−/−; *p19*−/−	No rescue of p53-dependent apoptosis observed	(Tsai et al. 2002a)

Lee et al. 1996) (Table 1). Combining *p107* deficiency with *p130* deficiency, results in perinatal death with defects in endochondral bone development (Cobrinik et al. 1996). On a 129Sv Balb/c background, *p130*-deficient embryos die and *p107*-deficient animals exhibit growth and myeloproliferative defects (LeCouter et al. 1998a,b), again suggesting that strain-specific mod-

Table 2 Phenotypes of Rb+/− mice lacking various cell cycle regulators

Genotype	Phenotype	References
Rb+/−	Neuroendocrine tumorigenesis in intermediate lobe of the pituitary	(Harrison et al. 1995; Hu et al. 1994; Jacks et al. 1992)
	Additional tumorigenesis in thyroid, anterior lobe of the pituitary, and adrenal gland	(Williams et al. 1994a; Nikitin et al. 1999)
Rb+/−; p107−/−	Neuroendocrine and non-endocrine tumorigenesis in chimaeras	(Dannenberg et al. 2004)
Rb+/−; p130−/−	No pituitary or thyroid tumorigenesis but other endocrine tumors at low frequency in chimaeras	(Dannenberg et al. 2004)
Rb+/−; E2f1−/−	Decreased neuroendocrine tumorigenesis and increased survival	(Yamasaki et al. 1998)
Rb+/−; E2f3−/−	Decreased pituitary tumorigenesis, but worsened thyroid tumors	(Ziebold et al. 2003)
Rb+/−; E2f4−/−	Decreased neuroendocrine tumorigenesis and increased survival	(Lee et al. 2002)
Rb+/−; p21−/−	Increased neuroendocrine tumorigenesis, including pheochromacytomas, and decreased survival	(Brugarolas et al. 1998)
Rb+/−; p27−/−	Increased neuroendocrine tumorigenesis with worsened thyroid tumors and decreased survival	(Park et al. 1999)
Rb+/−; p53−/−	Increased neuroendocrine tumorigenesis and decreased survival	(Williams et al. 1994a)
Rb+/−; p19−/−	Increased neuroendocrine tumorigenesis and decreased survival	(Tsai et al. 2002b)
Rb+/− (129Sv)	Increased tumorigenesis in the intermediate lobe of the pituitary, and greatly decreased survival	(Leung et al. 2004)
Rb+/− (C57BL/6)	Increased neuroendocrine tumorigenesis in the anterior pituitary and thyroid glands with increased survival	(Leung et al. 2004)

ifiers regulate the severity of phenotypes resulting from inactivation of *Rb* family members. While the constitutive inactivation of multiple *Rb* family members is required for the immortalization of primary mouse embryonic fibroblasts (MEFs) (Dannenberg et al. 2000; Sage et al. 2000; Peeper et al. 2001), the spontaneous loss of only *Rb* is sufficient to reverse cellular senescence (Sage et al. 2003). Conditional loss of *Rb* impairs the development of the cerebellum on a *p107*-deficient background (Marino et al. 2003) and produces medulloblastomas on a *p53*-deficient background (Marino et al. 2000). Chimaeras generated with ES cells that are *Rb*-deficient and either *p107*- or *p130*-deficient are highly tumor prone, demonstrating that pRB family members act in concert to suppress tumorigenesis in a wide variety of tissues in the mouse (Dannenberg et al. 2004) (Table 2). Chimaeras generated with ES cells that are Rb+/− and either *p107*- or *p130*-deficient develop tumors, but suggest that *p107* is a more effective tumor suppressor that *p130* (Dannenberg et al. 2004). The absence of retinoblastoma in Rb+/− mice prompted criticism of modeling human cancer in the mouse, but continued efforts to generate inherited models of mouse retinoblastoma with *Rb* deficiency have been successful recently (see Sect. 7 below).

3.4
pRB Regulates Growth and Differentiation

In the following section, the best characterized effector of pRB, the E2F/DP transcription factor family, is reviewed (see Sect. 4). The ability of E2F/DP complexes to control the expression of most if not all cell cycle related genes strongly suggests that the E2F/DP family is a crucial, downstream pRB target for controlling growth and thereby suppressing tumorigenesis. However, beyond the preponderance of reports on E2F and the significance of E2F for the growth suppressive function of pRB, there are numerous ($\sim$ 110) other interactors of pRB that have been identified (Morris and Dyson 2001). While none of the E2F family members contains an LCE motif, a number of these pRB interactors (e.g. RBP1, RBP2, HDAC) do contain this motif or one similar to it.

The existence of low penetrance retinoblastoma mutations that encode pRB mutants capable of E2F interaction, demonstrate that repression of E2F activity alone is insufficient for tumor suppression (Sellers et al. 1998). Such pRB mutants fail to interact with and activate transcription factors important for differentiation, suggesting that differentiation is an important component of pRB's ability to suppress tumor formation. Additionally, *Rb* deficiency inhibits the differentiation of particular lineages (e.g. adipogenesis, myogenesis and osteogenesis) due to the inability of lineage-specific transcription factors (e.g. C/EBPα, MyoD, CBFA1) to be activated by pRB (Gu et al. 1993; Chen et al. 1996; Thomas et al. 2001). These studies suggest that it is the unique ability of pRB to coordinate cell cycle exit with the induction of differentiation that confers upon pRB its tumor suppressor function.

The interaction of pRB with Id2 (inhibitor of differentiation) is especially compelling (Table 1). Id2 normally acts as a dominant negative inhibitor of helix–loop–helix transcription factors; however, it is also able to antagonize pRB family function by direct interaction. Loss of *Id2* rescues the mid-gestational lethality of *Rb*-deficient embryos, minimally by rescuing the defects seen in neurogenesis and erythropoiesis (Lasorella et al. 2000). Rescue of erythropoiesis in *Rb*-deficiency occurs with loss of *Id2*, because Id2 normally inhibits PU.1, a transcription factor that stimulates the macrophage lineage, upon which developing erythroblasts are dependent for enucleation (Iavarone et al. 2004).

4
The E2F/DP Transcription Factor Family

In 1992, researchers identified a critical downstream effector of pRB, the E2F1 transcription factor, binding sites for which were found in the adenovirus E2 promoter and the promoters of many cell cycle regulated genes (reviewed in Trimarchi and Lees 2002; Attwooll et al. 2004). E2F1 stimulates entry into S-phase and cooperates with activated *Ras* for cellular transformation. Multiple E2F family members (E2F1–6) have been identified that recognize E2F binding sites in target promoters following hetero-dimerization to a DP family member (DP1 and DP2). More recently E2F7 has been identified that binds E2F sites independently of association with a DP subunit. E2F1, E2F2 and E2F3 interact preferentially with pRB, while E2F4 and E2F5 interact well with p107 and p130. E2F4 can interact with pRB at lower affinity. E2F6 and E2F7 do not interact with pRB family members, and are not competent for activating transcription.

4.1
E2F Target Genes and Repression

Interaction of E2F1–3/DP complexes with pRB converts these transcriptional activators to repressor complexes that are known to bind to target promoters and inhibit interaction with the basal transcription machinery. Interaction with pRB blocks the ability of E2F/DP complexes to induce its many target genes that promote numerous cellular processes (Ishida et al. 2001; Kalma et al. 2001; Ma et al. 2002; Ren et al. 2002; Stevaux and Dyson 2002; Weinmann et al. 2002; Wells et al. 2002). Classical E2F target genes encode products that promote S-phase entry (e.g. Orc1 and Mcm2–7) and DNA replication (e.g. Dhfr, Rnr, TK, TS and Polα). Many E2F target genes encode products involved in cell cycle progression (e.g. cyclins A and E, pRB, *p*107, E2F1–3) and apoptosis (e.g. p73, Apaf1, caspase). Recently however, E2F target genes have been identified, the products of which act in DNA repair (e.g. Msh2, Mlh1,

Rad51), cell cycle checkpoints (e.g. Mad2, Chk1, Bub3) and mitotic events (e.g. cyclin B, Cdc2, and Smc2). These more recently discovered targets help in understanding the aneuploidy found in human tumors, because following loss of *RB*, the deregulation of E2F activity increases expression of *Mad2*, a mitotic checkpoint gene that contributes to genomic instability (Hernando et al. 2004).

pRB binds HDAC family members while tethered through E2F to target promoters, inhibiting gene expression via chromatin remodeling. Interestingly, pRB and E2F family members are known to be acetylated in vivo, which changes the ability of these proteins to bind DNA, suggesting that HDAC association may modify pRB and E2F directly as well as histones in neighboring nucleosomes (Martinez-Balbas et al. 2000; Marzio et al. 2000; Chan et al. 2001; Brown and Gallie 2002; Nguyen et al. 2004). While the pRB tumor suppressor pathway is widely inactivated in human tumors, the E2F or DP genes themselves are not, an observation that may be attributed to at least two underlying causes. First, the frequent inactivation of pRB and its upstream regulators in human tumors leads in effect to deregulation of the E2F activity. Second, the bifunctional nature of E2F/DP complexes to act as activators and/or repressors and the wide range of E2F target genes suggests that deregulation of E2F or DP may have pleiotropic and opposing effects that do not favor tumor cell survival.

4.2
Mice Deficient in E2F Family Members

Mice lacking various E2F family members have been generated, and in general, the viable phenotypes of individual *E2f*-deficient mice probably reflect substantial functional redundancy of many E2F family members (Table 3). Loss of multiple E2F family members results in stronger phenotypes, occasionally in embryonic death, consistent with subsets of E2Fs having similar function. For example, while inactivation of *E2f1* leads to reduced adult survival with broad-range of tumors and tissue atrophy (Field et al. 1996; Yamasaki et al. 1996), the simultaneous inactivation of *E2f1* and *E2f2* results in high penetrance phenotypes that result in premature death, including diabetes, exocrine pancreatic failure, hematopoietic failure and leukemias (Zhu et al. 2001; Li et al. 2003a,b; Iglesias et al. 2004). This is a similar situation to that seen in MEFs, where loss of *E2f1* or *E2f2* does not result in a cell cycle defect and loss of *E2f3* leads to defective induction of numerous E2F target genes (Humbert et al. 2000b); however, the simultaneous inactivation of *E2f1*, *E2f2* and *E2f3* is required to block MEF proliferation (Wu et al. 2001). Inactivation of *Dp1* leads to embryonic lethality prior to E12.5, a sharp contrast to the weaker phenotypes resulting from inactivation of individual E2F family members (Kohn et al. 2003). *Dp1* deficiency cripples the development of all extra-embryonic lineages (visualized as early as E6.5), leading to placen-

Table 3 Phenotypes of mice lacking *E2F* or DP family members

Genotype	Phenotype	References
E2f1-/-	Tumor predisposition and decreased thymocyte apoptosis	(Field et al. 1996; Yamasaki et al. 1996)
	Atrophy of testes and thyroid gland	
E2f2-/-	Enhanced T cell proliferation and auto-immunity	(Murga et al. 2001)
E2f3-/-	Decreased viability and congestive heart failure on mixed background	(Cloud et al. 2002; Humbert et al. 2000b)
	Inviable on 129Sv strain. Decreased *E2F* target induction and MEF proliferation	
E2f4-/-	Erythropoietic and craniofacial defects resulting in juvenile death	(Humbert et al. 2000a; Rempel et al. 2000)
E2f5-/-	Abnormal choroids plexus and hydrocephaly leading to premature death	(Lindeman et al. 1998)
E2f6-/-	Homeotic transformations of the axial skeleton without premature death	(Storre et al. 2002)
Dp1-/-	Embryonic lethality prior to E12.5	(Kohn et al. 2003, 2004)
	Placental bypass required for late gestational E17.5 survival without obvious defects	
E2f1-/-; E2f2-/-	Hematopoietic abnormalities (e.g., tumors and megaloblastic anemia)	(Zhu et al. 2001; Li et al. 2003a,b; Iglesias et al. 2004)
	Insulin-dependent diabetes with exocrine pancreatic failure and auto-immunity	
E2f1-/-; E2f3-/-	No enhanced tumor predisposition, but worsened testicular atrophy and viability	(Cloud et al. 2002)
E2f4-/-; E2f5-/-	Embryonic lethality between E13.5 and birth	(Gaubatz et al. 2000)
E2f1-/-; E2f2-/-; E2f3-/-	No organismal phenotype reported, but MEFs do not proliferate	(Wu et al. 2001)

tal insufficiency that is responsible for the death of the embryo secondarily and that occurs well before the requirement for *Rb* in the placenta. Bypassing the extra-embryonic requirement for *Dp1* with the construction of chimaeras surprisingly allows *Dp1*-deficient ES cells to develop into most tissues of the embryo proper (Kohn et al. 2004).

E2F family members represent direct targets of pRB as judged by the reversal of *Rb*-deficient phenotypes with inactivation of E2Fs. Combination of *E2f1*-, or *E2f3*-deficiency rescues the mid-gestational lethality of *Rb*-deficiency and reduces the penetrance of neuroendocrine tumorigenesis in *Rb+/-* mice (Tsai et al. 1998; Yamasaki et al. 1998; Ziebold et al. 2001, 2003). Surprisingly, *E2f4* deficiency is able to fulfil the latter function, by diverting p107 and p130 repressive function to E2F1-3 (Lee et al. 2002). It is also quite likely that *E2f2* loss will rescue *Rb*-deficient lethality. Taken together, these studies demonstrate the functional significance of *E2F* family members as downstream effectors of pRB functions in development and tumorigenesis.

5
Cyclin-dependent Kinases and their Inhibitors

5.1
Deregulation of Cyclins, Cdks and CKIs in Human Tumors

As predicted from the early studies of the cell cycle, deregulation of cell cycle components (cyclins, Cdks and CKIs) is observed in human cancer. Indeed, direct genetic alterations and overexpression of numerous cyclins are common events in human neoplasia, and correlate with a poor prognosis of survival (Reed 2003). Cyclin D1 overexpression occurs in mantle (centrocytic) cell lymphomas and parathyroid adenomas following chromosomal translocation of the *CCND1* (also known as the *BCL1* or *PRAD1*) locus at Chr11q13 into the immunoglobulin heavy chain locus on Chr14 and chromosomal inversion, respectively (Motokura et al. 1991). *CCND1* amplification also occurs frequently in cancer of the breast, esophagus, bladder and pancreas. Cyclin D2 overexpression occurs in ovarian and testicular tumors, and cyclin D3 overexpression resulting from *CCND3* rearrangements occurs in lymphoid malignancies (Sicinska et al. 1996, 2003). Cyclin E overexpression or amplification of the *CCNE* locus on Chr19 is commonly observed in uterine, ovarian and breast cancer, as well as an array of other tumor types. Cyclin A overexpression resulting from random integration of HBV in the *CCNA* locus has been reported in a liver tumor (Wang et al. 1990). Finally, *CDK4* at Chr12q13 is amplified in a number of different tumor types, and mutation of *CDK4* rendering insensitive to $p16^{INK4}$ inhibition has been reported in melanomas (Zuo et al. 1996).

Inhibitors of Cdks or CKIs (cyclin-dependent kinase inhibitors) would be predicted to act as tumor suppressors that slow cell cycle progression. There are two families of CKIs, the *INK4* family (p16^{INK4A}, p15^{INK4B}, p18^{INK4C} and p19^{INK4D}) and the *CIP/KIP* family (p21^{CIP1}, p27^{KIP1} and p57^{KIP2}). *INK4A* and *INK4B* at Chr9p21 are frequently mutated in human cancer (see Sect. 6 below for the discussion of overlap between the *INK4A* and *ARF* locus), while *INK4C* and *INK4D* show no such pattern (Kamb et al. 1994). *CIP1* (also known as *WAF1/SDI1*) and *KIP1* are infrequently mutated in human cancer; however, decreased expression of p27 through increased ubiquitin-mediated degradation, commonly occurs in human tumors and also correlates with a poor prognosis (Pagano and Benmaamar 2003). *KIP2* is one of the imprinted genes at Chr 11p15 frequently deleted in Beckwith–Wiedemann syndrome (BWS), characterized overgrowth, midline-defects and increased predisposition to pediatric cancer, and a small subset of BWS patients have *KIP2* mutations (Hatada et al. 1996; O'Keefe et al. 1997).

Overexpression of other cell cycle components, such as the Cdc25A and Cdc25B phosphatases that are important for Cdk activation in G1, is also observed in human tumors (Kristjansdottir and Rudolph 2004). It is of note that overexpression of Cdk1, Cdk2 or Cdc25C is not commonly seen in human cancer, perhaps because of the requirement to decrease mitotic Cdk1 activity for mitotic exit. Nevertheless, these reports strongly suggest that deregulation of the cell cycle facilitates tumorigenesis in humans.

5.2
Mice Deficient in Cyclins, Cdks and CKIs

Surprisingly, most individual G1 and G1/S regulators of the cell cycle are largely (if not completely) dispensable for normal mouse embryonic development (Gladden and Diehl 2003; Roberts and Sherr 2003; Pagano and Jackson 2004; Sherr and Roberts 2004) (Table 4). Individual D-type cyclins are dispensable for development (Fantl et al. 1995; Sicinski et al. 1995, 1996; Sicinska et al. 2003). Similarly, mice lacking individual E-type cyclins are viable (Geng et al. 2003; Parisi et al. 2003). *Cdk2*-deficient mice are viable and are sterile due to meiotic defects (Berthet et al. 2003; Ortega et al. 2003). *Cdk4*-deficient mice are viable, but runted and develop diabetes (Rane et al. 1999; Tsutsui et al. 1999). *Cdk6*-deficient mice are viable and develop hematopoietic defects (Malumbres et al. 2004).

Mice lacking p21 are viable (Brugarolas et al. 1995; Deng et al. 1995), while mice lacking p27 are viable, but display organomegaly and pituitary adenomas (Fero et al. 1996; Kiyokawa et al. 1996; Nakayama et al. 1996). Mice lacking p57 die in late gestation with placental and mid-line closure defects (Yan et al. 1997; Zhang et al. 1997). The absence of p21 or p27 accelerates the development of neuroendocrine tumorigenesis in *Rb+/−* mice, demonstrating the importance of these factors as negative regulators of neoplasia (Brugaro-

Table 4 Phenotypes of mice lacking G1 cyclins and Cdks

Genotype	Phenotype	References
CcnD1-/-	Viable with retinal and mammary hypoplasia and neurological defects	(Fantl et al. 1995; Sicinski et al. 1995)
CcnD2-/-	Viable with gonadal hypoplasia	(Sicinski et al. 1996)
CcnD3-/-	Viable with defective T-cell expansion	(Sicinska et al. 2003)
CcnD2-/-; CcnD3-/-	Born, but premature death due to megaloblastic anemia	(Ciemerych et al. 2002)
CcnD1-/-; CcnD3-/-	Born, but premature death due to neurological abnormalities	(Ciemerych et al. 2002)
CcnD1-/-; CcnD2-/-	Born, but premature death due to cerebellar deficiency	(Ciemerych et al. 2002)
CcnD1-/-; CcnD2-/-; CcnD3-/-	Late gestational failure with anemia and cardiac defects	(Kozar et al. 2004)
CcnE1-/-	Viable with no obvious phenotype	(Geng et al. 2003; Parisi et al. 2003)
CcnE2-/-	Viable with reduced fertility due to testicular atrophy	(Geng et al. 2003; Parisi et al. 2003)
CcnE1-/-; CcnE2-/-	Mid-gestational lethality. Placental bypass required for survival to late gestation. Megakaryocyte and cardiovascular defects in late gestation	(Geng et al. 2003; Parisi et al. 2003)
Cdk2-/-	Viable, but infertile	(Berthet et al. 2003; Ortega et al. 2003)
Cdk4-/-	Viable, small, infertile and insulin-dependent diabetes	(Rane et al. 1999; Tsutsui et al. 1999)
Cdk6-/-	Viable with mild hematopoietic defects	(Malumbres et al. 2004)
Cdk4-/-; Cdk6-/-	Late gestational failure with anemia	(Malumbres et al. 2004)

las et al. 1998; Park et al. 1999). Mice lacking individual *Ink4* family members are all viable, but only mice with mutations at the complex *Ink4a/Arf* locus show strong tumor predisposition. Originally, mice lacking the shared exon (2) of this locus displayed increased tumor predisposition (Serrano et al. 1996); however, this has been attributed mainly to loss of *Arf*, because *Arf*-/- mice phenocopy mice lacking both transcripts (Kamijo et al. 1997). Inactivation of *Ink4a* alone leads to a mild tumor phenotype in response to carcinogen treatment (Krimpenfort et al. 2001; Sharpless et al. 2001).

The viability of single deficiency strains has been attributed largely to the presence of other family members that functionally compensate for the loss of one member, rather than the possibility that the anticipated function of a particular family of regulators is not essential. The requirement for mitotic cell cycle regulators (e.g. Cdk1, cyclin A2 or B1) is closer to our expectations; that is, their loss results in early embryonic lethality and peri-implantation defects at E4.5, presumably after maternal supplies of mRNA or protein have expired (Murphy et al. 1997; Brandeis et al. 1998; Malumbres and Barbacid, unpublished result).

Surprisingly, we are now learning that, in fact, entire families of G1 regulators (e.g. cyclins E1/E2, cyclins D1/D2/D3) can be lost without impairing the normal development of most embryonic tissues. Simultaneous loss of cyclins E1 and E2 results in embryonic death due to placental failure at mid-gestation with abnormal endoreduplication of trophoblast giant cells (Geng et al. 2003; Parisi et al. 2003). Rescue of placental insufficiency allows cyclin E1/2-deficient embryos to survive to late gestation, where cardiac defects and megakaryocyte abnormalities occur. The fact that *Rb*, *Dp1* or cyclin E1/2 deficiency results in placental insufficiency is intriguing, and it is tempting to speculate that these genes cooperate in a single placental function; however, the complexity of the extra-embryonic compartment makes this possibility unlikely.

Mice expressing only one D-type cyclin are viable, yet experience premature death due to megaloblastic anemia, neurological abnormalities and absence of cerebella with the presence of only cyclin D1, cyclin D2 or cyclin D3, respectively (Ciemerych et al. 2002) (Table 4). The simultaneous inactivation of all three D-type cyclins leads to normal development until mid-to-late gestation when defects in the hematopoietic compartment and heart contribute to embryonic death (Kozar et al. 2004). Inactivation of both Cdk4 and Cdk6 leads to normal development until late gestation when embryos die with anemia (Malumbres et al. 2004). These remarkable findings demonstrate that most tissues in the embryo develop without E-type cyclins or Cdk2, and similarly that development is normal in most tissues without D-type cyclins or Cdk4/Cdk6. Certainly, the inactivation of these families of G1 regulators results in a pronounced phenotype, but the emphasis should be placed on how restricted this phenotype is, rather than how important any regulator is in a particular susceptible site. Clearly, these mutant mouse phenotypes demonstrate that most of embryonic development does not require these subsets of cyclins or Cdks, while data from human tumors strongly suggest deregulation of these cell cycle components facilitate tumorigenesis. We propose alternative views to interpret these data in Sect. 8.

6
Links Between the pRB and p53 Tumor Suppressor Pathway

The earliest clue that the pRB and p53 tumor suppressor pathways were connected was the observation that tumor viruses (adenovirus, SV40 and HPV), belonging to three distinct DNA viral families, had evolved regulatory proteins to inactivate both of these barriers to tumor growth. Presumably, inactivation of the pRB pathway by adenovirus E1A, SV40-T or HPV-E7 is advantageous for these tumor viruses to promote cell cycle progression into S-phase, when the synthesis of viral genomes would be optimal, allowing the production of multiple virion particles. Likewise, inactivation of the p53 pathway by adenovirus E1B, SV40-T or HPV-E6 allows these DNA tumor viruses to escape the cell death program activated by the virus entry and unscheduled S-phase activity, thereby optimizing virus production. In the case of abortive infections, the inactivation of the pRB and p53 pathways facilitates entry into S-phase and avoidance of apoptosis, thereby promoting cellular transformation.

The next formal link between the pRB and p53 tumor suppressor pathways came from the discovery of p21. Multiple groups identified p21, as a p53-target gene (*WAF1*), a Cdk2-associated protein (p21 or *CIP1*), a Cdk2 inhibitor (*CAP20*), or a senescence cell-derived inhibitor (*SDI1*). Since phosphorylation by Cdk2 complexes (either cyclin A or E associated) normally inactivates pRB, the stabilization of p53 following DNA damage will induce p21 levels, thereby inhibiting cyclin/Cdk2 activity towards pRB and halting cell cycle progression.

One of the most intriguing links between the pRB and p53 tumor suppressor pathways is the complex genetic Chr9p21 locus that is mutated in approximately half of all human tumors and that contains two overlapping, tumor suppressor genes. One of the genes is *INK4A*, which encodes the p16 inhibitor of the Cdk4 and Cdk6 kinases, which act in concert with Cdk2 kinase complexes to inactivate pRB during G1-phase. The other gene at 9p21 is *ARF* that encodes the p14^{ARF} (human) or p19^{Arf} (mice) protein that counteracts the ability of Hdm2 (human) or mdm2 (mice) to degrade p53. The second and third exons of *INK4A* and *ARF* are shared but translated in different reading frames, while each of these genes has a unique first exon (exon 1α for *INK4A* and exon 1β for *ARF*). Mutations at Chr9p21 often remove these shared exons, simultaneously inactivating the pRB and p53 tumor suppressor pathways. Large deletions at Chr9p21 also delete the upstream *INK4B* locus (encoding the p15 inhibitor of Cdk4 and Cdk6 kinases) as well as the overlapping *INK4A* and *ARF* loci (Sherr 2001a,b).

Finally, E2F1 is a key link between the pRB and p53 tumor suppressor pathways. E2F1 is normally repressed by pRB; however, when released from pRB-mediated control, E2F1 indirectly induces p53 levels by inducing the

ARF. Furthermore, E2F1 directly induces p73 (one of two p53 homologues), Apaf1 and caspase-7 that induce or facilitate apoptosis (Irwin et al. 2000; Lissy et al. 2000; Moroni et al. 2001; Nahle et al. 2002; Pediconi et al. 2003). Thus, under the proper signals to relieve repression from pRB, E2F1 appears to be particularly efficient in inducing p53-dependent and p53-independent apoptosis in vitro. There are conflicting reports, however, on the requirement for E2F1 in thymic lymphomagenesis or embryonic lethality observed with *p53* deficiency (Wikonkal et al. 2003; Wloga et al. 2004), and our own work has shown that E2F1 is dispensable for these p53-dependent phenotypes.

7
Murine Models of Retinoblastoma

One of the fascinating, yet poorly understood aspects of tumorigenesis is the narrow tumor spectrum resulting from the inheritance of loss-of-function mutations in specific tumor suppressors. This is the case for inherited retinoblastoma patients in which germ-line *RB* mutations predispose to pediatric retinoblastomas and mainly osteosarcomas later in life, although all tissues are predisposed to become pRB-negative following loss of the remaining wild-type *RB* allele. *Rb*+/− mice (that are analogous to retinoblastoma patients with germ-line *RB* mutations) develop a narrow range of neuroendocrine tumors following LOH in those tissues. Curiously, retinoblastomas are not seen in these mice, and much effort has been expended to resolve this discrepancy and generate mouse retinoblastomas. Disruption of pRB family function using transgenic expression of SV40-T in the photoreceptor layer of the retina leads to retinoblastoma development (Windle et al. 1990; al-Ubaidi et al. 1992). When chimaeras are constructed using ES cells lacking *Rb* and *p107*, then retinoblastomas develop by 1–4 months (Robanus-Maandag et al. 1998). Similarly, chimaeras generated with ES cells lacking *Rb* and *p130* develop retinoblastomas as well (Dannenberg et al. 2004).

Very recently three groups have succeeded in generating mouse models of heritable retinoblastoma using various *Cre* transgenic lines to conditionally delete a floxed allele of *Rb* in animals lacking either *p107* or *p130* (reviewed in Dyer and Bremner 2005). Loss of *Rb* on a wild-type background using α-*Cre*, *Nestin-Cre* or *Chx10-Cre* leads to the loss of distinct neuronal cell types (photoreceptors for α-*Cre*, ganglion cells, photoreceptors and bipolar cells for *Nestin-Cre* or rod cells for *Chx10-Cre*) without the development of retinoblastoma. Notably, the level of *p107* increases or the level of hypophosphorylated *p107* increases with loss of *Rb* in the mouse retina (Zhang et al. 2004a). With the additional deletion of another pRB family member, conditional deletion of *Rb* leads to retinoblastoma in mice. Deletion of *Rb* using α-*Cre* (expressed by E10.5 in the peripheral retina by the *Pax6* al-

pha enhancer) leads to retinoblastoma by 1–2 months on a *p107*-deficient background (Chen et al. 2004). Previously, no retinoblastomas were reported using *IRBP-Cre* (expressed in a mosaic manner by E14.5 in photoreceptors) to conditionally delete *Rb* in a *p107*-deficient background (Vooijs et al. 2002). Deletion of *Rb* using *nestin-Cre* (expressed in a mosaic pattern in developing neurons and glia when the maternal allele is inherited) leads to retinoblastoma by 3 months on a *p130*-deficient background (MacPherson et al. 2004). On a wild-type background, conditional deletion of *Rb* using paternally inherited *nestin-Cre* results in perinatal lethality (MacPherson et al. 2003), while on a *p107*-deficient background, conditional deletion of *Rb* using maternally inherited *nestin-Cre* resulted in death prior to weaning (MacPherson et al. 2004). In contrast, deletion of *Rb* using *Chx10-Cre* (expressed in retinal progenitors) generates moderate hyperproliferative lesions by 4 months on a *p107*-deficient background (Zhang et al. 2004b).

Loss of *p53* is not commonly seen in human retinoblastoma and does not enhance the development of mouse retinoblastoma on a *p130*-deficient background using *Nestin-Cre* to delete *Rb* conditionally (MacPherson et al. 2004); however, loss of *p53* leads to invasive retinoblastoma 3–6 weeks in mice on a *p107*-deficient background using *Chx10-Cre* to delete *Rb* conditionally in retinal progenitors (Zhang et al. 2004b). Similarly, injection of a retrovirus encoding E1A (13S) that minimally inactivates all *Rb* family members (as well as p300) into newborn mice leads to retinoblastoma by 3.5 months on a *p53*-deficient background (Zhang et al. 2004b). Previously, transgenic expression of HPV-E7 in the photoreceptor layer induced retinoblastomas only a *p53*-deficient background (Howes et al. 1994). Using α-*Cre* to delete *Rb* in a *p107*-deficient background, it appears that bypassing terminal differentiation rather than cell death is the mechanism allowing retinoblastoma development (Chen et al. 2004). Thus, there is still conflicting evidence concerning the importance of overcoming p53-dependent apoptosis during the development of retinoblastoma. This may reflect the fact that there are multiple routes to retinoblastoma development in the mouse (Dyer and Bremner 2005).

8
Revising Cell Cycle Models

Clearly, the field was initially swept away by the fervor to generalize cell cycle models in yeast or tissue culture cells to the whole organism. Still, we should still be impressed that so much regulation has been conserved with regards to mechanisms of activating and inactivating Cdk activity. However, the best tact would be to recognize simply that all cell types are not equivalent and thus, not all cell cycles are equivalent. The fact that tumor predisposition is so narrow in patients with germ-line *RB* mutations (i.e. retinoblastomas and mainly osteosarcomas later in life) underscores the reality that we do not understand the basis for tissue-specific tumor susceptibility in mouse or in humans.

Since only a few tissues require the presence of a family of G1 or G1/S regulators during mouse development, a number of intriguing questions arise concerning the critical nature of G1 itself. First, are there undiscovered homologues (e.g. novel G1 cyclins or Cdks) that functionally compensate for the loss of subsets of G1 regulators family members? In this post-genomic era, it is unlikely that other cyclin-like genes with substantial homology to the D- or E-type cyclins have not been detected. However, unconventional use of known cyclins or Cdks appears to occur when the optimal regulator is not present. Knock-in of cyclin E into the cyclin D1 locus suppresses cyclin D1-deficient phenotypes, demonstrating that cyclin E and presumably Cdk2 form a kinase complex that can substitute for cyclin D1 and either Cdk4 or Cdk6 complexes (Geng et al. 1999). Likewise, it appears that Cdk2 can functionally replace Cdk4 and Cdk6 in G1 (Malumbres et al. 2004).

Second, are mitotic regulators substituting for G1 or G1/S regulators? Perhaps Cdk1 pinch-hits in the absence of Cdk2 or even Cdk4/6 analogous to the role of yeast Cdc2 or Cdc28 in G2/M and G1 in combination with different cyclins. Similarly, cyclin A may substitute for E-type cyclins during the G1/S transition. While the proper cyclin may help substrate selection and drive cell cycle progression optimally, it may not be required to do so, and another cyclin may perform these functions with slower kinetics, but to the same end. This substitute cyclin could function even better when overexpressed, as in the case of human tumors.

Third, is G1-phase an optional stage rather than an integral segment of the cell cycle? Consider that early embryonic cell cycles actually do not have a G1-phase. Rather, distinct G1 phases are added later with specific signaling cascades that activate distinct transcriptional programs for each cell type. While G1 and G1/S regulators can make the cell cycle progress, subsets of them are apparently not required to do so during development. It is possible that G1 and G1/S regulators are only required for the cell to switch to a new state. In this way, we could liken G1 and G1/S regulators to the clutch on a car, to invoke the often-used analogy of the car's engine to the cell cycle. A car that is in motion does not need a clutch, unless the driver tries to shift speeds, go into reverse or stop, when the clutch becomes the only way not to ruin the engine. Thus, the cell may use G1 and G1/S regulators to proliferate faster, to differentiate or to senesce, but may not require G1 or G1/S regulators to simply cycle, contrary to the well-accepted cell cycle paradigms. However, the mutant mouse phenotypes tell us that there are many cell types that simply do not require these G1 and G1/S regulators for normal proliferation or development.

Finally, there may be other ways to regulate cell cycle progression than by activating Cdk activity. One of the key targets of G1 and G1/S Cdk activity is the pRB, and non-phosphorylatable mutants of pRB suppress cell cycle progression better than wild-type pRB. However, *RB* is also regulated at the level of transcription, increasing during cellular differentiation and low penetrance *RB* mutations have been mapped into the *RB* promoter, demonstrating that decreasing the absolute levels of pRB predisposes patients to retinoblastomas. Furthermore, pRB is regulated by degradation and acetylation, both of which could lead to changes in cell cycle progression during development. Thus, it is possible that regulating the absolute levels of pRB may be an alternative to activating cyclin/Cdk activity to control cell cycle progression. Undoubtedly, results from mutant mouse models will continue to challenge our views on how changes in the cell cycle impact cancer for many years to come.

Acknowledgements This work was supported by the Charlotte Geyer Foundation and a grant from the NIH-NCI (#2R01-CA079646). Special thanks to P. Kaldis for his patience with this review. LY thanks M. and I. Pagano for their continual support and strength.

References

al-Ubaidi MR, Font RL, Quiambao AB, Keener MJ, Liou GI, Overbeek PA, Baehr W (1992) Bilateral retinal and brain tumors in transgenic mice expressing simian virus 40 large T antigen under control of the human interphotoreceptor retinoid-binding protein promoter. J Cell Biol 119:1681–1687

Attwooll C, Denchi EL, Helin K (2004) The *E2F* family: specific functions and overlapping interests. Embo J 23:4709–4716

Balmain A (2001) Cancer genetics: from Boveri and Mendel to microarrays. Nat Rev Cancer 1:77–82

Balmain A, Gray J, Ponder B (2003) The genetics and genomics of cancer. Nat Genet 33:238–244

Berthet C, Aleem E, Coppola V, Tessarollo L, Kaldis P (2003) Cdk2 knockout mice are viable. Curr Biol 13:1775–1785

Brandeis M, Rosewell I, Carrington M, Crompton T, Jacobs MA, Kirk J, Gannon J, Hunt T (1998) Cyclin B2-null mice develop normally and are fertile whereas cyclin B1-null mice die in utero. Proc Natl Acad Sci USA 95:4344–4349

Brown VD, Gallie BL (2002) The B-domain lysine patch of pRB is required for binding to large T antigen and release of *E2F* by phosphorylation. Mol Cell Biol 22:1390–401

Brugarolas J, Chandrasekaran C, Gordon JI, Beach D, Jacks T, Hannon GJ (1995) Radiation-induced cell cycle arrest compromised by p21 deficiency. Nature 377:552–557

Brugarolas J, Bronson RT, Jacks T (1998) p21 is a critical CDK2 regulator essential for proliferation control in Rb-deficient cells. J Cell Biol 141:503–514

Chan HM, Krstic-Demonacos M, Smith L, Demonacos C, La Thangue NB (2001) Acetylation control of the retinoblastoma tumour-suppressor protein. Nat Cell Biol 3:667–674

Chen D, Livne-bar I, Vanderluit JL, Slack RS, Agochiya M, Bremner R (2004) Cell-specific effects of RB or RB/*p*107 loss on retinal development implicate an intrinsically death-resistant cell-of-origin in retinoblastoma. Cancer Cell 5:539–551

Chen PL, Riley DJ, Chen Y, Lee WH (1996) Retinoblastoma protein positively regulates terminal adipocyte differentiation through direct interaction with C/EBPs. Genes Dev 10:2794–2804

Ciemerych MA, Kenney AM, Sicinska E, Kalaszczynska I, Bronson RT, Rowitch DH, Gardner H, Sicinski P (2002) Development of mice expressing a single D-type cyclin. Genes Dev 16:3277–3289

Clarke AR, Maandag ER, van Roon M, van der Lugt NM, van der Valk M, Hooper ML, Berns A, te Riele H (1992) Requirement for a functional Rb-1 gene in murine development. Nature 359:328–330

Classon M, Harlow E (2002) The retinoblastoma tumour suppressor in development and cancer. Nat Rev Cancer 2:910–917

Cloud JE, Rogers C, Reza TL, Ziebold U, Stone JR, Picard MH, Caron AM, Bronson RT, Lees JA (2002) Mutant mouse models reveal the relative roles of E2F1 and E2F3 in vivo. Mol Cell Biol 22:2663–2672

Cobrinik D, Lee MH, Hannon G, Mulligan G, Bronson RT, Dyson N, Harlow E, Beach D, Weinberg RA, Jacks T (1996) Shared role of the pRB-related p130 and p107 proteins in limb development. Genes Dev 10:1633–1644

Dannenberg JH, van Rossum A, Schuijff L, te Riele H (2000) Ablation of the retinoblastoma gene family deregulates G1 control causing immortalization and increased cell turnover under growth-restricting conditions. Genes Dev 14:3051–3064

Dannenberg JH, Schuijff L, Dekker M, van der Valk M, te Riele H (2004) Tissue-specific tumor suppressor activity of retinoblastoma gene homologs p107 and p130. Genes Dev 18:2952–2962

de Bruin A, Wu L, Saavedra HI, Wilson P, Yang Y, Rosol TJ, Weinstein M, Robinson ML, Leone G (2003) Rb function in extraembryonic lineages suppresses apoptosis in the CNS of Rb-deficient mice. Proc Natl Acad Sci USA 100:6546–6551

DeCaprio JA, Ludlow JW, Figge J, Shew JY, Huang CM, Lee WH, Marsilio E, Paucha E, Livingston DM (1988) SV40 large tumor antigen forms a specific complex with the product of the retinoblastoma susceptibility gene. Cell 54:275–283

Deng C, Zhang P, Harper JW, Elledge SJ, Leder P (1995) Mice lacking p21$^{CIP1/WAF1}$ undergo normal development, but are defective in G1 checkpoint control. Cell 82:675–684

Dyer MA, Bremner R (2005) The search for the retinoblastoma cell of origin. Nat Rev Cancer 5:91–101

Dyson N, Howley PM, Munger K, Harlow E (1989) The human papilloma virus-16 E7 oncoprotein is able to bind to the retinoblastoma gene product. Science 243:934–937

Ephrussi B, Davidson RL, Weiss MC, Harris H, Klein G (1969) Malignancy of somatic cell hybrids. Nature 224:1314–1316

Ewart-Toland A, Briassouli P, de Koning JP, Mao JH, Yuan J, Chan F, MacCarthy-Morrogh L, Ponder BA, Nagase H, Burn J, Ball S, Almeida M, Linardopoulos S, Balmain A (2003) Identification of Stk6/STK15 as a candidate low-penetrance tumor-susceptibility gene in mouse and human. Nat Genet 34:403–412

Fantl V, Stamp G, Andrews A, Rosewell I, Dickson C (1995) Mice lacking cyclin D1 are small and show defects in eye and mammary gland development. Genes Dev 9:2364–2372

Ferguson KL, Vanderluit JL, Hebert JM, McIntosh WC, Tibbo E, MacLaurin JG, Park DS, Wallace VA, Vooijs M, McConnell SK, Slack RS (2002) Telencephalon-specific Rb knockouts reveal enhanced neurogenesis, survival and abnormal cortical development. Embo J 21:3337–3346

Fero ML, Rivkin M, Tasch M, Porter P, Carow CE, Firpo E, Polyak K, Tsai LH, Broudy V, Perlmutter RM, Kaushansky K, Roberts JM (1996) A syndrome of multiorgan hyperplasia with features of gigantism, tumorigenesis, and female sterility in p27^{Kip1}-deficient mice. Cell 85:733–744

Field SJ, Tsai FY, Kuo F, Zubiaga AM, Kaelin WG, Jr, Livingston DM, Orkin SH, Greenberg ME (1996) E2F-1 functions in mice to promote apoptosis and suppress proliferation. Cell 85:549–561

Gaubatz S, Lindeman GJ, Ishida S, Jakoi L, Nevins JR, Livingston DM, Rempel RE (2000) E2F4 and E2F5 play an essential role in pocket protein-mediated G1 control. Mol Cell 6:729–735

Geng Y, Whoriskey W, Park MY, Bronson RT, Medema RH, Li T, Weinberg RA, Sicinski P (1999) Rescue of cyclin D1 deficiency by knockin cyclin E. Cell 97:767–777

Geng Y, Yu Q, Sicinska E, Das M, Schneider JE, Bhattacharya S, Rideout WM, Bronson RT, Gardner H, Sicinski P (2003) Cyclin E ablation in the mouse. Cell 114:431–443

Gladden AB, Diehl JA (2003) Cell cycle progression without cyclin E/CDK2: breaking down the walls of dogma. Cancer Cell 4:160–162

Gu W, Schneider JW, Condorelli G, Kaushal S, Mahdavi V, Nadal-Ginard B (1993) Interaction of myogenic factors and the retinoblastoma protein mediates muscle cell commitment and differentiation. Cell 72:309–324

Harrison DJ, Hooper ML, Armstrong JF, Clarke AR (1995) Effects of heterozygosity for the Rb-1t19neo allele in the mouse. Oncogene 10:1615–1620

Hatada I, Ohashi H, Fukushima Y, Kaneko Y, Inoue M, Komoto Y, Okada A, Ohishi S, Nabetani A, Morisaki H, Nakayama M, Niikawa N, Mukai T (1996) An imprinted gene p57^{KIP2} is mutated in Beckwith–Wiedemann syndrome. Nat Genet 14:171–173

Hernando E, Nahle Z, Juan G, Diaz-Rodriguez E, Alaminos M, Hemann M, Michel L, Mittal V, Gerald W, Benezra R, Lowe SW, Cordon-Cardo C (2004) Rb inactivation promotes genomic instability by uncoupling cell cycle progression from mitotic control. Nature 430:797–802

Howes KA, Ransom N, Papermaster DS, Lasudry JG, Albert DM, Windle JJ (1994) Apoptosis or retinoblastoma: alternative fates of photoreceptors expressing the HPV-16 E7 gene in the presence or absence of p53. Genes Dev 8:1300–1310

Hu N, Gutsmann A, Herbert DC, Bradley A, Lee WH, Lee EY (1994) Heterozygous Rb-1 delta 20/+ mice are predisposed to tumors of the pituitary gland with a nearly complete penetrance. Oncogene 9:1021–1027

Huang HJ, Yee JK, Shew JY, Chen PL, Bookstein R, Friedmann T, Lee EY, Lee WH (1988) Suppression of the neoplastic phenotype by replacement of the RB gene in human cancer cells. Science 242:1563–1566

Humbert PO, Rogers C, Ganiatsas S, Landsberg RL, Trimarchi JM, Dandapani S, Brugnara C, Erdman S, Schrenzel M, Bronson RT, Lees JA (2000) E2F4 is essential for normal erythrocyte maturation and neonatal viability. Mol Cell 6:281–291

Humbert PO, Verona R, Trimarchi JM, Rogers C, Dandapani S, Lees JA (2000) E2f3 is critical for normal cellular proliferation. Genes Dev 14:690–703

Iavarone A, King ER, Dai XM, Leone G, Stanley ER, Lasorella A (2004) Retinoblastoma promotes definitive erythropoiesis by repressing Id2 in fetal liver macrophages. Nature 432:1040–1045

Iglesias A, Murga M, Laresgoiti U, Skoudy A, Bernales I, Fullaondo A, Moreno B, Lloreta J, Field SJ, Real FX, Zubiaga AM (2004) Diabetes and exocrine pancreatic insufficiency in E2F1/E2F2 double-mutant mice. J Clin Invest 113:1398–1407

Irwin M, Marin MC, Phillips AC, Seelan RS, Smith DI, Liu W, Flores ER, Tsai KY, Jacks T, Vousden KH, Kaelin WG Jr (2000) Role for the p53 homologue p73 in E2F-1-induced apoptosis. Nature 407:645–648

Ishida S, Huang E, Zuzan H, Spang R, Leone G, West M, Nevins JR (2001) Role for E2F in control of both DNA replication and mitotic functions as revealed from DNA microarray analysis. Mol Cell Biol 21:4684–4699

Jacks T, Fazeli A, Schmitt EM, Bronson RT, Goodell MA, Weinberg RA (1992) Effects of an Rb mutation in the mouse. Nature 359:295–300

Kalma Y, Marash L, Lamed Y, Ginsberg D (2001) Expression analysis using DNA microarrays demonstrates that E2F-1 up-regulates expression of DNA replication genes including replication protein A2. Oncogene 20:1379–1387

Kamb A, Gruis NA, Weaver-Feldhaus J, Liu Q, Harshman K, Tavtigian SV, Stockert E, Day RS 3rd, Johnson BE, Skolnick MH (1994) A cell cycle regulator potentially involved in genesis of many tumor types. Science 264:436–440

Kamijo T, Zindy F, Roussel MF, Quelle DE, Downing JR, Ashmun RA, Grosveld G, Sherr CJ (1997) Tumor suppression at the mouse INK4a locus mediated by the alternative reading frame product p19^{ARF}. Cell 91:649–659

Kaye FJ (2002) RB and cyclin dependent kinase pathways: defining a distinction between RB and p16 loss in lung cancer. Oncogene 21:6908–6914

Kiyokawa H, Kineman RD, Manova-Todorova KO, Soares VC, Hoffman ES, Ono M, Khanam D, Hayday AC, Frohman LA, Koff A (1996) Enhanced growth of mice lacking the cyclin-dependent kinase inhibitor function of p27^{Kip1}. Cell 85:721–732

Knudson AG Jr (1971) Mutation and cancer: statistical study of retinoblastoma. Proc Natl Acad Sci USA 68:820–823

Knudson AG (2001) Two genetic hits (more or less) to cancer. Nat Rev Cancer 1:157–162

Kohn MJ, Bronson RT, Harlow E, Dyson NJ, Yamasaki L (2003) Dp1 is required for extraembryonic development. Development 130:1295–1305

Kohn MJ, Leung SW, Criniti V, Agromayor M, Yamasaki L (2004) Dp1 is largely dispensable for embryonic development. Mol Cell Biol 24:7197–7205

Kozar K, Ciemerych MA, Rebel VI, Shigematsu H, Zagozdzon A, Sicinska E, Geng Y, Yu Q, Bhattacharya S, Bronson RT, Akashi K, Sicinski P (2004) Mouse development and cell proliferation in the absence of D-cyclins. Cell 118:477–491

Krimpenfort P, Quon KC, Mooi WJ, Loonstra A, Berns A (2001) Loss of p16^{Ink4a} confers susceptibility to metastatic melanoma in mice. Nature 413:83–86

Kristjansdottir K, Rudolph J (2004) Cdc25 phosphatases and cancer. Chem Biol 11:1043–1051

Lasorella A, Noseda M, Beyna M, Yokota Y, Iavarone A (2000) Id2 is a retinoblastoma protein target and mediates signalling by Myc oncoproteins. Nature 407:592–598

LeCouter JE, Kablar B, Hardy WR, Ying C, Megeney LA, May LL, Rudnicki MA (1998) Strain-dependent myeloid hyperplasia, growth deficiency, and accelerated cell cycle in mice lacking the Rb-related p107 gene. Mol Cell Biol 18:7455–7465

LeCouter JE, Kablar B, Whyte PF, Ying C, Rudnicki MA (1998) Strain-dependent embryonic lethality in mice lacking the retinoblastoma-related p130 gene. Development 125:4669–4679

Lee EY, Chang CY, Hu N, Wang YC, Lai CC, Herrup K, Lee WH, Bradley A (1992) Mice deficient for Rb are nonviable and show defects in neurogenesis and haematopoiesis. Nature 359:288–294

Lee EY, Cam H, Ziebold U, Rayman JB, Lees JA, Dynlacht BD (2002) E2F4 loss suppresses tumorigenesis in Rb mutant mice. Cancer Cell 2:463–472

Lee MH, Williams BO, Mulligan G, Mukai S, Bronson RT, Dyson N, Harlow E, Jacks T (1996) Targeted disruption of *p107*: functional overlap between *p107* and Rb. Genes Dev 10:1621–1632

Leung SW, Wloga EH, Castro AF, Nguyen T, Bronson RT, Yamasaki L (2004) A dynamic switch in Rb+/– mediated neuroendocrine Tumorigenesis. Oncogene 23:3296–3307

Li FX, Zhu JW, Hogan CJ, DeGregori J (2003) Defective gene expression, S phase progression, and maturation during hematopoiesis in E2F1/E2F2 mutant mice. Mol Cell Biol 23:3607–3622

Li FX, Zhu JW, Tessem JS, Beilke J, Varella-Garcia M, Jensen J, Hogan CJ, DeGregori J (2003) The development of diabetes in E2f1/E2f2 mutant mice reveals important roles for bone marrow-derived cells in preventing islet cell loss. Proc Natl Acad Sci USA 100:12935–12940

Lindeman GJ, Dagnino L, Gaubatz S, Xu Y, Bronson RT, Warren HB, Livingston DM (1998) A specific, nonproliferative role for E2F-5 in choroid plexus function revealed by gene targeting. Genes Dev 12:1092–1098

Lipinski MM, Macleod KF, Williams BO, Mullaney TL, Crowley D, Jacks T (2001) Cell-autonomous and non-cell-autonomous functions of the Rb tumor suppressor in developing central nervous system. Embo J 20:3402–3413

Lissy NA, Davis PK, Irwin M, Kaelin WG, Dowdy SF (2000) A common E2F-1 and p73 pathway mediates cell death induced by TCR activation. Nature 407:642–645

Ma Y, Croxton R, Moorer RL Jr, Cress WD (2002) Identification of novel E2F1-regulated genes by microarray. Arch Biochem Biophys 399:212–224

Maandag EC, van der Valk M, Vlaar M, Feltkamp C, O'Brien J, van Roon M, van der Lugt N, Berns A, te Riele H (1994) Developmental rescue of an embryonic-lethal mutation in the retinoblastoma gene in chimeric mice. Embo J 13:4260–4268

Macleod KF, Hu Y, Jacks T (1996) Loss of Rb activates both p53-dependent and independent cell death pathways in the developing mouse nervous system. Embo J 15:6178–6188

MacPherson D, Sage J, Crowley D, Trumpp A, Bronson RT, Jacks T (2003) Conditional mutation of Rb causes cell cycle defects without apoptosis in the central nervous system. Mol Cell Biol 23:1044–1053

MacPherson D, Sage J, Kim T, Ho D, McLaughlin ME, Jacks T (2004) Cell type-specific effects of Rb deletion in the murine retina. Genes Dev 18:1681–1694

Malumbres M, Sotillo R, Santamaria D, Galan J, Cerezo A, Ortega S, Dubus P, Barbacid M (2004) Mammalian cells cycle without the D-type cyclin-dependent kinases Cdk4 and Cdk6. Cell 118:493–504

Marino S, Vooijs M, van Der Gulden H, Jonkers J, Berns A (2000) Induction of medulloblastomas in p53-null mutant mice by somatic inactivation of Rb in the external granular layer cells of the cerebellum. Genes Dev 14:994–1004

Marino S, Hoogervoorst D, Brandner S, Berns A (2003) Rb and *p107* are required for normal cerebellar development and granule cell survival but not for Purkinje cell persistence. Development 130:3359–3368

Martinez-Balbas MA, Bauer UM, Nielsen SJ, Brehm A, Kouzarides T (2000) Regulation of E2F1 activity by acetylation. Embo J 19:662–671

Marzio G, Wagener C, Gutierrez MI, Cartwright P, Helin K, Giacca M (2000) *E2F* family members are differentially regulated by reversible acetylation. J Biol Chem 275:10887–10892

Minna JD, Gazdar AF, Sprang SR, Herz J (2004) Cancer. A bull's eye for targeted lung cancer therapy. Science 304:1458–1461

Morgan DO (1997) Cyclin-dependent kinases: engines, clocks, and microprocessors. Annu Rev Cell Dev Biol 13:261–291

Morgenbesser SD, Williams BO, Jacks T, DePinho RA (1994) p53-dependent apoptosis produced by Rb-deficiency in the developing mouse lens. Nature 371:72–74

Moroni MC, Hickman ES, Denchi EL, Caprara G, Colli E, Cecconi F, Muller H, Helin K (2001) Apaf-1 is a transcriptional target for E2F and p53. Nat Cell Biol 3:552–558

Morris EJ, Dyson NJ (2001) Retinoblastoma protein partners. Adv Cancer Res 82:1–54

Motokura T, Bloom T, Kim HG, Juppner H, Ruderman JV, Kronenberg HM, Arnold A (1991) A novel cyclin encoded by a bcl1-linked candidate oncogene. Nature 350:512–515

Murga M, Fernandez-Capetillo O, Field SJ, Moreno B, Borlado LR, Fujiwara Y, Balomenos D, Vicario A, Carrera AC, Orkin SH, Greenberg ME, Zubiaga AM (2001) Mutation of E2F2 in mice causes enhanced T lymphocyte proliferation, leading to the development of autoimmunity. Immunity 15:959–970

Murphy M, Stinnakre MG, Senamaud-Beaufort C, Winston NJ, Sweeney C, Kubelka M, Carrington M, Brechot C, Sobczak-Thepot J (1997) Delayed early embryonic lethality following disruption of the murine cyclin A2 gene. Nat Genet 15:83–86

Nahle Z, Polakoff J, Davuluri RV, McCurrach ME, Jacobson MD, Narita M, Zhang MQ, Lazebnik Y, Bar-Sagi D, Lowe SW (2002) Direct coupling of the cell cycle and cell death machinery by E2F. Nat Cell Biol 4:859–864

Nakayama K, Ishida N, Shirane M, Inomata A, Inoue T, Shishido N, Horii I, Loh DY (1996) Mice lacking p27^{Kip1} display increased body size, multiple organ hyperplasia, retinal dysplasia, and pituitary tumors. Cell 85:707–720

Nguyen DX, Baglia LA, Huang SM, Baker CM, McCance DJ (2004) Acetylation regulates the differentiation-specific functions of the retinoblastoma protein. Embo J 23:1609–1618

Nikitin AY, Juarez-Perez MI, Li S, Huang L, Lee WH (1999) RB-mediated suppression of spontaneous multiple neuroendocrine neoplasia and lung metastases in Rb+/− mice. Proc Natl Acad Sci USA 96:3916–3921

O'Keefe D, Dao D, Zhao L, Sanderson R, Warburton D, Weiss L, Anyane-Yeboa K, Tycko B (1997) Coding mutations in p57^{KIP2} are present in some cases of Beckwith-Wiedemann syndrome but are rare or absent in Wilms tumors. Am J Hum Genet 61:295–303

Ortega S, Prieto I, Odajima J, Martin A, Dubus P, Sotillo R, Barbero JL, Malumbres M, Barbacid M (2003) Cyclin-dependent kinase 2 is essential for meiosis but not for mitotic cell division in mice. Nat Genet 35:25–31

Pagano M, Benmaamar R (2003) When protein destruction runs amok, malignancy is on the loose. Cancer Cell 4:251–256

Pagano M, Jackson PK (2004) Wagging the dogma; tissue-specific cell cycle control in the mouse embryo. Cell 118:535–538

Palmero I, Peters G (1996) Perturbation of cell cycle regulators in human cancer. Cancer Surv 27:351–367

Parisi T, Beck AR, Rougier N, McNeil T, Lucian L, Werb Z, Amati B (2003) Cyclins E1 and E2 are required for endoreplication in placental trophoblast giant cells. Embo J 22:4794–4803

Park MS, Rosai J, Nguyen HT, Capodieci P, Cordon-Cardo C, Koff A (1999) p27 and Rb are on overlapping pathways suppressing tumorigenesis in mice. Proc Natl Acad Sci USA 96:6382–6387

Pediconi N, Ianari A, Costanzo A, Belloni L, Gallo R, Cimino L, Porcellini A, Screpanti I, Balsano C, Alesse E, Gulino A, Levrero M (2003) Differential regulation of E2F1 apoptotic target genes in response to DNA damage. Nat Cell Biol 5:552–558

Peeper DS, Dannenberg JH, Douma S, te Riele H, Bernards R (2001) Escape from premature senescence is not sufficient for oncogenic transformation by Ras. Nat Cell Biol 3:198–203

Pharoah PD, Antoniou A, Bobrow M, Zimmern RL, Easton DF, Ponder BA (2002) Polygenic susceptibility to breast cancer and implications for prevention. Nat Genet 31:33–36

Rane SG, Dubus P, Mettus RV, Galbreath EJ, Boden G, Reddy EP, Barbacid M (1999) Loss of Cdk4 expression causes insulin-deficient diabetes and Cdk4 activation results in beta-islet cell hyperplasia. Nat Genet 22:44–52

Reed SI (2003) Ratchets and clocks: the cell cycle, ubiquitylation and protein turnover. Nat Rev Mol Cell Biol 4:855–864

Rempel RE, Saenz-Robles MT, Storms R, Morham S, Ishida S, Engel A, Jakoi L, Melhem MF, Pipas JM, Smith C, Nevins JR (2000) Loss of E2F4 activity leads to abnormal development of multiple cellular lineages. Mol Cell 6:293–306

Ren B, Cam H, Takahashi Y, Volkert T, Terragni J, Young RA, Dynlacht BD (2002) E2F integrates cell cycle progression with DNA repair, replication, and G2/M checkpoints. Genes Dev 16:245–256

Robanus-Maandag E, Dekker M, van der Valk M, Carrozza ML, Jeanny JC, Dannenberg JH, Berns A, te Riele H (1998) *p107* is a suppressor of retinoblastoma development in pRb-deficient mice. Genes Dev 12:1599–1609

Roberts JM, Sherr CJ (2003) Bared essentials of CDK2 and cyclin E. Nat Genet 35:9–10

Sage J, Mulligan GJ, Attardi LD, Miller A, Chen S, Williams B, Theodorou E, Jacks T (2000) Targeted disruption of the three Rb-related genes leads to loss of G1 control and immortalization. Genes Dev 14:3037–3050

Sage J, Miller AL, Perez-Mancera PA, Wysocki JM, Jacks T (2003) Acute mutation of retinoblastoma gene function is sufficient for cell cycle re-entry. Nature 424:223–228

Sellers WR, Novitch BG, Miyake S, Heith A, Otterson GA, Kaye FJ, Lassar AB, Kaelin WG Jr (1998) Stable binding to *E2F* is not required for the retinoblastoma protein to activate transcription, promote differentiation, and suppress tumor cell growth. Genes Dev 12:95–106

Serrano M, Lee H, Chin L, Cordon-Cardo C, Beach D, DePinho RA (1996) Role of the INK4a locus in tumor suppression and cell mortality. Cell 85:27–37

Sharpless NE, Bardeesy N, Lee KH, Carrasco D, Castrillon DH, Aguirre AJ, Wu EA, Horner JW, DePinho RA (2001) Loss of p16^{Ink4a} with retention of p19^{Arf} predisposes mice to tumorigenesis. Nature 413:86–91

Sherr CJ (1996) Cancer cell cycles. Science 274:1672–1677

Sherr CJ (2001) The INK4a/ARF network in tumour suppression. Nat Rev Mol Cell Biol 2:731–737

Sherr CJ (2001) Parsing Ink4a/Arf: "pure" p16-null mice. Cell 106:531–534

Sherr CJ, Roberts JM (2004) Living with or without cyclins and cyclin-dependent kinases. Genes Dev 18:2699–2711

Sicinski P, Donaher JL, Parker SB, Li T, Fazeli A, Gardner H, Haslam SZ, Bronson RT, Elledge SJ, Weinberg RA (1995) Cyclin D1 provides a link between development and oncogenesis in the retina and breast. Cell 82:621–630

Sicinski P, Donaher JL, Geng Y, Parker SB, Gardner H, Park MY, Robker RL, Richards JS, McGinnis LK, Biggers JD, Eppig JJ, Bronson RT, Elledge SJ, Weinberg RA

(1996) Cyclin D2 is an FSH-responsive gene involved in gonadal cell proliferation and oncogenesis. Nature 384:470–474

Sicinska E, Aifantis I, Le Cam L, Swat W, Borowski C, Yu Q, Ferrando AA, Levin SD, Geng Y, von Boehmer H, Sicinski P (2003) Requirement for cyclin D3 in lymphocyte development and T cell leukemias. Cancer Cell 4:451–461

Spike BT, Dirlam A, Dibling BC, Marvin J, Williams BO, Jacks T, Macleod KF (2004) The Rb tumor suppressor is required for stress erythropoiesis. Embo J 23:4319–4329

Stevaux O, Dyson NJ (2002) A revised picture of the E2F transcriptional network and RB function. Curr Opin Cell Biol 14:684–691

Storre J, Elsasser HP, Fuchs M, Ullmann D, Livingston DM, Gaubatz S (2002) Homeotic transformations of the axial skeleton that accompany a targeted deletion of E2f6. EMBO Rep 3:695–700

Tetsu O, McCormick F (1999) Beta-catenin regulates expression of cyclin D1 in colon carcinoma cells. Nature 398:422–426

Thomas DM, Carty SA, Piscopo DM, Lee JS, Wang WF, Forrester WC, Hinds PW (2001) The retinoblastoma protein acts as a transcriptional coactivator required for osteogenic differentiation. Mol Cell 8:303–316

Trimarchi JM, Lees JA (2002) Sibling rivalry in the E2F family. Nat Rev Mol Cell Biol 3:11–20

Tsai KY, Hu Y, Macleod KF, Crowley D, Yamasaki L, Jacks T (1998) Mutation of E2f-1 suppresses apoptosis and inappropriate S phase entry and extends survival of Rb-deficient mouse embryos. Mol Cell 2:293–304

Tsai KY, MacPherson D, Rubinson DA, Crowley D, Jacks T (2002) ARF is not required for apoptosis in Rb mutant mouse embryos. Curr Biol 12:159–163

Tsai KY, MacPherson D, Rubinson DA, Nikitin AY, Bronson R, Mercer KL, Crowley D, Jacks T (2002) ARF mutation accelerates pituitary tumor development in Rb+/– mice. Proc Natl Acad Sci USA 99:16865–16870

Tsutsui T, Hesabi B, Moons DS, Pandolfi PP, Hansel KS, Koff A, Kiyokawa H (1999) Targeted disruption of CDK4 delays cell cycle entry with enhanced $p27^{Kip1}$ activity. Mol Cell Biol 19:7011–7019

Vooijs M, te Riele H, van der Valk M, Berns A (2002) Tumor formation in mice with somatic inactivation of the retinoblastoma gene in interphotoreceptor retinol binding protein-expressing cells. Oncogene 21:4635–4645

Wang J, Chenivesse X, Henglein B, Brechot C (1990) Hepatitis B virus integration in a cyclin A gene in a hepatocellular carcinoma. Nature 343:555–557

Weinmann AS, Yan PS, Oberley MJ, Huang TH, Farnham PJ (2002) Isolating human transcription factor targets by coupling chromatin immunoprecipitation and CpG island microarray analysis. Genes Dev 16:235–244

Wells J, Graveel CR, Bartley SM, Madore SJ, Farnham PJ (2002) The identification of E2F1-specific target genes. Proc Natl Acad Sci USA 99:3890–3895

Wikonkal NM, Remenyik E, Knezevic D, Zhang W, Liu M, Zhao H, Berton TR, Johnson DG, Brash DE (2003) Inactivating E2f1 reverts apoptosis resistance and cancer sensitivity in Trp53-deficient mice. Nat Cell Biol 5:655–660

Williams BO, Remington L, Albert DM, Mukai S, Bronson RT, Jacks T (1994) Cooperative tumorigenic effects of germline mutations in Rb and p53. Nat Genet 7:480–484

Williams BO, Schmitt EM, Remington L, Bronson RT, Albert DM, Weinberg RA, Jacks T (1994) Extensive contribution of Rb-deficient cells to adult chimeric mice with limited histopathological consequences. Embo J 13:4251–4259

Windle JJ, Albert DM, O'Brien JM, Marcus DM, Disteche CM, Bernards R, Mellon PL (1990) Retinoblastoma in transgenic mice. Nature 343:665–669

Wloga EH, Criniti V, Yamasaki L, Bronson RT (2004) Lymphomagenesis and female-specific lethality in p53-deficient mice occur independently of E2f1. Nat Cell Biol 6:565–567

Wu L, Timmers C, Maiti B, Saavedra HI, Sang L, Chong GT, Nuckolls F, Giangrande P, Wright FA, Field SJ, Greenberg ME, Orkin S, Nevins JR, Robinson ML, Leone G (2001) The E2F1-3 transcription factors are essential for cellular proliferation. Nature 414:457–462

Wu L, de Bruin A, Saavedra HI, Starovic M, Trimboli A, Yang Y, Opavska J, Wilson P, Thompson JC, Ostrowski MC, Rosol TJ, Woollett LA, Weinstein M, Cross JC, Robinson ML, Leone G (2003) Extra-embryonic function of Rb is essential for embryonic development and viability. Nature 421:942–947

Yamasaki L, Jacks T, Bronson R, Goillot E, Harlow E, Dyson NJ (1996) Tumor induction and tissue atrophy in mice lacking E2F-1. Cell 85:537–548

Yamasaki L, Bronson R, Williams BO, Dyson NJ, Harlow E, Jacks T (1998) Loss of E2F-1 reduces tumorigenesis and extends the lifespan of Rb1+/− mice. Nat Genet 18:360–364

Yan Y, Frisen J, Lee MH, Massague J, Barbacid M (1997) Ablation of the CDK inhibitor p57^{Kip2} results in increased apoptosis and delayed differentiation during mouse development. Genes Dev 11:973–983

Zachariae W, Nasmyth K (1999) Whose end is destruction: cell division and the anaphase-promoting complex. Genes Dev 13:2039–2058

Zacksenhaus E, Jiang Z, Chung D, Marth JD, Phillips RA, Gallie BL (1996) pRb controls proliferation, differentiation, and death of skeletal muscle cells and other lineages during embryogenesis. Genes Dev 10:3051–3064

Zhang P, Liegeois NJ, Wong C, Finegold M, Hou H, Thompson JC, Silverman A, Harper JW, DePinho RA, Elledge SJ (1997) Altered cell differentiation and proliferation in mice lacking p57^{KIP2} indicates a role in Beckwith-Wiedemann syndrome. Nature 387:151–158

Zhang J, Gray J, Wu L, Leone G, Rowan S, Cepko CL, Zhu X, Craft CM, Dyer MA (2004) Rb regulates proliferation and rod photoreceptor development in the mouse retina. Nat Genet 36:351–360

Zhang J, Schweers B, Dyer MA (2004) The first knockout mouse model of retinoblastoma. Cell Cycle 3:952–959

Zhu JW, Field SJ, Gore L, Thompson M, Yang H, Fujiwara Y, Cardiff RD, Greenberg M, Orkin SH, DeGregori J (2001) E2F1 and E2F2 determine thresholds for antigen-induced T-cell proliferation and suppress tumorigenesis. Mol Cell Biol 21:8547–8564

Ziebold U, Reza T, Caron A, Lees JA (2001) E2F3 contributes both to the inappropriate proliferation and to the apoptosis arising in Rb mutant embryos. Genes Dev 15:386–391

Ziebold U, Lee EY, Bronson RT, Lees JA (2003) E2F3 loss has opposing effects on different pRB-deficient tumors, resulting in suppression of pituitary tumors but metastasis of medullary thyroid carcinomas. Mol Cell Biol 23:6542–6552

Zuo L, Weger J, Yang Q, Goldstein AM, Tucker MA, Walker GJ, Hayward N, Dracopoli NC (1996) Germline mutations in the p16^{INK4a} binding domain of CDK4 in familial melanoma. Nat Genet 12:97–99

zur Hausen H (2002) Papillomaviruses and cancer: from basic studies to clinical application. Nat Rev Cancer 2:342–350

Senescence and Cell Cycle Control

Hiroaki Kiyokawa

Department of Molecular Pharmacology and Biological Chemistry,
Feinberg School of Medicine, Northwestern University, 303 E. Chicago Avenue,
Chicago, IL 60611, USA
kiyokawa@northwestern.edu

Abstract In response to various stresses, such as telomere shortening during continuous proliferation, oxidative stress, DNA damage and aberrant oncogene activation, normal cells undergo cellular senescence, which is a stable postmitotic state with particular morphology and metabolism. Signaling that induces senescence involves two major tumor suppressor cascades, i.e., the INK4a-Rb pathway and the ARF-p53 pathway. Diverse stimuli upregulate these interacting pathways, which orchestrate exit from the cell cycle. Recent studies have provided insights into substantial differences in senescence-inducing signals in primary cells of human and rodent origins. This review is focused on recent advances in understanding the roles of the tumor-suppressive pathways in senescence.

1
Senescence

Senescence was originally defined as an "irreversible" state of cell cycle arrest that reflects consumed proliferative capacity of the cell (Hayflick and Moorhead 1961). In eukaryotic cells each chromosome shortens from telomeres during every round of DNA replication (Smogorzewska and de Lange 2004; Campisi 2001). The structure of telomeres with repetitive sequences functions as a cap to prevent chromosome end fusions and genomic instability (de Lange 1998; Sharpless and DePinho 2004). While germ cells express telomerase, which resynthesizes the telomeric repeats to maintain the chromosomal length, most human somatic cells do not express telomerase. In somatic cells proliferating continuously, attrition of telomeres beyond a threshold triggers a response leading to "replicative" senescence. Recent studies have indicated that telomere attrition provokes DNA damage-responsive signaling pathways (d'Adda et al. 2003; Gire et al. 2004; Herbig et al. 2004). In addition, senescent cells exhibit a particular flat morphology with enlarged cytoplasm, and also express particular biochemical markers, such as senescence associated β-galactosidase activity (Dimri et al. 1995). "Premature" senescence or stasis could be triggered in cells without telomere attrition by ectopic oncogene activation, DNA damage, oxidative stress and other stressful conditions (Serrano et al. 1997; Chang et al. 2002; Chen and Ames 1994). These two forms of senescence are morphologically indistinguishable, and are likely to depend on common signaling pathways leading to cell cycle arrest. The pathways that play key roles in senescence induction, i.e., the ARF-p53-p21^{Cip1} pathway and the p16^{INK4a}-Rb pathway, are major tumor-suppressor cascades (Sherr and DePinho 2000). Thus, it has been postulated that senescence is a potent tumor-suppressive mechanism, like programmed cell death or apoptosis (Lowe and Sherr 2003). While there have been interesting discussions on whether cellular senescence is involved in organismal aging (Pelicci 2004), this review focuses on the tumor-suppressor pathways and cell cycle control during cellular senescence.

2
Role of the p53 Pathway in Senescence

The tumor-suppressor p53 plays a key role in induction of senescence as evidenced by studies using mouse embryonic fibroblasts (MEFs) from p53-deficient (knockout) mice (Livingstone et al. 1992; Lowe et al. 1994). Under standard culture conditions, MEFs from wild-type embryos proliferate up to 15–25 population doublings, followed by induction of senescence. In contrast, p53-null MEFs continue to proliferate without obvious cell cycle arrest or senescence-like morphology. The same immortal phenotype is seen in wild-type MEFs expressing dominant-negative p53 mutants, short interfering RNA (siRNA) against p53, or the papilloma virus oncoprotein E6, which facilitates p53 degradation (Munger and Howley 2002). Thus, loss of p53 function is sufficient to abrogate the senescence checkpoint in MEFs (Sharpless and DePinho 2002). The immortalization step dependent on p53 perturbation renders mouse cells susceptible for activated Ras-induced malignant transformation. In contrast, Ras activation in wild-type MEFs induces premature senescence, as already described (Serrano et al. 1997). Although the role of Ras in senescece induction remains to be fully understood, Ras activates the external transcribed spacer transcription factors (ets) through the mitogen-activated protein (MAP) kinase pathway, which may upregulate p16^{INK4a} (see Sect. 4 for details). Furthermore, Ras activation leads to accumulation of intracellular reactive oxygen species [ROS] (see Sect. 5 for details). These cellular changes are apparently involved in Ras-induced senescence response. Human diploid fibroblasts (HDFs) taken from p53-heterozygous patients with Li–Fraumeni syndrome show prolonged replicative life spans in culture, associated with loss of heterozygosity (Boyle et al. 1998). Studies using HDFs, keratinocytes, and mammary epithelial cells suggest that loss of p53 is not sufficient for cooperating with Ras activation to transform human cells (Drayton and Peters 2002). It has been described that a combination of SV40 large T-antigen (T-Ag), activated Ras and the telomerase catalytic subunit (hTERT) can transform HDFs and human keratinocytes (Hahn et al. 1999). T-Ag inactivates the p53- and Rb-dependent senescence-inducing pathways and hTERT eliminates telomere-mediated signaling, allowing cells to undergo transformation in response to Ras activation. Interestingly, human mammary epithelial cells seem to have higher requirements for transformation, being insensitive to the combinatory treatment (Hahn et al. 1999). These data suggest substantial diversity of senescence control in different types of cells and in different species.

Activation of p53 could result in one of the two cell fates, senescence or apoptosis. Loss of p53 function could contribute to immortalization and enhanced survival, both of which are hallmarks of cancer cells. Thus, p53 is a multifuctional tumor suppressor, playing a central role in preventing malignant transformation. While the proapoptotic function of p53 involves a number of genes, the only known mediator of its prosenescent function is

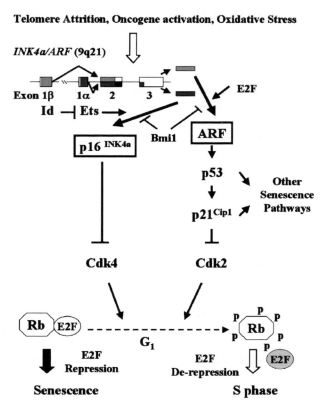

Fig. 1 The p16^{INK4a}-Rb and ARF-p53 tumor-suppressor pathways control senescence. The INK4a/ARF locus on human chromosome 9q21 encodes two tumor-suppressor proteins, p16^{INK4a} and p14^{Arf} (p19^{ARF} in mice). While p16^{INK4a} directly inhibits Cdk4, p14^{Arf} induces p21^{Cip1}, which inhibits Cdk2. Cdk inhibition results in repression of the E2F target genes via reduced phosphorylation of the retinoblastoma (*Rb*) family proteins

the Cdk inhibitor p21^{Cip1} (Fig. 1) (Sharpless and DePinho 2002; Vousden and Prives 2005; Xiong et al. 1993; Harper et al. 1993). In an early study, p21^{Cip1} was isolated as a senescence-related gene (Noda et al. 1994). During induction of senescence, p53 transactivates the p21^{Cip1} gene (el Deiry et al. 1993), and p21^{Cip1} protein binds to and inhibits G1-regulatory cyclin-dependent kinases (Cdk), especially Cdk2 in complex with cyclins E and A (Sherr and Roberts 1999). In addition to Cdk inhibition, p21^{Cip1} appears to affect expression and functions of various proteins, which may lead cells to senescence in an orchestrated manner (Roninson 2002). Importantly, forced expression of p21^{Cip1} results in cellular accumulation of ROS with undefined mechanisms (Macip et al. 2002). ROS could cause DNA damage, which then activates p53-dependent pathways, possibly forming a positive feedback. Disruption of p21^{Cip1} in HDFs prolongs the replicative life span in culture (Brown et al. 1997). These observations suggest a central role for p21^{Cip1} in senescence. However, studies using knockout mice provide a conflicting view. p21^{Cip1}-null MEFs undergo senescence normally (Pantoja and Serrano 1999), and p21^{Cip1}-null mice display only limited susceptibility to spontaneous tumorigenesis (Martin-Caballero et al. 2001). Thus, p21^{Cip1} may not be essential for senescence of mouse cells. An alternative possibility is that p21^{Cip1}-null mice undergo developmental adaptation to the absence of p21$^{Cip1/Waf1}$, for which other Cdk inhibitors and possibly p130 (Coats et al. 1999) may compensate. Studies using acute disruption of p21^{Cip1} by conditional gene targeting will be informative for better understanding of the role of p21^{Cip1} in senescence of mouse cells.

3
Role of the Rb Pathway in Senescence

The Rb-family pocket binding proteins, i.e., Rb, p107 and p130, also play critical roles in cell fate determination between senescence and immortalization. These proteins bind to the E2F family transcription factors and maintain the repressor function of E2F (Hatakeyama and Weinberg 1995; Stevaux and Dyson 2002). Phosphorylation of the pocket binding proteins by Cdk complexes, such as cyclin D/Cdk4 (or Cdk6) and cyclin E/Cdk2, results in dissociation of the proteins from E2F complexes, and is thought to mediate derepression or transactivation of E2F target genes (Fig. 1). Senescence of MEFs induced by p53 overexpression depends on the repressor activity of E2F (Rowland et al. 2002), suggesting that the senescence-inducing signal from p53 converges to the Rb/E2F pathway. This signaling crosstalk could result from p21^{Cip1} inhibition of Cdk2 and possibly Cdk4/6. Inactivation of these pocket binding proteins by the papillomavirus E7 oncoprotein, together with telomerase activation by hTERT expression, has been shown to immortalize primary human epithelial cells (Kiyono et al. 1998), although immortalization of this system may involve additional mutations. In MEFs, disruption of Rb, p107 and p130 results in increased proliferation with a shortened G1 phase and immortalization (Dannenberg et al. 2000), whereas MEFs with any one of the three proteins still exhibit senescence in culture (Sage et al. 2000). These observations suggest that the pocket binding proteins have overlapping functions in controlling senescence. However, acute disruption of Rb by the Cre-Loxp recombination system has been demonstrated to immortalize MEFs (Sage et al. 2003). This apparent discrepancy suggests that germline disruption of one or two pocket binding proteins leads to developmental adaptation, which helps cells retain the senescence checkpoint by a compensatory mechanism. For instance, p107 expression is upregulated in Rb-null MEFs. Interestingly, the same study showed that Rb disruption induced cell cycle reentry in a small fraction of apparently senescent cells, suggesting that Rb plays a key role in maintenance of the postmitotic status in senescent cells. Indeed, Rb, but not p107 or p130, is found in the senescence-associated heterochromatic foci (SAHF) (Narita et al. 2003), which may play a critical role in long-term transcriptional repression specific in senescent cells. Interestingly, similar cell cycle reentry from senescence has been described in MEFs infected with lentivirus for anti-p53 short hairpin RNA (Dirac and Bernards 2003). While these studies intriguingly suggest that senescence may not necessarily be an irreversible process, this notion awaits further investigations on regulation of SAHF and other characteristics of senescence.

Immortalization requires aberrant activation of the cell cycle machinery. Cdk4 activation plays a key role when cells overcome the senescence checkpoint, presumably via phosphorylation of the pocket binding proteins. MEFs with targeted $Cdk4^{R24C}$ mutation, which express a constitutively active Cdk4 protein insensitive to the INK4 inhibitors, exhibit an immortal phenotype in culture (Rane et al. 2002). Activated Ras is sufficient to transform $Cdk4^{R24C}$ MEFs, and mice with the $Cdk4^{R24C}$ mutation spontaneously develop various tumors, such as endocrine and skin tumors (Sotillo et al. 2001; Rane et al. 2002). In contrast, MEFs from Cdk4-null mice are resistant to immortalization induced by a dominant negative p53 mutant (DNp53) or disruption of the INK4a/ARF locus (Zou et al. 2002). Cdk4-null MEFs undergo transformation poorly in response to Ras plus DNp53 or Myc. Consistent with the resistance to transformation, Cdk4-null mice are refractory to skin carcinogenesis in response to the *keratin-5-Myc* transgene or the tumor initiator 7,12-dimethylbenz[4a4]anthracene plus the tumor promoter 12-O-tetradecanoylphorbol-13-acetate (Rodriguez-Puebla et al. 2002; Miliani de Marval et al. 2004). Genetic alterations that activate Cdk4, such as overexpression of Cdk4 or D-type cyclins and deletion of the Cdk4 inhibitor p16^{INK4a}, are observed in the majority of human cancers (Ortega et al. 2002). Thus, derepression of E2F target genes as a consequence of Cdk4 activation seems to be required for cellular immortalization. While

Cdk2- and Cdk6-null mice have been generated (Berthet et al. 2003; Malumbres et al. 2004; Ortega et al. 2003), it has not been determined yet whether Cdk6 or Cdk2 plays an indispensable role in immortalization, similarly to Cdk4.

4
The Role of the INK4A/ARF Locus in Senescence

The INK4A/ARF locus was originally identified as a major tumor-suppressor locus *MTS1* (multiple tumor suppressor-1) on human chromosome 9q21 (Kamb et al. 1994). This locus is very frequently deleted in a variety of human cancers. Importantly, the locus encodes two proteins, p16^{INK4a} and p14^{ARF} (or p19^{ARF} in the case of mice), which cooperate for induction of senescence (Fig. 1) (Quelle et al. 1995; Carnero et al. 2000; Sherr and DePinho 2000). While the coding sequences of these proteins partly overlap owing to the use of alternative reading frames, each product has its unique promoter and first exon. p16^{INK4a} is an specific inhibitor of Cdk4 and Cdk6 (Serrano et al. 1993). p14^{ARF} stabilizes p53 by interfering with ubiquitin-dependent degradation by the ring-finger protein MDM2 (Pomerantz et al. 1998; Stott et al. 1998; Zhang et al. 1998). p53 stabilization leads to induction of p21^{Cip1} expression. Therefore, p16^{INK4a} and p14^{ARF} are positive regulators of the Rb- and p53-dependent pathways, respectively. Both proteins are upregulated during the senescence process. A variety of stimuli could be involved in control of the p16^{INK4a} and p14^{ARF} promoters, although the regulation of the INK4a/ARF locus is yet to be completely elucidated. The expression of activated oncogenes, such as Ras (Serrano et al. 1997), Raf (Zhu et al. 1998) or MEK (Lin et al. 1998), upregulates the p16^{INK4a} promoter. It has been described that the transactivation involves the Ets family transcription factors (Ohtani et al. 2001). Ets-1 can directly transactivate p16^{INK4a} and induce senescence, and this regulation is abrogated by Id-1 as a specific inhibitor. Bmi-1, a transcriptional repressor in the polycomb group protein family, inhibits the senescence-associated induction of p16^{INK4a} (Jacobs et al. 1999). The polycomb proteins and the antagonizing trithorax group proteins are involved in transcriptional control during development (Park et al. 2004). In particular, Bmi-1 has been shown to be a critical factor for maintenance of the proliferative capacity in stem cells. Overexpression of Bmi-1 abrogates the induction of both p16^{INK4a} and p19^{ARF} in MEFs, slowing down the senescence process. In HDFs, Bmi-1 expression represses p16^{INK4a} and slows down senescence until telomere erosion causes crisis. Expression of a dominant negative mutant of Bmi-1, which lacks the ring-finger domain, induces p16^{INK4a} expression in HDFs, accelerating replicative senescence (Itahana et al. 2003). Another polycomb protein, Cbx7, can also downregulate the INK4a/ARF locus (Gil et al. 2004). Thus, Bmi-1 plays a role in the timing of senescence induction by controlling p16^{INK4a} (and p19^{ARF} in MEFs), while the detail of the promoter regulation requires further investigations. In mouse cells, p19^{ARF} expression is increased by senescence-inducing signals such as oncogene activation or cellular stresses. The ARF promoter contains E2F-binding sites, and can be activated by overexpression of E2F-1 or c-myc (DeGregori et al. 1997; Bates et al. 1998; Zindy et al. 1998; Dimri et al. 2000). Repression of the ARF promoter by E2F, especially E2F-3, suggests that the Rb pathway could function upstream for the p53 pathway, as a feedback or crosstalk mechanism that controls immortalization vs. senescence.

In contrast to p19^{ARF} in MEFs, p14^{ARF} is minimally upregulated when HDFs undergo replicative or premature senescence (Bates et al. 1998; Palmero et al. 1998; Ferbeyre et al. 2000; Wei et al. 2001). These data imply that mouse cells depend more on p19^{ARF} for senescence, compared with human cells. This hypothesis is consistent with observations that MEFs lacking p19^{ARF} with intact p16^{INK4a} expression are immortal in culture (Kamijo et al. 1997), so are MEFs lacking both p19^{ARF} and p16^{INK4a} (Serrano et al. 1996). On the other hand, MEFs lacking p16^{INK4a} with intact p19^{ARF} undergo senescence apparently normally (Sharpless et al. 2001; Krimpenfort et al. 2001). These studies suggest that p19^{ARF} is essential and p16^{INK4a} is dispensable for senescence of primary mouse cells (Krimpenfort et al. 2001). However, it is noteworthy that an antisense RNA construct against p16^{INK4a} can extend proliferative life span in primary wild-type MEFs (Carnero et al. 2000). This may suggest that acute loss of p16^{INK4a} in MEFs impacts on senescence control more significantly than germline knockout of the gene does. In contrast to p16^{INK4a}-null MEFs, HDFs from patients with mutations of the INK4A/ARF locus that disrupt only p16^{INK4a} expression show a prolonged proliferative life span in culture, followed by arrest between the senescence (M1) and crisis (M2) checkpoints (Brookes et al. 2002, 2004). Therefore, germline disruption of p16^{INK4a} in human cells appears to affect senescence more significantly than in mouse cells.

5
Mouse Cells vs. Human Cells: Roles of Reactive Oxygen Species and Telomere Attrition

Oxidative stress triggered by ROS plays a major role in cellular senescence, as well as organismal aging (Itahana et al. 2004). Intracellular accumulation of ROS, for instance, by treatment with H_2O_2 induces premature senescence. In contrast, culturing cells under 3–5% oxygen reduces ROS levels and prolongs replicative life span in HDFs. This hypoxic condition is closer to physiological conditions cells in vivo are exposed to than in atmospheric oxygen (approximately 20%). It is noteworthy that MEFs do not undergo replicative senescence under 3% oxygen (Parrinello et al. 2003). MEFs cultured in atmospheric oxygen exhibit characteristics of oxidative damage to DNA, which is less obvious in HDFs. These observations suggest that mouse cells are more sensitive to oxidative stress than human cells. In other words, senescence of mouse cells induced by continuous culture largely depends on oxidative stress-responsive pathways (Fig. 2a). Ras activation is known to result in ROS accumulation. ROS could activate p53 function in a DNA damage-dependent manner, through activation of ataxia-telangiectasia mutated (ATM)/ATM and Rad3-related (ATR) kinases. ROS can also upregulate p19^{ARF} effectively in primary mouse cells. Furthermore, the stress-responsive MAP kinase p38^{MAPK} may play a key role in mediating ROS signals to induce senescence, especially by upregulating p16^{INK4a} (Iwasa et al. 2003). In addition, a protein named seladin-1 was isolated in a genetic screen for regulators of Ras-induced premature senescence in rat embryonic fibroblasts (Wu et al. 2004). Seladin-1 may function as an ROS effector that facilitates p53 stabilization and consequently p21^{Cip1} induction.

In contrast to the ROS-dependent senescence of mouse cells, human cells undergo replicative senescence largely in a manner dependent on telomere shortening (Fig. 2b). Telomere attrition after every round of DNA replication could be an intrinsic mechanism that counts cumulative numbers of cell division. When telomeres get shorter than a threshold level, DNA damage-responsive pathways are activated to induce senescence. A recent report demonstrated that telomere shortening triggers senescence in human fibroblasts through the ATM-dependent DNA damage-responsive pathway leading to activation of p53 and induction of p21^{Cip1} (Herbig et al. 2004).

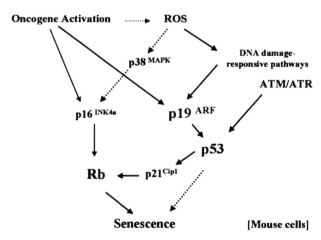

Fig. 2 a Senescence-inducing pathways in primary mouse cells. Reactive oxygen species (*ROS*) play a critical role in induction of senescence in cultured mouse cells. p19^{ARF} is predominantly upregulated during senescence by oncogene activation and/or oxidative stress

In human cells undergoing replicative senescence, p16^{INK4a} could be induced independently of telomeres or DNA damage. Nonetheless, a recent study provided evidence that p16^{INK4a} significantly contributes to p53-independent response to telomere attrition in HDFs (Jacobs and de Lange 2004). While the mechanism of p16^{INK4a} upregulation in telomere-mediated senescence remains to be clarified, p16^{INK4a} levels could affect cellular sensitivity to premature or replicative senescence by somehow affecting oncogene- and damage-responsive pathways. HDFs with high p16^{INK4a} levels undergo premature senescence in response to activated Ras, whereas HDFs with low p16^{INK4a} levels do not show Ras-induced senescence (Benanti and Galloway 2004). Compared with telomeres in human cells (5–15 kb), telomeres in mouse cells (40–60 kb) are remarkably longer, which may be associated with detectable levels of telomerase in many somatic cell types in mice. Therefore, telomere attrition is unlikely to function as a rate-limiting factor for induction of senescence in mouse cells under normal conditions. Studies using knockout mice deficient for the telomerase RNA component (mTR) showed that mTR−/− mice develop significant telomere attrition and chromosomal instability only after five to six generations of breeding (Rudolph et al. 1999). Interestingly, mTR−/− mice at the fifth generation show resistance to carcinogen-induced skin tumorigenesis in a p53-dependent manner (Gonzalez-Suarez et al. 2002), suggesting the critical role of the p53 pathway in senescence in this engineered mouse model for telomere shortening.

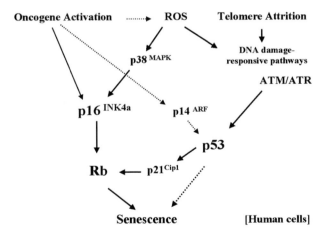

Fig. 2 b Senescence-inducing pathways in primary human cells. Unlike mouse cells, telomere attrition during each round of DNA replication plays a key role in triggering senecence-induction pathways in human cells. Premature senescence induced by oncogene activation involves oxidative stress with ROS accumulation. While the role of p16^{INK4a} in human cell senescence is well established, the role of p14^{Arf} in human cells seems more complex

6
Conclusions

Cellular senescence is an orchestrated program in response to a cue of various stresses. Senescence is also an intrinsic tumor-suppressor mechanism, forming a checkpoint barrier against malignant transformation. Diverse senescence-inducing signals converge to the $p16^{INK4a}$-Rb and ARF-p53 tumor-suppressor pathways. A number of studies using genetically engineered mouse and human cells have revealed how these pathways interact with each other to execute the senescence program. p53 plays a key role in mediating DNA damage-responsive signals elicited by telomere attrition or ROS. During the senescence program, the E2F repressive action of Rb could be sustained through Cdk inhibition as a consequence of p53-dependent upregulation of $p21^{Cip1}$, as well as independent upregulation of $p16^{INK4a}$. However, fundamental differences in senescence between mouse and human models exist. ARF plays a more prominent role in mouse cells than in human cells. In contrast, senescence-associated $p16^{INK4a}$ upregulation is generally more robust in human models than in mouse models. This is also the case for the tumor-suppressor role of $p16^{INK4a}$. Further investigations are needed for better understanding the role of senescence in tumor suppression and possibly in age-related pathological changes.

Acknowledgements I apologize to many colleagues for being unable to cite their papers critical for the field. I thank Nissim Hay, Rob Costa, Pradip Raychaudhuri, Oscar Colamonici, David Ucker and Xianghong Zou for helpful discussions, and the National Institutes of Health, the Department of Defense and the American Cancer Society for grant support for my research.

References

Bates S, Phillips AC, Clark PA, Stott F, Peters G, Ludwig RL, Vousden KH (1998) $p14^{ARF}$ links the tumour suppressors RB and p53. Nature 395:124–125

Benanti JA, Galloway DA (2004) Normal human fibroblasts are resistant to RAS-induced senescence. Mol Cell Biol 24:2842–2852

Berthet C, Aleem E, Coppola V, Tessarollo L, Kaldis P (2003) Cdk2 knockout mice are viable. Curr Biol 13:1775–1785

Boyle JM, Mitchell EL, Greaves MJ, Roberts SA, Tricker K, Burt E, Varley JM, Birch JM, Scott D (1998) Chromosome instability is a predominant trait of fibroblasts from Li-Fraumeni families. Br J Cancer 77:2181–2192

Brookes S, Rowe J, Ruas M, Llanos S, Clark PA, Lomax M, James MC, Vatcheva R, Bates S, Vousden KH, Parry D, Gruis N, Smit N, Bergman W, Peters G (2002) INK4a-deficient human diploid fibroblasts are resistant to RAS-induced senescence. EMBO J 21:2936–2945

Brookes S, Rowe J, Gutierrez DA, Bond J, Peters G (2004) Contribution of $p16^{INK4a}$ to replicative senescence of human fibroblasts. Exp Cell Res 298:549–559

Brown JP, Wei W, Sedivy JM (1997) Bypass of senescence after disruption of $p21^{CIP1/WAF1}$ gene in normal diploid human fibroblasts. Science 277:831–834

Campisi J (2001) Cellular senescence as a tumor-suppressor mechanism. Trends Cell Biol 11:S27–S31

Carnero A, Hudson JD, Price CM, Beach DH (2000) p16^{INK4a} and p19^{ARF} act in overlapping pathways in cellular immortalization. Nat Cell Biol 2:148–155

Chang BD, Swift ME, Shen M, Fang J, Broude EV, Roninson IB (2002) Molecular determinants of terminal growth arrest induced in tumor cells by a chemotherapeutic agent. Proc Natl Acad Sci USA 99:389–394

Chen Q, Ames BN (1994) Senescence-like growth arrest induced by hydrogen peroxide in human diploid fibroblast F65 cells. Proc Natl Acad Sci USA 91:4130–4134

Coats S, Whyte P, Fero ML, Lacy S, Chung G, Randel E, Firpo E, Roberts JM (1999) A new pathway for mitogen-dependent cdk2 regulation uncovered in p27^{Kip1}-deficient cells. Curr Biol 9:163–173

d'Adda dF, Reaper PM, Clay-Farrace L, Fiegler H, Carr P, Von Zglinicki T, Saretzki G, Carter NP, Jackson SP (2003) A DNA damage checkpoint response in telomere-initiated senescence. Nature 426:194–198

Dannenberg JH, van Rossum A, Schuijff L, te Riele H (2000) Ablation of the retinoblastoma gene family deregulates G1 control causing immortalization and increased cell turnover under growth-restricting conditions. Genes Dev 14:3051–3064

de Lange T (1998) Telomeres and senescence: ending the debate. Science 279:334–335

De Gregori J, Leone G, Miron A, Jakoi L, Nevins JR (1997) Distinct roles for E2F proteins in cell growth control and apoptosis. Proc Natl Acad Sci USA 94:7245–7250

Dimri GP, Lee X, Basile G, Acosta M, Scott G, Roskelley C, Medrano EE, Linskens M, Rubelj I, Pereira-Smith O, Peacocke M, Campisi J (1995) A biomarker that identifies senescent human cells in culture and in aging skin in vivo. Proc Natl Acad Sci USA 92:9363–9367

Dimri GP, Itahana K, Acosta M, Campisi J (2000) Regulation of a senescence checkpoint response by the E2F1 transcription factor and p14^{ARF} tumor suppressor. Mol Cell Biol 20:273–285

Dirac AM, Bernards R (2003) Reversal of senescence in mouse fibroblasts through lentiviral suppression of p53. J Biol Chem 278:11731–11734

Drayton S, Peters G (2002) Immortalisation and transformation revisited. Curr Opin Genet Dev 12:98–104

El-Deiry WS, Tokino T, Velculescu VE, Levy DB, Parsons R, Trent JM, Lin D, Mercer WE, Kinzler KW, Vogelstein B (1993) WAF1, a potential mediator of p53 tumor suppression. Cell 75:817–825

Ferbeyre G, de Stanchina E, Querido E, Baptiste N, Prives C, Lowe SW (2000) PML is induced by oncogenic ras and promotes premature senescence. Genes Dev 14:2015–2027

Gil J, Bernard D, Martinez D, Beach D (2004) Polycomb CBX7 has a unifying role in cellular lifespan. Nat Cell Biol 6:67–72

Gire V, Roux P, Wynford-Thomas D, Brondello JM, Dulic V (2004) DNA damage checkpoint kinase Chk2 triggers replicative senescence. EMBO J 23:2554–2563

Gonzalez-Suarez E, Flores JM, Blasco MA (2002) Cooperation between p53 mutation and high telomerase transgenic expression in spontaneous cancer development. Mol Cell Biol 22:7291–7301

Hahn WC, Stewart SA, Brooks MW, York SG, Eaton E, Kurachi A, Beijersbergen RL, Knoll JH, Meyerson M, Weinberg RA (1999) Inhibition of telomerase limits the growth of human cancer cells. Nat Med 5:1164–1170

Harper JW, Adami GR, Wei N, Keyomarsi K, Elledge SJ (1993) The p21 Cdk-interacting protein Cip1 is a potent inhibitor of G1 cyclin-dependent kinases. Cell 75:805–816

Hatakeyama M, Weinberg RA (1995) The role of Rb in cell cycle control. Prog Cell Cycle Res 1:9–19

Hayflick L, Moorhead PS (1961) The serial cultivation of human diploid cell strains. Exp Cell Res 25:585–621

Herbig U, Jobling WA, Chen BP, Chen DJ, Sedivy JM (2004) Telomere shortening triggers senescence of human cells through a pathway involving ATM, p53, and p21^{CIP1}, but not p16^{INK4a}. Mol Cell 14:501–513

Itahana K, Zou Y, Itahana Y, Martinez JL, Beausejour C, Jacobs JJ, Van Lohuizen M, Band V, Campisi J, Dimri GP (2003) Control of the replicative life span of human fibroblasts by p16 and the polycomb protein Bmi-1. Mol Cell Biol 23:389–401

Itahana K, Campisi J, Dimri GP (2004) Mechanisms of cellular senescence in human and mouse cells. Biogerontology 5:1–10

Iwasa H, Han J, Ishikawa F (2003) Mitogen-activated protein kinase p38 defines the common senescence-signalling pathway. Genes Cells 8:131–144

Jacobs JJ, de Lange T (2004) Significant role for p16^{INK4a} in p53-independent telomere-directed senescence. Curr Biol 14:2302–2308

Jacobs JJ, Kieboom K, Marino S, DePinho RA, Van Lohuizen M (1999) The oncogene and Polycomb-group gene bmi-1 regulates cell proliferation and senescence through the INK4a locus. Nature 397:164–168

Kamb A, Gruis NA, Weaver-Feldhaus J, Liu Q, Harshman K, Tavtigian SV, Stockert E, Day RS III, Johnson BE, Skolnick MH (1994) A cell cycle regulator potentially involved in genesis of many tumor types. Science 264:436–440

Kamijo T, Zindy F, Roussel MF, Quelle DE, Downing JR, Ashmun RA, Grosveld G, Sherr CJ (1997) Tumor suppression at the mouse INK4a locus mediated by the alternative reading frame product p19^{ARF}. Cell 91:649–659

Kiyono T, Foster SA, Koop JI, McDougall JK, Galloway DA, Klingelhutz AJ (1998) Both Rb/p16^{INK4a} inactivation and telomerase activity are required to immortalize human epithelial cells. Nature 396:84–88

Krimpenfort P, Quon KC, Mooi WJ, Loonstra A, Berns A (2001) Loss of p16^{INK4a} confers susceptibility to metastatic melanoma in mice. Nature 413:83–86

Lin AW, Barradas M, Stone JC, van Aelst L, Serrano M, Lowe SW (1998) Premature senescence involving p53 and p16 is activated in response to constitutive MEK/MAPK mitogenic signaling. Genes Dev 12:3008–3019

Livingstone LR, White A, Sprouse J, Livanos E, Jacks T, Tlsty TD (1992) Altered cell cycle arrest and gene amplification potential accompany loss of wild-type p53. Cell 70:923–935

Lowe SW, Sherr CJ (2003) Tumor suppression by INK4a-ARF: progress and puzzles. Curr Opin Genet Dev 13:77–83

Lowe SW, Jacks T, Housman DE, Ruley HE (1994) Abrogation of oncogene-associated apoptosis allows transformation of p53-deficient cells. Proc Natl Acad Sci USA 91:2026–2030

Macip S, Igarashi M, Fang L, Chen A, Pan ZQ, Lee SW, Aaronson SA (2002) Inhibition of p21-mediated ROS accumulation can rescue p21-induced senescence. EMBO J 21:2180–2188

Malumbres M, Sotillo R, Santamaria D, Galan J, Cerezo A, Ortega S, Dubus P, Barbacid M (2004) Mammalian cells cycle without the D-type cyclin-dependent kinases Cdk4 and Cdk6. Cell 118:493–504

Martin-Caballero J, Flores JM, Garcia-Palencia P, Serrano M (2001) Tumor susceptibility of p21$^{Waf1/Cip1}$-deficient mice. Cancer Res 61:6234–6238

Miliani de Marval PL, Macias E, Rounbehler R, Sicinski P, Kiyokawa H, Johnson DG, Conti CJ, Rodriguez-Puebla ML (2004) Lack of cyclin-dependent kinase 4 inhibits c-myc tumorigenic activities in epithelial tissues. Mol Cell Biol 24:7538–7547

Munger K, Howley PM (2002) Human papillomavirus immortalization and transformation functions. Virus Res 89:213–228

Narita M, Nunez S, Heard E, Narita M, Lin AW, Hearn SA, Spector DL, Hannon GJ, Lowe SW (2003) Rb-mediated heterochromatin formation and silencing of E2F target genes during cellular senescence. Cell 113:703–716

Noda A, Ning Y, Venable SF, Pereira-Smith OM, Smith JR (1994) Cloning of senescent cell-derived inhibitors of DNA synthesis using an expression screen. Exp Cell Res 211:90–98

Ohtani N, Zebedee Z, Huot TJ, Stinson JA, Sugimoto M, Ohashi Y, Sharrocks AD, Peters G, Hara E (2001) Opposing effects of Ets and Id proteins on p16^{INK4a} expression during cellular senescence. Nature 409:1067–1070

Ortega S, Malumbres M, Barbacid M (2002) Cyclin D-dependent kinases, INK4 inhibitors and cancer. Biochim Biophys Acta 1602:73–87

Ortega S, Prieto I, Odajima J, Martin A, Dubus P, Sotillo R, Barbero JL, Malumbres M, Barbacid M (2003) Cyclin-dependent kinase 2 is essential for meiosis but not for mitotic cell division in mice. Nat Genet 35:25–31

Palmero I, Pantoja C, Serrano M (1998) p19^{ARF} links the tumour suppressor p53 to Ras. Nature 395:125–126

Pantoja C, Serrano M (1999) Murine fibroblasts lacking p21 undergo senescence and are resistant to transformation by oncogenic Ras. Oncogene 18:4974–4982

Park IK, Morrison SJ, Clarke MF (2004) Bmi1, stem cells, and senescence regulation. J Clin Invest 113:175–179

Parrinello S, Samper E, Krtolica A, Goldstein J, Melov S, Campisi J (2003) Oxygen sensitivity severely limits the replicative lifespan of murine fibroblasts. Nat Cell Biol 5:741–747

Pelicci PG (2004) Do tumor-suppressive mechanisms contribute to organism aging by inducing stem cell senescence? J Clin Invest 113:4–7

Pomerantz J, Schreiber-Agus N, Liegeois NJ, Silverman A, Alland L, Chin L, Potes J, Chen K, Orlow I, Lee HW, Cordon-Cardo C, DePinho RA (1998) The INK4a tumor suppressor gene product, p19^{ARF}, interacts with MDM2 and neutralizes MDM2's inhibition of p53. Cell 92:713–723

Quelle DE, Zindy F, Ashmun RA, Sherr CJ (1995) Alternative reading frames of the INK4a tumor suppressor gene encode two unrelated proteins capable of inducing cell cycle arrest. Cell 83:993–1000

Rane SG, Cosenza SC, Mettus RV, Reddy EP (2002) Germ line transmission of the Cdk4^{R24C} mutation facilitates tumorigenesis and escape from cellular senescence. Mol Cell Biol 22:644–656

Rodriguez-Puebla ML, Miliani de Marval PL, LaCava M, Moons DS, Kiyokawa H, Conti JC (2002) CDK4 deficiency inhibits skin tumor development but does not affect normal keratinocyte proliferation. Am J Pathol 161:405–411

Roninson IB (2002) Oncogenic functions of tumour suppressor p21$^{Waf1/Cip1/Sdi1}$: association with cell senescence and tumour-promoting activities of stromal fibroblasts. Cancer Lett 179:1–14

Rowland BD, Denissov SG, Douma S, Stunnenberg HG, Bernards R, Peeper DS (2002) E2F transcriptional repressor complexes are critical downstream targets of p19^{ARF}/p53-induced proliferative arrest. Cancer Cell 2:55–65

Rudolph KL, Chang S, Lee HW, Blasco M, Gottlieb GJ, Greider C, DePinho RA (1999) Longevity, stress response, and cancer in aging telomerase-deficient mice. Cell 96:701–712

Sage J, Mulligan GJ, Attardi LD, Miller A, Chen S, Williams B, Theodorou E, Jacks T (2000) Targeted disruption of the three Rb-related genes leads to loss of G1 control and immortalization. Genes Dev 14:3037–3050

Sage J, Miller AL, Perez-Mancera PA, Wysocki JM, Jacks T (2003) Acute mutation of retinoblastoma gene function is sufficient for cell cycle re-entry. Nature 424:223–228

Serrano M, Hannon GJ, Beach D (1993) A new regulatory motif in cell-cycle control causing specific inhibition of cyclin D/CDK4. Nature 366:704–707

Serrano M, Lee H, Chin L, Cordon-Cardo C, Beach D, DePinho RA (1996) Role of the INK4a locus in tumor suppression and cell mortality. Cell 85:27–37

Serrano M, Lin AW, McCurrach ME, Beach D, Lowe SW (1997) Oncogenic ras provokes premature cell senescence associated with accumulation of p53 and p16^{INK4a}. Cell 88:593–602

Sharpless NE, DePinho RA (2002) p53: Good cop/bad cop. Cell 110:9–12

Sharpless NE, DePinho RA (2004) Telomeres, stem cells, senescence, and cancer. J Clin Invest 113:160–168

Sharpless NE, Bardeesy N, Lee KH, Carrasco D, Castrillon DH, Aguirre AJ, Wu EA, Horner JW, DePinho RA (2001) Loss of p16^{INK4a} with retention of p19^{ARF} predisposes mice to tumorigenesis. Nature 413:86–91

Sherr CJ, DePinho RA (2000) Cellular senescence: mitotic clock or culture shock? Cell 102:407–410

Sherr CJ, Roberts JM (1999) CDK inhibitors: positive and negative regulators of G1-phase progression. Genes Dev 13:1501–1512

Smogorzewska A, de Lange T (2004) Regulation of telomerase by telomeric proteins. Annu Rev Biochem 73:177–208

Sotillo R, Dubus P, Martin J, de La CE, Ortega S, Malumbres M, Barbacid M (2001) Wide spectrum of tumors in knock-in mice carrying a Cdk4 protein insensitive to INK4 inhibitors. EMBO J 20:6637–6647

Stevaux O, Dyson NJ (2002) A revised picture of the E2F transcriptional network and RB function. Curr Opin Cell Biol 14:684–691

Stott FJ, Bates S, James MC, McConnell BB, Starborg M, Brookes S, Palmero I, Ryan K, Hara E, Vousden KH, Peters G (1998) The alternative product from the human CDKN2A locus, p14^{ARF}, participates in a regulatory feedback loop with p53 and MDM2. EMBO J 17:5001–5014

Vousden KH, Prives C (2005) p53 and prognosis: new insights and further complexity. Cell 120:7–10

Wei W, Hemmer RM, Sedivy JM (2001) Role of p14^{ARF} in replicative and induced senescence of human fibroblasts. Mol Cell Biol 21:6748–6757

Wu C, Miloslavskaya I, Demontis S, Maestro R, Galaktionov K (2004) Regulation of cellular response to oncogenic and oxidative stress by Seladin-1. Nature 432:640–645

Xiong Y, Hannon GJ, Zhang H, Casso D, Kobayashi R, Beach D (1993) p21 is a universal inhibitor of cyclin kinases. Nature 366:701–704

Zhang Y, Xiong Y, Yarbrough WG (1998) ARF promotes MDM2 degradation and stabilizes p53: ARF-INK4a locus deletion impairs both the Rb and p53 tumor suppression pathways. Cell 92:725–734

Zhu J, Woods D, McMahon M, Bishop JM (1998) Senescence of human fibroblasts induced by oncogenic Raf. Genes Dev 12:2997–3007

Zindy F, Eischen CM, Randle DH, Kamijo T, Cleveland JL, Sherr CJ, Roussel MF (1998) Myc signaling via the ARF tumor suppressor regulates p53-dependent apoptosis and immortalization. Genes Dev 12:2424–2433

Zou X, Ray D, Aziyu A, Christov K, Boiko AD, Gudkov AV, Kiyokawa H (2002) Cdk4 disruption renders primary mouse cells resistant to oncogenic transformation, leading to ARF/p53-independent senescence. Genes Dev 16:2923–2934

Mouse Models of Cell Cycle Regulators: New Paradigms

Eiman Aleem[1,2] · Philipp Kaldis[1] (✉)

[1]National Cancer Institute, Mouse Cancer Genetics Program, NCI-Frederick,
Bldg. 560/22-56, 1050 Boyles Street, Frederick, MD 21702-1201, USA
kaldis@ncifcrf.gov

[2]Department of Zoology, Faculty of Science, University of Alexandria, Alexandria, Egypt

Abstract In yeast, a single cyclin-dependent kinase (Cdk) is able to regulate diverse cell cycle transitions (S and M phases) by associating with multiple stage-specific cyclins. The evolution of multicellular organisms brought additional layers of cell cycle regulation in the form of numerous Cdks, cyclins and Cdk inhibitors to reflect the higher levels of organismal complexity. Our current knowledge about the mammalian cell cycle emerged from early experiments using human and rodent cell lines, from which we built the current textbook model of cell cycle regulation. In this model, the functions of different cyclin/Cdk complexes were thought to be specific for each cell cycle phase. In the last decade, studies using genetically engineered mice in which cell cycle regulators were targeted revealed many surprises. We discovered the in vivo functions of cell cycle proteins within the context of a living animal and whether they are essential for animal development. In this review, we discuss first the textbook model of cell cycle regulation, followed by a global overview of data obtained from different mouse models. We describe the similarities and differences between the phenotypes of different mouse models including embryonic lethality, sterility, hematopoietic, pancreatic, and placental defects. We also describe the role of key cell cycle regulators in the development of tumors in mice, and the implications of these data for human cancer. Furthermore, animal models in which two or more genes are ablated revealed which cell cycle regulators interact genetically and functionally complement each other. We discuss for example the interaction of cyclin D1 and p27 and the compensation of Cdk2 by Cdc2. We also focus on new functions discovered for certain cell cycle regulators such as the regulation of S phase by Cdc2 and the role of p27 in regulating cell migration. Finally, we conclude the chapter by discussing the limitations of animal models and to what extent can the recent findings be reconciled with the past work to come up with a new model for cell cycle regulation with high levels of redundancy among the molecular players.

1
Introduction

"The next ten years will reveal whether we have the commitment for the hard experiments that will be needed to challenge current dogma, overturn it when necessary, and move on to a deeper understanding of the cell cycle."

Andrew W. Murray, 2004

The ability of the cell to reproduce is a defining feature of our existence. The cell reproduces (proliferates) through a complex regulatory process called the cell cycle. The process of cell proliferation is tightly linked with differentiation, senescence, and apoptosis. A hallmark of cancer cells is that the normal balance of these processes is perturbed. The process of maintaining active proliferation is especially important for cancer cells. Therefore, uncovering the mechanisms of regulation of normal cell proliferation sets the ground for understanding the deregulated proliferation characteristic of a cancer cell regardless of the type of cancer or where/how it originated.

After completing one round of cell division, every cell in metazoans has to decide whether it will re-enter the cell cycle, exit the cell cycle and enter a quiescence state, and every quiescent cell has to similarly decide to stay quiescent, enter a state of terminal differentiation, or re-enter the cycle. All these crucial decisions are made by a set of information processors (cell cycle machinery) that integrate extracellular and intracellular signals to coordinate cell cycle events.

In order that a cell can produce an exact duplicate of itself, it has to perform four tasks in a highly ordered fashion: first to grow in size, replicate its DNA (S phase), equally segregate the duplicated DNA (M phase) and finally divide into two equal daughter cells (Mitchison and Creanor 1971). Since the two daughter cells must have the same genetic composition, the parent cell needs to replicate the genome only one single time per cycle followed by equal segregation of the replicated chromosomes into daughter cells. This is a crucial task for the cell cycle: to coordinate DNA replication (S phase) and cell division (M phase) in a well-balanced temporal sequence. The molecular core machinery controlling the eukaryotic cell cycle consists of a family of serine/threonine protein kinases called cyclin-dependent kinases (Cdks). These are catalytic subunits, which are activated by association with regulatory subunits called cyclins. The activity of Cdk/cyclin complexes is further regulated by Cdk-inhibitors (CKIs), phosphorylation and dephosphorylation, ubiquitin-mediated degradation, transcriptional regulation, substrate recognition, and subcellular localization.

Our knowledge of the events regulating the cell cycle emerged primarily from experiments performed in yeast, frogs, and mammalian cell lines. While the information gained from these experimental systems has provided the foundation of our current knowledge of cell cycle regulation, it did not reveal how these regulators function in the development and homeostasis of a whole animal. Hence came the importance of generating animal models, in which cell cycle genes are ablated or functionally altered in the mouse using knockout and transgenic techniques and allowed to study the effects of such genetic manipulations on the mouse as an integrated in vivo system. Such in vivo models underscored the redundancy of cell cycle genes within the context of a living animal and brought many surprises and some new concepts contradicting the textbook hypotheses upon which the current cell cycle model has been built.

The goal of this chapter is to discuss the textbook cell cycle model based on yeast and cultured mammalian cell lines, the similarities and differences between mouse models of cell cycle regulators, the functional complementation between mammalian Cdks, and the collective new paradigms emerging from these studies. A detailed background covering the history of Cdc2, Cdk2 and cyclin E is given because we found it necessary as a link to the conclusion of this chapter. We also discuss how the new paradigms emerging from mouse models reflect the complexity of higher mammals but at the same time prove that the molecular machinery operating the cell cycle is highly conserved and in higher organisms could be as simple as that of the single celled yeast. Furthermore, because misregulation of the cell cycle is a hallmark of cancer, the implications of these new paradigms to cancer and cancer therapy are discussed.

2
History of the Cell Cycle model

2.1
The Concept of Mammalian Cell Cycle Regulation

The textbook cell cycle model (Morgan 1997; Sherr and Roberts 1999) was based on lessons from yeast and cultured mammalian cells, as we will see below and can be summarized as follows: several Cdk/cyclin complexes drive cell cycle progression in higher organisms, and it has been believed that their functions are confined to specific stages in the cell cycle. For example, in early G1, Cdk4/Cdk6 in complex with cyclin D receive the environmental cues and transfer these signals to start the cell division cycle. They initiate phosphorylation of the retinoblastoma protein (Rb). In late G1, Cdk2 in complex with cyclin E completes phosphorylation of Rb. At this point the cell is committed to complete the cycle and passes the "restriction point" (Pardee 1974). DNA replication takes place in S phase. Cdk2 is the only Cdk known to regulate G1/S phase transition and progression through S phase in association with cyclin E and later with cyclin A. Mitosis is then initiated by Cdc2/cyclin B complexes, also known as M phase promoting factor (MPF). Cdc2/cyclin A complexes also contribute to the preparation for mitosis in G2 phase (Edgar and Lehner 1996; Nigg 1995).

2.2
Lessons from Yeast

In yeast, a single Cdk, which is the product of the *CDC28* gene in the budding yeast *Saccharomyces cerevisiae* (Hartwell et al. 1974; Lorincz and Reed 1984; Reed 1980) or the *cdc2+* gene in the fission yeast *Schizosaccharomyces pombe* (Nurse and Bissett 1981) is able to regulate diverse cell cycle transitions (S and M phases) by associating with multiple stage-specific cyclins (reviewed in Morgan 1997). In *S. cerevisiae*, the G1 function of Cdc28 requires three G1 cyclins (Cln1–3) with overlapping functions. Another set of six cyclins (Clb1–6) controls entry into S phase (Clb5/Clb6) and into mitosis (Clb1–4) (Nasmyth 1996). In *S. pombe* cdc2 and three cyclins encoded by *cdc13*, *cig1* and *cig2* control cell cycle progression (Fisher and Nurse 1995). Activation of cdc2 complexed with cdc13 brings the onset of mitosis (Booher et al. 1989; Moreno et al. 1989), and degradation of cdc13 leads to inactivation of the protein kinase, which is a prerequisite to mitotic exit (King et al. 1995). The role of cdc13 and its relationship to cdc2 is therefore analogous to cyclin B/Cdc2 in higher eukaryotes as described below. Cig2 is the major partner of cdc2 in G1 phase (Fisher and Nurse 1996; Mondesert et al. 1996). However, cig 2 is not essential for the initiation of S phase, but the G1/S transition is delayed when *cig2* is deleted (Fisher and Nurse 1996). It has been found that in the absence of cig2, cdc13, which was thought to be acting exclusively as a mitotic cyclin, is able to control S phase entry. The onset of S phase is severely compromised in a *cig2Δcdc13Δ* double mutant (Fisher and Nurse 1996; Mondesert et al. 1996) and is completely blocked in a *cig1Δcig2Δcdc13Δ* triple mutant (Fisher and Nurse 1996). In *cig1Δcig2Δ* double mutant, the only remaining cyclin, cdc13, and its associated cdc2 kinase activity undergo a single oscillation during the cell cycle, peaking in mitosis (Fisher and Nurse 1996), and this single oscillation of cdc2/cdc13 protein kinase activity can bring about the onset of both S phase and mitosis (Stern and Nurse 1996). In yeast, a very low kinase activity at the end of mitosis followed by a moderate kinase activity at the G1-S transition was proposed to bring about S phase (Stern and Nurse 1996). Maintenance of this moderate kinase activity through the G2 phase blocks re-initiation of replication, and a further increase of kinase activity was thought to induce mitosis. Furthermore, premature loss of cdc2/cdc13 kinase activity at G2 phase by deleting *cdc13* (Hayles et al. 1994) or overexpressing rum1, a specific inhibitor for cdc2/cdc13 (Correa-Bordes and Nurse 1995; Moreno and Nurse 1994) [equivalent to p27^{Kip1} in mammals] causes re-replication and no mitosis, leading to an increase of DNA content up to 32–64C. A situation similar to this occurs when the S phase kinase-associated protein 2 (Skp2), which is required for ubiquitin-mediated degradation of p27 at S and G2 phases (Carrano et al. 1999; Sutterluty et al. 1999), is ablated from mice (Nakayama et al. 2000). *Skp2−/−* cells show large nuclei and polyploidy, and are unable to enter mitosis. This is because p27 (with similar function to rum1 in *S. pombe*) strongly inhibits the mitotic Cdk in mice; Cdc2, as we later showed in Aleem et al. (2005) and also in Nakayama et al. (2004).

From the yeast model we learn that in the fission yeast a single Cdk (cdc2) and a single cyclin (cdc13) can solely regulate the different phases of the cell cycle depending on the levels of associated-kinase activity. The question is whether we can apply the same concept to higher eukaryotes. Can mammals including mice and humans survive with a single Cdk and a single cyclin? And how can they achieve this given the higher level of complexity of the mammalian cell cycle?

2.3
Human Cdc2, Cdk2 and Cyclin E

In higher organisms such as in mammals there are functional homologues of cdc2 or Cdc28 and specialized S and M phase Cdks have replaced the single Cdk of yeast. The discovery of more than 10 Cdc2-related proteins in vertebrates led to the concept that the higher eukaryotic cell cycle involved complex combinations of Cdks and cyclins. It raised also a number of questions: How many of these cyclin/Cdk complexes are essential for viability? Do these Cdk/cyclin complexes differ in the proteins they phosphorylate (i.e., their substrates) or rather in when and where they are expressed in the cell cycle? How much functional overlap is there between different cyclin/Cdk complexes?

The first Cdk to be identified was the human homologue of the fission yeast cdc2, which has been cloned by expressing a human cDNA library in fission yeast and selecting for clones that complemented the function of a defective mutant yeast cdc2 (Lee and Nurse 1987). Human Cdc2 encodes a 34 kDa protein similar to that of yeast. Because of the structural similarity between human and yeast Cdc2 and because the human *CDC2* gene was able to carry out all the functions of the *S. pombe* cdc2, it has been reasonably assumed that Cdc2 performs a similar role in controlling the human cell cycle. Taking the yeast model into consideration, researchers have suggested that human Cdc2 regulates two points in the cell cycle: one analogous to "Start" in late G1 of yeast, which is called the Restriction (R) point in mammals. The R-point designates a certain time at late G1 in which cells become independent on the presence of growth factors and are committed to complete one round of cell cycle, also known as the point of "no return" (Pardee 1974). The second point is in late G2 at the initiation of mitosis, similar to the maturation promotion factor (MPF) detected in vertebrate eggs. In the language of higher eukaryotes these predictions for the function of Cdc2 described in 1987 by Lee and Nurse can be interpreted as follows: Cdc2 regulates G1/S transition (a function assigned to Cdk2/cyclin E) and it regulates entry into M phase (a function assigned to Cdc2/cyclin B). We will discuss below how these predictions may be indeed correct after almost 20 years of research in the field of cell cycle from 1987 until 2006. It is relevant to mention here that microinjection of antibodies against human Cdc2 arrested cells in G2 phase (Riabowol et al. 1989) and a temperature sensitive mutation in human *CDC2* gene arrested cells at the G2/M phase at the non-permissive temperature and this arrest could be suppressed by expression of the wild type human CDC2 (Th'ng et al. 1990).

The second human Cdk to be characterized was Cdk2 (short for *c*ell *d*ivision *k*inase 2, later renamed as cyclin-dependent kinase 2), which has been identified by complementation of a *cdc28-4* mutant in *S. cerevisiae*, using a human cDNA expression library (Elledge and Spottswood 1991). Human Cdk2 could perform all the functions of the Cdc28 protein in budding yeast, was found to encode a 33 kDa protein, and is 66% identical to human Cdc2. This suggested that Cdk2 is distinct from Cdc2 and performs different functions in the cell cycle. This notion has been corroborated by in vitro experiments with Xenopus egg extracts in which depletion of Cdk2 interfered with DNA synthesis but depletion of Cdc2 did not affect DNA synthesis but blocked mitosis (Fang and Newport 1991). In addition, Cdk2 mRNA levels increase upon entry into the cell cycle before the mRNA of Cdc2. Nevertheless, both Cdc2 and Cdk2 associate with cyclin A (Elledge et al. 1992).

Human cyclin E was isolated by complementation of a triple *cln* deletion in *S. cerevisiae* (Koff et al. 1991) indicating its role in G1 phase. Similarly, genes encoding cyclin C and cyclin D were discovered by screening human and Drosophila cDNA libraries for genes that could complement mutations in the *S. cerevisiae CLN* genes, which encode G1 cyclins (Lahue et al. 1991; Xiong et al. 1991). Two human genes were identified that could interact with cyclin E to perform START in yeast containing a defective *cdc28* mutation. One was human Cdk2 and the other human Cdc2 (Koff et al. 1991). Recombinant cyclin E was shown to bind and activate Cdk2 and Cdc2 in extracts from a human B cell line (MANCA cells) synchronized in early G1 (Koff et al. 1992) and allowed to progress into S phase (Marraccino et al. 1992). Furthermore, cyclin E-associated kinase activity increased during G1, was maximal just as cells entered S phase and it peaks before cyclin A-associated kinase activity (Koff et al. 1992). It was absent in G1 and first detected as cells entered S phase. This report emphasized the role of cyclin E in the activation of Cdk2 and the regulation of G1 by cyclin E/Cdk2 complex (Koff et al. 1992). Although these results hinted that Cdc2 interacted with cyclin E in human G1 cells (Koff et al. 1991, 1992), most of the attention of cell cycle studies later focused on the association between Cdk2 and cyclin E and identified Cdk2 to be the only Cdk that binds to cyclin E in mammalian cells at the beginning of S phase to induce the initiation of DNA synthesis.

2.4
G1 Phase in Mammalian Cultured Cells

In the first half of the 1990s, it was shown that in mammalian cells, Cdc2 associates mainly with cyclin A and B, Cdk2 with cyclin E and A, Cdk4 and Cdk6 with the D-type cyclins (Draetta and Beach 1988; Dulic et al. 1992; Koff et al. 1992; Lees et al. 1992; Matsushime et al. 1992; Meyerson et al. 1992; Pines and Hunter 1990; Rosenblatt et al. 1992; Tsai et al. 1991, 1993; Xiong et al. 1992). Many studies employing overexpression of cyclins or Cdks or the use of dominant negative mutations in Cdks in cultured human or rodent cells contributed significantly to the development of the textbook cell cycle model. Van den Heuvel generated dominant-negative mutations for all Cdks (van den Heuvel and Harlow 1993). When expressed at high levels in human cells, dominant negative mutations inactivate the functions of the wild type protein (its kinase activity in this case) by competing for essential interacting molecules including cyclins (Herskowitz 1987). These mutants were unable to rescue *cdc28* mutations at the non-permissive temperature (36 °C) unlike the wild type. When Cdk2^{D145N} was expressed in four different human cell lines (U2OS, Saos-2, C33A cervical carcinoma cells and T98G glioblastoma cells), an increase in G1 population occurred. When Cdc2^{D146N} was expressed it led to increased G2/M population. Transfection of wild type Cdk2 and Cdc2 did not affect the cell cycle distribution, and the effects of mutant kinases could be overcome by co-expression of the corresponding wild type kinase. These experiments indicated a specific inhibition of Cdc2 and Cdk2 kinase activities at a specific timing in the cell cycle and underscored the concept that Cdk2 and Cdc2 each functions in a cell cycle phase-specific manner. However, Cdc2^{D146N} had no effect in C33A cells unlike the other cell lines. This may indicate that the role of Cdc2 differs from one cell line to another. Another line of evidence supporting this idea is that, in spite of the fact that expression of Cdk2^{D145N} in the above mentioned four cell lines did result in a G1 block, it did not cause a G1 arrest in colon cancer cells (Tetsu and Mc-Cormick 2003). However, in the early 1990s the dominant direction driving cell cycle research in higher eukaryotes was to prove that multiple cyclin/Cdk complexes regulate different phases in the cell cycle and this reflects the complexity of the organism, even if one or more observations did not match the emerging concept of multiplicity and specificity of Cdks.

A rescue of the cell cycle block induced by dominant negative forms of Cdk2 and Cdc2 was attempted by overexpressing cyclins A, B1, B2, C, D1, D3, and E (Hinds et al. 1992). Cyclin D1 could rescue the Cdk2^{D145N} G1 block but cyclins E and A were less efficient in rescuing the inhibition, and no effects were observed when cyclins B1, B2, C and D3 were cotransfected with the Cdk2 mutant. In contrast, a reduction of the Cdc2^{D146N} effect was observed when either cyclin B1 or B2 was contransfected. These results were limited by the amount of expressed cyclin and did not support the earlier observations in yeast that when G1 cyclins are overexpressed in yeast, the duration of G1 decreases and this results in small cell size during exponential growth (Cross 1988; Hadwiger et al. 1989; Nash et al. 1988; Wittenberg et al. 1990). Accordingly, overexpression of cyclin E should have rescued the G1-block induced by Cdk2^{D145N}, especially that when human cyclin E was over-expressed in Rat-1 fibroblasts and in primary human fibroblasts the duration of G1 was shorter than control cells (Ohtsubo and Roberts 1993). The amount of cyclin E-associated kinase activity was also increased in cells overexpressing cyclin E but this was not sufficient to initiate DNA replication. Similar experiments to overexpress cyclin A and B in the same cells did not result in changes in the kinetics of G1 control. Similarly, overexpression of cyclins D1 and D2 in a mouse macrophage cell line did not affect G1 phase duration (in Ohtsubo and Roberts 1993), but it did partially rescue the Cdk2^{D145N} block in the four human cell lines described above indicating that various cell lines could differ dramatically in their response to overexpression or other type of experimental manipulations. Another interesting finding supporting this is that although the Cdk2^{D145N} effect could be rescued in Saos-2 cells by overexpressing cyclin D1, these cells do not express endogenous cyclin D1. This means that enforced expression of a cyclin, which is not naturally expressed in a certain cell line could result in an interesting phenotype that resulted only by artificial means.

Cdk3 can also complement *cdc28* mutations in yeast, similar to Cdk2 (Meyerson et al. 1992). Cdk3^{D145N} mutants were tested in the same manner and found to induce G1 arrest similar to Cdk2 in Saos-2 and C22A cells (van den Heuvel and Harlow 1993). However, expression of wild type Cdk2 could not rescue the Cdk3^{D145N} G1 block and in the converse experiment wild type Cdk3 could not rescue the Cdk2^{D145N} block. This suggested a specific role for Cdk3 in the G1/S transition that is not redundant with the function of Cdk2. Moreover, Cdk3/cyclin E complexes were found to promote S phase entry in quiescent cells as efficiently as can Cdk2/cyclin E complexes (Connell-Crowley et al. 1998). In the same report, transfection of wild type or mutant forms of Cdk4, Cdk5 or Cdk6 had no effect on cell cycle distribution in the four human cell lines (van den Heuvel and Harlow 1993). These observations coupled with the fact that Cdk3 was shown to be the only kinase in addition to Cdk2 and Cdc2 that could rescue the yeast *cdc28* mutations suggested that only Cdk2, Cdc2, and Cdk3 are the essential Cdks in the mammalian cell cycle.

The notion that Cdks have phase-specific functions during the mammalian cell cycle had been widely accepted for many years until another stage of cell cycle research emerged using mouse models lacking one or more cell cycle genes. Genetic targeting of cell cycle regulatory proteins in the mouse determined which cell cycle gene is essential for the development of a whole animal. It also revealed additional levels of cell cycle regulation present in the context of a living animal and which could not be uncovered otherwise in cultured cells. Two shocking results strongly contradicted the long accepted fact that Cdk2, Cdk3, and Cdc2 are the only essential Cdks in the mammalian cell cycle: Ye et al. (2001) demonstrated that most species of the laboratory mouse *Mus musculus* have a natural mutation that results in replacement of Trp-187 with a stop codon resulting in a null allele. In contrast, Cdk3 from two wild type mice species lack this mutation. The data suggested that Cdk3 is not required for the development of the mouse and that any functional roles played by Cdk3 in the G1/S phase transition is redundant with another Cdk, most likely Cdk2. These results left only Cdc2 and Cdk2 as the only two essential Cdks in the regulation of the mammalian cell cycle. Another surprise in the history of cell cycle research was uncovered when three separate laboratories (Barbacid, Kaldis and McCormick) questioned the role of Cdk2 as a master regulator of entry into and progression through S phase. The genetic targeting of Cdk2 in the mouse (Berthet et al. 2003; Ortega et al. 2003) revealed that Cdk2 is not essential for the development or for the mitotic cell cycle. Because there were no other Cdks known to operate during S phase but Cdk2, these results raised the question of whether there is another yet unknown kinase, which compensates the loss of Cdk2 or whether any of the other known Cdks can also regulate S phase. If the second possibility is true, then it challenges the idea that Cdks are independent classes; their functions are cell cycle phase-specific.

3
Mouse Models of Cell Cycle Regulators

The advantages of mouse models over in vitro studies is that it highlights the functions of a particular cell cycle regulator as it is in a living animal on the organismal and cellular levels. The first clear cut answer a knockout mouse can provide is whether a particular gene is essential or not for the life and development of this mouse, so if the phenotype is lethal it indicates that the function of this gene is unique and cannot be compensated by similar molecules. We will present data from these mouse models *according to the phenotypes* of different mouse models rather than listing the phenotype of each mouse model in a consequential manner. We will focus on the mouse models for Cdks, cyclins, and the Cdk inhibitors. We will not describe the phenotypes of the Rb/E2F mouse models in details because it is presented elsewhere in this book (see chapter by L. Yamasaki, and chapter by Dannenberg and Te Riele).

3.1
Targeting of Individual Cell Cycle Regulators Results in Embryonic Lethality

The cell cycle model predicted that mice lacking Cdk2, Cdk3 or Cdc2 would be embryonic lethal due to their specific functions. Regarding mouse models of cyclins, only mice lacking cyclin A2, cyclin B1 and cyclin F (not discussed here) display a lethal phenotype.

3.1.1
Cyclin A2

Cyclin A is particularly interesting among the cyclins because it activates two different Cdks; Cdk2 in S phase and Cdc2 in the G2/M phase. While in human

(Yang et al. 1997), mice (Sweeney et al. 1996) and *Xenopus* (Howe et al. 1995; Minshull et al. 1990) there are two types of cyclin A: cyclin A1 and cyclin A2, there is only one essential cyclin A gene in *Drosophila* (Knoblich and Lehner 1993; Lehner and O'Farrell 1989). Cyclin A1 is only expressed in meiosis; i.e., restricted mainly to the male and female germ cells, very early embryos, and in the brain (Ravnik and Wolgemuth 1996), whereas cyclin A2 is present in proliferating somatic cells. The only essential function of cyclin A1 in mice is in spermatogenesis (Liu et al. 1998). In contrast, cyclin A2 is essential in mice and disruption of its gene causes early embryonic lethality [$\approx$E5.5] (Murphy et al. 1997). *Cyclin A2-/-* embryos reach the blastocyst stage, but die soon after implantation (Murphy et al. 1997). This indicates that cyclin A2 is dispensable for the early preimplantation development. It is possible that at this stage of development other proteins may replace the functions of cyclin A2, for example cyclin B3, which shares homology with the A-type cyclins. Unlike cyclin E1 and E2, and the D-type cyclins, which can compensate for the deficiency of each other, cyclin A1 cannot compensate the loss of cyclin A2 in postnatal and adult cells because of the restricted expression of cyclin A1 in germ cells and early embryos. Similar to the essential role of cyclin A2 in vivo, it was shown to have a non-redundant role in both S and M phase progression in cultured mammalian cells (Furuno et al. 1999; Pagano et al. 1992; Resnitzky et al. 1995).

3.1.2
Cyclin B1

The B-type cyclins are known for their important role in regulation of M phase progression. In mammals, the family so far contains three B-type cyclins: B1 (Chapman and Wolgemuth 1992; Pines and Hunter 1989), B2 (Chapman and Wolgemuth 1993) and B3 (Gallant and Nigg 1994; Lozano et al. 2002; Nguyen et al. 2002). Cyclins B1 and B2 associate with Cdc2, while cyclin B3 was shown to interact with Cdk2 but not with Cdc2 (Nguyen et al. 2002). However, we could recently detect Cdk2 by immunoblotting in cyclin B1 immunoprecipitates from thymus lysates in mice (Aleem et al. 2005). Nevertheless, the biological meaning of cyclin B1/Cdk2 complexes remains to be elucidated. It is relevant to mention that cyclin B3 shares characteristics of both A- and B-type cyclins (Nieduszynski et al. 2002) and like cyclin A it is localized exclusively in the cell nucleus (Gallant and Nigg 1994). Cyclin B1 and B2 are expressed in the majority of proliferating cells; however, cyclin B1 associates with microtubules while cyclin B2 localizes with the intracellular membranes (Jackman et al. 1995; Ookata et al. 1993). Furthermore, cyclin B1, but not B2 translocates into the nucleus at the end of the G2 phase, suggesting that they play two different functions during cell cycle progression (Pines and Hunter 1991; Toyoshima et al. 1998; Yang et al. 1998). It has been demonstrated that nuclear cyclin B1/Cdc2 complexes are responsible for nuclear envelope breakdown, chromosome condensation and mitotic spindle assembly, while cytoplasmic cyclin B2/Cdc2 complexes functions in the mitotic reorganization of the Golgi apparatus (Draviam et al. 2001).

Cyclin B1 is an essential cell cycle gene; its deletion in mice caused embryonic lethality before day E10 (Brandeis et al. 1998). However, neither the exact timing when the embryos die, nor the reason of lethality in cyclin B1-deficient mice has been determined. It is interesting to note that in spite of the fact that both cyclin B1 and B2 are ubiquitously expressed their functions are not redundant and cyclin B2 cannot compensate for the loss of cyclin B1.

3.2
Sterility

Sterility is the most common phenotype observed when cell cycle regulators are ablated in mice. Mice lacking Cdk2, Cdk4, cyclin D2, cyclin A1, cyclin E2, $p27^{Kip1}$, and $p18^{INK4c}/p19^{INK4d}$ double knockout mice, share this phenotype. Sterility may be partial or complete, either in males or females or in both genders; however, males seem to be more susceptible to this phenotype than females as we will see from the examples below. Germ cells undergo mitotic divisions, meiotic reduction divisions, and morphogenetic differentiation as they progress from the primordial germ cell to the haploid gamete. The process of spermatogenesis in mammals as described is characterized by a sequence of at least two mitotic divisions starting from day 7 post partum (pp) that lead to the development of type A and type B spermatogonia (Zindy et al. 2001). Type B undergoes premeiotic replication and enters meiosis as primary spermatocytes. Segregation of homologous chromosomes occurs at the end of meiosis I, and resulting secondary spermatocytes then proceed through a second meiotic division generating haploid germ cells. These differentiate to form round spermatids and mature spermatozoa (spermiogenesis). The first round of spermatogenesis is followed by additional waves to allow continuous sperm production. In males, follicle-stimulating hormone (FSH) stimulates Sertoli cells, whose number determines the size of the testis (Sharpe 1989), and lutenizing hormone (LH) stimulates interstitial Leydig cells to produce testosterone (Hedger and de Kretser 2000). The situation in females is different: female germ cells proliferate by mitosis and enter meiosis in the embryo, arresting in prophase of meiosis I. These oocytes remain arrested until puberty when a pool of oocytes are recruited to grow and complete the first meiotic division, only to arrest again at metaphase II until fertilization triggers the resumption of meiosis (Peters 1969). Cdks control both the mitotic and meiotic divisions. The role of different cell cycle regulators in regulation of meiosis is illustrated in the following mouse models (for details see chapter by Rajesh and Pittman).

3.2.1
Cdk2

It has long been believed that the only Cdk, which binds to and is activated by cyclin E is Cdk2 and cyclin E/Cdk2 complexes are essential components of the cell cycle machinery (see Sect. 2.4). Cyclin E/Cdk2 complexes phosphorylate several targets such as Rb (Furstenthal et al. 2001; Harbour et al. 1999; Lundberg and Weinberg 1998), p27 (Sheaff et al. 1997; Vlach et al. 1997), Cdc25A (Hoffmann et al. 1994), as well as proteins involved in DNA replication (Arata et al. 2000; Krude et al. 1997; Zou and Stillman 2000), centrosome duplication such as nucleophosmin and CP110 (Chen et al. 2002; Okuda et al. 2000), $p220^{NPAT}$ required for histone biosynthesis (Ma et al. 1999), E2F5 (Morris et al. 2000) and p300/CBP (Ait-Si-Ali et al. 1998; Felzien et al. 1999; Perkins et al. 1997). In contrast to regulation of the G1/S transition in the mitotic cell division, a new role for Cdk2 in the regulation of meiosis has been uncovered when Cdk2 was ablated in mice (Berthet et al. 2003; Ortega et al. 2003). Mice lacking Cdk2 showed complete sterility in males and females. Males displayed reduced testicular size and the only stages of spermatogenesis observed were the spermatogonia, and females showed also ovarian atrophy and few or abnormal follicles (Berthet et al. 2003; Ortega et al. 2003). Indeed, Cdk2 has been shown in an earlier study to be highly expressed in all spermatocytes, notably in cells undergoing the meiotic reduction divisions (Ravnik and Wolgemuth 1999). However, males and females show differential requirement for Cdk2 at distinct stages of meiotic prophase I. Whereas in male germ cells, Cdk2 is required for synaptonemal complex formation during the pachytene stage, female germ cells progress further to the dictyate stage, at time at which they undergo apoptosis in the absence of Cdk2 (Ortega et al. 2003). Furthermore, Cdk2 is localized in the telomeric ends of chromosomes from leptotene to diplotene stages of meiosis (Ashley et al. 2001). The meiotic substrates of Cdk2 are largely unknown. Other loci whose inactivation leads to phenotypes similar to that of *Cdk2-/-* mice include those encoding SYCP3 (Yuan et al. 2000). Lack of Cdk2 causes perturbed distribution of SYCP3 in male and female germ cells. Thus, Cdk2 may promote proper dynamics of SYCP3, either by direct phosphorylation or by phosphorylating other proteins involved in this process (Ortega et al. 2003).

3.2.2
Cyclin A1

Cyclin A1 protein is present only in male germ cells, prior to or during the first, but not the second meiotic division (Ravnik and Wolgemuth 1999). *Cyclin A1-/-* mice are developmentally normal, demonstrating that it is not required for embryonic and postnatal somatic cell divisions (Liu et al. 1998). The most pronounced phenotype is male sterility whereas females are fertile. Lack of cyclin A1 resulted in an abrupt arrest of spermatogenesis during late meiotic prophase in *cyclin A1-/-* males (Liu et al. 1998). The histological structure of the *cyclin A1-/-* testis resembles that of the *Cdk2-/-* testis. In addition, *cyclin A1-/-* seminiferous tubules have also early primary spermatocytes, which appeared normal. Furthermore, nuclei at mid/late pachytene stages with normal synapsed chromosomes, but not mid-diplotene nuclei with desynapsing synaptonemal complexes, were detected in the testes of *cyclin A1-/-* testes. Similarly, meiotic metaphase chromosomes were not observed confirming that spermatogenesis did not progress beyond the diplotene stage (Liu et al. 1998). In addition, numerous spermatocytes were found to undergo apoptosis in the testes of *cyclin A1-/-* testes. High percentage of apoptosis in the testes was also detected in *Cdk2-/-* mutants (Berthet et al. 2003; Ortega et al. 2003). Interestingly, histone H1 kinase activity of Cdc2 was reduced by 80% in the testes of adult *cyclin A1-/-* mice compared to *cyclin A1+/-* controls, whereas Cdk2 activity only moderately declined (Liu et al. 1998). Whether this result indicates that Cdc2 is the main catalytic partner of cyclin A1 in the testis remains to be further studied, because (Liu et al. 1998) immuno-depleted cyclin A1 protein from wild type testicular extracts and did not find alterations in the levels of Cdc2 or Cdk2 activities.

3.2.3
Cyclin E1 and Cyclin E2

Two E-type cyclins have been described; cyclin E1 and E2, which are targets of E2F/DP-1 mediated transcription. E-type cyclins are largely dispensable for mouse development (Geng et al. 2003; Parisi et al. 2003). Mice lacking either cyclin E1 or E2 are viable; however, the double knockout mice deficient for both genes died at E10.5–11.5 (Geng et al. 2003; Parisi et al. 2003). These results will be further discussed below. However, in contrast to the complete sterility of Cdk2 mice, mice lacking cyclin E1 are normal and fertile and only males lacking cyclin E2 show partial sterility; about 50% of the *cyclin E2-/-* males are sterile, showing reduced testicular size and reduced sperm counts as compared to the wild type males as well as abnormal meiotic figures within the spermatocyte layers and the presence of multinuclear giant cells within the seminiferous epithelium (Geng et al. 2003). This phenotype is different from the phenotype of *Cdk2-/-* males that do not show any stages of spermatogenic maturation after spermatogonia. This may reflect different causes of sterility in both mouse models. *Cyclin E2-/-* females on the other hand develop normally and are fully fertile (Geng et al. 2003).

3.2.4
Cyclin D2

The mammalian D-type cyclin family consists of three members: cyclin D1, D2, and D3. These proteins are encoded by separate genes but they show substantial amino acid similarity and are expressed in a highly overlapping fashion in all proliferating cells (Sherr and Roberts 1999). The D-type cyclins bind to and activate Cdk4 and Cdk6, and phosphorylate members of the Rb family (Rb, p130, p107). This in turn leads to the release of E2F transcription factors and activation of transcription of E2F-responsive genes (Sherr and Roberts 1999; 2004). The D-type cyclins are considered the sensors of mitogenic stimuli linking the extracellular environment to the cell cycle core machinery (Matsushime et al. 1991). Mice lacking individual D-type cyclins have been generated and they were all viable (Fantl et al. 1995; Sicinska et al. 2003; Sicinski et al. 1995, 1996). The only mouse model with fertility problems is the *cyclin D2-/-* mouse (Sicinski et al. 1996). Cyclin D2-deficient females are sterile owing to the inability of ovarian granulosa cells to proliferate normally in response to follicle-stimulating hormone (FSH), but oocyte development is not affected. In ovarian granulosa cells, cyclin D2 is specifically induced by FSH via a cyclic-AMP-dependent pathway, indicating that expression of the various D-type cyclins is under control of distinct intracellular signalling pathways. In contrast *cyclin D2-/-* males are fertile but display hypoplastic testes and decreased sperm counts (Sicinski et al. 1996). Furthermore, the same group found that some human ovarian and testicular tumors contain high levels of cyclin D2 messenger RNA, which is consistent with the notion that cyclin D2 is important for these compartments (Sicinski et al. 1996).

3.2.5
Cdk4

Cdk4 and Cdk6 are the main partners of the D-type cyclins and the main function of these complexes is phosphorylating Rb, thus inactivating its S phase-inhibitory action (reviewed in Sherr and Roberts 1999, 2004). The ablation of Cdk4 in mice resulted in sterility in both males and females (Rane et al. 1999; Tsutsui et al. 1999). However, in contrast to the complete sterility caused by Cdk2 deficiency, only 10–20% of male *Cdk4-/-* mice were fertile whereas all female mutants were infertile. Furthermore, the limited number of males, which produced an offspring, had a small number of litter (3–6 pups), and over a short period of time (2–3 months of age). The defective spermatogenesis in *Cdk4-/-* males was manifest by reduced testicular mass (75% smaller than the wild type testis), degenerated seminiferous tubules with severe reduction of spermatozoa in older males and numerous apoptotic bodies (Rane et al. 1999; Tsutsui et al. 1999) and with reduced expression of developmental markers such as *Myb11, Hsp70-2, Mos, transition protein 1, Stah2, protamine 1* and *2* (Rane et al. 1999). Sterility of *Cdk4-/-* females was attributed to defects in the formation of corpus luteum not in the development of granulosa cells (Rane et al. 1999; Tsutsui et al. 1999). Mutant females also had very low levels of progesterone (secreted by corpus luteum) and of FSH, as well as defects in ovulation detected by prolonged estrus cycle (Rane et al. 1999). Transplantation of wild type ovaries in *Cdk4-/-* females did not result in offspring. In contrast, the reciprocal ovarian transplant in which wild type females received *Cdk4-/-* ovaries resulted in *Cdk4+/-* offspring when these females were mated with wild type males (Rane et al. 1999). These results indicated that lack of Cdk4 causes female sterility that is not due to developmental abnormalities of their reproductive organs, but due to defects in the endocrine hypothalamic-pituitary axis.

3.2.6
p19^{INK4d} and p18^{INK4c}

The INK4 family of Cdk inhibitors includes four 15 to 19-kDa polypeptides (p16^{INK4a}, p15^{INK4b}, p18^{INK4c}, and p19^{INK4d}). The INK4 family is one of two distinct families of inhibitors that block the activity of G1 Cdks: the other being the Cip/Kip family, which includes three members (p21^{Cip1}, p27^{Kip1} and p57^{Kip2}) (reviewed in Sherr and Roberts 1999, 2004). The INK4 proteins exclusively bind to and inhibit the cyclin D-dependent catalytic subunits Cdk4 and Cdk6, while the Cip/Kip family binds to all Cdk/cyclin complexes with preferential inhibition of cyclin E- and A/Cdk2. The INK4 family of inhibitors are structurally redundant but are differentially expressed during mouse development (Zindy et al. 1997). p18^{INK4c} and p19^{INK4d} are widely expressed during mouse embryogenesis while p16^{INK4a} and p15^{INK4b} expression are not detected before birth. Mice lacking individual or combined members of the INK4 family have been generated (Franklin et al. 1998; Krimpenfort et al. 2001; Latres et al. 2000; Serrano et al. 1996; Sharpless et al. 2001; Zindy et al. 2001); however, only mice lacking p19^{INK4d} displayed gonadal problems (Zindy et al. 2000), and mice lacking both p18^{INK4c} and p19^{INK4d} are infertile (Zindy et al. 2001). Deletion of p19^{INK4d} in the mouse does not affect mouse development. *p19-/-* mice did not develop tumors and cells of different lineages isolated from these mice showed no remarkable proliferative disorders. However, males studied at 7 to 14 weeks of age showed marked testicular atrophy associated with increased apoptosis of germ cells and reduced sperm counts, although they remained fertile (Zindy et al. 2000). p19^{INK4d} is expressed in the testis in germ cells undergoing meiosis and during differentiation from spermatids to spermatozoa. This pattern of expression differs from that of its target Cdk4, which is expressed in spermatogonia and in early stage primary spermatocytes but does not contribute to later stages of germ cell development (Rhee and Wolgemuth 1995). This implies that p19^{INK4d}

may prepare cells for meiosis by downregulating Cdk4 activity. Although mice deficient for p19^{INK4d} or for p18^{INK4c} are fertile, *p18-/-p19-/-* double knockout male – but not female – mice are all infertile (Zindy et al. 2001). This result indicates that both p19^{INK4d} and p18^{INK4c} cooperate in regulating spermatogenesis but not oogenesis. The expression of p19^{INK4d} and p18^{INK4c} in the seminiferous tubules of postnatal wild type mice is largely confined to postmitotic spermatocytes undergoing meiosis. Their combined loss is associated with delayed exit of spermatogonia from the mitotic cell cycle leading to the retarded appearance of meiotic cells that do not properly differentiate and instead undergo apoptosis at an increased frequency. Furthermore, the double knockout mice as well as *p18^{INK4c}-/-* mice develop hyperplasia of interstitial testicular Leydig cells, which produce reduced levels of testosterone (75% less than wild type levels). This defect in testosterone production is not due to defects in the production of lutenizing hormone (LH) from the anterior pituitary, because these animals produce normal LH levels. It was found that Leydig cells in both *p18^{INK4c}-/-* and the double knockout animals fail to differentiate and produce testosterone as indicated by severe reduction in the levels of the Leydig cells differentiation marker P450scc, in comparison to its levels in wild type and *p19^{INKd}-/-* mice. But despite Leydig cell hyperplasia, the double knockout mice have small testes with tubular atrophy, reduced sperm counts and the residual spermatozoa have reduced viability and motility, leading to sterility. It was also found that *p19^{INK4d}-/-* and *p18^{INK4c}-/-p19^{INK4d}-/-* double knockout mice produce elevated levels of FSH, but the functional significance of this observation remains unknown (Zindy et al. 2001).

3.2.7
p27^{Kip1}

p27 was initially discovered as a Cdk-inhibitory activity induced by extracellular anti-mitogenic signals (Firpo et al. 1994; Koff et al. 1993; Polyak et al. 1994; Slingerland et al. 1994). When members of the CIP/KIP family of Cdk inhibitors (i.e. p21, p27 and p57) are overexpressed in cell lines they cause cell cycle arrest due to their inhibitory activity on cyclin/Cdk complexes essential for G1 progression and S phase entry. Cdk2 complexes were known to be major targets of p27. Ablation of p27 in mice did not have an effect on embryonic development and the most characteristic feature of these animals is multiorgan hyperplasia (Fero et al. 1996; Kiyokawa et al. 1996; Nakayama et al. 1996) [see below]. Both male and female mice deficient for p27 demonstrated testicular and ovarian hyperplasia; however, only females were sterile. Although *p27-/-* male mice were reported to be fertile, we observed that they require much more time to impregnate females than *p27+/-* males (Aleem and Kaldis, unpublished data). Indeed, it has been demonstrated that in the adult testes of mice deficient for p27, there is 50% increase in the number of type A spermatogonia in epithelial stage VIII compared to that of the wild type testes. Furthermore, there was a significant number of prelepotene spermatocytes failing to enter meiotic prophase that were not detected in the wild type testes (Beumer et al. 1999). These results suggested an indirect role for p27 in maintaining the normal spermatogenic process because p27 is known to be expressed in Sertoli cells only in the adult testis (Beumer et al. 1999). *p27-/-* females were capable of mating and some mice vaginal plugs were formed but there were no pregnancies to full term. Some embryos were isolated at day 3.5 pc and transferred to the oviducts of pseudopregnant normal female mice. These embryos could develop to full term indicating that ovulation and fertilization occurred in the absence of p27. Nevertheless, there are two main problems with *p27-/-* females that contribute to their sterility: the absence of a corpus luteum, and a disordered estrus cycle. Corpus luteum formation plays an important role for maintenance of pregnancy by secreting progesterone and other factors. Granulosa cells are the somatic components of the ovarian follicles. They differentiate into progesterone-producing luteal cells after ovulation. p27 is highly expressed in corpora lutea of control animals, but undetectable in granulose cells of the follicles; therefore p27 may prevent the differentiation of proliferating granulosa cells to nonproliferating luteal cells. The luteal phase defect in female mice deficient for p27 was not due to lack of circulating gonadotropins because the levels of FSH and LH were comparable in both knockout and wild type females. The administration of superphysiologic levels of gonadotropins induced ovulation, differentiation of corpora lutea, and early development of viable embryos in knockout females, and these embryos implanted but did not develop to term. Estrus is an indicator of endocrine function. *p27-/-* females had prolonged estrus cycle, which may reflect a defect in endocrine signaling between the pituitary and ovary, especially that these mice develop pituitary tumors.

3.3
Mouse Models with Hematopoietic Defects

3.3.1
Cdk6

Cdk6 and Cdk4 are closely related proteins in terms of biochemical properties. Cdk6 is expressed in most mammalian tissues, but is preferentially expressed in hematopoietic cells, and is most abundant in lymphoid organs (Meyerson et al. 1992; Meyerson and Harlow 1994). When Cdk6 is ablated in mice, hematopoietic tissues such as spleen and thymus display decreased cellularity. For example, the number of megakaryocytes in spleens from *Cdk6-/-* mice was reduced to one third of those present in wild type spleens (Malumbres et al. 2004). In addition, peripheral blood of *Cdk6-/-*

mice had reduced numbers of red blood cells. There is also delayed G1 progression in lymphocytes but not in MEFs from *Cdk6-/-* mice (Malumbres et al. 2004). Therefore, Cdk6 is not essential for proliferation of any specific cell lineage, however it is a regulator of the proliferative response of T lymphocytes upon mitogenic stimuli and is required for the expansion of differentiated populations. In agreement with this is the case in MEL erythroleukemia cells – transformed erythroid precursor cells blocked at the proerythroblast stage – in which differentiation requires inhibition of Cdk6 but not Cdk4 (Matushansky et al. 2000).

3.3.2
Cyclin D3

Cyclin D3 is expressed in nearly all proliferating cells (Bartkova et al. 1998), and its function is mostly redundant with other D-type cyclins in most cell types except T lymphocytes. Ablation of cyclin D3 in the mouse resulted in failure of the normal expansion of T lymphocytes (Sicinska et al. 2003). The process of T cell development involves the following sequential stages: $CD4^-CD8^-$ (double negative) cells, $CD4^+CD8^+$ (double positive) then $CD4^-CD8^+$ or $CD4^+CD8^-$ (single positive). The double negative population is subdivided in turn into DN-1, DN-2, DN-3 and DN-4. The proliferation of thymocytes during DN-1 to DN-3 is cytokine-dependent, and then immature lymphocytes rearrange β chains of their T cell receptor (TCR) and assemble the pre-TCR, which drives proliferation that become cytokine independent (Fehling et al. 1995). Signals from pre-TCR drive expansion of the DN-4 and of "immature single positive" (ISP) cells, which differentiate into double positive thymocytes and arrest their proliferation. *Cyclin D3-/-* mice are normal and fertile, however, they have hypoplastic thymi and sevenfold fewer thymocytes than wild type littermates (Sicinska et al. 2003). Cytokine-dependent proliferation of thymocytes from *cyclin D3-/-* mice (i.e., DN-1 to DN-3) was similar to that of wild type littermates; however, the pre-TCR-driven expansion of the DN-4 and ISP thymocytes was reduced in cyclin D3-deficient mice. Because *cyclin D3-/-* mice expressed normal levels of TCRβ and other pre-TCR components, this indicated that cyclin D3 functions downstream of the pre-TCR in driving proliferation of immature T lymphocytes. Indeed, cyclin D3 protein was found to be strongly induced at the DN-4 and ISP stages in wild type mice. On the other hand, cyclin D2 expression was high at the cytokine-dependent stages (DN-1 to DN-3) and disappeared after TCRβ rearrangement took place (Sicinska et al. 2003). Further studies by the same group using mice lacking $p56^{LCK}$ – a proto-oncogene tyrosine kinase downstream of pre-TCR- and intercrosses between *$p56^{LCK}$-/-* and *cyclin D3-/-* mice identified cyclin D3 as the major downstream target of the pre-TCR/$p56^{LCK}$ pathway (Sicinska et al. 2003). Therefore, cyclin D3 has a very specific role in transmitting pre-TCR-dependent mitogenic signals in immature T cells. Furthermore, the critical role of the D-type cyclins in hematopoietic cells was underscored by generation of mice lacking cyclin D2 and D3 – i.e., expressing only D1 – (Ciemerych et al. 2002) and the triple knockout mice lacking D1, D2, and D3 (Kozar et al. 2004). Embryos lacking cyclin D2 and D3 die at E18.5 due to severe megaloblastic anemia (Ciemerych et al. 2002). The double mutant embryos revealed normal morphogenesis in all tissues except developing livers. Because fetal livers are the major source of erythropoiesis at this stage of development, this defect was reflected in significant reduction in number but increase in size of mature red blood cells in peripheral blood of the embryos. This megaloblastic feature is caused by impaired division of erythroid precursors (Ciemerych et al. 2002). This indicated that proper division of erythroid precursors requires cyclin D2 or D3 or both and this specialized function of D2 and D3 cannot be compensated by upregulation of cyclin D1. Targeting of the three cyclin D family members resulted in severe megaloblastic anemia similar to *cyclin D2-/- D3-/-* embryos and multi-lineage hematopoietic failure, but also revealed proliferative failure of myocardial cells, a defect in a new compartment that was not seen before in mouse models lacking different combinations of cyclin D family members. This defect resulted in abnormal heart development (Kozar et al. 2004).

3.4
Mouse Models with Pancreatic Defects

The pancreatic islets are an endocrine organ secreting insulin (β cells), glucagon (α cells), somatostatin (δ cells) and other peptide hormones (Slack 1995). The islets play an important role in regulating glucose homeostasis. Therefore, regulation of the adult β cell mass is important for preserving insulin levels. Insufficient insulin secretion and inadequate β cell growth are central components of the pathogenesis of diabetes (Bell and Polonsky 2001; Butler et al. 2003; Yoon et al. 2003). Several mechanisms have been proposed to explain how new β cells are formed, including replication of preexisting cells and neogenesis from putative precursors (Bonner-Weir et al. 2004; Dor et al. 2004). Many factors regulate β cell growth and function including the insulin/IGF signaling pathway through IRS-2. The mitogen signal is then received by the D-type cyclins, which activate Cdk4/6. In addition to inactivating pRb, cyclin D/Cdk4 complexes promote cell cycle progression also by activating cyclin E and cyclin A/Cdk2 complexes in late G1 and S phase through sequestering their inhibitor p27 (reviewed in Sherr and Roberts 1999, 2004). Indeed, p27 is a principal cell cycle inhibitor in β cells, as it accumulates in the nucleus of β cells from obese mice, inhibiting compensatory β cell expansion (Uchida et al. 2005).

3.4.1
Cdk4

Ablation of Cdk4 in mice did not affect embryogenesis and mice are viable but display growth retardation and reproductive dysfunction (Rane et al. 1999; Tsutsui et al. 1999). Cdk4 also regulates the expansion of pancreatic islets because 80% of mice deficient for Cdk4 develop diabetes mellitus associated with progressive degeneration of pancreatic islets by six weeks of age. At three weeks of age, *Cdk4-/-* pancreas had already fewer islets with disorganized cellularity and apoptotic cells (Rane et al. 1999; Tsutsui et al. 1999). There was no compensatory upregulation of Cdk6 expression in the pancreas of *Cdk4-/-* mice compared to the levels of Cdk6 in wild type mice. In contrast, pancreatic expression of Cdk6 is lower in *Cdk4-/-* mice (Rane et al. 1999; Tsutsui et al. 1999). In addition, constitutively active Cdk4^{R24C} renders Cdk4 insensitive to inhibition by p16^{INK4a} and expanded the mass of functional β cells (Marzo et al. 2004). Moreover, islet specific rescue of Cdk4 disruption prevents diabetes (Martin et al. 2003). Therefore, Cdk4 is indispensable for the postnatal pancreatic β cells and consequently required for maintenance of glucose homeostasis (Mettus and Rane 2003; Rane et al. 1999; Tsutsui et al. 1999).

3.4.2
Cyclin D2

Cdk4 in association with one of the D-type cyclins operates to maintain functional pancreatic β cells. Kushner et al. (2005) showed that cyclin D2 and D1 are essential for postnatal islet growth. In adult mouse islets, basal levels of cyclin D2 mRNA expression were detected; cyclin D1 at lower levels but cyclin D3 was not detected. Prenatal islet development appeared normal in *cyclin D2-/-* mice, but β cell proliferation, adult mass and glucose tolerance were decreased in adult *cyclin D2-/-* mice causing glucose intolerance that progressed to diabetes by 12 weeks of age (Georgia and Bhushan 2004), i.e., later than the diabetes caused by Cdk4 deficiency at six weeks of age. *Cyclin D1+/-* mice do not develop diabetes but when crossed with *cyclin D2-/-* mice, the *cyclin D1+/-cyclin D2-/-* mice of the C57BL/6 sv129 mixed genetic background develop a much more severe islet growth deficiency and diabetes. They die of diabetes complications by four months of age. This indicates that although diabetes was not detected in *cyclin D1-/-* mice, cyclin D1 seems to partially compensate for cyclin D2 in regulation of β cell proliferation because when *cyclin D2-/-* mice lose one allele of cyclin D1 the appearance of the diabetes phenotype is accelerated (Kushner et al. 2005). This compensatory mechanism is specific to cyclin D1 because disruption of one allele of cyclin D3 does not worsen the diabetes phenotype of *cyclin D2-/-* mice.

3.5
Placental Defects and Endoreduplication

Ablation of cell cycle genes in mice sometimes leads to extra-embryonic defects that compromise development and causes embryonic lethality, in spite of the fact that the major organs in the embryo are indistinguishable from wild type organs. For example, mice lacking Rb, DP1, and cyclin E die from placental defects (see below). Placental defects cause poor exchange of metabolites and oxygen leading to secondary phenotypes such as developmental delay and yolk sac abnormalities. In general, when the mouse embryo reaches the blastocyst stage, two cellular lineages are distinguishable: the inner cell mass that will give rise to the embryo proper; and the trophectoderm that will form extraembryonic tissues (reviewed in Ciemerych and Sicinski 2005). Mammalian trophoblasts, which contribute to the placenta, are very prominent in wild type mouse placentas due to the giant size of trophoblast cell nuclei. Trophoblast giant cells undergo repeated rounds of DNA synthesis without intervening mitoses, a process called endoreplication or endoreduplication. Endoreduplication gives rise to cells with giant nuclei containing extra copies of genomic DNA up to 1000 N. In addition to the trophoblast giant cells in mammals, megakaryocytes that produce platelets become polyploid by endoreduplication up to 128 N as part of their differentiation program (reviewed in Zimmet and Ravid 2000). Megakaryocyte ploidy has been found to be associated with overexpression of cyclin D3 (Zimmet et al. 1997).

3.5.1
Cyclin E

Mice deficient for both cyclin E1 and E2 die at E11.5 due to placental dysfunction (Geng et al. 2003; Parisi et al. 2003). Mutant placentas have an overall normal structure but the nuclei of trophoblast giant cells show marked reduction in DNA content indicating that cyclin E deficient embryos fail to undergo endoreduplication. Embryonic lethality could be rescued by providing mutant embryos with wild type extraembryonic tissues (Geng et al. 2003) through "tetraploid blastocyst complementation" (Eggan et al. 2001; Tanaka and Kanagawa 1997). The rescued embryos died of lung abnormalities caused by the technique not by cyclin E deficiency. As mentioned above, endoreduplication occurs also in megakaryocytes and indeed cyclin E-deficient mice show reduced DNA content in megakaryocytes as a result of failed endoreduplication (Geng et al. 2003; Parisi et al. 2003). Therefore, cyclin E is dispensable for development of the embryo proper but required for endoreduplication. Cyclin E was postulated to cause loading of MCM proteins onto DNA replication origins during endoreplicative cycles of *Drosophila melanogaster* salivary glands (Su and O'Farrell 1998). In agreement with this, *cyclin E1-/-E2-/-* MEFs fail to re-enter the cell cycle after quiescent, G0 state induced by serum starvation (Geng et al. 2003; Parisi et al. 2003) despite normal induction of cyclin A and cyclin A-associated kinase activity and normal phosphorylation of Rb. In the absence of cyclin E, mutant MEFs fail to load MCM proteins onto their DNA replication origins (Geng et al. 2003). In quiescent G0 state, unlike continuously dividing cells, MCM and CDC6 are displaced from chromatin and must be reloaded during cell cycle reentry (Madine et al. 2000). Defective binding of MCM to replication origins in the absence of cyclin E can also cause the defects in endoreduplication. This function seems to be specific for cyclin E and is carried out equally by cyclin E1 and E2.

3.5.2
DP-1 and Rb

E2F transcription factors carry their functions after heterodimerization with members of the DP family; DP-1 and DP-2 (Helin et al. 1993; Wu et al. 1995; Zhang and Chellappan 1995). During mouse development high DP-1 expression is observed in both the embryo proper and the extraembryonic tissues, and it remains in adult tissues to be ubiquitously expressed but at lower levels (Gopalkrishnan et al. 1996; Kohn et al. 2003; Tevosian et al. 1996; Wu et al. 1995). Disruption of DP-1 resulted in embryonic lethality at E12.5 and examination of earlier embryos (E9.5–10.5) revealed that they show severe developmental delay (Kohn et al. 2003). DP-1 ablation resulted in perturbed development of the ectoplacental cone, and affected the trophectoderm giant cells, which displayed DNA replication failure. Additional experiments by injecting *DP-1-/-* ES cells into wild type blastocyts and generating chimeric embryos revealed that DP-1 is dispensable for the embryo proper (Kohn et al. 2004).

Mice lacking Rb die *in utero* at E12–E15 (mid gestation) due to severe anemia. In addition mutant embryos revealed defects in lens development, massive apoptosis in the central nervous system (CNS) and peripheral nervous system (PNS) and abnormal S phase entry of postmitotic neurons. In addition, there is a significant increase in immature nucleated erythrocytes (Clarke et al. 1992; Jacks et al. 1992; Lee et al. 1992; Morgenbesser et al. 1994). Using in vitro erythroid differentiating culture experiments, researchers have shown that Rb is essential for cell cycle exit and terminal differentiation of erythroid cells (Clark et al. 2004). Many phenotypes of Rb-deficiency could be ascribed to placental abnormalities. Abnormal expansion and differentiation of trophoblast cells caused the failure of the labyrinth development in Rb-deficient placentas (Wu et al. 2003). Chimeric mice composed of *Rb-/-* embryos and wild type placentas overcame many of the developmental abnormalities described above and the embryos developed to full term, even the development of erythroid lineage was entirely rescued. No visible abnormalities in the nervous system were detected in chimeric mice (Wu et al. 2003).

Collectively speaking, ablation of three different cell cycle genes DP-1, Rb or cyclin E (E1 and E2) resulted in a similar phenotype, which is embryonic lethality mainly due to defects in the extraembryonic tissues. However, the defects in extraembryonic tissues among the three mouse models have different causes related to specific roles of each cell cycle gene. For example, the defect in loading of MCM proteins is specific to cyclin E deficiency.

3.5.3
Skp2

As we described above, endoreduplication occurs normally in mammals within certain cell types such as trophoblast giant cells of the placenta and megakaryocytes, but it can also be observed in cells due to a defect in the cell cycle regulation of these cells. For example inhibition of Cdc2 kinase activity (the mitotic machinery) using a potent Cdc2 inhibitor butyrolactone I (Kitagawa et al. 1993) leads to nuclear enlargement and centrosome duplication. The DNA content of butyrolactone I-treated cells increases in multiples of 2C, a characteristic of endoreplication (Nakayama et al. 2004). Skp2 is an F-box protein and a substrate recognition component of an Skp1-Cullin-F-box protein (SCF) ubiquitin ligase. Skp2 binds to p27 and mediates its ubiquitylation and degradation by the proteasome (Carrano et al. 1999; Sutterluty et al. 1999; Tsvetkov et al. 1999). Skp2 also targets free cyclin E (not in complex with Cdk2) for ubiquitination (Nakayama et al. 2000). Both p27 and free cyclin E accumulate to high levels in *Skp2-/-* cells (Nakayama et al. 2000; Nakayama et al. 2001). The most obvious cellular phenotype of *Skp2-/-* mice is the presence of markedly enlarged, polyploid nuclei and multiple centrosomes, suggesting impairment of the mechanism that prevents endoreplication. Therefore the genomic DNA content of the cell has increased without cell division. When both Skp2 and its target p27 are ablated in the mouse endoreduplication associated with Skp2 deficiency is rescued (Nakayama et al. 2004). Furthermore, the increase in p27 levels in *Skp2-/-* cells results in inhibition of the Cdc2 kinase activity and a consequent block of entry into M phase (Nakayama et al. 2004). This block in M phase entry is responsible for the endoreplication phenotype, thus implicating p27 as a potent inhibitor of Cdc2.

From the above sections we observe two different situations in which the process of endoreduplication contributed to the phenotype of the mouse model: (1) failure of the normal process of endoreduplication taking place in trophoblast giant cells of the mouse placenta as a result of the lack of both forms of cyclin E or the lack of DP-1 contributed significantly to embryonic lethality. (2) In the case of Skp2 deficiency abnormal endoreduplication emerged as a result of failure of cells to enter mitosis due to inhibitory effects of excess p27 on Cdc2.

4
Tumorigenesis in Mouse Models of Cell Cycle Regulators

One of the hallmarks of cancer cells is misregulation of cell proliferation. Cancer cells are able to evade the normal signals that stop cell division and thereby escape from the quiescent state (Sherr and Roberts 2004). In this regard, cell cycle regulators that are not essential for normal somatic cell cycles may be required for oncogenic transformation. Oncogenic Ras plus other oncogenes, such as Myc, adenovirus E1A, or dominant negative p53DN, can transform wild type MEFs. In contrast, MEFs lacking D-type or E-type cyclins resist such transformation (reviewed in Sherr and Roberts 2004). Similarly, *Cdk4-/-* MEFs are refractory to transformation by oncogenic Ras plus p53^{DN} (Zou et al. 2002), and they senesce rapidly in culture implicating Cdk4 in the regulation of the proliferative capacity of the cell. Therefore, cell cycle regulators may contribute to oncogenic transformation by promoting emergence from quiescence and/or allowing cells to avoid senescence. This raises the question of whether living animals lacking cell cycle regulators are also resistant to cancer. In this regard mouse models provided an indispensable tool. There are several examples of mouse models in which disruption of certain cell cycle genes resulted in spontaneous tumor formation and in increased susceptibility to cancer when treated with oncogenes such as p27, p18, p16/p19^{Arf}, and p53, thus underscoring the role of these genes as tumor suppressors (Donehower et al. 1992; Fero et al. 1996, 1998; Franklin et al. 1998; Kiyokawa et al. 1996; Nakayama et al. 1996; Serrano et al. 1996). In contrast, other mouse models displayed resistance or reduced susceptibility to tumors induced by oncogenes or other means, for example mice lacking cyclin D1, D2, D3, and Cdk4. Mouse models have also presented some intriguing cases such as the development of ovarian tumors in all female mice lacking Cdk2 (Berthet, Aleem, and Kaldis, unpublished). Cdk2 promotes cell proliferation, thus adopting an oncogene-like role. Therefore when Cdk2 is ablated; we expect reduction in susceptibility of the mouse to tumors but not otherwise. In this section we present some examples of mouse models illustrating the role of cell cycle regulators in tumor development.

4.1
Pituitary Tumors

Studies on knockout mice deficient in the G1 control pathways suggest that the pituitary gland is very sensitive to perturbation of G1 regulation. Ablation of each of the tumor suppressors p27^{Kip1} (Fero et al. 1996; Kiyokawa et al. 1996; Nakayama et al. 1996), p18^{INK4c} (Franklin et al. 1998; Latres et al. 2000) or one allele of Rb [Rb+/−] (Harrison et al. 1995; Hu et al. 1994; Jacks et al. 1992; Maandag et al. 1994; Williams et al. 1994b) individually resulted in pituitary tumors of the intermediate lobe in mice. This similarity in pituitary phenotype may reflect a similar role for the three tumor suppressors in regulating pituitary cell proliferation and differentiation. When combinations of these tumor suppressors were ablated, pituitary tumorigenesis was accelerated or exacerbated indicating functional cooperation of these proteins in regulating pituitary homeostasis. In contrast, eight-week-old *Cdk4-/-* mice showed hypoplastic anterior pituitaries (Moons et al. 2002).

4.1.1
p27 and Rb

Targeted disruption of p27 in mice results in adenomas in the intermediate lobe of the pituitary with 100% penetrance (Fero et al. 1996; Kiyokawa et al. 1996; Nakayama et al. 1996). Although the function of p27 seems to be specifically required for the intermediate lobe of the pituitary, a reduction in p27 expression is sufficient to sensitize somatotrophs of the anterior pituitary (cells secreting growth hormone) to the proliferative actions of excess growth hormone releasing hormone (GHRH), resulting in earlier and increased penetrance of hGHRH-induced pituitary tumors (Teixeira et al. 2000). *p27-/-* and *p27+/-* mice expressing a metallotheionin promotor-driven hGHRH transgene also show synergistic effects of hGHRG transgene expression and p27 deficiency on liver, spleen and ovarian growth (Teixeira et al. 2000).

Rb +/- mice develop more aggressive pituitary adenocarcinomas in the intermediate lobe after loss of heterozigosity [LOH] (Nikitin and Lee 1996). These tumors are associated with a reduction in Arf expression (Carneiro et al. 2003). In addition, loss of p27 mRNA and protein expression was detected in tumor cells compared to wild type cells of the pituitary from *Rb+/-* mice indicating a possible regulation of p27 mRNA by Rb in intermediate lobe melanophors (Park et al. 1999). Furthermore, *Rb+/-p27-/-* mice were generated and found to develop earlier and more aggressive pituitary adenocarcinomas in comparison to *Rb+/-* or *p27-/-* mice suggesting that there is functional cooperation between Rb and p27 to suppress tumor development. Additional studies by the same group demonstrated that p27 deficiency contributed to the tumor development in *Rb+/-* background by abrogating an Arf-dependent apoptotic response in *Rb-/-* tumor cells (Park et al. 1999). Another common phenotype found in both *p27-/-* and *Rb+/-* mice is the hyperplasia of adrenal medulla (Nakayama et al. 1996; Williams et al. 1994a), which developed later in *p27-/-* mice to pheocromocytomas (Aleem et al. 2005). Similarly, both genotypes develop small tumors of the thyroid C (chromagranin positive) cell origin, but *Rb+/-p27-/-* mice developed earlier and more aggressive thyroid C cell carcinoma (Park et al. 1999) suggesting that the cooperation between p27 and Rb in tumor development is not restricted to the pituitary. The proapoptotic tumor suppressor function of p27 is one of two mechanisms by which loss of p27 can accelerate tumor development; the primary tumor suppressor function of p27 is due to its inhibitory effects on cell cycle progression, as it had been shown in mouse models of prostate cancer associated with heterozygosity of *Pten+/–* (Di Cristofano et al. 2001) and colon cancer associated with heterozygosity of *Min* (Philipp-Staheli et al. 2002). When these mouse models are placed into a p27-deficient background, tumor development is accelerated and there is a greater fraction of cycling cells in the tumor (Carneiro et al. 2003).

The results from mouse genetic experiments suggest that Rb and p27 do not function in one linear pathway, but their functions overlap. Rb knockout leads to faster p27 degradation because Rb interacts with the N-terminus of Skp2 and interferes with Skp2-p27 interaction and inhibits ubiquitination of p27. Disruption of p27 function or expression of the Skp2 N-terminus prevents Rb from causing G1 arrest (Ji et al. 2004). Interestingly, Rb mutants defective for E2F binding retain full activity in inhibiting Skp2-p27 interaction and can induce G1 arrest. These timed Rb re-expression experiments using Rb-deficient Saos-2 tumor cells to study the temporal relationship between the cell cycle arrest and E2F repression effects of Rb demonstrated that after expression of Rb, the decrease in the percentage of cells in S phase occurred at least 8 hours before the decrease in the protein levels of Cdk2, cyclin E, cyclin A and E2F1. The timing of G1 arrest coincides with the induction of p27 by Rb and with the decrease in the activity of cyclin E and cyclin A/Cdk2 due to inhibition by p27, but not with the transcription-dependent decrease in their protein levels. These findings suggest that the Rb-mediated G1 arrest is a two-step process: a fast E2F-independent initiation of G1 arrest involving p27, and a slower maintenance of this arrest by E2F-dependent repression (Ji et al. 2004).

4.1.2
p27 and p18

The phenotype of mice deficient for $p18^{INK4c}$ (Franklin et al. 1998; Latres et al. 2000) resembles very much that of mice deficient for p27 (Fero et al. 1996; Kiyokawa et al. 1996; Nakayama et al. 1996). *p18–/–* mice develop gigantism and widespread organomegaly and hyperplastic spleen, thymus and pituitary. Loss of p18, like p27 leads to gradual progression of intermediate pituitary lobe hyperplasia in young mice to adenomas by 10 months of age with complete penetrance (Franklin et al. 1998). This phenotype is greatly accelerated in animals deficient for both Cdk inhibitors; *p27–/–p18–/–* mice died of pituitary adenomas by 3 months of age indicating a functional collaboration between p27 and p18 (Franklin et al. 1998). This acceleration in pituitary tumor development reflects an additive effect, which means that p27 and p18 are in two separate pathways controlling pituitary proliferation, probably through controlling the function of Rb.

4.1.3
Cdk4 and Cdk4$^{R24C/R24C}$

When pathways involving the three tumor suppressors (p27, p18, and Rb) regulating pituitary proliferation and differentiation are compared, Cdk4 is a common regulator or target. Cdk4 is a direct target of p18 and p27 but it lies upstream of Rb; Cdk4 phosphorylates Rb at several sites relieving its tumor suppressor capacity. *Cdk4-/-* mice display severe postnatal hypoplastic anterior pituitaries but no alterations in the prenatal pituitary reflecting an indispensable role for Cdk4 for postnatal proliferation of the anterior pituitary, specifically for somatotrophs and lactotrophs but not for gonadotrophs (Jirawatnotai et al. 2004). Furthermore, pituitary hyperplasia induced by transgenic expression of hGHRH is completely abrogated in *Cdk4-/-* background confirming the negative impact of Cdk4-deficiency on GHRH-induced hyperplasia (Jirawatnotai et al. 2004). The fact that the negative effect of Cdk4 deficiency on proliferation of anterior pituitary and pancreatic islets (described above) is postnatal and not prenatal supports the notion that this effect is through Rb because it has been shown that the Rb-pathway is dispensable for early embryonic development (Clarke et al. 1992; Jacks et al. 1992; Lee et al. 1992). In contrast to the hypoplastic pituitary phenotype in mice deficient for Cdk4, *Cdk4$^{R24C/R24C}$* mice developed a wide range of tumors including pituitary adenomas and carcinomas arising in the intermediate or the anterior lobes (Rane et al. 2002; Sotillo et al. 2001a).

4.1.4
Cdk4$^{R24C/R24C}$ and p27

The activation of an oncogene and the simultaneous inactivation of a tumor suppressor gene predisposes cells to cancer, however, this depends to a great extent on the genetic interaction and the location of these proteins relative to each other in a given pathway. Mouse models helped to a great extent to uncover these interactions. For example, we described above that lack of either p18 or p27 or the expression of active mutated Cdk4^{R24C} induces similar phenotypes in the mouse pituitary and we know that both inhibitors bind to and inhibit Cdk4. Therefore, if p18, p27 and Cdk4 act in a linear pathway we expect exacerbated pituitary phenotypes (i.e. additive effects) in mice harboring the Cdk4^{R24C} mutation plus being deficient for one of the inhibitors. This turned to be true only in the case of p27 because *Cdk4$^{R24C/R24C}$* mice develop pituitary tumors with complete penetrance and short latency in a *p27-/-* or *p27+/-* background but not in a *p18-/-* background indicating that p18 acts only as an inhibitor of Cdk4 but p18 and Cdk4^{R24C} do not cooperate in pituitary tumor development, unlike Cdk4^{R24C} and p27 (Sotillo et al. 2005).

4.2
Skin Cancer and Melanoma

To further characterize the role of Cdk4 in tumorigenesis, skin tumors were induced in mice using the DMBA/TPA protocol. Cdk4 deficiency resulted in 98% reduction in the number of skin tumors compared to wild type animals (Rodriguez-Puebla et al. 2002). However, lack of Cdk4 did not affect proliferation of normal keratinocytes suggesting that Cdk4 may be a valuable therapeutic target because of its requirement in tumor cell proliferation but not the corresponding normal tissue. Moreover, Cdk4 is regulated by Myc (Grandori and Eisenman 1997), and it mediates Myc-induced tumorigenesis in mice by sequestering p27 and p21 thereby indirectly activating Cdk2 (Miliani de Marval et al. 2004). Transgenic mice expressing Myc from the Keratin 5 promoter (K5-Myc) display epithelial neoplasia in the skin and oral mucosa (Rounbehler et al. 2002; Rounbehler et al. 2001). K5-Myc transgenic mice deficient for Cdk4 were generated and loss of Cdk4 in these mice results in complete inhibition of tumor development (Miliani de Marval et al. 2004).

In contrast to the decreased susceptibility of *Cdk4–/–* mice to skin tumors, *Cdk4*$^{R24C/R24C}$ mice show extraordinary susceptibility to skin carcinogenesis induced by DMBA/TPA protocol (Rane et al. 2002) indicating that an inhibitor-resistant Cdk4 protein can be considered a potent oncogene. Nevertheless, although this mutation predisposes humans to melanoma (Zuo et al. 1996) the incidence of melanoma in *Cdk4*$^{R24C/R24C}$ mice is very low, when these mice were subjected to treatment with DMBA/TPA they were highly susceptible to melanoma development. In addition, deletion of p18^{INK4c} but not p15^{INK4b} confered proliferative advantage to melanocytic tumor growth (Sotillo et al. 2001b).

4.3
Breast Cancer

4.3.1
Cyclin D1 and Breast Tumors

Ablation of cyclin D1 in mice resulted in a narrow set of developmental abnormalities restricted to the retina and the nervous system (Fantl et al. 1995; Sicinski et al. 1995). However, the most obvious phenotype in *cyclin D1–/–* females is the failure of the mammary glands to undergo full lobuloalveolar development during the late stage of pregnancy. This defect is restricted to pregnancy-associated proliferation, because *cyclin D1–/–* mice developed normal mammary glands during sexual maturation (Fantl et al. 1995; Sicinski et al. 1995). The cooperation between cyclin D1 loss and overexpressed oncogenes in the induction and development of breast cancer was studied by crossing *cyclin D1–/–* mice with four different strains of mouse mammary tumor virus (MMTV)-oncogene transgenic mice (Yu et al. 2001) that overexpress the oncogenes v-Ha-*Ras* (Sinn et al. 1987), c-*neu* (Muller et al. 1988), c-*myc* (Stewart et al. 1984) and *Wnt-1* (Tsukamoto et al. 1988), respectively. Interestingly, these studies suggested that mice lacking cyclin D1 were resistant to breast cancer induced only by *Ras* and *neu* but not to that induced by *myc* or *Wnt-1*. Collectively, it is clear that the neu/Ras pathway is connected to the cell cycle machinery exclusively *via* the cyclin D1 promoter in mammary epithelial cells, but in other cell types they may signal also through cyclin D2 or D3 (Yu et al. 2001). Another line of evidence is the finding that when cyclin D1 is replaced by the human cyclin E driven by cyclin D1 promoter, these "knockin" mice succumb to breast cancer when crossed with transgenic MMTV-*neu* mice at the same incidence as *cyclin D1+/+MMTV-neu* mice (Yu et al. 2001). Furthermore, cyclin D1 plays an important role in breast cancer formation in humans too. The cyclin D1 gene is amplified in up to 20% of human breast cancers (Dickson et al. 1995), while cyclin D1 protein is overexpressed in over 50% of human mammary carcinomas (Gillett et al. 1994; McIntosh et al. 1995). This overexpression can be detected at the earliest stage of breast cancer progression such as ductal carcinoma in situ (Weinstat-Saslow et al. 1995). Once acquired, overexpression of cyclin D1 is maintained in all stages of the disease including the metastatic lesions (Bartkova et al. 1994; Gillett et al. 1996).

4.3.2
Cyclin E

Similar to cyclin D1, cyclin E is overexpressed in human cancers, particularly breast cancer. It is associated with increased tumor aggressiveness and poor patient outcome (Fukuse et al. 2000; Hwang and Clurman 2005; Keyomarsi et al. 2002). A series of human primary breast cancers was divided into two subtypes; one is characterized by high cyclin D1 and elevated Rb phosphorylation, and the second is characterized by high cyclin E, but low cyclin D1 and lack of corresponding Rb phosphorylation (Loden et al. 2002). These breast tumors with high cyclin E showed also perturbations in p53, p27, and Bcl-2. These results indicate that cyclin D and E have different mechanisms to inactivate the Rb pathway and thereby achieve unrestrained growth of breast tumor cells (Loden et al. 2002). Overexpression of a cyclin E transgene induces breast carcinomas in mice (Bortner and Rosenberg 1997). In contrast to the cooperation between Ras and Neu with cyclin D1 to induce breast cancer (see Sect. 4.3.1), cyclin E expression increases in breast tumors arising in transgenic mice carrying MMTV-c-*Myc* but not the MMTV-v-Ha-*Ras* oncogene (Geng et al. 2001b), reflecting that c-*Myc*-dependent pathways may control the expression of cyclin E. Furthermore, *cyclin E1–/–E2–/–* MEFs are resistant to oncogenic transformation by c-*Myc*, H-*Ras* plus c-*Myc*, or H-*Ras* plus p53^{DN} and show reduced susceptibility for oncogenic transformation by H-*Ras* plus *E1A* (Geng et al. 2003). At the first thought, one may conclude that the mechanism by which cyclin E contributes to tumorigenesis is by continuous activation of Cdk2, the sole catalytic partner of cyclin E (according to the textbook cell cycle model). However, this is not the most important effect of cyclin E overexpression because cyclin E mutants defective to form an active kinase complex with Cdk2 are unable to drive cells from G1 into S phase but can still malignantly transform rat embryo fibroblasts in cooperation with H-*Ras* confirming that the oncogenic activity of cyclin E is independent of Cdk2 activation and p27 binding (Geisen and Moroy 2002). As to the mechanisms of oncogenicity of cyclin E; increased cyclin E stability achieved through mutating threonine 393, the phosphorylation site which is necessary to trigger its ubiquitination, causes genetic instability, and is oncogenic by itself, and *cyclin E^{T393A}*-/- mice had greatly increased susceptibility to Ras-induced lung cancer (Loeb et al. 2005).

4.4
Ovarian Tumors

Of all mouse models of cell cycle regulators, mice lacking Cdk2 (Berthet et al. 2003; Ortega et al. 2003) or cyclin E1/E2 (Geng et al. 2003; Parisi et al. 2003) showed the most unexpected phenotype. These mouse models challenged the cell cycle model described above by proving that continuous somatic cell proliferation can progress without Cdk2 or cyclin E. In tumor cells, most Cdks are overexpressed or their kinase activities are increased. For example, Cdk2, Cdk4 and Cdc2 are overexpressed on the protein level in primary colorectal carcinoma tissue than in adjacent normal tissue (Kim et al. 1999). Similarly, Cdk2 and Cdk4 overexpression was detected in early pancreatic intraepithelial neoplasias and progressively increased in carcinomas (Al-Aynati et al. 2004). Therefore, because of their functions in driving cell proliferation Cdks may be considered oncogenes, as we described above in case of Cdk4. Hence, loss of Cdk2 should either reduce the susceptibility of the animal to spontaneous or carcinogen-induced tumors or if it acts as a weak oncogene, its loss may not affect tumor development either positively or negatively. But it is intriguing to find that female mice lacking Cdk2 develop ovary tumors (cystadenomas) after 1.5–2 years of their life (Berthet, Aleem, and Kaldis, unpublished observation). Furthermore, this phenotype is accelerated when p27 is ablated on a *Cdk2*-/- background; the earliest ovary tumor appeared in a 7-month old female and by 10 months old, all *p27*-/-*Cdk2*-/- females had ovary tumors, reflecting a collaborative function between Cdk2 and p27 in ovary tumor development (Aleem et al. 2005). Cdk2 is essential for the meiotic cell cycle as all mice lacking Cdk2 are sterile and ovaries from *Cdk2*-/- mice show aberrant or no follicles (Berthet et al. 2003; Ortega et al. 2003). Because the ovary tumors are accompanied by granulosa cell hyperplasia, the role of Cdk2 in granulosa cell proliferation and differentiation should be further studied.

5
New Functions for Old Players

Results from mouse models revealed new concepts of cell cycle regulation. When cell cycle regulation remains intact after ablation of a particular gene, the search for compensatory mechanisms starts immediately. It is critical for the cell to preserve its integrity and it must switch on or upregulates the function of other proteins when a certain protein is off. For example, when three forms of the same D-type cyclins exist such as cyclin D1, D2, and D3, the cell may redistribute or upregulate the expression of one or the other D-type cyclin in the absence of the third one. In this case we do not describe this as taking up a new function because the compensation that occurred for example by D2 or D3 for D1 served to rescue the original general function of the D-type cyclins; that is, the reception of environmental cues and the consequent activation of Cdk4/6. However, if the compensatory mechanism in the mouse model involves a new function that was not known before for a cell cycle protein and which occurs in wild type as well as in knockout cells, then we may identify a new function for a known player. The following example illustrates this latter concept: the regulation of G1/S transition by Cdc2 (Aleem et al. 2005), known primarily for its mitotic function in association with cyclin B. The second example identifies an unexpected function for a tumor suppressor: the regulation of cell migration by p27.

5.1
Cdc2 Regulates S Phase Entry

As we described above, *Cdk2-/-* and *cyclin E-/-* mouse models are one of the recent surprises contradicting established cell cycle models. The only cyclin/Cdk complex known to operate during the fundamental stages of cell cycle, i.e., G1/S transition and S phase is the complex formed by cyclin E and Cdk2. In *Cdk2-/-* cells, there was no cyclin E activity detected above background levels (Berthet et al. 2003; Ortega et al. 2003). Therefore, we expected that when the expression of cyclin E in *cyclin E1-/-E2-/-* mice is lost, we should get similar phenotypes to the *Cdk2-/-* mice, but actually the phenotype of *cyclin E1-/-E2-/-* mice is more severe. *Cyclin E1-/-E2-/-* mice show defective endoreplication of megakaryocytes and trophoblasts leading to embryonic lethality and MEFs deficient for cyclin E1 and E2 fail to re-enter the cell cycle after quiescence and are resistant to oncogenic transformation (Geng et al. 2003; Parisi et al. 2003). These results suggest that cyclin E performs some physiological functions independent of Cdk2 activation. Indeed, (Matsumoto and Maller 2004) showed that cyclin E may initiate S phase entry through an undefined role in centrosomes rather than by activating Cdk2, and previously it has been demonstrated that the oncogenic potential of cyclin E is independent of activating Cdk2 (Geisen and Moroy 2002). How could the different phenotypes of *Cdk2-/-* and cyclin *E1-/-E2-/-* be reconciled? Several reviews discussed different scenarios such as the possibility that cyclin D/Cdk4 may compensate for cyclin E/Cdk2, especially that it does associate with the pre-RC (Gladden and Diehl 2003). A second alternative was that an unidentified kinase can partner with cyclin E and promote S phase entry in the absence of Cdk2 or another possibility that is not mutually exclusive with this is that a non-Cdk kinase might be able to compensate completely for Cdk2 especially in human tumor cells such as Aurora kinase (Hinds 2003). This issue has been resolved recently when we demonstrated that Cdc2 compensates for Cdk2 function and can control the G0/G1 and G1/S transition in *Cdk2-/-* MEFs as well as in wild type MEFs (Aleem et al. 2005), and reviewed in (Bashir and Pagano 2005; Kaldis and Aleem 2005; Welcker and Clurman 2005). Cdk2 was thought to be the major target of p27 (discussed above). If this is true, then the phenotypic defects of *p27-/-* mice should be modified by the loss of Cdk2. We generated *p27-/-Cdk2-/-* mice that displayed most of the defects in *p27-/-* mice such as pituitary intermediate lobe adenomas, enlarged thymi due to increased thymocyte proliferation, sterility of both genders similar to *Cdk2-/-* mice and accelerated development of ovary tumors (as we discussed in Sect. 4.4). We also found an increase in the proportion of thymocytes in mitosis and in S phase in both *p27-/-* and *p27-/-Cdk2-/-* mutants suggesting that p27 has other targets than Cdk2 that are able to promote entry into S phase and M phase (Aleem et al. 2005). The fact that Cdk2 is not the only target of p27 was supported by the finding that ectopic expression of p27 or p21 efficiently inhibits cell cycle progression in *Cdk2-/-* MEFs (Martin et al. 2005). In our study we could detect an upregulation in the kinase activity associated with cyclin E and Cdc2 in both *p27-/-* and *p27-/-Cdk2-/-* cells. Therefore, cyclin E associates with a catalytic subunit and this kinase activity is susceptible to p27 inhibition. Immunoprecipitation experiments revealed that cyclin E does bind and activate Cdc2, therefore providing a possible explanation of how DNA replication takes place in the absence of Cdk2. Our results do not exclude the possibility that cyclin E displays additional kinase independent functions. Cyclin E activity was not detected in *Cdk2-/-* cells previously (Berthet et al. 2003; Ortega et al. 2003) because cyclin E/Cdc2 activity in *Cdk2-/-* cells is low in the presence of p27 (Aleem et al. 2005). In *Cdk2-/-* cells, the amount of p27 that is normally bound to Cdk2 is probably redistributed to cyclin E/Cdc2 complexes. Silencing of Cdc2 caused a more pronounced inhibition in S phase entry after quiescence in *Cdk2-/-* compared to wild type MEFs (Aleem et al. 2005), indicating that Cdc2 becomes essential for S phase entry in the absence of Cdk2. These results suggest that Cdc2 not only plays an essential role in mitosis but also in S phase. Although our findings proves a functional overlap between Cdk2 and Cdc2 there is a considerable difference in their ability to compensate for each other; Cdk2-deficient mice are viable whether this is due to the full compensation by Cdc2 alone or in combination with other Cdks, however, *Cdc2-/-*

embryos die at a very early stage of embryonic development (Martin et al. 2005) indicating that Cdk2 does not compensate as efficiently for Cdc2 functions. It would be intriguing to define the functions of Cdc2 that cannot be compensated by Cdk2.

5.2
p27 Regulates the Rho Pathway

Although p27 is known for its negative regulation of cell proliferation through its inhibitory effects on Cdk2 and Cdc2, a new interesting function of p27 in regulation of cell migration has been recently discovered (McAllister et al. 2003). It is well established that a decrease or loss of p27 protein or its sequestration in the cytoplasm away from Cdks in the nucleus increases the incidence of tumor formation induced by different oncogenic events (Bloom and Pagano 2003; Philipp-Staheli et al. 2001; Slingerland and Pagano 2000). Similarly, in humans a low abundance of p27 is associated with tumor aggressiveness and patient mortality for a variety of tumors, characteristics that are related to tumor invasion and metastasis. In other words, the functions of p27 are modulated by changes in its abundance and in its subcellular localization, with its relocalization in the cytoplasm triggered by its phosphorylation on serine 10 (Ishida et al. 2002; Rodier et al. 2001). From the studies by McAllister (2003) and Besson (2004) it has been shown that when p27 is found in the cytoplasm it positively influences cell migration. Cell migration is very important for tumor metastasis and is regulated by Rho proteins, which coordinate the cytoskeletal remodeling that underlies changes in cell adhesion and migration (reviewed in Collard 2004). Guanine nucleotide exchange factors (GEFs) activate Rho proteins by promoting the replacement of guanine diphosphate (GDP) with guanine triphosphate (GTP). RhoA promotes the formation of actin stress fibres and focal adhesions through the recruitment and activation of its effectors including the Rho-kinases, ROCK1 and ROCK2 (Amano et al. 1997; Leung et al. 1996; Ridley and Hall 1992). Although Rho activity is important for motility, its role in the formation of stress fibers and focal adhesions can lead to an inhibition of cell migration by increasing cellular adhesion to its substrate and preventing focal adhesion turnover (Ren et al. 2000; Vial et al. 2003). p27 inhibits RhoA activation by interfering with its interaction with GEFs, thus promoting cell migration (Besson et al. 2004). Consistent with this, they found also that the motility of *p27–/–* fibroblasts decreases dramatically in comparison to that of wild type cells because a reduction in p27 levels increases RhoA activity and consequently ROCK activity, thereby reducing migration. Re-expression of wild type p27 or a truncated form of p27 that cannot bind to cyclin/Cdk complexes restores migration, which in the case of tumor cells may promote tumor progression and invasiveness (Besson et al. 2004). Taken together, it is apparent that p27 has dual functions, one as a tumor suppressor and another as an oncogene depending on its subcellular localization. When localized to the nucleus, p27 acts as a tumor suppressor and binds to cyclin/Cdk complexes through its amino-terminal region and inhibits cell proliferation. When p27 is phosphorylated and translocated into the cytoplasm it binds to RhoA through its carboxy terminus and interferes with RhoA activation by GEFs (Collard 2004).

6
Genetic Interaction and Functional Complementation of Cell Cycle Regulators

In the previous sections we described the phenotypes of several mouse models deficient for one or more cell cycle regulators. We can now distinguish between four types of mouse models: (a) mice lacking only one gene to answer whether this gene is "essential" or not, and if not we can learn its specific function. For example, we learned from *Cdk2–/–* mice that Cdk2 is not essential but it is required in meiosis. (b) Double knockout mice (e.g., cyclin D1 and p27) to investigate the genetic interaction and functional complementation of these genes as we will see below. (c) Mouse models lacking all isoforms of a particular protein, which is known to have an essential function (e.g., cyclin D1, D2, D3 triple knockout). (d) The fourth type of mouse models is when a certain gene can be replaced by another one (e.g., knockin of cyclin E into cyclin D1 locus). We learned that when a certain gene is lost in the mouse, compensation of the loss either through employing the closest family member, or through bypassing this particular pathway and switching to another mechanism is often observed.

6.1
Interactions of Cyclin D1 and p27

Cyclin D1 and p27 are different in their cell cycle regulatory functions; the former has a stimulatory role whereas the latter has an inhibitory role. These opposing functions are reflected in the phenotypes of mice lacking cyclin D1 and p27, respectively. While mice deficient for cyclin D1 show dwarfism, exhibit retinal and pregnant mammary gland hypoplasia, and reduced susceptibility to tumor formation (Fantl et al. 1995; Sicinski et al. 1995), *p27–/–* mice display gigantism, multiorgan hyperplasia with dysregulation in the retinal structure and increased susceptibility to tumorigenesis (Fero et al. 1996; Kiyokawa et al. 1996; Nakayama et al. 1996). In addition to their Cdk-dependent role, the D-type cyclins may control the activity of cyclin E indirectly through titration of p27 (Sherr and Roberts 1999). When both cyclin D1 and p27 are ablated in the same animal, correction of some of the defects to the wild type phenotype occurs providing evidence of genetic interaction between cyclin D1 and p27 and that they are counteracting forces acting in the same pathway (Geng et al. 2001a; Tong and Pollard 2001). *Cyclin D1–/– p27–/–* mice develop normally and do not exhibit premature mortality unlike *cyclin D1–/–* mice that are very small and some die within three weeks after birth (Fantl et al. 1995; Sicinski et al. 1995). Furthermore, ablation of p27 in *cyclin D1–/–* mice fully corrects the retinal phenotype of cyclin D1 deficiency and restores normal retinal development. On the molecular level, the rescue of cyclin D1 retinal hypoplasia in double mutants is not dependent on other D-type cyclins because the level of expression of D2 and D3 is barely detectable in retinas of wild type, *p27–/–*, and double mutants. However, Cdk2 kinase activity is markedly increased in the retinas of *p27–/–* and *cyclin D1–/–p27–/–* mice, but whether this increase in Cdk2 activity is responsible for the rescue of the retinal phenotype remains to be elucidated. From our recent report (Aleem et al. 2005), we can speculate that Cdc2 kinase activity may be also increased in the same way because p27 inhibits both Cdk2 and Cdc2. It is possible that the ablation of p27 in *cyclin D1–/–* mice results in an overall increase in cell proliferation due to increased Cdk2 and Cdc2 activities and this rescues the cyclin D1-associated hypoplasia. Similarly, the mammary epithelial phenotype of *cyclin D1–/–* animals is corrected and the epithelium undergoes full lobuloalveolar development (Geng et al. 2001a; Tong and Pollard 2001). On the other hand, ablation of cyclin D1 does not rescue any of the phenotypes of *p27–/–* mice. Importantly, the rate of development of pituitary adenomas is not reduced in *cyclin D1–/–p27–/–* mice (Geng et al. 2001a; Tong and Pollard 2001). These findings suggest that p27 functions may be downstream of cyclin D1.

6.2
Functional Complementation of Cdc2 and Cdk2 in G1/S Phase Transition

In Sect. 5.1, we described our recent results identifying a new function for Cdc2 in regulating S phase entry in *Cdk2–/–* animals. This is an example of functional complementation of two different Cdks. We already know that both Cdc2 and Cdk2 bind cyclin A, and we demonstrated that Cdc2 binds to and is activated by cyclin E, and Cdk2 binds cyclin B1, meaning that both Cdc2 and Cdk2 have the same cyclin partners. Furthermore, the same inhibitor, p27, inhibits both Cdc2 and Cdk2 (Aleem et al. 2005). As a matter of fact, Cdc2 has been shown before to bind to cyclin E in human cells (Koff et al. 1991) and Cdc2 is known to regulate S phase entry in yeast as we described in details in section 2.2. In addition, another report demonstrated that in *Xenopus* egg extracts, depletion of both Cdk2 and Cdc2 using p13^{suc1} beads inhibits DNA replication and replacement of either Cdk2 or Cdc2 proteins individually stimulates S phase initiation (Chevalier et al. 1995). In this experimental set up, both *Xenopus* and human Cdc2 mRNA stimulates S phase while mutant forms fail to restore this activity, illustrating that the restoration of S phase entry is not an artifact of the system used. Therefore, Cdc2 carries out the G2/M and the G1/S transition, thereby performing a unique function of its own (in G2/M) in addition to sharing a function with Cdk2 (in G1/S). This raises two questions: why have we overlooked Cdc2 function in G1/S transition until now? And why does the multicellular organism need so many Cdks if Cdc2 can regulate both S and M phases? The answer to the first question as discussed (Bashir and Pagano 2005) is that Cdk2 activity is high during G1/S and this might obscure the comparatively low Cdc2 G1/S activity. The second reason is that the G1/S Cdc2 activity is considered negligible compared with the high Cdc2 activity in G2/M phase, which led us to believe that the only function of Cdc2 is during G2/M. As long as Cdc2 is active during both S and M phases, there should be some mechanism to stop DNA replication until mitosis is completed and stop mitosis until DNA replication has properly taken place. This is to ensure a correct cell division resulting in identical daughter cells. Full activation of Cdc2 is prevented during S and G2 phase by the pathway involving ATR, Chk1 and Cdc25 phosphatases (Bartek et al. 2004). In addition, the inhibition of Cdc2 by p27 (Aleem et al. 2005; Nakayama et al. 2004) may be another mechanism to coordinate both S and M phase machineries (Pagano 2004). Furthermore, subcellular localization is another important level of regulation to control the activity of cyclin B/Cdc2 (Moore et al. 2003) and determine their specificity (reviewed also by Takizawa and Morgan 2000). In vertebrates, cyclin B is predominantly cytoplasmic (Moore et al. 2002; Pines and Hunter 1991; Yang et al. 1998) and it has a nuclear export signal (NES) between residues 108-117 (Hagting et al. 1998; Toyoshima et al. 1998). Cyclin B/Cdc2 complexes display S phase promoting abilities that can be unmasked by relocating it from the cytoplasm to the nucleus and moderately stimulating its activity (Moore et al. 2003). The authors generated a cyclin B1 fusion protein that retains the Cdk-activating portion of cyclin B1 but lacks its NES and accumulates in the nucleus, and they introduced the nuclear localization sequence (NLS) of cyclin E into the *N*-terminus of cyclin B1. They added this GST-cyclin E/B1 chimera to *Xenopus* egg extracts lacking endogenous cyclin E and detected substantial DNA synthesis. They experimented with different concentrations of the cyclin E/B1 fusion protein and found that only on adding the appropriate concentration of cyclin E-B1 a gradual increase of Cdc2 activity took place that first triggered replication then entry into mitosis, which terminated further DNA synthesis. Cyclin B/Cdc2 acts to suppress its inhibitors, Wee1 and Myt1 kinases, and it stimulates its activators, the Cdc25 phosphatases (Mueller et al. 1995). Wee1 and Myt1 suppress the kinase activity of low concentrations of cyclin B/Cdc2 until cyclin B levels rise beyond a critical threshold value that fires the autocatalytic activation mechanism. To overcome the inhibitory effects of endogenous Wee1 and Myt1, a variant of Cdc25B phosphatase containing only the catalytic domain was added (Moore et al. 2003). Collectively speaking, cyclin B/Cdc2 activity can trigger

DNA replication if a suitable level of activity is targeted to the nucleus. However, cyclin E/Cdk2 is unable to provoke entry into mitosis, even when present in large amounts (Furuno et al. 1999; Moore et al. 2002; Strausfeld et al. 1996). In a similar fashion, the major mitotic B-type cyclins in fission and budding yeast are able to compensate for the absence of their S phase promoting counterparts (Amon et al. 1994; Fisher and Nurse 1996; Haase and Reed 1999; Miller and Cross 2001).

To answer the second question of why the cell needs so many Cdks given the high degree of flexibility and redundancy among them, we believe that this reflects the complexity of higher animals. As in a soccer game, there are main players and reserve ones, who attend the whole game and their role is to replace the main players in the event of injury or towards the end of the game when one of the main players is exhausted. In analogy, we can imagine Cdk2 as a main player during G1/S transition with its activity peaking during this time and Cdc2 as a reserve player, whose activity is kept low but is fully redundant with Cdk2. During G2/M transition the control is tighter and we do not know of any other Cdk that can completely replace the activity of cyclin B/Cdc2 during M phase. This raises the question of why the regulation of M phase is much tighter than S phase? The reason could be that M phase is the final stage of the cell division cycle after which a cell gives rise directly to two daughter cells and there is no return, so may be the regulation at this point is very tight to ensure that if there are any defects that escaped the checkpoints in G1, S or G2 they will be caught up in M phase and prevented from being passed to the offspring.

6.3
Functional Cooperation Between Cdk2, Cdk4 and p27

The genetic targeting of Cdk4 in mice exhibited the functional cooperation between Cdk4, Cdk2, and p27 as has been reviewed in (Aleem et al. 2004; Sherr and Roberts 1999) and elsewhere in this book. In summary, *Cdk4–/–* MEFs exhibit a four hour delay in S phase entry after quiescence induced by serum starvation. This coincides with a marked decrease in the activity of Cdk2 (Tsutsui et al. 1999). However, the S phase delay is restored when p27 is ablated from *Cdk4–/–* mice and *Cdk4–/–p27–/–* MEFs re-enter the cell cycle with almost normal kinetics after quiescence. p27 binds both Cdk4 and Cdk2 but at any given time most of p27 is in complex with cyclin D/Cdk4, and this complex is active because p27 has been shown to serve as an assembly factor for cyclin D and Cdk4 (Cheng et al. 1999). However, in the case of *Cdk4–/–* MEFs, most of p27 will be redistributed to cyclin E and cyclin A/Cdk2, thereby inhibiting these complexes and causing the delay in S phase entry. It is, therefore, conceivable that the deletion of p27 from *Cdk4–/–* mice relieved Cdk2 inhibition and S phase progressed with normal kinetics. It is also noteworthy to mention that *Cdk2–/–* MEFs exhibit a 4-hour delay in S phase entry similar to *Cdk4-/-* MEFs, and this delay is restored by ectopic expression of wild type Cdk2 (Berthet et al. 2003). The sequestration of p27 by Cdk4 from Cdk2 represent a functional cooperation between these three cell cycle proteins to ensure proper G1/S transition and S phase progression. In other situations, as in single cyclin D embryos, i.e., embryos lacking two of the three D-type cyclins, the levels of p27 in different organs are reduced on the posttranscriptional level. However, the down-regulation of p27 in several organs did not result in increase in the kinase activity of Cdk2 as would be expected, further supporting that p27 might have other targets.

6.4
Compensation Between the D-type Cyclins

The three D-type cyclins are expressed often in mutually exclusive cell types during mouse development. In addition to their growth promoting functions, they play non-redundant roles in promoting cell differentiation of specific cellular compartments. Nevertheless, ablation of individual or combinations of the D-type cyclins does not affect the development and morphogenesis of major organs in the embryo (reviewed in Ciemerych and Sicinski 2005; Sherr and Roberts 2004). This indicates that there is functional complementation between cyclin D1, D2, and D3. If one or two members of the family are absent, the third D-type cyclin compensates their loss and its expression is reactivated or upregulated in the tissues, in which it is normally not expressed or expressed at low levels. For example, developing spleens of wild type animals express cyclins D2 and D3 but minimal levels of cyclin D1, but in spleens of *cyclin D2-/-D3-/-* animals, cyclin D1 is greatly upregulated. Therefore, in single cyclin embryos the tissue specific pattern of expression of D-type cyclins is lost, and the remaining cyclin becomes ubiquitously expressed in all tissues (Ciemerych et al. 2002). This indicates that the functions of D-type cyclins can be interchanged. *Cyclin D1-/-D2-/-D3-/-* triple knockout embryos developed relatively normally until mid gestation and died prior to day E17.5. The majority of cell types in these embryos could proliferate relatively normally in the absence of all the D-type cyclins indicating that proliferation could occur in a cyclin D-independent fashion. The proliferative failure was limited to myocardial cells and to hematopoietic stem cells, demonstrating that D-type cyclins are required specifically in these cell lineages Kozar et al. (2004). Interestingly, the very same compartments require Cdk4 and Cdk6 as the *Cdk4-/-Cdk6-/-* embryos (Malumbres et al. 2004) show similar phenotypes to the *cyclin D1-/-D2-/-D3-/-* animals. These findings provided genetic evidence that the main function of the D-type cyclins in development is to activate their catalytic partners Cdk4 and Cdk6 but in contrast to the textbook model of cell cycle regulation, the cyclin D-Cdk4/6 complex is not essential for proliferation of most cell types. The surprise from this mouse model is that the D-type cyclins were thought to be critical as sensors to extracellular mitogenic stimuli but *cyclin D1-/-D2-/-D3-/-* cells behave exactly like wild type cells in response to serum requirements and can reenter the cell cycle efficiently after quiescence by serum starvation, however, the mechanism of mitogenic response in the triple knockout cells is still unknown at this point. As to the mechanism by which the triple knockout cells proliferate, (Kozar et al. 2004) demonstrated that *cyclin D1-/-D2-/-D3-/-* MEFs were resistant to p16^{INK4a} inhibition but remained sensitive to inhibition by the Cdk2- and Cdc2-inhibitor p27, indicating that the mechanism of proliferation is independent of Cdk4/6. Furthermore, knockdown of Cdk2 using shRNA strongly inhibited proliferation in asynchronous *cyclin D1-/-D2-/-D3-/-* MEFs, as well as inhibited their re-entry into the cell cycle after serum starvation, whereas it had minimal effects on wild type cells (Kozar et al. 2004). Collectively, these results demonstrated that in the event of total loss of cyclin D, Cdk4 and Cdk6 become functionless and the cell shifts to Cdk2-dependent proliferation. Nevertheless, although proliferation proceeded using Cdk2 in the triple knockout cells, the Rb sites, which are phosphorylated by cyclin D/Cdk remained hypophosphorylated but this did not affect proliferation indicating that the cyclin E and cyclin A-driven phosphorylation of Rb appears to suffice for the functional inactivation of Rb (Kozar et al. 2004).

6.5
Interactions Between Cdk4 and Cdk6

Mice deficient for both Cdk4 and Cdk6 die *in utero* or hours after birth (Malumbres et al. 2004). However, this is not due to defects in embryonic development but as a result of severe anemia due to limited proliferation of erythroid progenitors. Given the fact that both Cdk4 and Cdk6 are known to be the Cdks responsible for initiating cell cycle progression in response to mitogenic stimuli, it is surprising that serum starved MEFs are able to re-enter the cell cycle upon mitogenic stimulation with normal kinetics (Malumbres et al. 2004). In addition, *Cdk4-/-Cdk6-/-* MEFs proliferate well and have a similar doubling time to wild type MEFs at P2 but a significantly longer time at P4 (72 versus 52 hours). This is reflected in a drop in the percentage of cells entering S phase from 28% at P2 to 14% at P4, which is not observed in the wild type MEFs. Furthermore, *Cdk4-/-Cdk6-/-* MEFs display features of senescence at P4-P5, in contrast to the wild type MEFs, which display the same features at a later passage number. However, all *Cdk4-/-Cdk6-/-* MEFs become immortalized upon continuous culture following 3T3 protocol. This phenotype of *Cdk4-/-Cdk6-/-* MEFs is similar to *Cdk2-/-* MEFs that stop growing in culture at P4-P5. However, not all *Cdk2-/-* MEFs can be immortalized and the few ones, which become immortal require longer passaging than their wild type counterparts (Aleem, Berthet, and Kaldis, unpublished observation). This result indicates that there are other Cdks that can compensate the loss of Cdk4 and Cdk6 and bind to the D-type cyclins to respond to extracellular mitogenic signals in MEFs, but not in the hematopoietic cells in which Cdk6 is preferentially expressed. Silencing of Cdk2 reduced the proliferation of *Cdk4-/-Cdk6-/-* MEFs in comparison to wild type MEFs. These results and the resistance of *Cdk4-/-Cdk6-/-* MEFs to inhibition by p16 but not to inhibition by p21 follow a similar pattern to the D-type triple knockout cells (Kozar et al. 2004) and indicate that Cdk2 may compensate the loss of Cdk4/Cdk6 and the D-type cyclins. To study the synergy between Cdk6 and Cdk2, mice deficient for both Cdk6 and Cdk2 were generated and displayed phenotypes similar to those observed in Cdk2 and Cdk6 single mutants (Malumbres et al. 2004) Therefore, there is no synergism between Cdk6 and Cdk2 deficiencies in vivo.

6.6
Cyclin E Can Functionally Compensate for Cyclin D1

Both cyclin E and D1 are G1 cyclins but they share limited sequence similarity [only 38% in the most conserved "cyclin box" domain] (Inaba et al. 1992; Xiong et al. 1992). A knockin of cyclin E into the cyclin D1 locus was generated, in which the coding exons of cyclin D1 were deleted and replaced with the human cyclin E cDNA (Geng et al. 1999). The knockin (KI) mice do not express cyclin D1 but express cyclin E in its place under control of the mouse cyclin D1 promoter, consequently cyclin E was targeted to tissues normally expressing cyclin D1. In these mice cyclin E could rescue the neurologic abnormalities, the retinal and mammary gland hypoplasia in *cyclin D1-/-* mice. On the molecular level, immunoprecipitation of human cyclin E from extracts of the KI-mice pulled down Cdk2, but neither Cdk4 nor Cdk6 indicating that cyclin E did not associate with the cyclin D-catalytic partners Cdk4 and Cdk6 but exclusively with its normal partner Cdk2. In addition, the human cyclin E containing complexes did not bring down significant amounts of p27 in contrast to cyclin D1 immunoprecipitates from wild type extracts, and they did not rescue the decrease in Rb phosphorylation seen in *cyclin D1-/-* tissues. Therefore, the cyclin E-D1 knockin mouse model provides another example, in which one cell cycle regulator can compensate the loss of another one not by fully replacing its function, in this situation full restoration of Rb phosphorylation, but possibly by bypassing this function (Geng et al. 1999).

7
Implications of Data from Cell Cycle Mouse Models to Human Cancer

Cancer is the outcome of perturbations in many pathways including cell cycle control. Cyclins, Cdks, and Cdk inhibitors are either targets for genetic modifications in cancer or are disrupted indirectly by other oncogenic events. We reviewed above the impact of mutations in cell cycle genes on tumor development in the mouse, but the most important question raised is: how about the relevance of these data to human cancer? Can every genetic alteration that induces tumors in mice give rise to the same phenotype in humans? Is our current knowledge of cell cycle regulation sufficient to transfer the lab data to clinical trials? When we find that a certain Cdk is not essential for somatic cell proliferation, does it necessarily mean tumor cells do not require it? There are a lot of pathways known to be altered in human cancer; in particular cyclin D-Cdk4/6-INK4-Rb-E2F pathway is altered in about 80% of human cancer (Ortega et al. 2002). In addition, more than 50% of human neoplasia shows alterations in p53 and its downstream effectors. We will not discuss either the p53 or Rb pathways because they have been extensively reviewed elsewhere, instead we will focus on Cdk2 in human cancer.

7.1
Cdk2 in Human Tumors and in Tumor Cell Lines

As we have described at the beginning of this chapter, the textbook model of cell cycle regulation has assigned Cdk2 the function of a master regulator of G1/S transition and progression through S phase, through Rb phosphorylation, and p27 phosphorylation leading to p27 degradation (Sherr and Roberts 1999). Indeed, pharmacologic inhibitors of Cdk2 are in development and several are in clinical testing. Examples of Cdk2 inhibitors include flavopiridol, UCN-01, roscovitine, and olomucine (reviewed in Fischer and Gianella-Borradori 2003; Wadler 2002). Nevertheless, the majority of these inhibitors are not specific for Cdk2 but can inhibit also other Cdks. It is also highly likely that the effects of Cdk2 inhibition depend on the cell type. The fact that Cdk2 is not required for normal mitotic cell division does not necessarily mean that it is not required in tumor cells. The first report that questioned the requirement of Cdk2 in certain tumor cell lines was by Tetsu and McCormick (2003). They demonstrated that inhibition of Cdk4 through inhibition of the MEK/MAP kinase pathway is sufficient to induce G1 arrest in colon cancer cell lines. In contrast, inhibition of Cdk2 by p27 redistribution or ectopic p27 expression has no effect on proliferation since these cell lines were chosen for this criterium. They also used dominant negative Cdk2, antisense nucleotides and siRNA against Cdk2 and found that Cdk2 activity may be dispensable in this particular setting. Osteosarcomas and Rb-negative cervical cancer cells continued to proliferate in spite of Cdk2 inhibition (Tetsu and McCormick 2003). Results from *Cdk2–/–* mice further corroborated the dispensability of Cdk2 for somatic cell proliferation. However, the requirement of Cdks in tumor cells may differ from normal cells in the sense that cell proliferation and differentiation in response to physiologic levels of growth factors in vivo might be able to proceed with only low levels of net G1 Cdk activity, whereas an increased threshold of sustained mitogenic signals invokes a requirement for these G1 phase regulators in order for transformation to occur (Sherr and Roberts 2004). Mutations in Cdk2 have not been often described in human cancers but alterations in its expression have been associated with malignancy. For example, the expression of Cdk2 and its partner cyclin E has been studied in the progression of colorectal carcinoma (Li et al. 2001). The authors studied normal mucosa, hyperplastic polyps, adenomas, adenocarcinomas in adenomas, primary cancers, lymph node and hepatic metastases. They found that the expression of Cdk2 increased significantly mainly from the primary to the lymph node metastatic foci. In another study using colorectal carcinoma from 23 patients (AJCC/UICC Stage II–III), about 30% had increased Cdk2 activity in cancer tissue compared with adjacent normal tissue (Kim et al. 1999). The increased expression of Cdk2 is not limited to colorectal cancer but was also observed in other cancers such as epithelial ovarian tumors (Sui et al. 2001), in which Cdk2 expression was increased gradually from benign, to borderline, and finally to malignant tumors. In fact, this pattern of expression does not mean that Cdk2 is required in tumors but maybe its expression is increased in order to cope with the higher rate of proliferation of malignant cells versus benign ones and in this situation the data from cell lines can be reconciled with those from human tumors. If this is the case, then Cdk2 may not be the optimal therapeutic target in human cancer and possibly if we inhibit Cdk2 pharmacologically, tumor cells will shift to other Cdks to compensate for Cdk2 inhibition.

8
Conclusions

Genetic targeting of cell cycle components in the mouse contributed significantly to our understanding of the cell cycle machinery. Analysis of mice lacking a particular cell cycle gene provides clear-cut answers to questions of whether a gene is essential or not for animal development, and whether it is required for the proliferation and differentiation of a specific cellular compartment. For example, D-type cyclins and Cdk4/6 are specifically required for the hematopoietic system (Kozar et al. 2004; Malumbres et al. 2004). Mouse phenotypes provided also genetic evidence of interaction and cooperation between particular cell cycle proteins. Combined ablation of p107 and p130 led to more pronounced phenotypes than targeting each gene individually, revealing overlapping functions for these proteins in development (Cobrinik et al. 1996). Ablation of p27 in cyclin D1 background rescued the defects of cyclin D1 deficiency reflecting that p27 lies downstream of cyclin D1 (Geng et al. 2001a; Tong and Pollard 2001), however ablation of p27 did not rescue the phenotype of Cdk2-deficient mice indicating that p27 and Cdk2 are not in a linear pathway (Aleem et al. 2005; Martin et al. 2005). As a matter of fact, the mild phenotype of Cdk2-deficient mice (Berthet et al.

2003; Ortega et al. 2003) but not *cyclin E1-/-E2-/-* mice (Geng et al. 2003; Parisi et al. 2003) was one of the most unexpected findings in this field, raising the possibility that cyclin E may activate other kinase partners, but the undetectable cyclin E-associated kinase activity in *Cdk2-/-* cells led to the conclusion that cyclin E may have kinase-independent functions. The resolution of this dilemma came from analysis of *p27-/-Cdk2-/-* mice, in which we demonstrated that cyclin E activates Cdc2 that can in turn regulate S phase entry in *Cdk2-/-* and wild type cells, and that both Cdk2 and Cdc2 are major targets of p27 (Aleem et al. 2005). In summary, analysis of mouse models of cell cycle regulators is very informative but limitations remain. For example, the mutant animals develop in the absence of a given protein. The animal, hence, may activate compensatory mechanisms that allow cell proliferation to occur in the long-term absence of this particular gene. We do not know whether animals will switch on the same compensatory mechanisms in the event of acute sudden loss of a given protein. Indeed, an acute loss of Rb in primary cells has different phenotypic consequences from germ-line ablation of Rb function (Sage et al. 2003). The same limitation is valid when we apply data from our analysis of knockout mice to human tumors. The tumor cell has its own program of regulation of cell cycle, apoptosis and differentiation and the fact that normal cell proliferation can proceed in the absence of a certain cell cycle protein does not mean it is not required by tumor cells.

Taken together, analyses of genetically targeted mice contributed significantly to our understanding of the regulation of mammalian cell cycle and led to the re-consideration of the current cell cycle model and the design of a new one (Fig. 1) based on the fact that the mammalian cell cycle is not much different from the yeast cell cycle in the fact that it appears to have only one essential Cdk, which is Cdc2. Cdc2 can regulate progression through all phases of the cell cycle in complex with different cyclins and p27 is an essential component regulating the kinase activities of Cdk4/6, Cdk2, and Cdc2. The second conclusion is that different cyclin/Cdk complexes and the high degree of redundancy between them, reflects the complexity of animals and

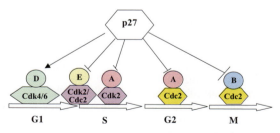

Fig. 1 A new model for mammalian cell cycle regulation showing Cdk2/Cdc2 in complex with cyclin E controlling the G1/S transition, and p27 as a master regulator of all cyclin/Cdk complexes in different phases of the cell cycle

revealed that these complexes cannot operate in stage-specific linear pathways but they are members of a large network.

Acknowledgements We thank Cyril Berthet for discussion, support, and communication of unpublished results. We are grateful to all members of the Kaldis lab for their support. We are also indebted to Nancy Jenkins and Neal Copeland and all members of the Mouse Cancer Genetics Program, at the NCI-Frederick for their thoughtful discussions and continuous support. Our thanks extend also to Hiroaki Kiyokawa, Northwestern, and Michele Pagano, NYU, for support and discussions. The work in our laboratory was supported by the Intramural Research Program of the NIH, National Cancer Institute, Center for Cancer Research.

References

Ait-Si-Ali S, Ramirez S, Barre FX, Dkhissi F, Magnaghi-Jaulin L, Girault JA, Robin P, Knibiehler M, Pritchard LL, Ducommun B, Trouche D, Harel-Bellan A (1998) Histone acetyltransferase activity of CBP is controlled by cycle-dependent kinases and oncoprotein E1A. Nature 396:184–186

Al-Aynati MM, Radulovich N, Ho J, Tsao MS (2004) Overexpression of G1-S cyclins and cyclin-dependent kinases during multistage human pancreatic duct cell carcinogenesis. Clin Cancer Res 10:6598–6605

Aleem E, Berthet C, Kaldis P (2004) Cdk2 as a master of S phase entry: fact or fake? Cell Cycle 3:35–37

Aleem E, Kiyokawa H, Kaldis P (2005) Cdc2-cyclin E complexes regulate the G1/S phase transition. Nat Cell Biol 7:831–836

Amano M, Chihara K, Kimura K, Fukata Y, Nakamura N, Matsuura Y, Kaibuchi K (1997) Formation of actin stress fibers and focal adhesions enhanced by Rho-kinase. Science 275:1308–1311

Amon A, Irniger S, Nasmyth K (1994) Closing the cell cycle circle in yeast: G2 cyclin proteolysis initiated at mitosis persists until the activation of G1 cyclins in the next cycle. Cell 77:1037–1050

Arata Y, Fujita M, Ohtani K, Kijima S, Kato JY (2000) Cdk2-dependent and -independent pathways in E2F-mediated S phase induction. J Biol Chem 275:6337–6345

Ashley T, Walpita D, de Rooij DG (2001) Localization of two mammalian cyclin dependent kinases during mammalian meiosis. J Cell Sci 114:685–693

Bartek J, Lukas C, Lukas J (2004) Checking on DNA damage in S phase. Nat Rev Mol Cell Biol 5:792–804

Bartkova J, Lukas J, Muller H, Lutzhoft D, Strauss M, Bartek J (1994) Cyclin D1 protein expression and function in human breast cancer. Int J Cancer 57:353–361

Bartkova J, Lukas J, Strauss M, Bartek J (1998) Cyclin D3: requirement for G1/S transition and high abundance in quiescent tissues suggest a dual role in proliferation and differentiation. Oncogene 17:1027–1037

Bashir T, Pagano M (2005) Cdk1: the dominant sibling of Cdk2. Nat Cell Biol 7:779–781

Bell GI, Polonsky KS (2001) Diabetes mellitus and genetically programmed defects in beta-cell function. Nature 414:788–791

Berthet C, Aleem E, Coppola V, Tessarollo L, Kaldis P (2003) Cdk2 knockout mice are viable. Curr Biol 13:1775–1785

Besson A, Gurian-West M, Schmidt A, Hall A, Roberts JM (2004) p27^{Kip1} modulates cell migration through the regulation of RhoA activation. Genes Dev 18:862–876

Beumer TL, Kiyokawa H, Roepers-Gajadien HL, van den Bos LA, Lock TM, Gademan IS, Rutgers DH, Koff A, de Rooij DG (1999) Regulatory role of p27^{Kip1} in the mouse and human testis. Endocrinology 140:1834–1840

Bloom J, Pagano M (2003) Deregulated degradation of the cdk inhibitor p27 and malignant transformation. Semin Cancer Biol 13:41–47

Bonner-Weir S, Toschi E, Inada A, Reitz P, Fonseca SY, Aye T, Sharma A (2004) The pancreatic ductal epithelium serves as a potential pool of progenitor cells. Pediatr Diabetes 5 Suppl 2:16–22

Booher RN, Alfa CE, Hyams JS, Beach DH (1989) The fission yeast cdc2/cdc13/suc1 protein kinase: regulation of catalytic activity and nuclear localization. Cell 58:485–497

Bortner DM, Rosenberg MP (1997) Induction of mammary gland hyperplasia and carcinomas in transgenic mice expressing human cyclin E. Mol Cell Biol 17:453–459

Brandeis M, Rosewell I, Carrington M, Crompton T, Jacobs AM, Kirk J, Gannon J, Hunt T (1998) Cyclin B2-null mice develp normally and are fertile whereas cyclin B1-null mice die in utero. Proc Natl Acad Sci USA 95:4344–4349

Butler AE, Janson J, Bonner-Weir S, Ritzel R, Rizza RA, Butler PC (2003) Beta-cell deficit and increased beta-cell apoptosis in humans with type 2 diabetes. Diabetes 52:102–110

Carneiro C, Jiao MS, Hu M, Shaffer D, Park M, Pandolfi PP, Cordon-Cardo C, Koff A (2003) p27 deficiency desensitizes Rb-/- cells to signals that trigger apoptosis during pituitary tumor development. Oncogene 22:361–369

Carrano AC, Eytan E, Hershko A, Pagano M (1999) SKP2 is required for ubiquitin-mediated degradation of the CDK inhibitor p27. Nat Cell Biol 1:193–199

Chapman DL, Wolgemuth DJ (1992) Identification of a mouse B-type cyclin which exhibits developmentally regulated expression in the germ line. Mol Reprod Dev 33:259–269

Chapman DL, Wolgemuth DJ (1993) Isolation of the murine cyclin B2 cDNA and characterization of the lineage and temporal specificity of expression of the B1 and B2 cyclins during oogenesis, spermatogenesis and early embryogenesis. Development 118:229–240

Chen Z, Indjeian VB, McManus M, Wang L, Dynlacht BD (2002) CP110, a cell cycle-dependent CDK substrate, regulates centrosome duplication in human cells. Dev Cell 3:339–350

Cheng A, Ross KE, Kaldis P, Solomon MJ (1999) Dephosphorylation of cyclin-dependent kinases by type 2C protein phosphatases. Genes Dev 13:2946–2957

Chevalier S, Tassan JP, Cox R, Philippe M, Ford C (1995) Both cdc2 and cdk2 promote S phase initiation in Xenopus egg extracts. J Cell Sci 108:1831–1841

Ciemerych MA, Kenney AM, Sicinska E, Kalaszczynska I, Bronson RT, Rowitch DH, Gardner H, Sicinski P (2002) Development of mice expressing a single D-type cyclin. Genes Dev 16:3277–3289

Ciemerych MA, Sicinski P (2005) Cell cycle in mouse development. Oncogene 24:2877–2898

Clark AJ, Doyle KM, Humbert PO (2004) Cell-intrinsic requirement for pRb in erythropoiesis. Blood 104:1324–1326

Clarke PR, Leiss D, Pagano M, Karsenti E (1992) Cyclin A- and cyclin B-dependent protein kinases are regulated by different mechanisms in *Xenopus* egg extraacts. EMBO J 11:1751–1761

Cobrinik D, Lee MH, Hannon G, Mulligan G, Bronson RT, Dyson N, Harlow E, Beach D, Weinberg RA, Jacks T (1996) Shared role of the pRB-related p130 and p107 proteins in limb development. Genes Dev 10:1633–1644

Collard JG (2004) Cancer: Kip moving. Nature 428:705–708

Connell-Crowley L, Elledge SJ, Harper JW (1998) G1 cyclin-dependent kinases are sufficient to initiate DNA synthesis in quiescent human fibroblasts. Curr Biol 8:65–68

Correa-Bordes J, Nurse P (1995) p25^{rum1} orders S phase and mitosis by acting as an inhibitor of the p34^{cdc2} mitotic kinase. Cell 83:1001–1009

Cross FR (1988) DAF1, a mutant gene affecting size control, pheromone arrest, and cell cycle kinetics of Saccharomyces cerevisiae. Mol Cell Biol 8:4675–4684

Di Cristofano A, De Acetis M, Koff A, Cordon-Cardo C, Pandolfi PP (2001) Pten and p27^{KIP1} cooperate in prostate cancer tumor suppression in the mouse. Nat Genet 27:222–224

Dickson C, Fantl V, Gillett C, Brookes S, Bartek J, Smith R, Fisher C, Barnes D, Peters G (1995) Amplification of chromosome band 11q13 and a role for cyclin D1 in human breast cancer. Cancer Lett 90:43–50

Donehower LA, Harvey M, Slagle BL, McArthur MJ, Montgomery CA Jr, Butel JS, Bradley A (1992) Mice deficient for p53 are developmentally normal but susceptible to spontaneous tumours. Nature 356:215–221

Dor Y, Brown J, Martinez OI, Melton DA (2004) Adult pancreatic beta-cells are formed by self-duplication rather than stem-cell differentiation. Nature 429:41–46

Draetta G, Beach D (1988) Activation of cdc2 protein kinase during mitosis in human cells: cell cycle-dependent phosphorylation and subunit rearrangement. Cell 54:17–26

Draviam VM, Orrechia S, Lowe M, Pardi R, Pines J (2001) The localization of human cyclins B1 and B2 determines CDK1 substrate specificity and neither enzyme requires MEK to disassemble the Golgi apparatus. J Cell Biol 152:945–958

Dulic V, Lees E, Reed SI (1992) Association of human cyclin E with a periodic G1-S phase protein kinase. Science 257:1958–1961

Edgar BA, Lehner CF (1996) Developmental control of cell cycle regulators: a fly's perspective. Science 274:1646–1652

Eggan K, Akutsu H, Loring J, Jackson-Grusby L, Klemm M, Rideout WM III, Yanagimachi R, Jaenisch R (2001) Hybrid vigor, fetal overgrowth, and viability of mice derived by nuclear cloning and tetraploid embryo complementation. Proc Natl Acad Sci USA 98:6209–6214

Elledge SJ, Richman R, Hall FL, Williams RT, Lodgson N, Harper JW (1992) *CDK2* encodes a 33-kDa cyclin A-associated protein kinase and is expressed before *CDC2* in the cell cycle. Proc Natl Acad Sci USA 89:2907–2911

Elledge SJ, Spottswood MR (1991) A new human p34 protein kinase, CDK2, identified by complementation of a *cdc28* mutation in *Saccharomyces cerevisiae*, is a homolog of *Xenopus* Eg1. EMBO J 10:2653–2659

Fang F, Newport JW (1991) Evidence that the G1-S and G2-M transitions are controlled by different cdc2 proteins in higher eukaryotes. Cell 66:731–742

Fantl V, Stamp G, Andrews A, Rosewell I, Dickson C (1995) Mice lacking cyclin D1 are small and show defects in eye and mammary gland development. Genes Dev 9:2364–2372

Fehling HJ, Krotkova A, Saint-Ruf C, von Boehmer H (1995) Crucial role of the pre-T-cell receptor alpha gene in development of alpha beta but not gamma delta T cells. Nature 375:795–798

Felzien LK, Farrell S, Betts JC, Mosavin R, Nabel GJ (1999) Specificity of cyclin E-Cdk2, TFIIB, and E1A interactions with a common domain of the p300 coactivator. Mol Cell Biol 19:4241–4246

Fero ML, Randel E, Gurley KE, Roberts JM, Kemp CJ (1998) The murine gene p27^{Kip1} is haplo-insufficient for tumour suppression. Nature 396:177–180

Fero ML, Rivkin M, Tasch M, Porter P, Carow CE, Firpo E, Polyak K, Tsai LH, Broudy V, Perlmutter RM, Kaushansky K, Roberts JM (1996) A syndrome of multiorgan hyperplasia with features of gigantism, tumorigenesis, and female sterility in p27^{Kip1}-deficient mice. Cell 85:733–744

Firpo EJ, Koff A, Solomon MJ, Roberts JM (1994) Inactivation of a Cdk2 inhibitor during interleukin 2-induced proliferation of human T lymphocytes. Mol Cell Biol 14:4889–4901

Fischer PM, Gianella-Borradori A (2003) CDK inhibitors in clinical development for the treatment of cancer. Expert Opin Investig Drugs 12:955–970

Fisher D, Nurse P (1995) Cyclins of the fission yeast Schizosaccharomyces pombe. Semin Cell Biol 2:73–78

Fisher D, Nurse P (1996) A single fission yeast mitotic cyclin B p34^{cdc2} kinase promotes both S-phase and mitosis in the absence of G1 cyclins. EMBO J 15:850–860

Franklin DS, Godfrey VL, Lee H, Kovalev GI, Schoonhoven R, Chen-Kiang S, Su L, Xiong Y (1998) CDK inhibitors p18^{INK4c} and p27^{Kip1} mediate two separate pathways to collaboratively suppress pituitary tumorigenesis. Genes Dev 12:2899–2911

Fukuse T, Hirata T, Naiki H, Hitomi S, Wada H (2000) Prognostic significance of cyclin E overexpression in resected non-small cell lung cancer. Cancer Res 60:242–244

Furstenthal L, Swanson C, Kaiser BK, Eldridge AG, Jackson PK (2001) Triggering ubiquitination of a CDK inhibitor at origins of DNA replication. Nat Cell Biol 3:715–722

Furuno N, den Elzen N, Pines J (1999) Human cyclin A is required for mitosis until mid prophase. J Cell Biol 147:295–306

Gallant P, Nigg EA (1994) Identification of a novel vertebrate cyclin: cyclin B3 shares properties with both A- and B-type cyclins. EMBO J 13:595–605

Geisen C, Moroy T (2002) The oncogenic activity of cyclin E is not confined to Cdk2 activation alone but relies on several other, distinct functions of the protein. J Biol Chem 277:39909–39918

Geng Y, Whoriskey W, Park MY, Bronson RT, Medema RH, Li T, Weinberg RA, Sicinski P (1999) Rescue of cyclin D1 deficiency by knockin cyclin E. Cell 97:767–777

Geng Y, Yu Q, Sicinska E, Das M, Bronson RT, Sicinski P (2001a) Deletion of the p27^{Kip1} gene restores normal development in cyclin D1-deficient mice. Proc Natl Acad Sci USA 98:194–199

Geng Y, Yu Q, Whoriskey W, Dick F, Tsai KY, Ford HL, Biswas DK, Pardee AB, Amati B, Jacks T, Richardson A, Dyson N, Sicinski P (2001b) Expression of cyclins E1 and E2 during mouse development and in neoplasia. Proc Natl Acad Sci USA 98:13138–13143

Geng Y, Yu Q, Sicinska E, Das M, Schneider JE, Bhattacharya S, Rideout WM, Bronson RT, Gardner H, Sicinski P (2003) Cyclin E ablation in the mouse. Cell 114:431–443

Georgia S, Bhushan A (2004) Beta cell replication is the primary mechanism for maintaining postnatal beta cell mass. J Clin Invest 114:963–968

Gillett C, Fantl V, Smith R, Fisher C, Bartek J, Dickson C, Barnes D, Peters G (1994) Amplification and overexpression of cyclin D1 in breast cancer detected by immunohistochemical staining. Cancer Res 54:1812–1817

Gillett C, Smith P, Gregory W, Richards M, Millis R, Peters G, Barnes D (1996) Cyclin D1 and prognosis in human breast cancer. Int J Cancer 69:92–99

Gladden AB, Diehl JA (2003) The cyclin D1-dependent kinase associates with the pre-replication complex and modulates RB-MCM7 binding. J Biol Chem 278:9754–9760

Gopalkrishnan RV, Dolle P, Mattei MG, La Thangue NB, Kedinger C (1996) Genomic structure and developmental expression of the mouse cell cycle regulatory transcription factor DP1. Oncogene 13:2671–2680

Grandori C, Eisenman RN (1997) Myc target genes. Trends Biochem Sci 22:177–181

Haase SB, Reed SI (1999) Evidence that a free-running oscillator drives G1 events in the budding yeast cell cycle. Nature 401:394–397

Hadwiger JA, Wittenberg C, Richardson HE, de Barros Lopes M, Reed SI (1989) A family of cyclin homologs that control the G1 phase in yeast. Proc Natl Acad Sci USA 86:6255–6259

Hagting A, Karlsson C, Clute P, Jackman M, Pines J (1998) MPF localization is controlled by nuclear export. EMBO J 17:4127–4138

Harbour JW, Luo RX, Dei Santi A, Postigo AA, Dean DC (1999) Cdk phosphorylation triggers sequential intramolecular interactions that progressively block Rb functions as cells move through G1. Cell 98:859–869

Harrison DJ, Hooper ML, Armstrong JF, Clarke AR (1995) Effects of heterozygosity for the Rb-1t19neo allele in the mouse. Oncogene 10:1615–1620

Hartwell LH, Culotti J, Pringle JR, Reid BJ (1974) Genetic control of the cell division cycle in yeast. Science 183:46–51

Hayles J, Fisher D, Woollard A, Nurse P (1994) Temopral order of S phase ans mitosis in fission yeast is determined by the state of the $p34^{cdc2}$-mitotic B cyclin complex. Cell 78:813–822

Hedger M, de Kretser D (2000) Leydig cell function and its regulation. Results Probl Cell Differ 28:69–110

Helin K, Wu CL, Fattaey AR, Lees JA, Dynlacht BD, Ngwu C, Harlow E (1993) Heterodimerization of the transcription factors E2F-1 and DP-1 leads to cooperative trans-activation. Genes Dev 7:1850–1861

Herskowitz I (1987) Functional inactivation of genes by dominant negative mutations. Nature 329:219–222

Hinds PW (2003) Cdk2 dethroned as master of S phase entry. Cancer Cell 3:305–307

Hinds PW, Mittnacht S, Dulic V, Arnold A, Reed SI, Weinberg RA (1992) Regulation of retinoblastoma protein functions by ectopic expression of human cyclins. Cell 70:993–1006

Hoffmann I, Draetta G, Karsenti E (1994) Activation of the phosphatase activity of human cdc25A by a cdk2-cyclin E dependent phosphorylation at the G1/S transition. EMBO J 13:4302–4310

Howe JA, Howell M, Hunt T, Newport JW (1995) Identification of a developmental timer regulating the stability of embryonic cyclin A and a new somatic A-type cyclin at gastrulation. Genes Dev 9:1164–1176

Hu N, Gutsmann A, Herbert DC, Bradley A, Lee WH, Lee EY (1994) Heterozygous Rb-1 delta 20/+ mice are predisposed to tumors of the pituitary gland with a nearly complete penetrance. Oncogene 9:1021–1027

Hwang HC, Clurman BE (2005) Cyclin E in normal and neoplastic cell cycles. Oncogene 24:2776–2786

Inaba T, Matsushime H, Valentine M, Roussel MF, Sherr CJ, Look AT (1992) Genomic organization, chromosomal localization, and independent expression of human cyclin D genes. Genomics 13:565–574

Ishida N, Hara T, Kamura T, Yoshida M, Nakayama K, Nakayama KI (2002) Phosphorylation of p27^{Kip1} on serine 10 is required for its binding to CRM1 and nuclear export. J Biol Chem 277:14355–14358

Jackman M, Firth M, Pines J (1995) Human cyclins B1 and B2 are localized to strikingly different structures: B1 to microtubules, B2 primarily to the Golgi apparatus. EMBO J 14:1646–1654

Jacks T, Fazeli A, Schmitt EM, Bronson RT, Goodell MA, Weinberg RA (1992) Effects of an Rb mutation in the mouse. Nature 359:295–300

Ji P, Jiang H, Rekhtman K, Bloom J, Ichetovkin M, Pagano M, Zhu L (2004) An Rb-Skp2-p27 pathway mediates acute cell cycle inhibition by Rb and is retained in a partial-penetrance Rb mutant. Mol Cell 16:47–58

Jirawatnotai S, Aziyu A, Osmundson EC, Moons DS, Zou X, Kineman RD, Kiyokawa H (2004) Cdk4 is indispensable for postnatal proliferation of the anterior pituitary. J Biol Chem 279:51100–51106

Kaldis P, Aleem E (2005) Cell cycle sibling rivalry: Cdc2 vs. Cdk2. Cell Cycle 4:1491–1494

Keyomarsi K, Tucker SL, Buchholz TA, Callister M, Ding Y, Hortobagyi GN, Bedrosian I, Knickerbocker C, Toyofuku W, Lowe M, Herliczek TW, Bacus SS (2002) Cyclin E and survival in patients with breast cancer. N Engl J Med 347:1566–1575

Kim JH, Kang MJ, Park CU, Kwak HJ, Hwang Y, Koh GY (1999) Amplified CDK2 and cdc2 activities in primary colorectal carcinoma. Cancer 85:546–553

King RW, Peters J-M, Tugendreich S, Rolfe M, Hieter P, Kirschner MW (1995) A 20S complex containing CDC27 and CDC16 catalyzes the mitosis-specific conjugation of ubiquitin to cyclin B. Cell 81:279–288

Kitagawa M, Okabe T, Ogino H, Matsumoto H, Suzuki-Takahashi I, Kokubo T, Higashi H, Saitoh S, Taya Y, Yasuda H, et al. (1993) Butyrolactone I, a selective inhibitor of cdk2 and cdc2 kinase. Oncogene 8:2425–2432

Kiyokawa H, Kineman RD, Manova-Todorova KO, Soares VC, Hoffman ES, Ono M, Khanam D, Hayday AC, Frohman LA, Koff A (1996) Enhanced growth of mice lacking the cyclin-dependent kinase inhibitor function of p27^{Kip1}. Cell 85:721–732

Knoblich JA, Lehner CF (1993) Synergistic action of Drosophila cyclins A and B during the G2-M transition. EMBO J 12:65–74

Koff A, Cross F, Fisher A, Schumacher J, Leguellec K, Philippe M, Roberts JM (1991) Human cyclin E, a new cyclin that interacts with two members of the CDC2 gene family. Cell 66:1217–1228

Koff A, Giordano A, Desai D, Yamashita K, Harper JW, Elledge S, Nishimoto T, Morgan DO, Franza BR, Roberts JM (1992) Formation and activation of a cyclin E-cdk2 complex during the G1 phase of the human cell cycle. Science 257:1689–1694

Koff A, Ohtsuki M, Polyak K, Roberts JM, Massague J (1993) Negative regulation of G1 in mammalian cells: inhibition of cyclin E-dependent kinase by TGF-beta. Science 260:536–539

Kohn MJ, Bronson RT, Harlow E, Dyson NJ, Yamasaki L (2003) Dp1 is required for extra-embryonic development. Development 130:1295–1305

Kohn MJ, Leung SW, Criniti V, Agromayor M, Yamasaki L (2004) Dp1 is largely dispensable for embryonic development. Mol Cell Biol 24:7197–7205

Kozar K, Ciemerych MA, Rebel VI, Shigematsu H, Zagozdzon A, Sicinska E, Geng Y, Yu Q, Bhattacharya S, Bronson RT, Akashi K, Sicinski P (2004) Mouse development and cell proliferation in the absence of D-cyclins. Cell 118:477–491

Krimpenfort P, Quon KC, Mooi WJ, Loonstra A, Berns A (2001) Loss of p16^{Ink4a} confers susceptibility to metastatic melanoma in mice. Nature 413:83–86

Krude T, Jackman M, Pines J, Laskey RA (1997) Cyclin/Cdk-dependent initiation of DNA replication in a human cell-free system. Cell 88:109–119

Kushner JA, Ciemerych MA, Sicinska E, Wartschow LM, Teta M, Long SY, Sicinski P, White MF (2005) Cyclins D2 and D1 are essential for postnatal pancreatic beta-cell growth. Mol Cell Biol 25:3752–3762

Lahue EE, Smith AV, Orr-Weaver TL (1991) A novel cyclin gene from Drosophila complements CLN function in yeast. Genes Dev 5:2166–2175

Latres E, Malumbres M, Sotillo R, Martin J, Ortega S, Martin-Caballero J, Flores JM, Cordon-Cardo C, Barbacid M (2000) Limited overlapping roles of p15^{INK4b} and p18^{INK4c} cell cycle inhibitors in proliferation and tumorigenesis. EMBO J 19:3496–3506

Lee EY, Chang CY, Hu N, Wang YC, Lai CC, Herrup K, Lee WH, Bradley A (1992) Mice deficient for Rb are nonviable and show defects in neurogenesis and haematopoiesis. Nature 359:288–294

Lee MG, Nurse P (1987) Complementation used to clone a human homologue of the fission yeast cell cycle control gene cdc2. Nature 327:31–35

Lees E, Faha B, Dulic V, Reed SI, Harlow E (1992) Cyclin E/cdk2 and cyclin A/cdk2 kinases associate with p107 and E2F in a temporally distinct manner. Genes Dev 6:1874–1885

Lehner CF, O'Farrell PH (1989) Expression and function of Drosophila cyclin A during embryonic cell cycle progression. Cell 56:957–968

Leung T, Chen XQ, Manser E, Lim L (1996) The p160 RhoA-binding kinase ROK alpha is a member of a kinase family and is involved in the reorganization of the cytoskeleton. Mol Cell Biol 16:5313–5327

Li JQ, Miki H, Ohmori M, Wu F, Funamoto Y (2001) Expression of cyclin E and cyclin-dependent kinase 2 correlates with metastasis and prognosis in colorectal carcinoma. Hum Pathol 32:945–953

Liu D, Matzuk MM, Sung WK, Guo Q, Wang P, Wolgemuth DJ (1998) Cyclin A1 is required for meiosis in the male mouse. Nat Genet 20:377–380

Loden M, Stighall M, Nielsen NH, Roos G, Emdin SO, Ostlund H, Landberg G (2002) The cyclin D1 high and cyclin E high subgroups of breast cancer: separate pathways in tumorogenesis based on pattern of genetic aberrations and inactivation of the pRb node. Oncogene 21:4680–4690

Loeb KR, Kostner H, Firpo E, Norwood T, K DT, Clurman BE, Roberts JM (2005) A mouse model for cyclin E-dependent genetic instability and tumorigenesis. Cancer Cell 8:35–47

Lorincz AT, Reed SI (1984) Primary structure homology between the product of yeast cell division control gene CDC28 and vertebrate oncogenes. Nature 307:183–185

Lozano JC, Perret E, Schatt P, Arnould C, Peaucellier G, Picard A (2002) Molecular cloning, gene localization, and structure of human cyclin B3. Biochem Biophys Res Commun 291:406–413

Lundberg AS, Weinberg RA (1998) Functional inactivation of the retinoblastoma protein requires sequential modification by at least two distinct cyclin-cdk complexes. Mol Cell Biol 18:753–761

Ma T, Zou N, Lin BY, Chow LT, Harper JW (1999) Interaction between cyclin-dependent kinases and human papillomavirus replication-initiation protein E1 is required for efficient viral replication. Proc Natl Acad Sci USA 96:382–387

Maandag EC, van der Valk M, Vlaar M, Feltkamp C, O'Brien J, van Roon M, van der Lugt N, Berns A, te Riele H (1994) Developmental rescue of an embryonic-lethal mutation in the retinoblastoma gene in chimeric mice. EMBO J 13:4260–4268

Madine MA, Swietlik M, Pelizon C, Romanowski P, Mills AD, Laskey RA (2000) The roles of the MCM, ORC, and Cdc6 proteins in determining the replication competence of chromatin in quiescent cells. J Struct Biol 129:198–210

Malumbres M, Sotillo R, Santamaria D, Galan J, Cerezo A, Ortega S, Dubus P, Barbacid M (2004) Mammalian cells cycle without the D-type cyclin-dependent kinases Cdk4 and Cdk6. Cell 118:493–504

Marraccino RL, Firpo EJ, Roberts JM (1992) Activation of the p34^{CDC25} protein kinase at the start of S phase in the human cell cycle. Mol Biol Cell 3:389–401

Martin A, Odajima J, Hunt SL, Dubus P, Ortega S, Malumbres M, Barbacid M (2005) Cdk2 is dispensable for cell cycle inhibition and tumor suppression mediated by p27^{Kip1} and p21^{Cip1}. Cancer Cell 7:591–598

Martin J, Hunt SL, Dubus P, Sotillo R, Nehme-Pelluard F, Magnuson MA, Parlow AF, Malumbres M, Ortega S, Barbacid M (2003) Genetic rescue of Cdk4 null mice restores pancreatic beta-cell proliferation but not homeostatic cell number. Oncogene 22:5261–5269

Marzo N, Mora C, Fabregat ME, Martin J, Usac EF, Franco C, Barbacid M, Gomis R (2004) Pancreatic islets from cyclin-dependent kinase 4/R24C (Cdk4) knockin mice have significantly increased beta cell mass and are physiologically functional, indicating that Cdk4 is a potential target for pancreatic beta cell mass regeneration in Type 1 diabetes. Diabetologia 47:686–694

Matsumoto Y, Maller JL (2004) A centrosomal localization signal in cyclin E required for Cdk2-independent S phase entry. Science 306:885–888

Matsushime H, Ewen ME, Storm DK, Kato J-y, Hanks SK, Roussel MF, Sherr CJ (1992) Identification and properties of an atypical catalytic subunit (p34^{PSK-J3}/cdk4) for mammalian D type G1 cyclins. Cell 71:323–334

Matsushime H, Roussel MF, Ashmun RA, Sherr CJ (1991) Colony-stimulating factor 1 regulates novel cyclins during the G1 phase of the cell cycle. Cell 65:701–713

Matushansky I, Radparvar F, Skoultchi AI (2000) Reprogramming leukemic cells to terminal differentiation by inhibiting specific cyclin-dependent kinases in G1. Proc Natl Acad Sci USA 97:14317–14322

McAllister SS, Becker-Hapak M, Pintucci G, Pagano M, Dowdy SF (2003) Novel p27^{Kip1} C-terminal scatter domain mediates Rac-dependent cell migration independent of cell cycle arrest functions. Mol Cell Biol 23:216–228

McIntosh GG, Anderson JJ, Milton I, Steward M, Parr AH, Thomas MD, Henry JA, Angus B, Lennard TW, Horne CH (1995) Determination of the prognostic value of cyclin D1 overexpression in breast cancer. Oncogene 11:885–891

Mettus RV, Rane SG (2003) Characterization of the abnormal pancreatic development, reduced growth and infertility in Cdk4 mutant mice. Oncogene 22:8413–8421

Meyerson M, Enders GH, Wu C-L, Su L-K, Gorka C, Nelson C, Harlow E, Tsai L-H (1992) A family of human cdc2-related protein kinases. EMBO J 11:2909–2917

Meyerson M, Harlow E (1994) Identification of G$_1$ kinase activity for cdk6, a novel cyclin D partner. Mol Cell Biol 14:2077–2086

Miliani de Marval PL, Macias E, Rounbehler R, Sicinski P, Kiyokawa H, Johnson DG, Conti CJ, Rodriguez-Puebla ML (2004) Lack of cyclin-dependent kinase 4 inhibits c-myc tumorigenic activities in epithelial tissues. Mol Cell Biol 24:7538–7547

Miller ME, Cross FR (2001) Cyclin specificity: how many wheels do you need on a unicycle? J Cell Sci 114:1811–1820

Minshull J, Golsteyn R, Hill CS, Hunt T (1990) The A- and B-type cyclin associated cdc2 kinases in Xenopus turn on and off at different times in the cell cycle. EMBO J 9:2865–2875

Mitchison JM, Creanor J (1971) Induction synchrony in the fission yeast. Schizosaccharomyces pombe. Exp Cell Res 67:368–374

Mondesert O, McGowan C, Russel P (1996) Cig2, a B-type cyclin, promotes the onset of S in *Schizosaccharomyces pombe*. Mol Cell Biol 16:1527–1533

Moons DS, Jirawatnotai S, Parlow AF, Gibori G, Kineman RD, Kiyokawa H (2002) Pituitary hypoplasia and lactotroph dysfunction in mice deficient for cyclin-dependent kinase-4. Endocrinology 143:3001–3008

Moore JD, Kirk JA, Hunt T (2003) Unmasking the S-phase-promoting potential of cyclin B1. Science 300:987–990

Moore JD, Kornbluth S, Hunt T (2002) Identification of the nuclear localization signal in Xenopus cyclin E and analysis of its role in replication and mitosis. Mol Biol Cell 13:4388–4400

Moreno S, Hayles J, Nurse P (1989) Regulation of p34^{cdc2} protein kinase during mitosis. Cell 58:361–372

Moreno S, Nurse P (1994) Regulation of progression through the G1 phase of the cell cycle by the rum1+ gene. Nature 367:236–242

Morgan DO (1997) Cyclin-dependent kinases: engines, clocks, and microprocessors. Annu Rev Cell Dev Biol 13:261–291

Morgenbesser SD, Williams BO, Jacks T, DePinho RA (1994) p53-dependent apoptosis produced by Rb-deficiency in the developing mouse lens. Nature 371:72–74

Morris L, Allen KE, La Thangue NB (2000) Regulation of E2F transcription by cyclin E-Cdk2 kinase mediated through p300/CBP co-activators. Nat Cell Biol 2:232–239

Mueller PR, Coleman TR, Dunphy WG (1995) Cell cycle regulation of a *Xenopus* Wee1-like kinase. Mol Biol Cell 6:119–134

Muller WJ, Sinn E, Pattengale PK, Wallace R, Leder P (1988) Single-step induction of mammary adenocarcinoma in transgenic mice bearing the activated c-neu oncogene. Cell 54:105–115

Murphy M, Stinnakre M-G, Senamaud-Beaufort C, Winston NJ, Sweeney C, Kubelka M, Carrington M, Brechot C, Sobczak-Thepot J (1997) Delayed early embryonic lethality following disruption of the murine cyclin A2 gene. Nat Genet 15:83–86

Nakayama K, Ishida N, Shirane M, Inomata A, Inoue T, Shishido N, Horii I, Loh DY, Nakayama K (1996) Mice lacking p27^{Kip1} display increased body size, multiple organ hyperplasia, retinal dysplasia, and pituitary tumors. Cell 85:707–720

Nakayama K, Nagahama H, Minamishima YA, Matsumoto M, Nakamichi I, Kitagawa K, Shirane M, Tsunematsu R, Tsukiyama T, Ishida N, Kitagawa M, Nakayama K, Hatakeyama S (2000) Targeted disruption of Skp2 results in accumulation of cyclin E and p27^{Kip1}, polyploidy and centrosome overduplication. EMBO J 19:2069–2081

Nakayama K, Nagahama H, Minamishima YA, Miyake S, Ishida N, Hatakeyama S, Kitagawa M, Iemura S, Natsume T, Nakayama KI (2004) Skp2-mediated degradation of p27 regulates progression into mitosis. Dev Cell 6:661–672

Nakayama KI, Hatakeyama S, Nakayama K (2001) Regulation of the cell cycle at the G1-S transition by proteolysis of cyclin E and p27^{Kip1}. Biochem Biophys Res Commun 282:853–860

Nash R, Tokiwa G, Anand S, Erickson K, Futcher AB (1988) The WHI1+ gene of Saccharomyces cerevisiae tethers cell division to cell size and is a cyclin homolog. EMBO J 7:4335–4346

Nasmyth K (1996) Viewpoint: putting the cell cycle in order. Science 274:1643–1645

Nguyen TB, Manova K, Capodieci P, Lindon C, Bottega S, Wang XY, Refik-Rogers J, Pines J, Wolgemuth DJ, Koff A (2002) Characterization and expression of mammalian cyclin B3, a prepachytene meiotic cyclin. J Biol Chem 277:41960–41969

Nieduszynski CA, Murray J, Carrington M (2002) Whole-genome analysis of animal A- and B-type cyclins. Genome Biol 3: RESEARCH0070

Nigg EA (1995) Cyclin-dependent protein kinases: key regulators of the eukaryotic cell cycle. Bioessays 17:471–480

Nikitin A, Lee WH (1996) Early loss of the retinoblastoma gene is associated with impaired growth inhibitory innervation during melanotroph carcinogenesis in Rb+/− mice. Genes Dev 10:1870–1879

Nurse P, Bissett Y (1981) Gene required in G1 for commitment to cell cycle and in G2 for control of mitosis in fission yeast. Nature 292:558–560

Ohtsubo M, Roberts JM (1993) Cyclin-dependent regulation of G1 in mammalian fibroblasts. Science 259:1908–1912

Okuda M, Horn HF, Tarapore P, Tokuyama Y, Smulian AG, Chan PK, Knudsen ES, Hofmann IA, Snyder JD, Bove KE, Fukasawa K (2000) Nucleophosmin/B23 is a target of CDK2/cyclin E in centrosome duplication. Cell 103:127–140

Ookata K, Hisanaga S, Okumura E, Kishimoto T (1993) Association of p34^{cdc2}/cyclin B complex with microtubules in starfish oocytes. J Cell Sci 105:873–881

Ortega S, Malumbres M, Barbacid M (2002) Cyclin D-dependent kinases, INK4 inhibitors and cancer. Biochim Biophys Acta 1602:73–87

Ortega S, Prieto I, Odajima J, Martin A, Dubus P, Sotillo R, Barbero JL, Malumbres M, Barbacid M (2003) Cyclin-dependent kinase 2 is essential for meiosis but not for mitotic cell division in mice. Nat Genet 35:25–31

Pagano M (2004) Control of DNA synthesis and mitosis by the Skp2-p27-Cdk1/2 axis. Mol Cell 14:414–416

Pagano M, Pepperkok R, Verde F, Ansorge W, Draetta G (1992) Cyclin A is required at two points in the human cell cycle. EMBO J 11:961–971

Pardee AB (1974) A restriction point for control of normal animal cell proliferation. Proc Natl Acad Sci USA 71:1286–1290

Parisi T, Beck AR, Rougier N, McNeil T, Lucian L, Werb Z, Amati B (2003) Cyclins E1 and E2 are required for endoreplication in placental trophoblast giant cells. EMBO J 22:4794–4803

Park MS, Rosai J, Nguyen HT, Capodieci P, Cordon-Cardo C, Koff A (1999) p27 and Rb are on overlapping pathways suppressing tumorigenesis in mice. Proc Natl Acad Sci USA 96:6382–6387

Perkins ND, Felzien LK, Betts JC, Leung K, Beach DH, Nabel GJ (1997) Regulation of NF-kappaB by cyclin-dependent kinases associated with the p300 coactivator. Science 275:523–527

Peters H (1969) The development of the mouse ovary from birth to maturity. Acta Endocrinol (Copenh) 62:98–116

Philipp-Staheli J, Kim KH, Payne SR, Gurley KE, Liggitt D, Longton G, Kemp CJ (2002) Pathway-specific tumor suppression. Reduction of p27 accelerates gastrointestinal tumorigenesis in Apc mutant mice, but not in Smad3 mutant mice. Cancer Cell 1:355–368

Philipp-Staheli J, Payne SR, Kemp CJ (2001) p27^{Kip1}: regulation and function of a haploin-sufficient tumor suppressor and its misregulation in cancer. Exp Cell Res 264:148–168

Pines J, Hunter T (1989) Isolation of a human cyclin cDNA: evidence for cyclin mRNA and protein regulation in the cell cycle and for interaction with p34^{cdc2}. Cell 58:833–846

Pines J, Hunter T (1990) Human cyclin A is adenovirus E1A-associated protein p60 and behaves differently from cyclin B. Nature 346:760–763

Pines J, Hunter T (1991) Human cyclins A and B1 are differentially located in the cell and undergo cell cycle-depenent nuclear transport. J Cell Biol 115:1–17

Polyak K, Kato JY, Solomon MJ, Sherr CJ, Massague J, Roberts JM, Koff A (1994) p27^{Kip1}, a cyclin-Cdk inhibitor, links transforming growth factor-beta and contact inhibition to cell cycle arrest. Genes Dev 8:9–22

Rane SG, Cosenza SC, Mettus RV, Reddy EP (2002) Germ line transmission of the Cdk4^{R24C} mutation facilitates tumorigenesis and escape from cellular senescence. Mol Cell Biol 22:644–656

Rane SG, Dubus P, Mettus RV, Galbreath EJ, Boden G, Reddy EP, Barbacid M (1999) Loss of Cdk4 expression causes insulin-deficient diabetes and Cdk4 activation results in β-islet cell hyperplasia. Nat Genet 22:44–52

Ravnik SE, Wolgemuth DJ (1996) The developmentally restricted pattern of expression in the male germ line of a murine cyclin A, cyclin A2, suggests roles in both mitotic and meiotic cell cycles. Dev Biol 173:69–78

Ravnik SE, Wolgemuth DJ (1999) Regulation of meiosis during mammalian spermatogenesis: the A-type cyclins and their associated cyclin-dependent kinases are differentially expressed in the germ-cell lineage. Dev Biol 207:408–418

Reed SI (1980) The selection of S. cerevisiae mutants defective in the start event of cell division. Genetics 95:561–577

Ren XD, Kiosses WB, Sieg DJ, Otey CA, Schlaepfer DD, Schwartz MA (2000) Focal adhesion kinase suppresses Rho activity to promote focal adhesion turnover. J Cell Sci 113:3673–3678

Resnitzky D, Hengst L, Reed SI (1995) Cyclin A-associated kinase activity is rate limiting for entrance into S phase and is negatively regulated in G1 by p27^{Kip1}. Mol Cell Biol 15:4347–4352

Rhee K, Wolgemuth DJ (1995) Cdk family genes are expressed not only in dividing but also in terminally differentiated mouse germ cells, suggesting their possible function during both cell division and differentiation. Dev Dyn 204:406–420

Riabowol K, Draetta G, Brizuela L, Vandre D, Beach D (1989) The cdc2 kinase is a nuclear protein that is essential for mitosis in mammalian cells. Cell 57:393–401

Ridley AJ, Hall A (1992) The small GTP-binding protein rho regulates the assembly of focal adhesions and actin stress fibers in response to growth factors. Cell 70:389–399

Rodier G, Montagnoli A, Di Marcotullio L, Coulombe P, Draetta GF, Pagano M, Meloche S (2001) p27 cytoplasmic localization is regulated by phosphorylation on Ser10 and is not a prerequisite for its proteolysis. EMBO J 20:6672–6682

Rodriguez-Puebla ML, Miliani de Marval PL, LaCava M, Moons DS, Kiyokawa H, Conti CJ (2002) Cdk4 deficiency inhibits skin tumor development but does not affect normal keratinocyte proliferation. Am J Pathol 161:405–411

Rosenblatt J, Gu Y, Morgan DO (1992) Human cyclin-dependent kinase 2 is activated during the S and G$_2$ phases of the cell cycle and associates with cyclin A. Proc Natl Acad Sci USA 89:2824–2828

Rounbehler RJ, Rogers PM, Conti CJ, Johnson DG (2002) Inactivation of E2f1 enhances tumorigenesis in a Myc transgenic model. Cancer Res 62:3276–3281

Rounbehler RJ, Schneider-Broussard R, Conti CJ, Johnson DG (2001) Myc lacks E2F1's ability to suppress skin carcinogenesis. Oncogene 20:5341–5349

Sage J, Miller AL, Perez-Mancera PA, Wysocki JM, Jacks T (2003) Acute mutation of retinoblastoma gene function is sufficient for cell cycle re-entry. Nature 424:223–228

Serrano M, Lee H, Chin L, Cordon-Cardo C, Beach D, DePinho RA (1996) Role of the INK4a locus in tumor suppression and cell mortality. Cell 85:27–37

Sharpe RM (1989) Possible role of elongated spermatids in control of stage-dependent changes in the diameter of the lumen of the rat seminiferous tubule. J Androl 10:304–310

Sharpless NE, Bardeesy N, Lee KH, Carrasco D, Castrillon DH, Aguirre AJ, Wu EA, Horner JW, DePinho RA (2001) Loss of p16^{Ink4a} with retention of p19^{Arf} predisposes mice to tumorigenesis. Nature 413:86–91

Sheaff RJ, Groudine M, Gordon M, Roberts JM, Clurman BE (1997) Cyclin E-CDK2 is a regulator of p27^{Kip1}. Genes Dev 11:1464–1478

Sherr CJ, Roberts JM (1999) CDK inhibitors: positive and negative regulators of G1-phase progression. Genes Dev 13:1501–1512

Sherr CJ, Roberts JM (2004) Living with or without cyclins and cyclin-dependent kinases. Genes Dev 18:2699–2711

Sicinska E, Aifantis I, Le Cam L, Swat W, Borowski C, Yu Q, Ferrando AA, Levin SD, Geng Y, von Boehmer H, Sicinski P (2003) Requirement for cyclin D3 in lymphocyte development and T cell leukemias. Cancer Cell 4:451–461

Sicinski P, Donaher JL, B. PS, Li T, Fazeli A, Gardner H, Haslam SZ, Bronson RT, Elledge SJ, Weinberg RA (1995) Cyclin D1 provides a link between development and oncogenesis in the retina and breast. Cell 82:621–630

Sicinski P, Donaher JL, Geng Y, Parker SB, Gardner H, Park MY, Robker RL, Richards JS, McGinnis LK, Biggers JD, Eppig JJ, Bronson RT, Elledge SJ, Weinberg RA (1996) Cyclin D2 is an FSH-responsive gene involved in gonadal cell proliferation and oncogenesis. Nature 384:470–474

Sinn E, Muller W, Pattengale P, Tepler I, Wallace R, Leder P (1987) Coexpression of MMTV/v-H-*Ras* and MMTV/c-myc genes in transgenic mice: synergistic action of oncogenes in vivo. Cell 49:465–475

Slack JM (1995) Developmental biology of the pancreas. Development 121:1569–1580

Slingerland J, Pagano M (2000) Regulation of the cdk inhibitor p27 and its deregulation in cancer. J Cell Physiol 183:10–17

Slingerland JM, Hengst L, Pan CH, Alexander D, Stampfer MR, Reed SI (1994) A novel inhibitor of cyclin-Cdk activity detected in transforming growth factor beta-arrested epithelial cells. Mol Cell Biol 14:3683–3694

Sotillo R, Dubus P, Martin J, de la Cueva E, Ortega S, Malumbres M, Barbacid M (2001a) Wide spectrum of tumors in knock-in mice carrying a Cdk4 protein insensitive to INK4 inhibitors. EMBO J 20:6637–6647

Sotillo R, Garcia JF, Ortega S, Martin J, Dubus P, Barbacid M, Malumbres M (2001b) Invasive melanoma in Cdk4-targeted mice. Proc Natl Acad Sci USA 98:13312–13317

Sotillo R, Renner O, Dubus P, Ruiz-Cabello J, Martin-Caballero J, Barbacid M, Carnero A, Malumbres M (2005) Cooperation between Cdk4 and p27^{Kip1} in tumor development: a preclinical model to evaluate cell cycle inhibitors with therapeutic activity. Cancer Res 65:3846–3852

Stern B, Nurse P (1996) A quantitative model for the cdc2 control of S phase and mitosis in fission yeast. Trends Genet 12:345–350

Stewart TA, Pattengale PK, Leder P (1984) Spontaneous mammary adenocarcinomas in transgenic mice that carry and express MTV/myc fusion genes. Cell 38:627–637

Strausfeld UP, Howell M, Descombes P, Chevalier S, Rempel RE, Adamczewski J, Maller JL, Hunt T, Blow JJ (1996) Both cyclin A and cyclin E have S-phase promoting (SPF) activity in Xenopus egg extracts. J Cell Sci 109:1555–1563

Su TT, O'Farrell PH (1998) Size control: cell proliferation does not equal growth. Curr Biol 8: R687–689

Sui L, Dong Y, Ohno M, Sugimoto K, Tai Y, Hando T, Tokuda M (2001) Implication of malignancy and prognosis of p27^{Kip1}, Cyclin E, and Cdk2 expression in epithelial ovarian tumors. Gynecol Oncol 83:56–63

Sutterluty H, Chatelain E, Marti A, Wirbelauer C, Senften M, Muller U, Krek W (1999) p45^{SKP2} promotes p27^{Kip1} degradation and induces S phase in quiescent cells. Nat Cell Biol 1:207–214

Sweeney C, Murphy M, Kubelka M, Ravnik SE, Hawkins CF, Wolgemuth DJ, Carrington M (1996) A distinct cyclin A is expressed in germ cells in the mouse. Development 122:53–64

Takizawa CG, Morgan DO (2000) Control of mitosis by changes in the subcellular location of cyclin B1-Cdk1 and Cdc25C. Curr Opin Cell Biol 12:658–665

Tanaka H, Kanagawa H (1997) Influence of combined activation treatments on the success of bovine nuclear transfer using young or aged oocytes. Anim Reprod Sci 49:113–123

Teixeira LT, Kiyokawa H, Peng XD, Christov KT, Frohman LA, Kineman RD (2000) p27^{Kip1}-deficient mice exhibit accelerated growth hormone-releasing hormone (GHRH)-induced somatotrope proliferation and adenoma formation. Oncogene 19:1875–1884

Tetsu O, McCormick F (2003) Proliferation of cancer cells despite CDK2 inhibition. Cancer Cell 3:233–245

Tevosian SG, Paulson KE, Bronson R, Yee AS (1996) Expression of the E2F-1/DP-1 transcription factor in murine development. Cell Growth Differ 7:43–52

Th'ng JP, Wright PS, Hamaguchi J, Lee MG, Norbury CJ, Nurse P, Bradbury EM (1990) The FT210 cell line is a mouse G2 phase mutant with a temperature-sensitive CDC2 gene product. Cell 63:313–324

Tong W, Pollard JW (2001) Genetic evidence for the interactions of cyclin D1 and p27^{Kip1} in mice. Mol Cell Biol 21:1319–1328

Toyoshima F, Moriguchi T, Wada A, Fukuda M, Nishida E (1998) Nuclear export of cyclin B1 and its possible role in the DNA damage-induced G2 checkpoint. EMBO J 17:2728–2735

Tsai LH, Harlow E, Meyerson M (1991) Isolation of the human cdk2 gene that encodes the cyclin A- and adenovirus E1A-associated p33 kinase. Nature 353:174–177

Tsai LH, Lees E, Faha B, Harlow E, Riabowol K (1993) The cdk2 kinase is required for the G1-to-S transition in mammalian cells. Oncogene 8:1593–602

Tsukamoto AS, Grosschedl R, Guzman RC, Parslow T, Varmus HE (1988) Expression of the int-1 gene in transgenic mice is associated with mammary gland hyperplasia and adenocarcinomas in male and female mice. Cell 55:619–625

Tsutsui T, Hesabi B, Moons DS, Pandolfi PP, Hansel KS, Koff A, Kiyokawa H (1999) Targeted disruption of CDK4 delays cell cycle entry with enhanced p27^{Kip1} activity. Mol Cell Biol 19:7011–7019

Tsvetkov LM, Yeh KH, Lee SJ, Sun H, Zhang H (1999) p27^{Kip1} ubiquitination and degradation is regulated by the SCF^{Skp2} complex through phosphorylated Thr187 in p27. Curr Biol 9:661–664

Uchida T, Nakamura T, Hashimoto N, Matsuda T, Kotani K, Sakaue H, Kido Y, Hayashi Y, Nakayama KI, White MF, Kasuga M (2005) Deletion of Cdkn1b ameliorates hyperglycemia by maintaining compensatory hyperinsulinemia in diabetic mice. Nat Med 11:175–182

van den Heuvel S, Harlow E (1993) Distinct roles for cyclin-dependent kinases in cell cycle control. Science 262:2050–2054

Vial D, Oliver C, Jamur MC, Pastor MV, da Silva Trindade E, Berenstein E, Zhang J, Siraganian RP (2003) Alterations in granule matrix and cell surface of focal adhesion kinase-deficient mast cells. J Immunol 171:6178–6186

Vlach J, Hennecke S, Amati B (1997) Phosphorylation-dependent degradation of the cyclin-dependent kinase inhibitor p27^{Kip1}. EMBO J 16:5334–5344

Wadler S (2002) Perspectives for cancer therapies with cdk2 inhibitors. Drug Res Updates 4:347–367

Weinstat-Saslow D, Merino MJ, Manrow RE, Lawrence JA, Bluth RF, Wittenbel KD, Simpson JF, Page DL, Steeg PS (1995) Overexpression of cyclin D mRNA distinguishes invasive and in situ breast carcinomas from non-malignant lesions. Nat Med 1:1257–1260

Welcker M, Clurman B (2005) Cell cycle: how cyclin E got its groove back. Curr Biol 15:R810–812

Williams BO, Morgenbesser SD, DePinho RA, Jacks T (1994a) Tumorigenic and developmental effects of combined germ-line mutations in Rb and p53. Cold Spring Harb Symp Quant Biol 59:449–457

Williams BO, Remington L, Albert DM, Mukai S, Bronson RT, Jacks T (1994b) Cooperative tumorigenic effects of germline mutations in Rb and p53. Nat Genet 7:480–484

Wittenberg C, Sugimoto K, Reed SI (1990) G1-specific cyclins of S. cerevisiae: cell cycle periodicity, regulation by mating pheromone, and association with the p34^{CDC28} protein kinase. Cell 62:225–237

Wu CL, Zukerberg LR, Ngwu C, Harlow E, Lees JA (1995) In vivo association of E2F and DP family proteins. Mol Cell Biol 15:2536–2546

Wu L, de Bruin A, Saavedra HI, Starovic M, Trimboli A, Yang Y, Opavska J, Wilson P, Thompson JC, Ostrowski MC, Rosol TJ, Woollett LA, Weinstein M, Cross JC, Robinson ML, Leone G (2003) Extra-embryonic function of Rb is essential for embryonic development and viability. Nature 421:942–947

Xiong Y, Connolly T, Futcher B, Beach D (1991) Human D-type cyclin. Cell 65:691–699

Xiong Y, Menninger J, Beach D, Ward DC (1992) Molecular cloning and chromosomal mapping of CCND genes encoding human D-type cyclins. Genomics 13:575–584

Yang J, Bardes ES, Moore JD, Brennan J, Powers MA, Kornbluth S (1998) Control of cyclin B1 localization through regulated binding of the nuclear export factor CRM1. Genes Dev 12:2131–2143

Yang R, Morosetti R, Koeffler HP (1997) Characterization of a second human cyclin A that is highly expressed in testis and in several leukemic cell lines. Cancer Res 57:913–920

Ye X, Zhu C, Harper JW (2001) A premature-termination mutation in the *Mus musculus* cyclin-dependent kinase 3 gene. Proc Natl Acad Sci USA 98:1682–1686

Yoon KH, Ko SH, Cho JH, Lee JM, Ahn YB, Song KH, Yoo SJ, Kang MI, Cha BY, Lee KW, Son HY, Kang SK, Kim HS, Lee IK, Bonner-Weir S (2003) Selective beta-cell loss and alpha-cell expansion in patients with type 2 diabetes mellitus in Korea. J Clin Endocrinol Metab 88:2300–2308

Yu Q, Geng Y, Sicinski P (2001) Specific protection against breast cancers by cyclin D1 ablation. Nature 411:1017–1021

Yuan L, Liu JG, Zhao J, Brundell E, Daneholt B, Hoog C (2000) The murine SCP3 gene is required for synaptonemal complex assembly, chromosome synapsis, and male fertility. Mol Cell 5:73–83

Zhang Y, Chellappan SP (1995) Cloning and characterization of human DP2, a novel dimerization partner of E2F. Oncogene 10:2085–2093

Zimmet J, Ravid K (2000) Polyploidy: occurrence in nature, mechanisms, and significance for the megakaryocyte-platelet system. Exp Hematol 28:3–16

Zimmet JM, Ladd D, Jackson CW, Stenberg PE, Ravid K (1997) A role for cyclin D3 in the endomitotic cell cycle. Mol Cell Biol 17:7248–7259

Zindy F, den Besten W, Chen B, Rehg JE, Latres E, Barbacid M, Pollard JW, Sherr CJ, Cohen PE, Roussel MF (2001) Control of spermatogenesis in mice by the cyclin D-dependent kinase inhibitors p18^{Ink4c} and p19^{Ink4d}. Mol Cell Biol 21:3244–3255

Zindy F, Soares H, Herzog K-H, Morgan J, Sherr CJ, Roussel MF (1997) Expression of INK4 inhibitors of cyclin D-dependent kinases during mouse brain development. Cell Growth Differ 8:1139–1150

Zindy F, van Deursen J, Grosveld G, Sherr CJ, Roussel MF (2000) INK4d-deficient mice are fertile despite testicular atrophy. Mol Cell Biol 20:372–378

Zou L, Stillman B (2000) Assembly of a complex containing Cdc45p, replication protein A, and Mcm2p at replication origins controlled by S-phase cyclin-dependent kinases and Cdc7p-Dbf4p kinase. Mol Cell Biol 20:3086–3096

Zou X, Ray D, Aziyu A, Christov K, Boiko AD, Gudkov AV, Kiyokawa H (2002) Cdk4 disruption renders primary mouse cells resistant to oncogenic transformation, leading to Arf/p53-independent senescence. Genes Dev 16:2923–2934

Zuo L, Weger J, Yang Q, Goldstein AM, Tucker MA, Walker GJ, Hayward N, Dracopoli NC (1996) Germline mutations in the p16^{INK4a} binding domain of CDK4 in familial melanoma. Nat Genet 12:97–99.

Control of Cell Proliferation and Growth by Myc Proteins

Sandra Bernard · Martin Eilers (✉)

Institute for Molecular Biology and Tumor Research, University of Marburg,
35033 Marburg, Germany
eilers@imt.uni-marburg.de

Abstract Myc proteins act as signal transducers that alter cell proliferation in dependence on signals from the extracellular environment. In normal cells, the expression of *MYC* genes is therefore under tight control by growth factor dependent signals. The enormous interest in the function of these proteins is motivated by the observation that the close control of *MYC* expression is disrupted in a large percentage of human tumors, leading to deregulated expression of Myc proteins. A large body of evidence shows that this deregulation is a major driving force of human tumorigenesis; in cells with deregulated Myc, proliferation often takes place in the complete absence of external stimuli. We will discuss current models to understand Myc function and also potential avenues to selectively interfere with the proliferation of Myc-transformed cells.

1
Introduction

MYC genes form a small multigene family; the family has attracted enormous attention, since the enhanced expression of one of its three members (*MYC*, *MYCL* or *MYCN*) contributes to multiple human tumors. Deregulation of expression can occur through diverse mechanisms, only some of which involve the mutation of *MYC* genes themselves. More frequently, mutations occur in human tumors in pathways that control the expression of *MYC* genes or that control the function of the encoded proteins. The full spectrum of such mutations remains to be elucidated, since the regulation of any member of the *MYC* gene family is complex. As a result, the precise percentage of human tumors in which *MYC* genes are activated and/or deregulated remains a matter of some debate; it is possible that the 'Myc pathway', as suggested for the E2F pathway, needs to be deregulated for any human tumor to emerge. Alternatively, there is evidence that at least some cells proliferate in a Myc-independent manner and therefore it is possible that tumors derived from such cells do not have an 'activated' *MYC* gene.

Exogenously introduced *MYC* genes that are expressed under the control of a strong promoter (to mimic the activation of *MYC* genes seen in human tumors) elicit a stereotype response in most cells into which they are introduced: they promote cell proliferation even in the absence of mitogenic signals, they often promote cell growth and they almost invariably promote apoptosis or at least sensitize cells to apoptotic stimuli. Keratinocytes are one of only a few examples of cells that respond differently to Myc: in skin, proliferation is tightly linked to adhesion to the basal lamina and Myc disrupts this adhesion; therefore, keratinocytes respond to Myc with premature differentiation and arrest of proliferation (Gandarillas and Watt 1997).

The frequent activation of *MYC* genes in tumorigenesis has suggested that Myc proteins might also have an important function in normal proliferation and growth and this suggestion has by now been tested in numerous experimental systems: Rat1 fibroblasts, in which both alleles of *c-myc* have been deleted, show a reduced rate of proliferation and cell growth (Mateyak et al. 1997; Mateyak et al. 1999). Similarly, mouse embryo fibroblasts that carry a floxed allele of *c-myc* arrest upon cre-mediated excision of *c-myc* (de Alboran et al. 2001). Importantly, deletion of the *mnt* gene, which encodes a member of the Mad family of antagonists of Myc, restores proliferation to *c-myc* deleted cells, suggesting that the balance of expression of members of the Myc/Max/Mad network of proteins (see Fig. 1) may dictate the response to loss of Myc (Walker et al. 2005).

In mice, the effects of deletion of the *c-myc* or *n-myc* genes in vivo are not uniform. Deletion of *c-myc* in mice results in embryonic lethality and certain cell types do not proliferate (de Alboran et al. 2001; Trumpp et al. 2001). In the hematopoietic lineage, stem cells continue to proliferate upon genetic ablation of c-myc, whereas differentiated (lineage positive) cells arrest (Wilson et al. 2004). While these data show that *c-myc* is not required for proliferation of hematopoietic stem cells, such cells also express *n-myc* and therefore it is possible that *n-myc* provides essential functions of Myc proteins in these cells. Similarly, postnatal proliferation of hepatocytes in the liver does not require c-Myc, but again the potential compensation by other *myc* family members is not completely clear (Baena et al. 2005). In the small intestine, deletion of *c-myc* leads to a transient failure to form normal numbers of crypts in the small intestine, but in long-term experiments the mice maintain a normal epithelium in the absence of c-Myc activity and without apparent compensation by N-Myc or L-Myc (Bettess et al. 2005). Both findings argue that at least some cell types can proliferate in the absence of functional Myc. There are also clear examples for a strict requirement for Myc function in cell proliferation: for example, deletion of *n-myc* in neuronal precursor cells leads to a dramatic loss of proliferative capacity of such cells (Knoepfler et al. 2002). As long as questions of redundancy between different members of the *myc* gene family and compensation by loss of antagonists such as *mnt* are not fully resolved, it seems to us that no definitive answer is possible as to whether Myc function is generally required for cell proliferation and cell growth or whether its essential role in proliferation is restricted to specific cell types.

Control of Cell Proliferation and Growth by Myc Proteins

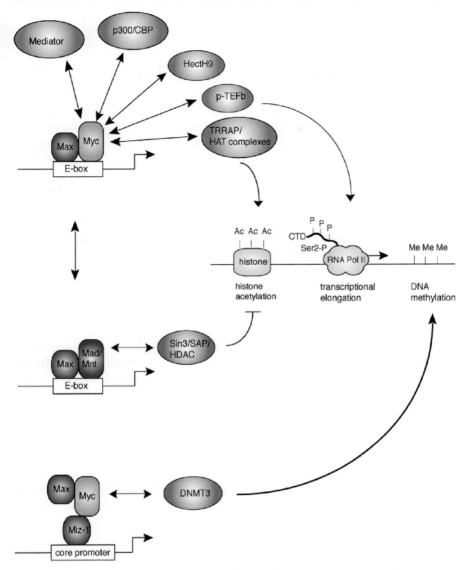

Fig. 1 Transcriptional regulatory complexes formed by Myc proteins and their co-factors. For details, see text. Almost certainly, Myc forms more than one repressive complex, so the Myc/Miz1 complex should be viewed as one well-understood example of a group of similar complexes

2
Mechanisms of Myc Action

Myc proteins act at least in part as transcription factors that activate and repress large groups of genes (see Fig. 2). They activate transcription as part of a binary complex together with an obligate partner protein, Max; the complex binds to a specific sequence, termed E-Box (CACGTG or related sequences), which is found in all genes that are activated by Myc. Interestingly, Myc proteins not only regulate protein-coding genes that are transcribed by polymerase II.

There are at least three exceptions: first, there is evidence that microRNAs can be target genes that are activated by Myc: for example, a microRNA targeting E2F1 is induced by Myc (O'Donnell et al. 2005). Since the *e2f1* gene is at the same time a target for transcriptional upregulation by Myc, the data suggest that there may be a fine tuning of expression of Myc target genes and proteins (Baudino et al. 2003). In the case of E2F1, there is evidence both that it mediates proliferation downstream of Myc (in B-lymphocytes) and that it mediates Myc-dependent apoptosis, so the microRNA-mediated regulation of E2F1 protein levels may ensure the correct balance between the two (Baudino et al. 2003; Leone et al. 2001).

The second exception are ribosomal RNA genes. Myc activates transcription through direct binding to canonical binding sites in the rDNA promoter (Arabi et al. 2005; Grandori et al. 2005; Poortinga et al. 2004). rDNA genes are transcribed by polymerase I in the nucleolus, arguing that at least a fraction of Myc proteins are localized in this compartment. Indeed, immunofluorescence experiments show that Myc can be found in the nucleolus, in particular after proteasome inhibition. The latter finding also suggests that Myc can be degraded in the nucleolus and at least one E3 ligase that targets Myc, Fbw7γ, is specifically localized in this compartment (Welcker et al. 2004a,b; Yada et al. 2004).

The third exception are several tRNA genes, demonstrating that Myc can also stimulate polymerase III-dependent transcription (Gomez-Roman et al.

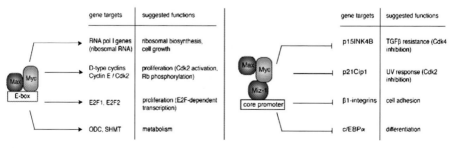

Fig. 2 Genetic targets of Myc in cell proliferation and their proposed functions

2003). There is clear evidence that this effect is direct and not mediated by activation of protein-coding genes that regulate polymerase III function; however, the mechanism of activation has not been clarified.

Myc proteins also act as transcriptional repressors; all current models suggest that this is mediated by protein/protein interactions with other transcription factors. One such factor is the Myc-interacting zinc finger protein, Miz1: through interaction with Miz1 (and likely additional proteins) Myc is recruited to non-canonical sites in the genome (Schneider et al. 1997; Seoane et al. 2002; Seoane et al. 2001; Staller et al. 2001; Wanzel et al. 2003). Free Miz1 acts as a transcriptional activator; in contrast, the Myc/Miz1 complex represses transcription from the same binding sites. Interestingly, at least one other oncogene, Bcl6, uses Miz1 as a 'platform' to repress transcription, suggesting that both Bcl6 and Myc may repress a common set of target genes (Phan et al. 2005).

Mechanistically, many questions remain open. Myc proteins interact with a number of potential co-activator and co-repressor proteins and, for some of them, functions in Myc-dependent activation and repression have been demonstrated. Some of these are shown in Fig. 1. Potentially, the best-understood interaction is that with TRRAP, since Myc recruits two distinct histone acetylases, Gcn5 and Tip60, to Myc/Max target sites in vivo through interaction with TRRAP (Bouchard et al. 2001; Frank et al. 2003; McMahon et al. 1998; McMahon et al. 2000). TRRAP binds to a highly conserved domain in the amino-terminus of Myc proteins (MycboxII), and Myc controls acetylation of its target genes in a MycboxII-dependent manner. Mutations in this domain impair both activation and repression of many, but not all, genes by Myc, and abolish transformation by Myc (Nikiforov et al. 2002). Together, the data strongly support the view that TRRAP-dependent stimulation of local histone acetylation is a key function in transcriptional activation by Myc. The precise role of most other interactions is less clear, mainly because genetic analyses are missing or have yielded unexpected results. For example, while mutations in Tip48 or Tip49 (which also bind to MycboxII) genetically interact with mutations in Myc in Drosophila, the pattern of genes that are regulated by either mutation show only very little overlap (Bellosta et al. 2005). Clearly, more work will be required to achieve a clear picture of which protein interactions of Myc contribute to which aspects of its transcriptional regulatory functions and how those link its biological properties.

This is most certainly true for an exciting link that emerges between ubiquitination of Myc and its transcriptional properties. There are two aspects to this link: first, phosphorylation of threonine 58 (T58) by GSK3 stimulates recognition by Fbw7γ, an SCF-type E3 ligase complex (Welcker et al. 2004b; Yada et al. 2004). T58 mutations are frequently found in human lymphomas, suggesting that this is one mechanism selecting for enhanced levels of Myc in human tumors. Surprisingly, however, these mutations also show an altered gene-regulatory behavior in that they fail to repress p21$^{\text{Cip1}}$, a target of the Myc/Miz1 complex and, potentially as an indirect result, fail to induced Bim1, a pro-apoptotic protein that is downstream of p21$^{\text{Cip1}}$ (Collins et al. 2005; Hemann et al. 2005). Consequently, such mutations allow lymphomas to develop, which retain wild-type p53 and ARF and therefore bypass a mutational requirement that exist for wild-type Myc. Does that mean that Fbw7γ is a co-factor of repression? Possibly the situation is very similar to that for Mdm2 and p53, where Mdm2-mediated histone ubiquitination contributes to repression of p53 function by Mdm2 (Minsky and Oren 2004). Similarly, ubiquitination of Myc by Skp2 has been shown to contribute to degradation of Myc and its gene-regulatory properties (Kim et al. 2003; von der Lehr et al. 2003).

Secondly, the ARF tumor suppressor inhibits transcriptional activation, but not repression by Myc (Qi et al. 2004). The mechanism of this repression is not completely clear and may involve direct interactions of both proteins. Alternatively, ARF acts through inhibiting the E3-ligase function of HectH9/p500ARF-BP1, a major ARF binding protein in human cells (Adhikary et al. 2005; Chen et al. 2005). HectH9 is also an E3-ligase for Myc and assembles a K63-linked poly-ubiquitin chain required for transcriptional activation by Myc. Both findings strongly suggest that ubiquitination of Myc is not solely used to control Myc levels, but to control its biological functions as well.

3
Targets

All available evidence suggests that Myc proteins affect both cell growth and cell proliferation through multiple pathways and that there is neither a single critical effector gene nor a small group of such genes. In cell culture, Myc can affect cell growth, Cdk2 activity, E2F-dependent transcription and intermediary metabolism and the effects on each pathway can be genetically separated from each other.

Myc has been shown to stimulate both polymerase I- and polymerase III-dependent transcription, leading to enhanced transcription of ribosomal RNA and tRNA genes and to stimulate transcription of many genes involved in all steps of protein translation (Arabi et al. 2005; Boon et al. 2001; Gomez-Roman et al. 2003; Grandori et al. 2005). While a single 'master' gene downstream of Myc may not exist, there can be little doubt that stimulation of protein translation and biogenesis is a major effector pathway of Myc. Undoubtedly, the stimulation of cell growth will also feed back and contribute to the stimulation of cell proliferation by Myc, although there is no evidence to suggest that this is the 'primary' function of Myc in stimulating cell proliferation in mammalian cells (Beier et al. 2000; Trumpp et al. 2001). Similar arguments hold for many enzymes of intermediary metabolism that are activated by Myc, such as lactate dehydrogenase A (LDH-A) and ornithine decarboxylase: induction of these genes contributes to and can be required for Myc-induced proliferation and tumorigenesis, but there is little evidence that the effects of Myc on cell proliferation are in any way mediated by the activation of these genes (Guo et al. 2005; Shim et al. 1997).

Several direct target genes of Myc contribute to the phosphorylation of the retinoblastoma protein and the regulation of E2F-dependent transcription. Cyclins D1 and D2 have been suggested to be downstream targets of the Myc protein (Bouchard et al. 1999; Kenney et al. 2003; Oliver et al. 2003; Osterhout et al. 1999; Perez-Roger et al. 1999), and detailed mechanistic studies have been published for cyclin D2 regulation by Myc (Bouchard et al. 2001, 2004). Also, Cdk4 is a target for transactivation by Myc (Hermeking et al. 2000) and the Cdk4 inhibitor, p15$^{\text{Ink4b}}$, is a direct target for repression by Myc (Seoane et al. 2001; Staller et al. 2001). In vivo, N-Myc is essential to maintain expression of cyclin D2 in neuronal precursor cells and the phenotypes of an N-Myc knockout are similar to those of a cyclin D2 knockout (Knoepfler et al. 2002; Kowalczyk et al. 2004). N-Myc is also required to suppress expression of p18$^{\text{Ink4c}}$ in vivo, but it is not absolutely clear whether this is a direct regulation (Knoepfler et al. 2002). Cdk4 is essential for Myc-induced tumorigenesis in skin (Miliani de Marval et al. 2004), and fibroblasts lacking D-type cyclins are resistant to transformation by Myc (Kozar et al. 2004), while fibroblasts expressing any single D-type cyclin can be transformed by Myc (Yu et al. 2005). Repression of p15$^{\text{Ink4b}}$ is critical for stimulation of proliferation by Myc in keratinocytes (Gebhardt et al. 2005). Taken together, the data strongly suggest that regulation of Cdk4 function is a critical function of Myc in tumorigenesis.

Most likely, the regulation of Cdk4 reflects at least in part a requirement for phosphorylation of the retinoblastoma protein or related pocket proteins in Myc-driven proliferation, since Myc leads to an activation of E2F-dependent transcription and distinct activating E2F proteins act downstream of Myc in controlling cell proliferation and tumorigenesis (Baudino et al. 2003; Jansen-Dürr et al. 1993; Leone et al. 2001). In addition, expression of E2F-target genes is closely correlated with expression of specific *MYC* genes in human tumors, arguing that Myc proteins act upstream of E2F-dependent transcription in human tumorigenesis (Hernando et al. 2004) [B. Samans and M. Eilers, unpublished observations].

The requirement for Cdk4 function may also reflect a need to sequester p27$^{\text{Kip1}}$ or p21$^{\text{Cip1}}$, leading to activation of cyclin E/Cdk2 kinase activity downstream of Myc (Leone et al. 1997; Steiner et al. 1995; Vlach et al. 1996). Indeed, transcription of both p27$^{\text{Kip1}}$ and p21$^{\text{Cip1}}$ is suppressed by enhanced expression of Myc, supporting the notion that Myc may regulate Cdk2 kinase activity (Herold et al. 2002; Seoane et al. 2002; Yang et al. 2001). Tissue-culture experiments suggest that activation of Cdk2 is required for Myc-induced S-phase entry (Rudolph et al. 1996). However, the recent findings that Cdk2 is dispensable for cell proliferation in vivo forces one to re-consider this issue. Rigorous experiments analyzing Myc-induced proliferation and/or transformation in cells lacking Cdk2 or cyclins E1 or E2 are currently missing. A related issue is whether Cdk2 activation is required for Myc-induced apoptosis and there are models suggesting that suppression of p21^{Cip1} by Myc favors Myc-dependent apoptosis and p53-dependent apoptosis upon DNA damage at the expense of cell-cycle arrest, potentially through upregulation of Bim1 (Collins et al. 2005; Hemann et al. 2005).

4
Checkpoints and Apoptosis

One of the most exciting questions in the field addresses the issues of what failsafe processes limit Myc-induced proliferation and transformation and which mutations disable such failsafe mechanisms during tumorigenesis.

There are multiple indications that activation of Myc causes multiple problems for the cells, the clearest being that Myc can induce apoptosis or dramatically sensitize the cells to apoptotic stimuli. Several mechanisms contribute to this effect, including repression of Bcl-xl, the induction of ARF, an activator of p53, and the regulation of Fas ligand. Blocking apoptosis by providing exogenous bcl$_{xl}$ dramatically accelerates Myc-induced tumorigenesis in vivo, demonstrating that apoptosis indeed provides a physiological barrier to Myc-dependent tumorigenesis.

There is also clear evidence that deregulation of Myc causes DNA damage: for example, cells in which Myc has been acutely activated stain positive for phosphorylated histone H2Ax and show foci of Mre11, indicative of double-strand breaks (Bartkova et al. 2005; Vafa et al. 2002); a potential explanation is provided by the observation that the repair of double-strand breaks occurs inefficiently in human fibroblasts that express *c-myc* (Karlsson et al. 2003). It has been argued that these are tissue-culture problems due to unphysiologically high concentrations of oxygen (Soucek and Evan 2002). This is formally possible: however, several Myc-induced tumors are genomically unstable and show complex karyotypes and enhanced rates of double-strand breaks, which suggests that deregulated expression of Myc can also promote genomic instability in vivo. Whether induction of DNA damage by Myc is a causal event in Myc-induced tumorigenesis is an open question: clearly, a transgenic situation can be engineered in mice where this is almost certainly not the case (Pelengaris et al. 2002). However, such an engineered situation is likely to bypass key steps of early tumorigenesis by activating Myc simultaneously in a large group of cells or even entire organs.

A second important issue is whether both induction of apoptosis and induction of DNA damage by Myc are causally related. Mediators of Myc-induced apoptosis include p53 and E2F1 and both have been implicated in

DNA damage responses (Hermeking and Eick 1994; Leone et al. 2001). However, activation of p53 by Myc is mediated by induction of ARF, at least in lymphoid cells and in mouse embryo fibroblasts (Zindy et al. 1998). While induction of ARF by Myc is not direct and the intermediate steps in ARF activation have not been resolved, little evidence exists to suggest that ARF expression is responsive to DNA damage. The situation is different in human fibroblasts, where induction of p53 by Myc occurs independently of p14^{ARF} and can be blocked by pharmacological inhibition of ATM and ATR; these findings suggest that Myc activates key components of the DNA damage pathway (Lindstrom and Wiman 2003). Whether this activation occurs as a response to DNA damage or as a response to alterations to chromatin structure (which activate ATM) is open. In vivo studies that measure the impact of ATM mutations on Myc-induced tumorigenesis will be required to unequivocally answer these important questions.

If indeed induction of DNA damage and apoptosis are hallmarks of Myc-dependent tumors in vivo, then tumors in which Myc has been activated should harbor secondary mutations that are required to suppress the checkpoint responses to damaged DNA. This would be an exciting class of mutations to explore, since restoration of checkpoints or targeting specific apoptotic pathways may lead to a synthetic lethal effect, selectively killing Myc-dependent tumor cells while sparing normal cells. At present, there are two indications that such mutations may indeed exist: one is the finding that cells lacking the Werner syndrome (WRN) helicase, a protein that has been implicated in DNA repair, undergo strongly accelerated senescence following expression of Myc (Grandori et al. 2003). A second indication is the recent identification of synthetic lethal interactions of deregulated Myc with the TRAIL-receptor pathway, which argue that targeting Myc degradation may lead to selective killing of tumor cells that express high levels of Myc (Rottmann et al. 2005; Wang et al. 2004). Clearly, the systematic identification of such synthetic lethal interactions holds a promise for the therapy of Myc-dependent tumors.

5
Conclusions

Myc proteins act at least in part as transcription factors that activate and repress large groups of genes; as a result, tumors that express high levels of Myc proteins have a distinct gene-expression profile. Many of the regulatory mechanisms that control Myc function and the transcriptional mechanisms through which Myc proteins act have been clarified over the last years. In contrast, questions remain about the identity of the genes that mediate Myc's various biological functions. Similarly, convincing evidence shows that Myc-induced proliferation is balanced by apoptosis, yet the precise mechanisms that recognize Myc-induced proliferation as aberrant remain to be elucidated.

Acknowledgements Work in the authors' laboratory is supported by the Deutsche Forschungsgemeinschaft, the European Community through the Framework 6 program, the Thyssen- and the Sander-Stiftung, AICR and the Deutsche Krebshilfe.

References

Adhikary S, Marinoni F, Hock A, Hulleman E, Popov N, Beier R, Bernard S, Quarto M, Capra M, Goettig S, Kogel U, Scheffner M, Helin K, Eilers M (2005) The ubiquitin ligase HectH9 regulates transcriptional activation by Myc and is essential for tumor cell proliferation. Cell 123:409–421

Arabi A, Wu S, Ridderstrale K, Bierhoff H, Shiue C, Fatyol K, Fahlen S, Hydbring P, Soderberg O, Grummt I, Larsson LG, Wright AP (2005) c-Myc associates with ribosomal DNA and activates RNA polymerase I transcription. Nat Cell Biol 7:303–310

Baena E, Gandarillas A, Vallespinos M, Zanet J, Bachs O, Redondo C, Fabregat I, Martinez AC, de Alboran IM (2005) c-Myc regulates cell size and ploidy but is not essential for postnatal proliferation in liver. Proc Natl Acad Sci USA 102:7286–7291

Bartkova J, Horejsi Z, Koed K, Kramer A, Tort F, Zieger K, Guldberg P, Sehested M, Nesland JM, Lukas C, Orntoft T, Lukas J, Bartek J (2005) DNA damage response as a candidate anti-cancer barrier in early human tumorigenesis. Nature 434:864–870

Baudino TA, Maclean KH, Brennan J, Parganas E, Yang C, Aslanian A, Lees JA, Sherr CJ, Roussel MF, Cleveland JL (2003) Myc-mediated proliferation and lymphomagenesis, but not apoptosis, are compromised by E2f1 loss. Mol Cell 11:905–914

Beier R, Burgin A, Kiermaier A, Fero M, Karsunky H, Saffrich R, Moroy T, Ansorge W, Roberts J, Eilers M (2000) Induction of cyclin E-cdk2 kinase activity, E2F-dependent transcription and cell growth by myc are genetically separable events. EMBO J 19:5813–5823

Bellosta P, Hulf T, Balla Diop S, Usseglio F, Pradel J, Aragnol D, Gallant P (2005) Myc interacts genetically with Tip48/Reptin and Tip49/Pontin to control growth and proliferation during Drosophila development. Proc Natl Acad Sci USA 102:11 799–11 804

Bettess MD, Dubois N, Murphy MJ, Dubey C, Roger C, Robine S, Trumpp A (2005) c-Myc is required for the formation of intestinal crypts but dispensable for homeostasis of the adult intestinal epithelium. Mol Cell Biol 25:7868–7878

Boon K, Caron HN, van Asperen R, Valentijn L, Hermus MC, van Sluis P, Roobeek I, Weis I, Voute PA, Schwab M, Versteeg R (2001) N-myc enhances the expression of a large set of genes functioning in ribosome biogenesis and protein synthesis. EMBO J 20:1383–1393

Bouchard C, Dittrich O, Kiermaier A, Dohmann K, Menkel A, Eilers M, Luscher B (2001) Regulation of cyclin D2 gene expression by the Myc/Max/Mad network: Myc-dependent TRRAP recruitment and histone acetylation at the cyclin D2 promoter. Genes Dev 15:2042–2047

Bouchard C, Marquardt J, Bras A, Medema RH, Eilers M (2004) Myc-induced proliferation and transformation require Akt-mediated phosphorylation of FoxO proteins. EMBO J 23:2830–2840

Bouchard C, Thieke K, Maier A, Saffrich R, Hanley-Hyde J, Ansorge W, Reed S, Sicinski P, Bartek J, Eilers M (1999) Direct induction of cyclin D2 by Myc contributes to cell cycle progression and sequestration of p27. EMBO J 18:5321–5333

Chen D, Kon N, Li M, Zhang W, Qin J, Gu W (2005) ARF-BP1/Mule is a critical mediator of the ARF tumor suppressor. Cell 121:1071–1083

Collins NL, Reginato MJ, Paulus JK, Sgroi DC, Labaer J, Brugge JS (2005) G1/S cell cycle arrest provides anoikis resistance through Erk-mediated Bim suppression. Mol Cell Biol 25:5282–5291

de Alboran IM, O'Hagan RC, Gartner F, Malynn B, Davidson L, Rickert R, Rajewsky K, DePinho RA, Alt FW (2001) Analysis of C-MYC function in normal cells via conditional gene-targeted mutation. Immunity 14:45–55

Frank SR, Parisi T, Taubert S, Fernandez P, Fuchs M, Chan HM, Livingston DM, Amati B (2003) MYC recruits the TIP60 histone acetyltransferase complex to chromatin. EMBO Rep 4:575–580

Gandarillas A, Watt FM (1997) c-Myc promotes differentiation of human epidermal stem cells. Genes Dev 11:2869–2882

Gebhardt A, Frye M, Herold S, Benitah SA, Braun K, Samans B, Watt F, Elsässer HP, Eilers M (2005) Myc regulates keratinocyte adhesion and differentiation via complex formation with Miz1. J Cell Biol 172:139–149

Gomez-Roman N, Grandori C, Eisenman RN, White RJ (2003) Direct activation of RNA polymerase III transcription by c-Myc. Nature 421:290–294

Grandori C, Gomez-Roman N, Felton-Edkins ZA, Ngouenet C, Galloway DA, Eisenman RN, White RJ (2005) c-Myc binds to human ribosomal DNA and stimulates transcription of rRNA genes by RNA polymerase I. Nat Cell Biol 7:311–318

Grandori C, Wu KJ, Fernandez P, Ngouenet C, Grim J, Clurman BE, Moser MJ, Oshima J, Russell DW, Swisshelm K, Frank S, Amati B, Dalla-Favera R, Monnat RJ Jr (2003) Werner syndrome protein limits MYC-induced cellular senescence. Genes Dev 17:1569–1574

Guo Y, Cleveland JL, O'Brien TG (2005) Haploinsufficiency for odc modifies mouse skin tumor susceptibility. Cancer Res 65:1146–1149

Hemann MT, Bric A, Teruya-Feldstein J, Herbst A, Nilsson JA, Cordon-Cardo C, Cleveland JL, Tansey WP, Lowe SW (2005) Evasion of the p53 tumour surveillance network by tumour-derived MYC mutants. Nature 436:807–811

Hermeking H, Eick D (1994) Myc-induced apoptosis is mediated by p53 protein. Science 265:2091–2093

Hermeking H, Rago C, Schuhmacher M, Li Q, Barrett JF, Obaya AJ, O'Connell BC, Mateyak MK, Tam W, Kohlhuber F, Dang CV, Sedivy JM, Eick D, Vogelstein B, Kinzler KW (2000) Identification of CDK4 as a target of c-MYC. Proc Natl Acad Sci USA 97:2229–2234

Hernando E, Nahle Z, Juan G, Diaz-Rodriguez E, Alaminos M, Hemann M, Michel L, Mittal V, Gerald W, Benezra R, Lowe SW, Cordon-Cardo C (2004) Rb inactivation promotes genomic instability by uncoupling cell cycle progression from mitotic control. Nature 430:797–802

Herold S, Wanzel M, Beuger V, Frohme C, Beul D, Hillukkala T, Syvaoja J, Saluz HP, Hänel F, Eilers M (2002) Negative regulation of the mammalian UV response by Myc through association with Miz-1. Mol Cell 10:509–521

Jansen-Dürr P, Meichle A, Steiner P, Pagano M, Finke K, Botz J, Wessbecher J, Draetta G, Eilers M (1993) Differential modulation of cyclin gene expression by MYC. Proc Natl Acad Sci USA 90:3685–3689

Karlsson A, Deb-Basu D, Cherry A, Turner S, Ford J, Felsher DW (2003) Defective double-strand DNA break repair and chromosomal translocations by MYC overexpression. Proc Natl Acad Sci USA 100:9974–9979

Kenney AM, Cole MD, Rowitch DH (2003) Nmyc upregulation by sonic hedgehog signaling promotes proliferation in developing cerebellar granule neuron precursors. Development 130:15–28

Kim SY, Herbst A, Tworkowski KA, Salghetti SE, Tansey WP (2003) Skp2 regulates myc protein stability and activity. Mol Cell 11:1177–1188

Knoepfler PS, Cheng PF, Eisenman RN (2002) N-myc is essential during neurogenesis for the rapid expansion of progenitor cell populations and the inhibition of neuronal differentiation. Genes Dev 16:2699–2712

Kowalczyk A, Filipkowski RK, Rylski M, Wilczynski GM, Konopacki FA, Jaworski J, Ciemerych MA, Sicinski P, Kaczmarek L (2004) The critical role of cyclin D2 in adult neurogenesis. J Cell Biol 167:209–213

Kozar K, Ciemerych MA, Rebel VI, Shigematsu H, Zagozdzon A, Sicinska E, Geng Y, Yu Q, Bhattacharya S, Bronson RT, Akashi K, Sicinski P (2004) Mouse development and cell proliferation in the absence of D-cyclins. Cell 118:477–491

Leone G, DeGregori J, Sears R, Jakoi L, Nevins JR (1997) Myc and Ras collaborate in inducing accumulation of active cyclin E/Cdk2 and E2F. Nature 387:422–426

Leone G, Sears R, Huang E, Rempel R, Nuckolls F, Park C-H, Giangrande P, Wu L, Saavedra HI, Field SJ, Thompson MA, Yang H, Fujiwara Y, Greenberg ME, Orkin S, Smith C, Nevins JR (2001) Myc requires distinct E2F activities to induce S phase and apoptosis. Mol Cell 8:105–113

Lindstrom MS, Wiman KG (2003) Myc and E2F1 induce p53 through p14^{ARF}-independent mechanisms in human fibroblasts. Oncogene 22:4993–5005

Mateyak MK, Obaya AJ, Sedivy JM (1999) c-Myc regulates cyclin D-cdk4 and -cdk6 activity but affects cell cycle progression at multiple independent points. Mol Cell Biol 19:4672–4683

Mateyak MK, Obaya AJ, Adachi S, Sedivy JM (1997) Phenotypes of c-Myc-deficient rat fibroblasts isolated by targeted homologous recombination. Cell Growth Differ 8:1039–1048

McMahon SB, Wood MA, Cole MD (2000) The essential cofactor TRRAP recruits the histone acetyltransferase hGCN5 to c-Myc. Mol Cell Biol 20:556–562

McMahon SB, van Buskirk HA, Dugan KA, Copeland TD, Cole MD (1998) The novel ATM-related protein TRRAP is an essential cofactor for the c-Myc and E2F oncoproteins. Cell 94:363–374

Miliani de Marval PL, Macias E, Rounbehler R, Sicinski P, Kiyokawa H, Johnson DG, Conti CJ, Rodriguez-Puebla ML (2004) Lack of cyclin-dependent kinase 4 inhibits c-myc tumorigenic activities in epithelial tissues. Mol Cell Biol 24:7538–7547

Minsky N, Oren M (2004) The RING domain of Mdm2 mediates histone ubiquitylation and transcriptional repression. Mol Cell 16:631–639

Nikiforov MA, Chandriani S, Park J, Kotenko I, Matheos D, Johnsson A, McMahon SB, Cole MD (2002) TRRAP-dependent and TRRAP-independent transcriptional activation by Myc family oncoproteins. Mol Cell Biol 22:5054–5063

O'Donnell KA, Wentzel EA, Zeller KI, Dang CV, Mendell JT (2005) c-Myc-regulated microRNAs modulate E2F1 expression. Nature 435:839–843

Oliver TG, Grasfeder LL, Carroll AL, Kaiser C, Gillingham CL, Lin SM, Wickramasinghe R, Scott MP, Wechsler-Reya RJ (2003) Transcriptional profiling of the Sonic hedgehog response: a critical role for N-myc in proliferation of neuronal precursors. Proc Natl Acad Sci USA 100:7331–7336

Osterhout DJ, Wolven A, Wolf RM, Resh MD, Chao MV (1999) Morphological differentiation of oligodendrocytes requires activation of Fyn tyrosine kinase. J Cell Biol 145:1209–1218

Pelengaris S, Khan M, Evan GI (2002) Suppression of myc-induced apoptosis in beta cells exposes multiple oncogenic properties of myc and triggers carcinogenic progression. Cell 109:321–334

Perez-Roger I, Kim S-H, Griffiths B, Sewing A, Land H (1999) Cyclins D1 and D2 mediate Myc-induced proliferation via sequestration of p27^{Kip1} and p21^{Cip1}. EMBO J 18:5310–5320

Phan RT, Saito M, Basso K, Niu H, Dalla-Favera R (2005) BCL6 interacts with the transcription factor Miz-1 to suppress the cyclin-dependent kinase inhibitor p21 and cell cycle arrest in germinal center B cells. Nat Immunol 6:1054–1060

Poortinga G, Hannan KM, Snelling H, Walkley CR, Jenkins A, Sharkey K, Wall M, Brandenburger Y, Palatsides M, Pearson RB, McArthur GA, Hannan RD (2004) MAD1 and c-MYC regulate UBF and rDNA transcription during granulocyte differentiation. EMBO J 23:3325–3335

Qi Y, Gregory MA, Li Z, Brousal JP, West K, Hann SR (2004) p19^{ARF} directly and differentially controls the functions of c-Myc independently of p53. Nature 431:712–717

Rottmann S, Wang Y, Nasoff M, Deveraux QL, Quon KC (2005) A TRAIL receptor-dependent synthetic lethal relationship between MYC activation and GSK3beta/FBW7 loss of function. Proc Natl Acad Sci USA 102:15 195–15 200

Rudolph B, Zwicker J, Saffrich R, Henglein B, Müller R, Ansorge W, Eilers M (1996) Activation of cyclin dependent kinases by Myc mediates transcriptional activation of cyclin A, but not apoptosis. EMBO J 15:3065–3076

Schneider A, Peukert K, Hänel F, Eilers M (1997) Association with the zinc finger protein Miz-1 defines a novel pathway for gene regulation by Myc. Curr Top Microbiol Immunol 224:137–149

Seoane J, Le HV, Massague J (2002) Myc suppression of the p21^{Cip1} Cdk inhibitor influences the outcome of the p53 response to DNA damage. Nature 419:729–734

Seoane J, Pouponnot C, Staller P, Schader M, Eilers M, Massague J (2001) TGFbeta influences Myc, Miz-1 and Smad to control the CDK inhibitor p15^{INK4b}. Nat Cell Biol 3:400–408

Shim H, Dolde C, Lewis BC, Wu CS, Dang G, Jungmann RA, Dalla-Favera R, Dang CV (1997) c-Myc transactivation of LDH-A: implications for tumor metabolism and growth. Proc Natl Acad Sci USA 94:6658–6663

Soucek L, Evan G (2002) Myc – is this the oncogene from hell? Cancer Cell 1:406–408

Staller P, Peukert K, Kiermaier A, Seoane J, Lukas J, Karsunky H, Moroy T, Bartek J, Massague J, Hanel F, Eilers M (2001) Repression of p15^{INK4b} expression by Myc through association with Miz-1. Nat Cell Biol 3:392–399

Steiner P, Philipp A, Lukas J, Godden-Kent D, Pagano M, Mittnacht S, Bartek J, Eilers M (1995) Identification of a Myc-dependent step during the formation of active G1 cyclin/cdk complexes. EMBO J 14:4814–4826

Trumpp A, Refaeli Y, Oskarsson T, Gasser S, Murphy M, Martin GR, Bishop JM (2001) c-Myc regulates mammalian body size by controlling cell number but not cell size. Nature 414:768–773

Vafa O, Wade M, Kern S, Beeche M, Pandita TK, Hampton GM, Wahl GM (2002) c-Myc can induce DNA damage, increase reactive oxygen species, and mitigate p53 function. A mechanism for oncogene-induced genetic instability. Mol Cell 9:1031–1044

Vlach J, Hennecke S, Alevizopoulos K, Conti D, Amati B (1996) Growth arrest by the cyclin-dependent kinase inhibitor p27^{Kip1} is abrogated by c-Myc. EMBO J 15:6595–6604

von der Lehr N, Johansson S, Wu S, Bahram F, Castell A, Cetinkaya C, Hydbring P, Weidung I, Nakayama K, Nakayama KI, Soderberg O, Kerppola TK, Larsson LG (2003)

The F-box protein Skp2 participates in c-Myc proteosomal degradation and acts as a cofactor for c-Myc-regulated transcription. Mol Cell 11:1189–1200

Walker W, Zhou ZQ, Ota S, Wynshaw-Boris A, Hurlin PJ (2005) Mnt-Max to Myc-Max complex switching regulates cell cycle entry. J Cell Biol 169:405–413

Wang Y, Engels IH, Knee DA, Nasoff M, Deveraux QL, Quon KC (2004) Synthetic lethal targeting of MYC by activation of the DR5 death receptor pathway. Cancer Cell 5:501–512

Wanzel M, Herold S, Eilers M (2003) Transcriptional repression by Myc. Trends Cell Biol 13:146–150

Welcker M, Orian A, Grim JA, Eisenman RN, Clurman BE (2004a) A nucleolar isoform of the Fbw7 ubiquitin ligase regulates c-Myc and cell size. Curr Biol 14:1852–1857

Welcker M, Orian A, Jin J, Grim JA, Harper JW, Eisenman RN, Clurman BE (2004b) The Fbw7 tumor suppressor regulates glycogen synthase kinase 3 phosphorylation-dependent c-Myc protein degradation. Proc Natl Acad Sci USA 101:9085–9090

Wilson A, Murphy MJ, Oskarsson T, Kaloulis K, Bettess MD, Oser GM, Pasche AC, Knabenhans C, Macdonald HR, Trumpp A (2004) c-Myc controls the balance between hematopoietic stem cell self-renewal and differentiation. Genes Dev 18:2747–2763

Yada M, Hatakeyama S, Kamura T, Nishiyama M, Tsunematsu R, Imaki H, Ishida N, Okumura F, Nakayama K, Nakayama KI (2004) Phosphorylation-dependent degradation of c-Myc is mediated by the F-box protein Fbw7. EMBO J 23:2116–2125

Yang W, Shen J, Wu M, Arsura M, FitzGerald M, Suldan Z, Kim DW, Hofmann CS, Pianetti S, Romieu-Mourez R, Freedman LP, Sonenshein GE (2001) Repression of transcription of the $p27^{Kip1}$ cyclin-dependent kinase inhibitor gene by c-Myc. Oncogene 20:1688–1702

Yu Q, Ciemerych MA, Sicinski P (2005) Ras and Myc can drive oncogenic cell proliferation through individual D-cyclins. Oncogene 24:7114–7119

Zindy F, Eischen CM, Randle DH, Kamijo T, Cleveland JL, Sherr CJ, Roussel MF (1998) Myc signaling via the ARF tumor suppressor regulates p53-dependent apoptosis and immortalization. Genes Dev 12:2424–2433

Results Probl Cell Differ (42)
P. Kaldis: Cell Cycle Regulation
DOI 10.1007/003/Published online: 20 December 2005
© Springer-Verlag Berlin Heidelberg 2005

Cell Cycle Regulation in Mammalian Germ Cells

Changanamkandath Rajesh · Douglas L. Pittman (✉)

Department of Physiology and Cardiovascular Genomics, Medical University of Ohio, Toledo, Ohio 43614, USA
dpittman@meduohio.edu

Abstract Meiosis is a unique form of cellular division by which a diploid cell produces genetically distinct haploid gametes. Initiation and regulation of mammalian meiosis differs between the sexes. In females, meiosis is initiated during embryo development and arrested shortly after birth during prophase I. In males, spermatogonial stem cells initiate meiosis at puberty and proceed through gametogenesis with no cell cycle arrest. Mouse genes required for early meiotic cell cycle events are being identified by comparative analysis with other eukaryotic systems, by virtue of gene knockout technology and by mouse mutagenesis screens for reproductive defects. This review focuses on mouse reproductive biology and describes the available mouse mutants with defects in the early meiotic cell cycle and prophase I regulatory events. These research tools will permit rapid advances in such medically relevant research areas as infertility, embryo lethality and developmental abnormalities.

1
Introduction

Meiosis is a developmental pathway used to reduce the chromosome number by one-half, resulting in the production of haploid gametes and permitting sexual reproduction (Fig. 1). Given that chromosomal aneuploidy is the leading genetic cause of pregnancy loss and birth defects, it is critical to dissect out meiotic cell cycle checkpoints and proper chromosome segregation (Hassold and Hunt 2001). However, understanding cell cycle regulation in mammalian germ cells has considerably lagged behind that of the mitotic cell cycle. Challenges to investigations of meiosis in mammals have included the long reproductive cycles, cell migration patterns, somatic/germ cell interaction and the inability to propagate germ cells in culture for extended time. Understanding mammalian meiosis has benefited considerably by general principles being conserved across species. Much of our knowledge is built upon studies in model organisms such as fungi, Arabidopsis and *Drosophila* (Engebrecht 2003; Page and Orr-Weaver 1997; Wilson and Yang 2004). Tools are now available in mouse genetics to make tremendous strides towards a thorough understanding of cell cycle regulatory genes during meiosis and gametogenesis.

Mammalian meiosis and differentiation take place in multicellular and hormonally regulated environments. During the gastrulation stage of mouse

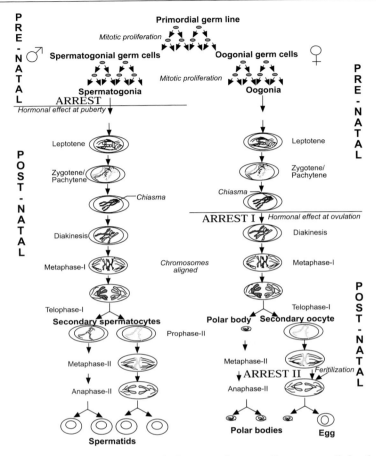

Fig. 1 Diagram illustrating the principle features of mammalian germ cell development. Following proliferation and migration to the genital ridge, primordial germ cells differentiate towards the spermatogonial or oogonial pathways, which differ in the number and timing of meiotic arrests. In male mice, spermatogenesis starts shortly after birth from the spermatogonia and once initiated, continues without interruption until haploid spermatid formation. Meiotic divisions of each spermatogonia generate four spermatids which mature by spermiogenesis to spermatozoa (not illustrated). Spermatogonia divide continuously after birth to maintain a constant supply of spermatocytes. In the females, oocyte development is initiated before birth and is arrested at the dictyate stage of meiotic prophase I. An oocyte resumes meiosis in response to hormonal stimuli only to be arrested a second time at metaphase II. This arrest is overcome if fertilization ensues, which triggers completion of the meiotic division. Cell division in females is unequal, generating three non-functional polar bodies and an egg. The stages of meiosis are the same in both sexes with the first division (meiosis I) resulting in reduction in the number of chromosomes per cell. The prophase I is sub-divided into five sub-stages (leptotene, zygotene, pachytene, diplotene and diakinesis) as illustrated. The second division (meiosis II) conserves the number of chromosomes in a mitosis-like process. Only a set of homologous chromosomes is illustrated in the figure

embryonic development, primordial germ cells (PGCs) begin to form at 7.0 days postconception (dpc) and migrate from the base of the allantois through the gut, up the dorsal mesentery and into the genital ridges (Anderson et al. 2000; Nagy 2003). In males, germ cells undergo a G1 mitotic arrest at 14.0 dpc (McLaren 2003). During the early stages of spermatogenesis, stem cells, termed type A spermatogonia, appear 3–5 days post-partum and undergo self-renewal to produce type B spermatogonia. The type B spermatogonia divide and generate primary diploid (2n) spermatocytes capable of initiating meiosis and forming four haploid (n) spermatids per spermatocyte, which eventually produce mature sperm (de Rooij and de Boer 2003). These events are supported by Sertoli cells in the seminiferous tubules of the testis, which are non-dividing diploid cells that form the blood-testis barrier. The entire process from spermatogonial stem cells to mature spermatozoa takes approximately 5 weeks (Handel 1987; Russell 1990).

In females, meiosis is initiated following PGC migration and undergoes a number of checkpoint regulated arrests (Mintz and Russell 1957). At 14.0 dpc in mouse, the peak number of oocytes are formed and meiosis begins. After birth, oocytes arrest during the first meiotic prophase, followed by a loss of more than one-third of the germ cell population by apoptosis (atresia). Shortly before ovulation, the meiotic cell cycle is activated and another, meiosis II, arrest ensues; activation and completion of oogenesis is then dependent upon fertilization (Fig. 1). The first meiotic division is asymmetrical, resulting in the production of a large secondary oocyte and a small non-functional polar body. The second meiotic division, if completed, results in four final haploid cells (three polar bodies and one egg). With recent evidence that oogonial stem cells may proliferate and replenish the follicle pool in adult mice, efforts to develop new technologies for identifying genes involved during cell cycle regulation is even more alluring (Johnson et al. 2004). The possible replenishment from the oogonial stem cells will be critical for addressing issues regarding infertility, preserving fertility and potentially prolonging the reproductive years.

The genetic program that triggers meiosis and maintains the cycle through checkpoints require some mechanisms not present during the mitotic cell cycle. Until the past decade, the availability of mouse models for meiosis studies were limited by the number of spontaneous meiosis mutants, most having pleiotropic phenotypes (Handel 1987). Currently, there are four primary approaches for identifying meiosis regulatory genes in the mouse: cell cycle genes having meiotic properties (Wolgemuth et al. 2004), homology searches and compiling encyclopedic compendiums (Critchlow et al. 2004), gene products associating with meiotic chromosomes (Heyting and Dietrich 1991) and phenotype-driven approaches (Reinholdt et al. 2004). This chapter reviews known mouse genes involved in cell cycle regulation during early meiosis, focusing on entry through meiosis I. Taking advantage of database and literature searches, 53 genes for which mouse mutants are avail-

Table 1 Gene knockouts affecting germ cell development in mouse

Gene	Protein Name	Effect on germ cell development	Refs.
A. Cellular Proliferation and Differentiation			
Ccna1	Cyclin A1	Arrest of germ cells prior to first meiotic division.	Liu et al. 1998
Cdc25b	Cell cycle protein	Cell cycle protein involved in activation of MPF. Deficiency leads to oocyte arrest at prophase → metaphase transition.	Lincoln et al. 2002; Spruck et al. 2003
Cdk4	Cyclin dependant kinase 4	Decreased number of spermatocytes and spermatogonia and defective luteal function in females.	Moons et al. 2002
Cdk2	Cyclin dependant kinase 2	Sterile males arrest at G_2/M transition stage of cell cycle, while females lack follicular development due to breakdown at the pachytene stage.	Ortega et al. 2003; Berthet et al. 2003
CyclinD2	Cell cycle regulating D-cyclin	Absence of germ cells due to lack of proliferation.	Sicinski et al. 1996
CyclinE2	Cyclin E2	Absence confers abnormal synaptonemal complex formation.	Geng et al. 2003; Parisi et al. 2003
Gcd/Pog	Proliferation of germ cells (POG)	Drastic depletion of germ cells in the developing ridges due to proliferation defects during PGC migration to genital ridges.	Agoulnik et al. 2002
p18[INK4C], *p19*[INK4D]	Cyclin D-dependant kinase inhibitors	Defects in transition from a pre-leptotene to the leptotene stage in spermatogenesis, results in decreased sperm numbers.	Zindy et al. 2001
p27[Kip1]	p27[Kip1] a Cdk inhibitor	Factor responsible for the G_1/G_0 arrest in oocytes and absence results in defects in pre-leptotene to leptotene transition. Female sterility from aberration in follicular maturation. Knockout male mice show a large number of pre-leptotene stage spermatocytes.	Beumer et al. 1999; Fero et al. 1996

Table 1 (continued)

Gene	Protein Name	Effect on germ cell development	Refs.
B. Transcriptional and Translational Factors			
Ahch	Dax1	A transcriptional factor that has a role in gonadal differentiation and sex determination. Lack of which leads to complete loss of germ cells due to progressive degeneration of germinal epithelium.	Yu et al. 1998
Cpeb	Cytoplasmic polyadenylation element binding protein	These RNA binding proteins regulate translation during oocyte maturation. The absence of which results in vestigial ovaries with immature oocytes arrested at the pachytene stage.	Tay and Richter 2001
Dazla	Dazla protein (Deleted in azoospermia phenotype)	RNA binding protein that affects translational control. Knockout results in lack of prenatal germ cells (females) or spermatogenic arrest by blockage of spermatogonial differentiation (male).	Ruggiu et al. 1997; Schrans-Stassen et al. 2001
Egr4	Early growth response protein (EGR) 4, NGFI-C, pAT133	A zinc finger transcriptional factor involved in cell differentiation and growth. In its absence incomplete block of germ cell maturation at mid-pachytene stage occurs. Also causes sperm with abnormal morphology.	Tourtellotte et al. 1999
Eif2s3y	Eukaryotic translation initiation factor 2, subunit 3, structural gene Y-linked	Involved in the early steps of protein synthesis, causes spermatogonial proliferation impairment in its absence.	Mazeyrat et al. 2001
Fox3a	Foxo transcription factors	Factor involved in metabolism, cellular stress response and aging. The absence leads to early activation of follicles leading to low functional follicles.	Castrillon et al. 2003
Hspa2	Heat shock-related 70 kDa protein 2 (HSP70-2)	Chaperone protein involved in protein folding, leads to spermatogenic arrest at metaphase stage in knockouts.	Dix et al. 1996

Table 1 (continued)

Gene	Protein Name	Effect on germ cell development	Refs.
Taf4b	TAF4B RNA polymerase II, TAFII 105	Transcription factor important in RNA polymerase II machinery. Impaired folliculogenesis is seen in knockouts leading to lack of mature follicles.	Freiman et al. 2001
Tiar	TIAR	An RNA-recognizing motif involved in splicing, transport, translation and stability of mRNA. Knockout results in decrease in the survival of germ cells at the genital ridge.	Beck et al. 1998
Tls/Fus	Translated in liposarcoma (TLS/FUS)	RNA-binding protein reported to contribute to N-terminal half of fusion proteins in liposarcomas and leukemias. Absence leads to increase in unpaired and mispaired chromosomal axes in pre-meiotic spermatocytes leading to apoptosis.	Kuroda et al. 2000
Trcp1	F-box protein to accumulation β-Trcp1	Required for mitotic progression. Lack of gene leads of spermatocytes arrested at metaphase-I.	Guardavaccaro et al. 2003
C. Cell Signalling			
Bmp15	Bone morphogenetic protein 15 (growth differentiation factor 9b)	Growth factor required for ovarian function, shows defects in ovarian folliculogenesis and ovulation.	Yan et al. 2001
Cit-k	CIT-K (citron kinase)	A serine/threonine kinase interacting with Rho. Knockouts show embryonic and postnatal loss of germ cells leading to complete lack of spermatocytes. Female effects not studied.	Cunto et al. 2002
Cks2	CKS2 (mammalian homolog of the yeast Cdk1-binding protein)	Essential for the biological function of cyclin dependent kinases by binding to its catalytic subunit. The lack of the gene results in germ cell arrest at metaphase-I.	Spruck et al. 2003

Table 1 (continued)

Gene	Protein Name	Effect on germ cell development	Refs.
Crem	Cyclic AMP response element modulator (CREM)	Binds to the cAMP response element (CRE), loss of which leads to interruption of spermatogenesis at early haploid stage (Round spermatids).	Blendy et al. 1996
Gja1/Cx43	Connexin 43	These intracellular signaling molecules in their absence cause impaired folliculogenesis resulting in a germ cell deficiency within the gonads.	Juneja et al. 1999
Gja4/Cx37	Connexin 37	Intracellular signaling molecules located at the gap junctions preventing the maturation of the follicles beyond meiotic competence in their absence.	Simon et al. 1997
Madh1/5	MAD homolog Smad 1/5	Signal mediators for bone morphogenetic proteins. Loss leads to greatly reduced or absence of primordial germ cells in embryos.	Chang and Matzuk 2001; Tremblay et al. 2001
Steel[panda]	Steel factor	Serves as a ligand for tyrosine kinase in *c-kit* protooncogene, knockout shows folliculogenesis defects and reduced germ cell number.	Huang et al. 1993
D. Cytoplasmic and Apoptotic Factors			
Apaf1	APAF1 (apoptotic protease activating factor 1)	The protein has a role in cytochrome c-mediated apoptosis, absence of which leads to spermatogonial degeneration.	Honarpour et al. 2000
AR	Androgen receptor (AR)	Sertoli cell specific knockout of this receptor resulted in spermatogenic arrest predominantly at the diplotene stage. Also results in low serum testosterone levels leading to azoospermia and infertility.	Chang et al. 2004

Table 1 (continued)

Gene	Protein Name	Effect on germ cell development	Refs.
Bcl2/Bclx	B-cell leukemia/ lymphoma 2/X	Anti-apoptotic protein controlling cell survival. The absence of which leads to severe loss of the primordial germ cells and hence absence of spermatogonia in testis and depletion of follicles in the post-natal ovary.	Ratts et al. 1995; Rucker et al. 2000
Bcl2l2 (Bclw)	Bcl2 like 2	A cytoplasmic protein that promotes cell survival and has a role in maintenance of reproductive germ cells. The knockout leads to progressive loss of germ cells, Sertoli cells and Leydig cells and cause extensive testicular degradation.	Ross et al. 1998
Bsg	Basigin	A protein belonging to the immunoglobulin superfamily important for pre-implantation development and spermatogenesis. The knockout mice show azoospermia due to metaphase-I arrest.	Toyama et al. 1999; Igakura et al. 1998
Casp2	Caspase 2	Intracellular death effectors, the absence of which significantly increase the primordial follicles in the ovary due to the absence of cell death.	Bergeron et al. 1998
c-mos	c-mos, proto-oncogene product MOS	An essential part of cytostatic factor (CSF), lack of which results in oocyte maturation from second meiotic metaphase arrest, without any activation, resulting in ovarian teratomas.	Colledge et al. 1994
Miwi	Miwi (Murine homolog of piwi- P-element induced wimpy testis)	Cytoplasmic protein expressed in spermatocytes and spermatids, absence of which can lead to spermatogenic arrest at beginning of spermatid stage.	Deng and Lin 2002
Tlf/Trf2	TATA box binding protein-like factor (TLF/TRF2)	Physiological function not known but absence results in arrest during spermiogenesis at transition from round spermatids, leading to apoptosis.	Martianov et al. 2001

Table 1 (continued)

Gene	Protein Name	Effect on germ cell development	Refs.
E. Prophase-I Regulation			
Atm	Ataxia telangiectasia	A nuclear protein with a role in cell cycle and DNA repair. Disrupted cell division leading to lack of ovarian follicles and spermatids.	Barlow et al. 1996
Brca1	BRCA1	Absence leads to prophase I arrest of spermatocytes accompanied by p53 dependent and independent apoptosis.	Xu et al. 2003
Brca2	BRCA2	Plays role in meiotic recombination, chromosome pairing and synapsis during spermatogenesis. Absence of Brca 2 leads to prophase-I arrest of spermatocytes.	Sharan et al. 2004
Dmc1	Disrupted meiotic cDNA (DMC) 1	Gene involved in homologous recombination, lack of which leads to defects in chromosomal synapsis.	Pittman et al. 1998; Yoshida et al. 1998
Ercc1/Xpf	DNA repair proteins	Involved in recombination, double strand break repair and repair of interstrand cross-links. The lack of Ercc1 leads to oocyte degeneration and low number of oocytes.	Hsia et al. 2003
Fkbp6	FK506 binding protein 6	Synaptonemal complex component essential for sex-specific fertility and chromosome pairing. Male knockouts have complete spermatogenesis block and death of meiotic spermatocytes.	Crackower et al. 2003
H2afx	Histone H2A.X	Facilitates specific DNA-specific DNA-repair complex assembly on damaged DNA. Absence leads to pachytene stage arrest of spermatocytes.	Celeste et al. 2002
Mei1	Meiosis defective 1	Homozygote mutants of both sexes are sterile due to meiotic arrest arising from defects in recombinational repair and chromosomal synapsis.	Libby et al. 2002; Libby et al. 2003
Mlh1	Mismatch repair enzyme	Non-viable oocyte or spermatocytes.	Edelmann et al. 1996

Table 1 (continued)

Gene	Protein Name	Effect on germ cell development	Refs.
Msh4/5	Mut S homologue 4/5	Post replicative DNA mismatch repair gene whose disruption leads to loss of oocytes and spermatocytes. Abnormal chromosomal pairing occurs at zygotene phase of prophase-I in spermatocytes.	Kneitz et al. 2000
Nbs1	Nijmegen breakage syndrome (NBS 1) protein	Component of Mre11 complex involved in DNA strand break repair. Female knockout mice are sterile due to oogenesis failure.	Kang et al. 2002
Siah1a	Ubiquitin ligase component	The protein is required for completion of meiosis I, defects are observed in synaptonemal complex formation and progression from metaphase I in knockouts.	Dickins et al. 2002
Sycp1	Synaptonemal complex protein 1 (SCP1)	Component of the transverse filament of the synaptonemal complex. Absence leads to male and female sterility with disruption of spermatogenesis primarily at pachytene.	de Vries et al. 2005
Scp3	Synaptonemal complex protein 3 (SCP3)	Component of axial/lateral element of the synaptonemal complex and is associated with the centromeres, lack of which leads to disruption of spermatogenesis at zygotene stage, massive cell death and female germ cell aneuploidy.	Yuan et al. 2000; Yuan et al. 2002
Spo11	Spo11 protein	Protein involved in initiation of genetic recombination. Arrest prior to pachytene stage of meiosis I in mutants.	Romanienko and Camerini-Otero 2000; Baudat et al. 2000

able are listed in Table 1. Though a strict classification is not possible due to the lack of in-depth knowledge concerning complete mechanisms, the genes are arranged into 5 categories based upon the cell cycle process affected: initiation and maintenance of meiosis, transcriptional and translational factors, cellular signaling, cytoplasmic and apoptotic factors and prophase I regulation.

2
Cell Cycle Regulatory Genes Required for Initiation and Maintenance of Meiosis

During germ cell development, spermatogonia or oogonia stem cells undergo a series of mitotic divisions leading to the formation of gametocytes that go through a final interphase and enter meiotic prophase I. The triggers for meiosis entry are not known, but like the mitotic cell cycle (G_1 to S, S progression and G_2 to M), transition is controlled, at least in part, by phosphorylation events catalyzed by cyclin dependant kinase (Cdk) complexes. In this section, Cdks, cyclins and Cdk inhibitors and activators are discussed.

The G_2/M transition in meiosis is controlled by the Cdc2 kinase, Cdk1 and the B-type cyclin complex, called the M-phase-promoting factor (MPF) (Choi et al. 1991; Masui and Markert 1971). The build up of cyclin B and its degradation at the end of the cell cycle is a hallmark of MPF activity guiding the initiation and termination of the cycle (Pines and Hunter 1989). MPF activity in oocytes peaks once at the time of meiotic initiation and again at the meiotic arrest in metaphase. The MPF is involved in several features of cell division such as disassembly of the nucleus, chromosome condensation, cytoskeletal rearrangements and transcriptional activity (Moreno and Nurse 1990).

At least 11 distinct Cdks (Cdk1-11) and 11 cyclins (cyclins A–J, T) exist in higher eukaryotes. Cyclin A1, *Ccna*, was one of the first cyclins demonstrated by targeted disruption in mouse as having a definite role in meiosis (Liu et al. 1998). Otherwise healthy males were sterile due to a block in spermatogonia before the first meiotic division, while females were fully fertile. The disruption did not have significant effect on either Cdk1 or Cdk2 kinase activity, suggesting that cyclin A1 affects downstream targets such as the MPF complex and other protein kinases or phosphatases for G_2/M transition. Additionally, cyclin B/Cdk1 activity was insufficient to bypass cyclin A1 for entry into meiosis (Liu et al. 2000).

In addition to mitotic cell cycle control, Cdk2 was implicated as having a role during meiosis by localization at telomeres during spermatogenesis when axial elements form and at synaptic sites corresponding to recombination events (Ashley et al. 2001). Surprisingly, *Cdk2* knockout mice were viable, apparently being compensated by another kinase, but male and female mice were sterile (Berthet et al. 2003; Ortega et al. 2003). Females lack follicular development, suggesting that meiotic arrest occurred at the early stages of meiosis. *Cdk2*-deficient oocytes developed up to the pachytene stage but break down at the dictyate stage, suggesting a failure to maintain chromosome synapsis. In spermatocytes, chromosomes also fail to synapse.

Cdk4 localizes along synaptonemal complexes of newly synapsed bivalents and disappear by mid-pachytene (Ashley et al. 2001). Cdk4 and its catalytic partner, cyclin D, are responsible for the initial phosphorylation of the Rb protein by regulating cyclin E/Cdk2 and lifting the S-phase inhibition by preventing binding of its inhibitor, p27 (Tsutsui et al. 1999). Cdk4/cyclin D2 have an important role in follicle stimulating hormone-induced proliferation of granulosa cells in females (Moons et al. 2002) and in decreased numbers of spermatogonia leading to sterility in males (Wolgemuth 2003).

Cdk1 function is regulated by phosphorylation at several sites. For example, human Cdk1 phosphorylation at Thr-14 and Tyr-15 are inhibitory while Thr-160 is activating (Pines 1999). The activating threonine phosphorylation, opens up a catalytic cleft, which is catalyzed by a Cdk activating kinase comprising Cdk7 and cyclin H. The phosphate is removed after cyclin degradation by a phosphatase KAP. The Cdc25 family of phosphatases are activators of Cdk complexes by removal of the phosphates from the ATP-binding site of Cdk1, but their function may differ between sexes. Females deficient for *Cdc25b* are sterile and oocytes arrest at prophase, but males are fertile (Lincoln et al. 2002).

Two classes of Cdk inhibitors modulate Cdk activity during the mitotic cell cycle. Members of the INK4 family ($p15^{INK4b}$, $p16^{INK4a}$, $p18^{INK4c}$, $p19^{INK4d}$) specifically bind and regulate cyclin D dependant kinases. The Kip/Cip family ($p21^{Cip1/Waf1}$, $p27^{Kip1}$ and $p57^{Kip2}$) binds and regulates cyclin A, D and E dependant kinases (Sherr and Roberts 1999). $p27^{Kip1}$ is an important factor guiding the G_1/G_0 arrest in germ cells as suggested by their high expression levels at 16 dpc. In adult males, $p27^{Kip1}$ expression is observed in the Sertoli cells and Leydig cells, probably to mediate spermatocyte development. $p27^{Kip1}$ knockout male mice show a large number of pre-leptotene stage spermatocytes indicating a role in regulation of spermatogonial proliferation and onset of meiotic prophase (Beumer et al. 1999). The pre-leptotene spermatocytes enter mitosis instead of meiosis, leading to speculations about their role in male germ cell tumorigenesis. In females, sterility results from aberration in follicular maturation and corpora lutea formation (Fero et al. 1996).

The $p19^{INK4d}$ gene is expressed during meiotic prophase and in postmeiotic spermatids (Zindy et al. 2001). The $p19^{INK4d}$ knockouts are fertile but display markedly lower sperm counts owing to testicular atrophy associated with apoptosis in seminiferous tubules. $p18^{INK4c}$ null mice are also fertile but develop Leydig cell hyperplasia and pituitary tumors. The $p19^{INK4d}$ and $p18^{INK4c}$ double mutant knockout mice have lower fertility due to a decrease of spermatocytes, while females were fertile. A delay in the mitosis to meiosis transition in double mutant males resulted in improper differentiation and higher rate of apoptosis.

In summary, phosphorylation by Cdk complexes are important for regulatory events leading to mammalian germ cell development. The interplay of various cyclins and their interacting kinases is responsible for guiding initiation, subsequent smooth progress through each stage and finally, termination of meiosis and gametogenesis. Inhibitors of Cdk activity also mediate meiosis and kinases and phosphatases govern the reorganization of the subcellular architecture for cell cycle progression.

3
Transcriptional and Translational Factors

Regulation of meiosis and gametogenesis also takes place at the level of transcription, which affects proliferation, differentiation and maturation of germ cells. The *Dazla* gene encodes an RNA-binding protein; the deficiency of which leads to complete absence of spermatocytes or oocytes (Ruggiu et al. 1997). The spermatogenic defect found in *Dazla*-deficient males is failure of spermatogonia progression, leading to accumulation and apoptosis of primordial spermatogonial cells in the seminiferous tubules (Schrans-Stassen et al. 2001). Other RNA-binding proteins affect oocyte maturation (*Cpeb*), pairing at the recombination stage in pre-meiotic spermatocytes (*Tls/Fus*) or stability of RNA at the genital ridges (*Tiar*). Genes encoding transcription factors such as Dax1 (*Ahch*) have been implicated with gonadal differentiation and sex determination. Loss of Dax1 affects female ovarian development, but males are sterile due to the progressive degeneration of the testicular germinal epithelium (Yu et al. 1998). Similarly, the transcription factors early growth response protein (*Egr4*), TAF4B RNA polymerase II (*Taf4b*) and Foxo (*Foxo3a*) affect gonadal differentiation and sex determination, germ cell maturation, folliculogenesis and the cellular stress response. These overall effects lead to the absence of functional gametes.

Impaired germ cell development also results from mutations in genes affecting translation or post-translational modification. Eukaryotic translation initiation factor 2 (*Eif2s3y*) is a Y-chromosome encoded subunit of the Eif2 translational initiation factor (also known as the spermatogonial proliferation factor or Spy). *Eif2s3y* is responsible for normal spermatogonic proliferation and progression to the meiotic cycle. The introduction of the transgenic *Eif2s3y* into XO males results in the release of spermatogonial arrest and spermatogenesis proceeds until the secondary spermatocyte stage prior to meiosis II initiation (Mazeyrat et al. 2001). Unpaired chromosomes lead to meiotic arrest and apoptosis at this stage. Heat shock-related protein 2 (Hspa2/Hsp70-2) is a chaperone protein affecting protein folding and is part of the synaptonemal complex. *Hspa2*-deficient males are infertile due to lack of post-meiotic spermatids and mature sperm. Females, which lack expression of *Hspa2* in pachytene oocytes, are fertile. *Hspa2* is also linked to mechanisms inhibiting spermatocyte apoptosis that increases in the knockouts, probably due to the incorrect folding and transportation of DNA repair and recombination proteins (Dix et al. 1996).

Germ cell development is regulated at both transcriptional and translational levels by specific transcription factors and RNA binding proteins. Genes involved in translation and post-translational modifications also play a role in the progression of meiosis and prevent stage specific arrests during gametogenesis.

4
Cell Signaling

Cell signaling proteins act as mediators of various cell cycle regulators, including Cdk activity. The CKS (Cdc28 kinase subunit) protein family is the mammalian homolog of the yeast Cdk binding proteins Cks1 and Suc1 and comprises two closely related orthologs, Cks1 and Cks2, which have 81% amino acid sequence similarity. In females, *Cks2* is expressed in developing oocytes, and *Cks1* is expressed in the surrounding granulosa cells. *Cks2*-deficient males and females are sterile due to the failure of germ cells to enter meiotic anaphase (Spruck et al. 2003). Cks1 has specific role in proteolysis of the Cdk inhibitor, p27^{Kip1} (Ganoth et al. 2001; Spruck et al. 2001). *Cks1*-deficient mice were histologically indistinguishable from p27^{Kip1} knockout mice except in the testis, where arrest occurs at the primary spermatocyte stage.

One gene regulating proliferation in germ cells is the transforming growth factor-β signaling superfamily protein. A knockout of the oocyte specific mouse homolog of growth differentiation factor-9 (GDF-9), bone morphogenetic protein-15 (BMP-15) confers infertility due to a block in folliculogenesis at the primary follicle stage. *Bmp15*-deficient males are fertile, but female mice show subfertility due to the defects in ovulation and fertilization defects (Yan et al. 2001).

Another class of signaling effectors are the Smad proteins, which are intracellular mediators of the TGF-β family of growth factors. Smad-deficient embryos die by 10.5 dpc due to impairment of allantois formation and failure to connect to the placenta (Tremblay et al. 2001). Importance of Smad-dependant signaling in primordial germ cell proliferation is suggested by the drastic reduction of PGCs in mutant embryos. Smad signaling acts as the second signal to the extra-embryonic signal from BMP that leads to the PGC formation (Chang and Matzuk 2001). Intracellular signaling molecules of gap junctions called connexins have been implicated in reproductive processes as germ cell and follicular development (Juneja et al. 1999). Follicles from *Gja4*-deficient mice (connexin 37) demonstrate inhibition of oocyte-granulosa cell communication, while other cellular signaling processes appear unaffected. Female infertility is due to this loss of communication resulting in abnormalities in follicular growth, control of luteinization and oocyte maturation. *Gja1* (connexin 43) is expressed in fully developed mouse oocytes and associated cumulus cells and is present in the gap junctions between Leydig, Sertoli and peritubular cells of males. *Gja1*-deficient mice have unusually small gonads due to the low number of germ cells. The Cx43 mediated gap junctional coupling plays an indispensable role in early germ cell development and in post-natal folliculogenesis. Other signaling molecules implicating effects in germ cell development are ligands for tyrosine kinase in the c-kit proto-oncogene, Steel factor (*Steel*^{panda}), cyclic AMP response element modulator (*Crem*) and serine/threonine Citron kinase (*Cit-k*).

Since cell signaling pathways affect all cell types, few of these cell effectors have been demonstrated to affect germ cell development pathways based on mouse knockout studies. The mouse mutants described in this section highlight indispensable roles in mediating cell cycle regulators, growth factors and intracellular communication pathways for germ cell development.

5
Cytoplasmic and Apoptotic Factors

During gametogenesis, apoptosis of germ cells normally occurs. Cytoplasmic factors, proto-oncogenes, intracellular effectors and extracellular factors aid in arresting cell development and subsequent degeneration or by preventing loss of germ cells. The proto-oncogene *c-mos* has been implicated in meiotic maturation (Colledge et al. 1994). Although *c-mos* mutant males were unaffected, *c-mos*$^{-/-}$ females were subfertile due to failure of oocyte arrest after ovulation. Germinal vesicle degeneration and extrusion of polar bodies sometimes led to parthenogenetic development of unfertilized eggs. Another cytoplasmic factor promoting cell survival is Bcl2-like 2 (*Bcl2l2/Bclw*) (Ross et al. 1998). *Bcl2l2* expression is limited to developing spermatids and Sertoli cells and loss of the gene results in gradual depletion of germ cells to develop a Sertoli-cell-only phenotype. Similar spermatogenic arrest before the spermatid stage is seen for a knockout of *Miwi*, a murine homolog of piwi family proteins, expressed primarily in spermatocyte and spermatid cytoplasm (Deng and Lin 2002).

The absence of certain transmembrane proteins also affects the development of germ cells. In Sertoli cell conditionally disrupted models, androgen receptor (*AR*) was found to specifically direct the progressive spermatogonial maintenance of the cell cycle (Chang et al. 2004). A knockout in the Basigin (*Bsg*), which belongs to the immunoglobulin superfamily, results in the arrest of spermatogenesis at metaphase of meiosis I. The lumen of the seminiferous tubules and duct systems of $Bsg^{-/-}$ were filled with round degenerated germ cells (Toyama et al. 1999).

Intracellular death effectors as caspases (*Casp2*) (Bergeron et al. 1998), apoptotic protease activating factor 1 (*Apaf1*) (Honarpour et al. 2000) or antiapoptotic proteins B-cell leukemia/lymphoma2/x (*Bcl2/Bclx*) (Ratts et al. 1995; Rucker et al. 2000) are responsible for preventing cell death to maintain germ cell numbers. The absence of these leads to degradation of spermatogonial cells (*Apaf1*), or loss of primordial germ cells in prenatal ovary (*Bcl2/Bclx*). *Casp2*-deficient mice have severely increased primordial follicles in the ovary due to the absence of cell death.

6
Cell Cycle Regulation during Prophase I

Prophase I is divided into five stages based upon chromosome appearance (Fig. 1): leptotene (long, thin chromosomes, no pairing), zygotene (homologous pairing and recombination initiated), pachytene (chromosomes shorter, synaptonemal complex formation and homologs paired), diplotene (homologs repel, visible chromosome exchange) and diakinesis (maximum chromosome condensation, nuclear envelope disappears). One of the major events during the leptotene to pachytene stages is the alignment process of homologous chromosomes, which requires genetic recombination and formation of a structure called the synaptonemal complex. Altered levels and positions of meiotic recombination are associated with chromosome nondisjunction (Hassold and Sherman 2000) and a cause of human infertility (Sun et al. 2004). Studies of some genes proposed to have a role during prophase I are complicated because of embryo lethality or pleiotropic phenotypes resulting from the mutations. The available mouse mutants defective in chromosome alignment and activation of prophase I cell cycle checkpoints are summarized in this section.

Initiation of genetic recombination is probably a DNA double strand break (DSB). At least four mammalian genes are predicted to be involved in generating DSBs (*Rad50, Mre11, Nbs1 and Spo11*) and mouse meiosis mutants are available for two of them, *Spo11* and *NbsI*. The Spo11 protein is expressed primarily during prophase I and is a member of the type II topoisomerase family (Gadelle et al. 2003; Keeney et al. 1999). In the knockout mice, both males and females are sterile. In males, the homologous chromosomes fail to synapse and cell cycle arrest occurs during pachytene, resulting in high levels of apoptosis. Females are born with oocytes which undergo degeneration. Two pieces of data are consistent with DSB initiation being defective in the *Spo11* mouse, absence of DSB intermediates and failure of a histone modified in the presence of DSBs, H2AX, to be phosphorylated (Zenvirth et al. 2003). For the second mutant, *Nbs1*-deficient males were fertile. In females, oogenesis was abolished but the phenotypes have not been fully characterized (Kang et al. 2002).

Genes necessary for repairing the DSBs in meiotic cells are part of the mammalian RAD51 homology dependent pathway. RAD51 binds to meiotic chromosomes during chromosome synapsis (Moens et al. 1997; Plug et al. 1996) and in several meiotic mutants, the number of foci increase, consistent with the proposed role of RAD51 for binding along the DSB and performing the DNA strand invasion search during homologous recombination. Due to the embryo lethality conferred by disruptions in the *Rad51* genes, a disruption in only one proposed late exchange gene, *Dmc1*, is currently available to study the role of homologous recombination during meiosis in mammals (Pittman et al. 1998; Yoshida et al. 1998). *Dmc1*-deficient males displayed an arrest of gamete development at the pachytene (spermatocyte) stage. Oocytes were present in the female fetus, but the chromosomes were unorganized, suggesting a failure in homologous chromosome pairing and synapsis. In both sexes, the differentiating cells arrested during the first meiotic division and were eliminated by apoptosis. The breast cancer susceptibility gene *Brca1* is also essential for recombination during spermatogenesis. *Brca1*-deficient males had defects during pachytene and increased apoptosis (Xu et al. 2003). Another group of genes involved in later steps for resolving the DSB intermediates are the mismatch DNA repair genes (Kolas and Cohen 2004). Mismatch repair (MMR) enzymes are involved in fixing mispaired bases and four mouse mutants in MMR genes have been generated. Targeted mutagenesis of the mouse *Pms2* gene resulted in chromosome synapsis defects in males but not females (Baker et al. 1995). A mutation in *Mlh1* also caused sex-specific defects; meiotic arrest occurred at the spermatocyte stage in males and exhibited premature chromosome separation (Baker et al. 1996; Edelmann et al. 1996). Yet, oocytes were able to complete the first meiotic division. Mice deficient in *Msh4* and *Msh5* were sterile and chromosome synapsis defects were also observed (Kneitz et al. 2000).

The synaptonemal complex (SC) is a zipper-like proteinaceous structure that unites homologous chromosomes during the zygotene stage (Page and Hawley 2004). Mouse mutants in two genes necessary for the SC to assemble have been generated. An *Sycp3* disruption resulted in male sterility due to chromosome synapsis defects and meiotic arrest, followed by massive apoptotic cell death (Yuan et al. 2000). Fertility was only slightly reduced in females. However, chromosome missegregation defects increased with maternal aging (Yuan et al. 2002). Recently, mutations in *Sycp3* were demonstrated to be a cause of human male infertility (Miyamoto et al. 2003) and SC defects have been associated with infertility and meiotic arrest at the zygotene stage (Judis et al. 2004). A second mouse line, deficient for *Sycp1* was infertile with phenotypes similar to *Sycp3* mutants observed in males. Unlike $Sycp3^{-/-}$ mice, female *Sycp1*-deficient mice were not fertile (de Vries et al. 2005).

The various genes affecting germ cell development at prophase I are linked to various stages of homologous recombination events: DSB formation, DSB repair, mismatch repair and organization of the synaptonemal complexes. The interplay of these gene products are necessary for chromosome alignment and foolproof development of the germ cells.

7
Future Perspectives

In this chapter, we have provided a summary of mouse models available for studies involving cell cycle regulation during the early events of meiosis. Thus far, more than 200 mouse models affecting fertility have been generated that affect the male or female at various stages of gonadal development, germ cell development, maturation, fertilization or embryo development (Naz and Rajesh 2005a,b). For a thorough understanding of cell cycle regulation during mammalian meiosis, more specific strategies are being developed. One approach is to generate double knockouts of the available meiotic mutants to further clarify functions in recognition of a cell cycle arrest. Examples are generation of *Spo11 Dmc1*, *Spo11 Msh5* and *Spo11 Atm* double mutant mice, providing insights into epistasis and signaling during prophase I (Di Giacomo et al. 2005). A second set of approaches to bypass embryo lethal phenotypes will be the generation of gametogenesis conditional gene disruptions, allelic variants and germline specific RNAi knockdowns in the genes proposed to be involved in meiosis cell cycle regulation (Prawitt et al. 2004; Chung et al. 2004). A complementary and perhaps less time consuming approach, is the mutagenesis efforts that specifically screen for sterility. *N*-ethyl-*N*-nitrosourea (ENU) is used to induce point mutations in spermatogonial stem cells or in ES cells used to generate offspring for phenotype screenings (http://reprogenomics.jax.org/index.html). Methods for ENU mutagenesis and identification of infertile mutants were recently reviewed (Reinholdt et al. 2004). The first mutant isolated and cloned using this strategy was *Mei1* (meiosis defective 1) and the phenotype of the mutants were similar to the *Spo11* knockout mice (Libby et al. 2002). The *Mei1* gene is unique to mammals, demonstrating the value of the mutagenesis and infertility screening strategies (Libby et al. 2003). Identification of a number of new genes and generation of combinations of double or even triple knockouts will help to decipher meiosis cell cycle checkpoint mechanisms (Reinholdt and Schimenti 2005). Finally, in vitro derived sperm cells and oocytes from ES cells will be useful to study early meiosis events (Hubner et al. 2003). Additionally, technology for deriving germ cells from human ES cells will permit studies of germ cell differentiation (Surani 2004). Not only will these approaches be used to identify the signals involved in regulating germ cell development, but also to develop contraceptives or generate treatments for infertility.

Cell Cycle Regulation in Mammalian Germ Cells   361

Acknowledgements We gratefully acknowledge support from the March of Dimes Birth Defects Foundation, a Helen and Harold McMaster Endowment, and the American Cancer Society.

References

Agoulnik AI, Lu B, Zhu Q, Truong C, Ty MT, Arango N, Chada KK, Bishop CE (2002) A novel gene, Pog, is necessary for primordial germ cell proliferation in the mouse and underlies the germ cell deficient mutation, gcd. Hum Mol Genet 11:3047–3053

Anderson R, Copeland TK, Scholer H, Heasman J, Wylie C (2000) The onset of germ cell migration in the mouse embryo. Mech Dev 91:61–68

Ashley T, Walpita D, de Rooij DG (2001) Localization of two mammalian cyclin dependent kinases during mammalian meiosis. J Cell Sci 114:685–693

Baker SM, Bronner CE, Zhang L, Plug AW, Robatzek M, Warren G, Elliott EA, Yu J, Ashley T, Arnheim N, Flavell RA, Liskay RM (1995) Male mice defective in the DNA mismatch repair gene PMS2 exhibit abnormal chromosome synapsis in meiosis. Cell 82:309–319

Baker SM, Plug AW, Prolla TA, Bronner CE, Harris AC, Yao X, Christie DM, Monell C, Arnheim N, Bradley A, Ashley T, Liskay RM (1996) Involvement of mouse Mlh1 in DNA mismatch repair and meiotic crossing over. Nat Genet 13:336–342

Bannister LA, Reinholdt LG, Munroe RJ, Schimenti JC (2004) Positional cloning and characterization of mouse mei8, a disrupted allele of the meiotic cohesin Rec8. Genesis 40:184–194

Barlow C, Hirotsune S, Paylor R, Liyanage M, Eckhaus M, Collins F, Shiloh Y, Crawley JN, Ried T, Tagle D, Wynshaw-Boris A (1996) Atm-deficient mice: A paradigm of ataxia telangiectasia. Cell 86:159–171

Baudat F, Manova K, Yuen JP, Jasin M, Keeney S (2000) Chromosome synapsis defects and sexually dimorphic meiotic progression in mice lacking Spo11. Mol Cell 6:989–998

Beck AR, Miller IJ, Anderson P, Streuli M (1998) RNA-binding protein TIAR is essential for primordial germ cell development. Proc Natl Acad Sci USA 95:2331–2336

Bergeron L, Perez GI, Macdonald G, Shi L, Sun Y, Jurisicova A, Varmuza S, Latham KE, Flaws JA, Salter JC, Hara H, Moskowitz MA, Li E, Greenberg A, Tilly JL, Yuan J (1998) Defects in regulation of apoptosis in caspase-2-deficient mice. Genes Dev 12:1304–1314

Berthet C, Aleem E, Coppola V, Tessarollo L, Kaldis P (2003) Cdk2 knockout mice are viable. Curr Biol 13:1775–1785

Beumer TL, Kiyokawa H, Roepers-Gajadien HL, van den Bos LA, Lock TM, Gademan IS, Rutgers DH, Koff A, de Rooij DG (1999) Regulatory role of $p27^{kip1}$ in the mouse and human testis. Endocrinology 140:1834–1840

Blendy JA, Kaestner KH, Weinbauer GF, Nieschlag E, Schutz G (1996) Severe impairment of spermatogenesis in mice lacking the CREM gene. Nature 380:162–165

Castrillon DH, Miao L, Kollipara R, Horner JW, DePinho RA (2003) Suppression of ovarian follicle activation in mice by the transcription factor Foxo3a. Science 301:215–218

Celeste A, Petersen S, Romanienko PJ, Fernandez-Capetillo O, Chen HT, Sedelnikova OA, Reina-San-Martin B, Coppola V, Meffre E, Difilippantonio MJ, Redon C, Pilch DR, Olaru A, Eckhaus M, Camerini-Otero RD, Tessarollo L, Livak F, Manova K, Bonner WM, Nussenzweig MC, Nussenzweig A (2002) Genomic instability in mice lacking histone H2AX. Science 296:922–927

Chang C, Chen YT, Yeh SD, Xu Q, Wang RS, Guillou F, Lardy H, Yeh S (2004) Infertility with defective spermatogenesis and hypotestosteronemia in male mice lacking the androgen receptor in Sertoli cells. Proc Natl Acad Sci USA 101:6876–6881

Chang H, Matzuk MM (2001) Smad5 is required for mouse primordial germ cell development. Mech Dev 104:61–67

Choi T, Aoki F, Mori M, Yamashita M, Nagahama Y, Kohmoto K (1991) Activation of p34^{cdc2} protein kinase activity in meiotic and mitotic cell cycles in mouse oocytes and embryos. Development 113:789–795

Chung SS, Cuzin F, Rassoulzadegan M, Wolgemuth DJ (2004) Primary spermatocyte-specific Cre recombinase activity in transgenic mice. Transgenic Res 13:289–294

Colledge WH, Carlton MB, Udy GB, Evans MJ (1994) Disruption of c-mos causes parthenogenetic development of unfertilized mouse eggs. Nature 370:65–68

Crackower MA, Kolas NK, Noguchi J, Sarao R, Kikuchi K, Kaneko H, Kobayashi E, Kawai Y, Kozieradzki I, Landers R, Mo R, Hui CC, Nieves E, Cohen PE, Osborne LR, Wada T, Kunieda T, Moens PB, Penninger JM (2003) Essential role of Fkbp6 in male fertility and homologous chromosome pairing in meiosis. Science 300:1291–1295

Critchlow HM, Payne A, Griffin DK (2004) Genes and proteins involved in the control of meiosis. Cytogenet Genome Res 105:4–10

Cunto FD, Imarisio S, Camera P, Boitani C, Altruda F, Silengo L (2002) Essential role of citron kinase in cytokinesis of spermatogenic precursors. J Cell Sci 115:4819–4826

de Rooij DG, de Boer P (2003) Specific arrests of spermatogenesis in genetically modified and mutant mice. Cytogenet Genome Res 103:267–276

de Vries FA, de Boer E, van den Bosch M, Baarends WM, Ooms M, Yuan L, Liu JG, van Zeeland AA, Heyting C, Pastink A (2005) Mouse Sycp1 functions in synaptonemal complex assembly, meiotic recombination, and XY body formation. Genes Dev 19:1376–1389

Deng W, Lin H (2002) Miwi, a murine homolog of piwi, encodes a cytoplasmic protein essential for spermatogenesis. Dev Cell 2:819–830

Di Giacomo M, Barchi M, Baudat F, Edelmann W, Keeney S, Jasin M (2005) Distinct DNA-damage-dependent and -independent responses drive the loss of oocytes in recombination-defective mouse mutants. Proc Natl Acad Sci USA 102:737–742

Dickins RA, Frew IJ, House CM, O'Bryan MK, Holloway AJ, Haviv I, Traficante N, de Kretser DM, Bowtell DD (2002) The ubiquitin ligase component Siah1a is required for completion of meiosis I in male mice. Mol Cell Biol 22:2294–2303

Dix DJ, Allen JW, Collins BW, Mori C, Nakamura N, Poorman-Allen P, Goulding EH, Eddy EM (1996) Targeted gene disruption of Hsp70-2 results in failed meiosis, germ cell apoptosis, and male infertility. Proc Natl Acad Sci USA 93:3264–3268

Edelmann W, Cohen PE, Kane M, Lau K, Morrow B, Bennett S, Umar A, Kunkel T, Cattoretti G, Chaganti R, Pollard JW, Kolodner RD, Kucherlapati R (1996) Meiotic pachytene arrest in MLH1-deficient mice. Cell 85:1125–1134

Engebrecht J (2003) Cell signaling in yeast sporulation. Biochem Biophys Res Commun 306:325–328

Fero ML, Rivkin M, Tasch M, Porter P, Carow CE, Firpo E, Polyak K, Tsai LH, Broudy V, Perlmutter RM, Kaushansky K, Roberts JM (1996) A syndrome of multiorgan hyperplasia with features of gigantism, tumorigenesis, and female sterility in p27^{Kip1}-deficient mice. Cell 85:733–744

Freiman RN, Albright SR, Zheng S, Sha WC, Hammer RE, Tjian R (2001) Requirement of tissue-selective TBP-associated factor TAFII105 in ovarian development. Science 293:2084–2087

Gadelle D, Filee J, Buhler C, Forterre P (2003) Phylogenomics of type II DNA topoisomerases. Bioessays 25:232–242

Ganoth D, Bornstein G, Ko TK, Larsen B, Tyers M, Pagano M, Hershko A (2001) The cell-cycle regulatory protein Cks1 is required for SCF^{Skp2}-mediated ubiquitinylation of p27. Nat Cell Biol 3:321–324

Geng Y, Yu Q, Sicinska E, Das M, Schneider JE, Bhattacharya S, Rideout WM, Bronson RT, Gardner H, Sicinski P (2003) Cyclin E ablation in the mouse. Cell 114:431–443

Guardavaccaro D, Kudo Y, Boulaire J, Barchi M, Busino L, Donzelli M, Margottin-Goguet F, Jackson PK, Yamasaki L, Pagano M (2003) Control of meiotic and mitotic progression by the F box protein beta-Trcp1 in vivo. Dev Cell 4:799–812

Handel MA (1987) Genetic control of spermatogenesis in mice. Results Probl Cell Differ 15:1–62

Hassold T, Hunt P (2001) To err (meiotically) is human: the genesis of human aneuploidy. Nat Rev Genet 2:280–291

Hassold T, Sherman S (2000) Down syndrome: genetic recombination and the origin of the extra chromosome 21. Clin Genet 57:95–100

Heikinheimo O, Gibbons WE (1998) The molecular mechanisms of oocyte maturation and early embryonic development are unveiling new insights into reproductive medicine. Mol Hum Reprod 4:745–756

Heyting C, Dietrich AJ (1991) Meiotic chromosome preparation and protein labeling. Methods Cell Biol 35:177–202

Honarpour N, Du C, Richardson JA, Hammer RE, Wang X, Herz J (2000) Adult Apaf-1-deficient mice exhibit male infertility. Dev Biol 218:248–258

Hsia KT, Millar MR, King S, Selfridge J, Redhead NJ, Melton DW, Saunders PT (2003) DNA repair gene Ercc1 is essential for normal spermatogenesis and oogenesis and for functional integrity of germ cell DNA in the mouse. Development 130:369–378

Huang EJ, Manova K, Packer AI, Sanchez S, Bachvarova RF, Besmer P (1993) The murine steel panda mutation affects kit ligand expression and growth of early ovarian follicles. Dev Biol 157:100–109

Hubner K, Fuhrmann G, Christenson LK, Kehler J, Reinbold R, De La Fuente R, Wood J, Strauss JF 3rd, Boiani M, Scholer HR (2003) Derivation of oocytes from mouse embryonic stem cells. Science 300:1251–1256

Igakura T, Kadomatsu K, Kaname T, Muramatsu H, Fan QW, Miyauchi T, Toyama Y, Kuno N, Yuasa S, Takahashi M, Senda T, Taguchi O, Yamamura K, Arimura K, Muramatsu T (1998) A null mutation in basigin, an immunoglobulin superfamily member, indicates its important roles in peri-implantation development and spermatogenesis. Dev Biol 194:152–165

Johnson J, Canning J, Kaneko T, Pru JK, Tilly JL (2004) Germline stem cells and follicular renewal in the postnatal mammalian ovary. Nature 428:145–150

Judis L, Chan ER, Schwartz S, Seftel A, Hassold T (2004) Meiosis I arrest and azoospermia in an infertile male explained by failure of formation of a component of the synaptonemal complex. Fertil Steril 81:205–209

Juneja SC, Barr KJ, Enders GC, Kidder GM (1999) Defects in the germ line and gonads of mice lacking connexin43. Biol Reprod 60:1263–1270

Kang J, Bronson RT, Xu Y (2002) Targeted disruption of NBS1 reveals its roles in mouse development and DNA repair. EMBO J 21:1447–1455

Keeney S, Baudat F, Angeles M, Zhou ZH, Copeland NG, Jenkins NA, Manova K, Jasin M (1999) A mouse homolog of the *Saccharomyces Cerevisiae* meiotic recombination DNA transesterase Spo11p. Genomics 61:170–182

Kneitz B, Cohen PE, Avdievich E, Zhu L, Kane MF, Hou H Jr, Kolodner RD, Kucherlapati R, Pollard JW, Edelmann W (2000) MutS homolog 4 localization to meiotic chromosomes is required for chromosome pairing during meiosis in male and female mice. Genes Dev 14:1085–1097

Kolas NK, Cohen PE (2004) Novel and diverse functions of the DNA mismatch repair family in mammalian meiosis and recombination. Cytogenet Genome Res 107:216–231

Kuroda M, Sok J, Webb L, Baechtold H, Urano F, Yin Y, Chung P, de Rooij DG, Akhmedov A, Ashley T, Ron D (2000) Male sterility and enhanced radiation sensitivity in TLS$^{-/-}$ mice. EMBO J 19:453–462

Libby BJ, De La Fuente R, O'Brien MJ, Wigglesworth K, Cobb J, Inselman A, Eaker S, Handel MA, Eppig JJ, Schimenti JC (2002) The mouse meiotic mutation mei1 disrupts chromosome synapsis with sexually dimorphic consequences for meiotic progression. Dev Biol 242:174–187

Libby BJ, Reinholdt LG, Schimenti JC (2003) Positional cloning and characterization of Mei1, a vertebrate-specific gene required for normal meiotic chromosome synapsis in mice. Proc Natl Acad Sci USA 100:15706–15711

Lincoln AJ, Wickramasinghe D, Stein P, Schultz RM, Palko ME, De Miguel MP, Tessarollo L, Donovan PJ (2002) Cdc25b phosphatase is required for resumption of meiosis during oocyte maturation. Nat Genet 30:446–449

Liu D, Liao C, Wolgemuth DJ (2000) A role for cyclin A1 in the activation of MPF and G2-M transition during meiosis of male germ cells in mice. Dev Biol 224:388–400

Liu D, Matzuk MM, Sung WK, Guo Q, Wang P, Wolgemuth DJ (1998) Cyclin A1 is required for meiosis in the male mouse. Nat Genet 20:377–380

Martianov I, Fimia GM, Dierich A, Parvinen M, Sassone-Corsi P, Davidson I (2001) Late arrest of spermiogenesis and germ cell apoptosis in mice lacking the TBP-like TLF/TRF2 gene. Mol Cell 7:509–515

Masui Y, Markert CL (1971) Cytoplasmic control of nuclear behavior during meiotic maturation of frog oocytes. J Exp Zool 177:129–145

Mazeyrat S, Saut N, Grigoriev V, Mahadevaiah SK, Ojarikre OA, Rattigan A, Bishop C, Eicher EM, Mitchell MJ, Burgoyne PS (2001) A Y-encoded subunit of the translation initiation factor Eif2 is essential for mouse spermatogenesis. Nat Genet 29:49–53

McLaren A (2003) Primordial germ cells in the mouse. Dev Biol 262:1–15

Mintz B, Russell ES (1957) Gene-induced embryological modifications of primordial germ cells in the mouse. J Exp Zool 134:207–237

Miyamoto T, Hasuike S, Yogev L, Maduro MR, Ishikawa M, Westphal H, Lamb DJ (2003) Azoospermia in patients heterozygous for a mutation in SYCP3. Lancet 362:1714–1719

Moens PB, Chen DJ, Shen Z, Kolas N, Tarsounas M, Heng HH, Spyropoulos B (1997) Rad51 immunocytology in rat and mouse spermatocytes and oocytes. Chromosoma 106:207–215

Moons DS, Jirawatnotai S, Tsutsui T, Franks R, Parlow AF, Hales DB, Gibori G, Fazleabas AT, Kiyokawa H (2002) Intact follicular maturation and defective luteal function in mice deficient for cyclin-dependent kinase-4. Endocrinology 143:647–654

Moreno S, Nurse P (1990) Substrates for p34^{cdc2}: In vivo veritas? Cell 61:549–551

Nagy A (2003) Manipulating the mouse embryo: a laboratory manual. Cold Spring Harbor Laboratory Press, Cold Spring Harbor, New York

Naz RK, Rajesh C (2005a) Gene knockouts that cause female infertility: search for novel contraceptive targets. Front Biosci 10:2447–2459

Naz RK, Rajesh P (2005b) Novel testis/sperm-specific contraceptive targets identified using gene knockout studies. Front Biosci 10:2430–2446

Nojimak H (2004) G1 and S-phase checkpoints, chromosome instability and cancer. In: Schonthal AH (ed) Checkpoint controls and cancer. Humana Press, New Jersey USA, p 3–49

Ortega S, Prieto I, Odajima J, Martin A, Dubus P, Sotillo R, Barbero JL, Malumbres M, Barbacid M (2003) Cyclin-dependent kinase 2 is essential for meiosis but not for mitotic cell division in mice. Nat Genet 35:25–31

Page AW, Orr-Weaver TL (1997) Stopping and starting the meiotic cell cycle. Curr Opin Genet Dev 7:23–31

Page SL, Hawley RS (2004) The genetics and molecular biology of the synaptonemal complex. Annu Rev Cell Dev Biol 20:525–558

Parisi T, Beck AR, Rougier N, McNeil T, Lucian L, Werb Z, Amati B (2003) Cyclins E1 and E2 are required for endoreplication in placental trophoblast giant cells. EMBO J 22:4794–4803

Pines J (1999) Four-dimensional control of the cell cycle. Nat Cell Biol 1:E73–79

Pines J, Hunter T (1989) Isolation of a human cyclin cDNA: evidence for cyclin mRNA and protein regulation in the cell cycle and for interaction with p34^{cdc2}. Cell 58:833–846

Pittman DL, Cobb J, Schimenti KJ, Wilson LA, Cooper DM, Brignull E, Handel MA, Schimenti JC (1998) Meiotic prophase arrest with failure of chromosome synapsis in mice deficient for Dmc1, a germline-specific RecA homolog. Mol Cell 1:697–705

Plug AW, Xu J, Reddy G, Golub EI, Ashley T (1996) Presynaptic association of Rad51 protein with selected sites in meiotic chromatin. Proc Natl Acad Sci USA 93:5920–5924

Prawitt D, Brixel L, Spangenberg C, Eshkind L, Heck R, Oesch F, Zabel B, Bockamp E (2004) RNAi knock-down mice: an emerging technology for post-genomic functional genetics. Cytogenet Genome Res 105:412–421

Ratts VS, Flaws JA, Kolp R, Sorenson CM, Tilly JL (1995) Ablation of bcl-2 gene expression decreases the numbers of oocytes and primordial follicles established in the post-natal female mouse gonad. Endocrinology 136:3665–3668

Reinholdt L, Ashley T, Schimenti J, Shima N (2004) Forward genetic screens for meiotic and mitotic recombination-defective mutants in mice. Methods Mol Biol 262:87–107

Reinholdt LG, Schimenti JC (2005) Mei1 is epistatic to Dmc1 during mouse meiosis. Chromosoma 114:127–134

Romanienko PJ, Camerini-Otero RD (2000) The mouse Spo11 gene is required for meiotic chromosome synapsis. Mol Cell 6:975–987

Ross AJ, Waymire KG, Moss JE, Parlow AF, Skinner MK, Russell LD, MacGregor GR (1998) Testicular degeneration in Bclw-deficient mice. Nat Genet 18:251–256

Rucker EB 3rd, Dierisseau P, Wagner KU, Garrett L, Wynshaw-Boris A, Flaws JA, Hennighausen L (2000) Bcl-x and Bax regulate mouse primordial germ cell survival and apoptosis during embryogenesis. Mol Endocrinol 14:1038–1052

Ruggiu M, Speed R, Taggart M, McKay SJ, Kilanowski F, Saunders P, Dorin J, Cooke HJ (1997) The mouse textitDazla gene encodes a cytoplasmic protein essential for gametogenesis. Nature 389:73–77

Russell LD (1990) Histological and histopathological evaluation of the testis. Cache River Press, Clearwater, Fl, xiv, p 286

Schrans-Stassen BH, Saunders PT, Cooke HJ, de Rooij DG (2001) Nature of the spermatogenic arrest in Dazl –/– mice. Biol Reprod 65:771–776

Sharan SK, Pyle A, Coppola V, Babus J, Swaminathan S, Benedict J, Swing D, Martin BK, Tessarollo L, Evans JP, Flaws JA, Handel MA (2004) BRCA2 deficiency in mice leads to meiotic impairment and infertility. Development 131:131–142

Sherr CJ, Roberts JM (1999) CDK inhibitors: positive and negative regulators of G1-phase progression. Genes Dev 13:1501–1512

Sicinski P, Donaher JL, Geng Y, Parker SB, Gardner H, Park MY, Robker RL, Richards JS, McGinnis LK, Biggers JD, Eppig JJ, Bronson RT, Elledge SJ, Weinberg RA (1996) Cyclin D2 is an FSH-responsive gene involved in gonadal cell proliferation and oncogenesis. Nature 384:470–474

Simon AM, Goodenough DA, Li E, Paul DL (1997) Female infertility in mice lacking connexin 37. Nature 385:525–529

Spruck C, Strohmaier H, Watson M, Smith AP, Ryan A, Krek TW, Reed SI (2001) A CDK-independent function of mammalian Cks1: targeting of SCF^{Skp2} to the CDK inhibitor p27^{Kip1}. Mol Cell 7:639–650

Spruck CH, de Miguel MP, Smith AP, Ryan A, Stein P, Schultz RM, Lincoln AJ, Donovan PJ, Reed SI (2003) Requirement of Cks2 for the first metaphase/anaphase transition of mammalian meiosis. Science 300:647–650

Sun F, Kozak G, Scott S, Trpkov K, Ko E, Mikhaail-Philips M, Bestor TH, Moens P, Martin RH (2004) Meiotic defects in a man with non-obstructive azoospermia: case report. Hum Reprod 19:1770–1773

Surani MA (2004) Stem cells: how to make eggs and sperm. Nature 427:106–107

Tay J, Richter JD (2001) Germ cell differentiation and synaptonemal complex formation are disrupted in CPEB knockout mice. Dev Cell 1:201–213

Tourtellotte WG, Nagarajan R, Auyeung A, Mueller C, Milbrandt J (1999) Infertility associated with incomplete spermatogenic arrest and oligozoospermia in Egr4-deficient mice. Development 126:5061–5071

Toyama Y, Maekawa M, Kadomatsu K, Miyauchi T, Muramatsu T, Yuasa S (1999) Histological characterization of defective spermatogenesis in mice lacking the basigin gene. Anat Histol Embryol 28:205–213

Tremblay KD, Dunn NR, Robertson EJ (2001) Mouse embryos lacking Smad1 signals display defects in extra-embryonic tissues and germ cell formation. Development 128:3609–3621

Tsutsui T, Hesabi B, Moons DS, Pandolfi PP, Hansel KS, Koff A, Kiyokawa H (1999) Targeted disruption of CDK4 delays cell cycle entry with enhanced p27^{Kip1} activity. Mol Cell Biol 19:7011–7019

Ward JO, Reinholdt LG, Hartford SA, Wilson LA, Munroe RJ, Schimenti KJ, Libby BJ, O'Brien M, Pendola JK, Eppig J, Schimenti JC (2003) Toward the genetics of mammalian reproduction: induction and mapping of gametogenesis mutants in mice. Biol Reprod 69:1615–1625

Wilson ZA, Yang C (2004) Plant gametogenesis: conservation and contrasts in development. Reproduction 128:483–492

Wolgemuth DJ (2003) Insights into regulation of the mammalian cell cycle from studies on spermatogenesis using genetic approaches in animal models. Cytogenet Genome Res 103:256–266

Wolgemuth DJ, Lele KM, Jobanputra V, Salazar G (2004) The A-type cyclins and the meiotic cell cycle in mammalian male germ cells. Int J Androl 27:192–199

Xu X, Aprelikova O, Moens P, Deng CX, Furth PA (2003) Impaired meiotic DNA-damage repair and lack of crossing-over during spermatogenesis in BRCA1 full-length isoform deficient mice. Development 130:2001–2012

Yan C, Wang P, DeMayo J, DeMayo FJ, Elvin JA, Carino C, Prasad SV, Skinner SS, Dunbar BS, Dube JL, Celeste AJ, Matzuk MM (2001) Synergistic roles of bone morphogenetic protein 15 and growth differentiation factor 9 in ovarian function. Mol Endocrinol 15:854–866

Yoshida K, Kondoh G, Matsuda Y, Habu T, Nishimune Y, Morita T (1998) The mouse RecA-like gene Dmc1 is required for homologous chromosome synapsis during meiosis. Mol Cell 1:707–718

Yu RN, Ito M, Saunders TL, Camper SA, Jameson JL (1998) Role of Ahch in gonadal development and gametogenesis. Nat Genet 20:353–357

Yuan L, Liu JG, Hoja MR, Wilbertz J, Nordqvist K, Hoog C (2002) Female germ cell aneuploidy and embryo death in mice lacking the meiosis-specific protein SCP3. Science 296:1115–1118

Yuan L, Liu JG, Zhao J, Brundell E, Daneholt B, Hoog C (2000) The murine SCP3 gene is required for synaptonemal complex assembly, chromosome synapsis, and male fertility. Mol Cell 5:73–83

Zenvirth D, Richler C, Bardhan A, Baudat F, Barzilai A, Wahrman J, Simchen G (2003) Mammalian meiosis involves DNA double-strand breaks with 3′ overhangs. Chromosoma 111:369–376

Zindy F, den Besten W, Chen B, Rehg JE, Latres E, Barbacid M, Pollard JW, Sherr CJ, Cohen PE, Roussel MF (2001) Control of spermatogenesis in mice by the cyclin D-dependent kinase inhibitors $p18^{Ink4c}$ and $p19^{Ink4d}$. Mol Cell Biol 21:3244–3255

Subject Index

20S particle, 149

A-box, 153
acetyltransferase, 186
ademomatous polyposis coli (Apc), 103
aging, 258, 263
Ago, 157
allantois, 343, 356
anaphase, 154, 160, 168
anaphase promoting complex/cyclosome (APC/C), 93, 95, 101, 149, 150, 153–156, 158, 160, 163, 164
aneuploidy, 93, 99, 103
APC^{Cdc20}, 153–155, 160, 168
APC^{Cdh1}, 153, 155, 162
Apc11, 152
Apc2, 152
apoptosis, 183, 187, 195–199, 205, 206, 271, 282, 285, 292, 312, 329, 332, 335–337, 345, 354, 355, 357–359
Archipelago, 157
ARF, 333, 334, 336
ARS, 32, 35
Ase1, 150
ATPase, 39–41, 166
atresia, 345
Aurora A, 150, 153
Aurora B, 94, 99, 101, 103, 104
Aurora C, 101, 103
Aurora kinase, 98, 103

bHLH, 209, 210
bipolar attachment, 153
bipolar chromosome attachment, 168
Bmi-1, 262
Bub1, 94, 95, 98, 101, 103, 104
Bub3, 94, 95, 97, 98, 104, 155
BubR1, 94, 95, 97, 98, 103, 104, 155, 164

c-myc, 33, 158, 160
cancer, 159, 183, 211, 227, 259, 271, 272, 276, 294, 295, 297–299, 310, 311
carcinogen, 229
Cdc16, 152
Cdc18, 158
Cdc2, 230, 271–273, 275–280, 282, 293, 300, 301, 303–305, 307, 309, 312, 313
Cdc20, 93, 95, 97, 98, 102, 104, 150, 152–155, 158, 160, 163, 164, 166, 168
Cdc23, 152
Cdc25, 164, 230, 354
Cdc25A, 150, 158, 164, 165
Cdc26, 152
Cdc27, 152
Cdc34, 151
Cdc4, 151, 157–160, 162, 163, 168
– haploinsufficiency of, 160
Cdc42, 162
Cdc45, 43–47
Cdc5/Plk, 150
Cdc53, 151, 152, 156
Cdc6, 36–42, 49, 150, 158
Cdc7, 44, 45
Cdh1, 152, 153, 158, 160, 162
Cdk, 5, 149, 154, 162, 353–356
– activation of, 5
– Cdk2 knockouts, 14
– Cdk3 knockouts, 14
– Cdk4 knockouts, 12
– Cdk6 knockouts, 12
– S phase, 159, 168
Cdk inhibitor (CKI), 159, 161, 162, 166, 168, 230
Cdk1, 149, 159, 161, 164, 165
Cdk2, 153, 159, 164, 165, 184, 185, 187, 188, 193, 196, 208, 232, 259–261, 271–273, 275–286, 289, 293, 294, 296, 298–301, 303–305, 307, 309–313, 334, 335, 353, 354

Cdk3, 278, 279
Cdk4, 184, 185, 188, 193, 196, 208, 232, 259–262, 273, 276, 278, 281, 284, 285, 287, 289, 290, 294, 297, 298, 300, 301, 307–313
Cdk6, 184, 232, 260–262, 273, 276, 278, 284, 285, 287, 290, 308–310
CDKN1A, 166
Cdks, 161, 271, 272, 275, 276, 278, 279, 281, 285, 300, 302, 303, 305, 307, 309–311
Cdt1, 36, 37, 40–42, 49, 161, 165
cell cycle, 77, 78, 150, 183–186, 188, 190, 192, 193, 195, 196, 198, 199, 205–207, 210, 271–282, 284–286, 289, 291–296, 298–301, 304, 307–313, 343, 345, 353–358, 360
– Cdc25A, 77, 78
– Cdc7, 78
– Cdk2, 78
– p21, 77
– p53, 77
– yeast, 151
cell cycle engine, 4, 8
– MPF, 4
cell cycle theory, 1, 19
– cell division, 2
– development of, 4
– future of, 23
cell division, 230, 272, 273, 275, 282, 293, 294, 305, 307, 311, 353
cell division cycle (cdc), 147
– mutants, 151, 152
cell growth, 6
– conservation of mass, 6
– mass increase, 22
cell lines, 271, 272, 276–278, 286, 311
cell migration, 271, 301, 303
CENP-E, 94, 97, 98, 101, 102
centrosomes, 293, 301
checkpoint, 4, 42, 46, 47, 93–95, 98, 99, 101–104, 163, 166, 238
– critical size threshold, 6
– DNA damage, 4, 164, 165
– replication, 164
– size threshold, 21
– spindle assembly, 163
– spindle integrity, 163
– stress response, 20
checkpoint signaling, 97, 101
chimaeras, 236

Chk1, 164, 165
Chk2, 164, 165
chromatin, 166
chromatin modulation, 66, 81
– 19S proteasome, 81
– histone H2AX, 81
– INO80, 81
– methylation, 81
– NuA4, 81
– ubiquitination, 81
chromosomal instability, 103
chromosomal passeneger complexes, 101
chromosome, 154, 343, 353, 355, 358, 359
chromosome instability, 159
Cin8/Kip1, 150
CKIs, 272
Cks1, 161
Clb2, 150
Clb5, 150
Cln1, 158, 159, 162
Cln2, 158, 159
cohesin, 154
– sister chromatid, 154
cohesion
– sister chromatid, 156
crisis, 262, 263
Ctd1, 158, 162
Cul1, 152, 156
Cullin, 151
cyclin, 5, 149, 154, 155, 161, 271–273, 275–277, 279, 280, 283, 284, 288–290, 294, 301, 304, 306–310, 312, 313, 353, 355
– cyclin D knockouts, 13
– cyclin E knockouts, 15
– expression of, 8
– G1, 159
– mitotic, 149, 152, 155
cyclin A, 150, 153, 155, 160, 240, 273, 275–277, 279, 280, 289, 291, 296, 305, 307, 309
cyclin A-Cdk2, 162
cyclin B, 150, 152, 153, 155
cyclin B-Cdk1, 154, 160
cyclin D, 184, 185, 188, 208, 210
– cyclin D knockouts, 13
cyclin D1, 9, 240, 271, 277, 284, 288, 290, 294, 298, 299, 301, 304, 308, 310, 312
cyclin D2, 281, 284, 288, 290, 298, 308
cyclin D3, 240, 243, 288, 290, 291

Subject Index

cyclin E, 8, 158, 159, 184, 185, 188, 190, 193, 196, 208, 240, 272, 273, 275–278, 282, 285, 289, 291–293, 296, 298–301, 304, 305, 307, 309–312
– cyclin E knockouts, 15
– regulation of, 10
– substrates of, 11, 18
cyclin-dependent kinase (Cdk), 36–38, 42–45, 48, 50, 51, 230

D-box, 152, 153
Dbf4, 45, 46, 150
deubiquitylating, 149
deubiquitylating enzymes, 167
development, 247, 271, 272, 276, 278, 279, 281, 283–286, 288, 290–292, 294–298, 300, 302, 304, 308–312
DHFR, 33, 34
differentiation, 183, 185, 192, 198, 206–210, 232, 271, 272, 281, 285–287, 291, 292, 294, 297, 300, 308, 311, 312
DNA damage, 37, 38, 45–47, 49, 65, 68, 69, 71, 72, 79, 163, 165, 257, 260, 264, 266
– aphidicolin, 72
– DNA replication interference, 66
– double-strand DNA breaks (DSBs), 66, 69, 71, 79
– HU (hydoxyurea), 72
– ionizing irradiation (IR), 79
– junctions, 71
– MMC (mitomycin C), 79
– MMS (methyl methanesulfonate), 72
– RPA-coated ssDNA, 68, 69
– RPA-ssDNA, 72
– single-stranded DNA (ssDNA), 67
– UV (ultraviolet light), 72
DNA polymerase α, 44
DNA polymerase ε, 40, 47
DNA repair, 66, 73, 77, 79, 80
– base excision repair, 80
– homologous recombination (HR), 76, 79
– mismatch repair, 73, 80
– NER (nucleotide excision repair), 73
– non-homologous end joining [NHEJ], 76, 79
– nucleotide excision repair, 80
DNA replication, 66, 78, 79, 165, 291, 305
– BLM, 79
– Claspin, 78
– Mcm2, 78

– Mus81, 79
– RPA, 78
Dpb11, 46, 47

E2F, 9, 153, 160, 183–186, 188–192, 194–200, 205, 208–210, 259–262, 266, 329, 334, 335
– E2F knockout, 17
– function of, 9, 21
E2F/DP, 236
E2F1, 185, 187, 188, 190, 191, 197, 198, 205, 206, 209
E2F2, 188, 190
E2F3, 188, 190
embryogenesis, 285, 290
Embryonic, 291
Emi1, 153, 158, 160
endoreduplication, 291, 293
Esp1, 154
external transcribed spacer, 258

F-box, 156, 157
F-box motif, 156
F-box protein, 151, 156–158, 161, 162
Far1, 158
Fbw7, 157–160, 162, 163
fertility, 345, 354, 360
folliculogenesis, 355–357

G1, 150
G1 arrest, 161
G1 checkpoint, 77
G1 cyclin, 162, 168
G1 phase, 273, 276, 277, 311
G2 phase, 273, 280, 306
G2/M checkpoint, 77
gametogenesis, 343
geminin, 37, 41–43, 50, 150
genomic instability, 160, 238, 257
Gic1,2, 158
Gic2, 162
GINS, 46, 47, 51
granulosa cells, 354, 356
GRB2, 18
growth, 329, 330, 334
Grr1, 157, 162
GSK3β, 159

hCdc4, 157
helicase, 31, 40, 41, 45, 48, 51
hematopoietic, 271, 287, 288, 308, 309, 312

histone deacetylases, 186
histone H3, 100
histone methyltransferases, 186
Hsl1, 150
human disease, 230
human papillomavirus (HPV), 228

Id2 (inhibitor of differentiation), 237
immortalization, 193, 195, 196, 258–262
INCENP, 99, 101
inhibitors, 98, 100, 103, 104, 271, 272, 279, 285, 286, 296, 297, 306, 310, 311
initiator, 32, 34, 35, 39
INK4A/ARF, 262, 263
interphase, 353
intra-S checkpoint, 77
Ipl1, 99

KEN-box, 153
kinase, 95, 98, 100, 101, 103
kinase activity, 273, 274, 276, 277, 282, 291, 293, 302, 304, 306, 307, 312
kinetochores, 93–95, 98, 101–103, 163
knockout mice, 12, 229, 260, 265

lamin B2, 33
leucine-rich repeat, 157, 161, 162
Leydig cells, 354
licensing, 37, 40–43, 46, 48, 51
localization, 272, 303, 306
LXCXE motif, 187

M phase, 272, 273, 275, 279, 280, 293, 302, 306, 307
M-phase-promoting factor, 353
Mad, 330
Mad1, 94, 95, 97, 98, 101
Mad2, 94, 95, 97, 98, 101, 104, 155, 164
Mad3, 94, 98, 155, 164
mammalian, 271–274, 276, 278, 279, 284, 287, 313
MAPK, 94, 102
maturation promoting factor (MPF), 230
MCAK, 100
Mcm10, 43, 44, 46, 48, 51
MCM2-7, 38–45, 48, 50, 51
MDM2, 165, 166, 187, 193, 198, 262
mediators, 75, 76
– 53BP1, 75, 76
– Brca1, 75, 76
– Claspin, 75, 76

– CtIP, 75
– Mcm7, 75
– Mdc1, 75, 76
– TopBP1, 75
MEFs, 287, 291, 294, 299, 301, 307, 309
meiosis, 156, 279, 281, 282, 285, 304, 343, 345, 353–360
meiosis I, 345
meiosis II, 345
meiotic division
– first, 156
– second, 156
meiotic prophase, 345, 353, 354
Met30, 157, 161
Met4, 158, 161, 163, 167
metaphase, 353, 357
metaphase-anaphase transition, 153
methionine, 163
MgcRacGTP, 100
microtubules, 95, 99
mitogen-activated protein kinase, 94, 102
mitosis, 93–95, 97–99, 101–104, 149, 159, 165, 166, 273, 275, 281, 293, 302, 305, 354
– premature entrance into, 160
mitotic checkpoint complex, 93
mitotic checkpoint complex (MCC), 94
mitotic divisions, 353
mitotic spindle, 154, 163
Miz1, 331, 333
mouse, 271, 272, 277–279, 281, 283–285, 288, 290–298, 300, 301, 304, 308, 310, 312, 343, 345, 353, 356–360
mouse embryonic fibroblasts (MEFs), 236
mouse models, 229
Mps1, 94, 101
mutant
– cdc, 157
Myc, 190, 192, 195–197, 208, 329–337

negative feedback loop, 166
Nek2A, 150
NLS, 190, 207
Notch, 160

oncogenes, 183
oncogenic transformation, 183, 195, 196, 200, 205, 208, 210
oocytes, 281, 345, 353–360
oogonial stem cells, 345
Orc1, 158

Subject Index

origin recognition complex, 32, 34–37
ovary, 286, 300, 302
ovulation, 345, 356, 357
oxidative stress, 257, 263–265

p107, 183, 185, 187–192, 194–196, 198, 199, 205–210, 232
p130, 158, 161, 162, 183, 187–192, 194–196, 198, 199, 205–210, 232
p14^{ARF}, 259, 263, 265
p15^{INK4B}, 241
p16^{INK4A}, 184, 190, 192–194, 196, 197, 199, 240, 241, 285, 290, 309
p18^{INK4C}, 241
p19^{ARF}, 193–199
p19^{INK4D}, 241
p21, 158, 161, 162, 166, 241, 333, 335
p27, 10, 158, 161, 162, 241, 271, 273, 282, 286, 289, 293–301, 303–305, 307, 309–313
– function of, 10, 18
– p27 knockout, 18
– regulation of, 11
p53, 160, 163, 165, 166, 185, 193–199, 205, 206, 210, 229, 257–262, 264, 266
p57^{KIP2}, 241
pachytene, 353, 354, 356, 358, 359
paclitaxel (Taxol), 104
PCNA, 166, 167
Pds1, 150, 154, 164, 168
phosphatases, 164
phosphodegron, 158–161, 164, 165, 168
– Skp2, 161
phosphorylation, 95, 98, 99, 102, 103, 158, 184, 188–190, 200, 208, 232, 272, 273, 282, 291, 299, 303, 309–311
pituitary, 284–286, 294–297, 301, 305
placenta, 291, 293
Plk, 160, 161
pocket proteins, 185, 187–190, 192, 195, 196, 199, 200, 205–207, 209
polo-like kinase, 160
polycomb, 262
polyploidy, 159
Pop1, 157
Pop2, 157
positive feedback loop, 161
pRb, 183–199, 201, 205–210
pre-initiation complex, 46, 48
pre-replication complex, 48

pre-replication complex assembly, 159
primordial germ cells, 343, 358
processing, 71, 73, 74
– Exo1, 72
– exonuclease I, 72
– MRN complex (Mre11-Rad50-Nbs1), 73, 74
proliferation, 271, 287, 288, 290, 294, 296–298, 300, 302, 303, 305, 308–312, 329, 330, 332, 334–337, 354–356
proliferative life span, 263
prometaphase/metaphase arrest, 163
promoters, 186, 188, 190, 194, 209
prophase, 343, 353, 354, 358–360
proteasome, 149, 165–167
– 19S, 149
– 19S regulatory component, 166
– 20S catalytic core, 166
– 26S, 149
protein-ubiquitin ligases, 147, 151, 156

quiescence, 272, 294, 301, 307, 308
quiescent, 272, 278, 291, 294

Rad23, 149
Rad53, 164
Ras, 258, 261, 262, 264, 265
retinoblastoma (Rb), 9, 183, 184, 187, 183, 192, 194, 199, 200, 206–209, 227, 231, 257–260, 262, 266, 273, 279, 282, 284, 291, 292, 294–297, 299, 309–312
– function of, 9, 22
– Rb knockout, 16
– regulation of, 10
Rbx1, 152, 156
redundancy, 271, 272, 307, 313
regulatory particle, 149
replication, 272, 273, 277, 281, 282, 289, 291, 292, 302, 305
– transcription and, 34
– viral, 43, 50
replication factors, phosphorylation of, 36, 37
replication origin, 31–36
replicator, 32, 35, 36
rereplication, 36, 38, 41, 49
restriction point, 7, 273
– cyclohexamide, 7
– early mRNAs, 9
– late mRNAs, 9
– START, 7

RhoA, 18
ring finger motifs, 152, 156
Roc1, 152, 156
ROS, 258, 260, 263–266
Rum1, 158, 159

S phase, 160, 168, 184, 189, 190, 192, 232, 271–273, 276, 278, 279, 284, 286, 289, 292, 296, 299, 301, 305, 307, 309, 311, 312
S-adenosyl methionine (SAM), 161
SAM, 163, 167
Scc1, 154
SCF, 11, 151, 152, 156, 158
SCF$^{\beta\text{-TrCP}}$, 160, 164, 165
SCF^{Cdc4}, 158–160, 168
SCF^{Grr1}, 162
SCF^{Met30}, 161, 163, 167
SCF$^{Pop1/Pop2}$, 159
SCF^{Skp2}, 161, 162, 165
securin, 150, 154, 155, 164, 168
seminiferous tubules, 345, 354, 355, 357
senescence, 185, 190, 193, 194, 196, 244, 257–266
sensors, 67
– 9-1-1 complex, 69, 71
– ATR-ATRIP, 67, 68, 71
– Hus1, 69
– Mec1-Ddc2, 67, 68
– Rad1, 69
– Rad17, 69
– Rad17 complex, 71
– Rad9, 69
separase, 154, 168
Sertoli cells, 345, 354, 357
serum starvation, 291, 307, 309
serum stimulation, 8
– Ras/Map kinase pathway, 18
– signal transduction, 8
Sic1, 151, 158, 159, 168
sister chromatid, 154, 156
skin carcinogenesis, 261
Skp1, 151, 156, 163
Skp2, 150, 157, 161, 162, 273, 293, 296
Sld2, 47
Sld3, 45, 46
Slimb, 157
SOS, 18
sperm, 345, 354, 356, 360
spermatocytes, 345, 353–355

spermatogenesis, 343, 353, 355, 357, 359
spermatogonia, 345, 353–355
spindle, 93–95, 97–99, 101–104, 154, 168
spindle assembly checkpoint, 93, 99, 101
substrate phosphorylation, 151
SUMO, 166
sumoylation, 166, 167
survivin, 100
Swe1, 158, 161
synapsis, 353, 359
synaptonemal complexes, 354, 359

telomerase, 257, 258, 260, 265
telomere, 80
Telomere attrition, 264
telomere erosion, 262
testis, 281, 282, 284–286, 345, 356
tetratricopeptide repeat, 152
transcription, 166, 355, 356
transcription factors, 183–185, 188, 190, 191, 208–210, 237, 332, 333, 337
transcriptional repressors, 333
transformation, 258, 259, 261, 266
transformed cells, 329
transgenic mice, 229
β-TrCP, 157, 158, 160, 161
tumor, 183, 184, 187, 188, 193, 195–197, 199, 200, 205–208, 210, 227, 271, 284–286, 294, 295, 297–300, 302, 303, 310–312, 329, 333, 335–337
tumor spectrum, 245
tumor suppressor, 183, 187, 205, 206, 257–259, 262, 266
tumorigenesis, 183, 185, 193, 195–197, 199, 200, 205, 207–210
tyrosine 15, 160

Uba domains, 149
Ubc10, 153
ubiquitin, 148, 149
– conjugating enzymes, 147
– ligase, 152
– ubiquitin-mediated proteolysis, 147, 149
ubiquitin conjugating enzyme, 151, 153
ubiquitin-proteasome pathway, 152
ubiquitylation, 153, 155, 161, 165
Ubl domain, 149

WD40 repeat, 157, 158, 160, 161
Wee1, 158, 160, 161, 230

Xkid, 150

Printing: Krips bv, Meppel
Binding: Stürtz, Würzburg